CAMBRIDGE MONOGRAPHS ON
MATHEMATICAL PHYSICS

General editors: P. V. Landshoff, D. W. Sciama, S. Weinberg

SUPERSTRING THEORY

Volume 2: Loop amplitudes, anomalies and phenomenology

TO OUR PARENTS

SUPERSTRING THEORY

Volume 2

Loop amplitudes, anomalies and phenomenology

MICHAEL B. GREEN

Queen Mary College, University of London

JOHN H. SCHWARZ

California Institute of Technology

EDWARD WITTEN

Princeton University

CAMBRIDGE
UNIVERSITY PRESS

PUBLISHED BY THE PRESS SYNDICATE OF THE UNIVERSITY OF CAMBRIDGE
The Pitt Building, Trumpington Street, Cambridge, United Kingdom

CAMBRIDGE UNIVERSITY PRESS
The Edinburgh Building, Cambridge CB2 2RU, UK   www.cup.cam.ac.uk
40 West 20th Street, New York, NY 10011-4211, USA   www.cup.org
10 Stamford Road, Oakleigh, Melbourne 3166, Australia
Ruiz de Alarcón 13, 28014 Madrid, Spain

First published 1987
Reprinted 1987 (three times)
First paperback edition with corrections 1988
Reprinted 1988, 1993, 1995, 1996, 1998, 1999

Printed in the United States of America

Typeset in Times

A catalog record for this book is available from the British Library

Library of Congress Cataloging in Publication Data is available

ISBN 0 521 35753 5 paperback

Contents

Preface

Recent years have brought a revival of work on string theory, which has been a source of fascination since its origins nearly twenty years ago. There seems to be a widely perceived need for a systematic, pedagogical exposition of the present state of knowledge about string theory. We hope that this book will help to meet this need. To give a comprehensive account of such a vast topic as string theory would scarcely be possible, even in two volumes with the length to which these have grown. Indeed, we have had to omit many important subjects, while treating others only sketchily. String field theory is omitted entirely (though the subject of chapter 11 is closely related to light-cone string field theory). Conformal field theory is not developed systematically, though much of the background material needed to understand recent papers on this subject is presented in chapter 3 and elsewhere. Our discussion of string propagation in background fields is limited to the bosonic theory, and multiloop diagrams are discussed only in very general and elementary terms. The omissions reflect a combination of human frailty and an attempt to keep the combined length of the two volumes from creeping too much over 1000 pages.

We hope that these two volumes will be useful for a wide range of readers, ranging from those who are motivated mainly by curiosity to those who actually wish to do research on string theory. The first volume is supposed to be self-contained. It gives a detailed introduction to the basic ideas of string theory, requiring as background only a moderate knowledge of particle physics and quantum field theory. The second volume delves into a number of more advanced topics, including a study of one-loop amplitudes, the low-energy effective field theory, and anomalies. There is also a substantial amount of mathematical background on differential and algebraic geometry, as well as their possible application to phenomenology.

We feel that the the two volumes should be suitable for use as textbooks in an advanced graduate-level course. The amount of material is probably more than can be covered in a one-year course. This should provide the instructor the luxury of emphasizing those topics he or she finds especially important while omitting others. Despite our best efforts, it is inevitable that a substantial number of misprints, notational inconsistencies and other errors have survived. We will be grateful if they are brought to our attention so that we can correct them in future editions.

We have benefitted greatly from the assistance of several people whom we are pleased to be able to acknowledge here. Kyle Gary worked with skill and diligence in typing substantial portions of the manuscript, as well as figuring out how to implement the formatting requirements of Cambridge University Press in TEX, the type-setting system that we have used. Marc Goroff brought his wealth of knowledge about computing systems to help solve a myriad of problems that arose in the course of this work. We also received help with computing systems from Paul Kyberd and Vadim Kaplunovsky. Patricia Moyle Schwarz put together the index and made useful comments on the manuscript. Harvey Newman set up communications links that enabled us to transfer files between Pasadena, Princeton and London. Judith Wallrich helped to compile the bibliography. Useful criticisms and comments on the text were offered by Čedomir Crnković, Chiara Nappi, Ryan Rohm and Larry Romans.

We would like to dedicate this book to our parents.

1986   Michael B. Green
                                                                  John H. Schwarz
                                                                  Edward Witten

8. One-Loop Diagrams in the Bosonic String Theory

Our discussions of string scattering amplitudes in the first volume of this book were limited to tree diagrams. These are the lowest-order approximations to string scattering amplitudes. In principle, quantum corrections to the tree level or classical results should be obtained by a perturbation expansion derived from string quantum field theory. Our present state of knowledge does not make this possible. Historically, loop diagrams were constructed by using unitarity to construct loop diagrams from tree diagrams. This unitarization of the tree diagrams led, in time, to the topological expansion, as sketched in chapter 1.

As has been explained in chapters 1 and 7, the tree amplitudes for on-mass-shell string states can be represented by functional integrals over Riemann surfaces that are topologically equivalent to a disk (for open strings) or a sphere (for closed strings). Higher-order corrections are identified with functional integrals over surfaces of higher genus. An important ingredient in the calculation of scattering amplitudes is the correlation function of vertex operators corresponding to the external particles emitted from the surface. The possible world-sheet topologies include surfaces with holes or 'windows' cut out (for type I theories, where the surfaces have boundaries) or 'handles' attached. For theories with oriented strings the surfaces must be orientable. Similarly, for theories containing only closed strings the surfaces must be closed.

As the genus of a surface increases, the power of the coupling constant that accompanies it also increases. For example, adding a handle to a surface is equivalent to adding a loop of closed strings (as in fig. 8.1) and increases the order of a diagram by a factor of κ^2, where κ is the gravitational coupling constant. Cutting a window out of a surface (which is only possible in theories that contain open strings) adds a boundary and hence it increases the number of internal open strings (fig. 8.2a). The order of the diagram is increased by $g^2 \sim \kappa$ for each window, where g is the Yang–Mills coupling constant. However, the presence of a window does not always correspond to adding a loop of open strings. For example,

1

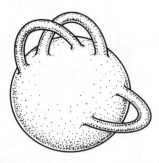

Figure 8.1. A handle added to a world sheet of arbitrary topology.

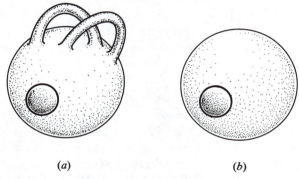

(a)   (b)

Figure 8.2. Cutting a window out of a world sheet adds a boundary. This increases the number of internal open-string propagators as seen in (a). Cutting a window out of a spherical world sheet results in a diagram that is topologically a disk, as shown in (b).

cutting a window out of a sphere is a modification of the (type I) closed-string tree amplitude, which gives a world sheet that is topologically a disk with external closed-string particles attached at interior points of the surface (fig. 8.2b). Type I superstring theory is based on unoriented open and closed strings and therefore also includes nonorientable surfaces.

This topological classification of diagrams in string theories is certainly strikingly different from the classification of Feynman diagrams in point-particle field theory. In string theories there are far fewer diagrams to consider at each order in perturbation theory, and there is no meaningful separation of diagrams into tadpoles, mass insertions, vertex corrections, etc. At the one-loop level, the analysis of world-sheet path integrals is tractable. In fact, one-loop diagrams can be generated by the same operator methods that we used for tree diagrams in chapter 7. Beyond the one-loop level, the analysis of world-sheet path integrals involves somewhat esoteric mathematics, which we will not explore in this book.

In the bosonic theory calculations based on the covariant operator formalism require the same mathematical manipulations as those that arise in light-cone gauge, at least when the external on-shell states are taken to have vanishing + components of momentum. Given Lorentz invariance, amplitudes for external particles with momenta restricted in this way completely determine the amplitudes provided that there are not too many external states. Although we use the covariant method in most of this chapter, very similar techniques also apply to the light-cone gauge method in this special frame.

8.1 Open-String One-Loop Amplitudes

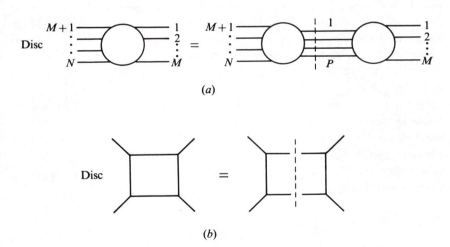

(a)

(b)

Figure 8.3. (a) Unitarity equates the discontinuity of a scattering amplitude (with M incoming and N outgoing particles) across a threshold cut (due to P intermediate particles) to the product of $M \to P$ and $P \to N$ scattering amplitudes integrated over intermediate state phase space. (b) At one loop, unitarity relates the discontinuity of a loop diagram to the integral of the product of two tree diagrams over the phase space for the intermediate on-shell two-particle states.

In point-particle theories the one-loop diagrams can be determined by unitarity in terms of tree diagrams without using the apparatus of second-quantized field theory. Unitarity requires that scattering amplitudes should have suitable branch cuts as a function of the Lorentz-invariant quantities formed out of the external momenta. These cuts arise from the regions of momentum space in which intermediate states are on their mass shells. For example, fig. 8.3a depicts the unitarity equation for an amplitude with M incoming and N outgoing particles. A given set of P intermediate on-shell physical states contributes to the discontinuity across the branch cut an amount that is proportional to the product of the amplitude for $M \to P$ particles multiplied by the amplitude for $P \to N$ particles integrated over the accessible phase space for the intermediate particles.

When expanded as a power series in the coupling constant this nonlinear equation relates the discontinuity of a one-loop amplitude to the product of two tree amplitudes. In this case, illustrated in fig. 8.3b, the number of intermediate states, P, is two. In particular, the form of the one-loop amplitude, including its normalization, is determined in terms of the tree diagrams up to an arbitrary entire function of these invariants. In the case of ordinary field theory, the arbitrary entire function corresponds to the arbitrariness associated with the renormalization procedure. In gauge invariant field theories, it is also necessary to avoid including in loop diagrams the contributions of timelike or longitudinally polarized gauge mesons. These contributions can be removed by going to a light cone or unitary gauge, or can be canceled by correctly including the Faddeev-Popov ghosts.

Similar considerations apply to the construction of the one-loop amplitudes in string theories from the tree diagrams. In this case the requirement of Regge behavior at high energies eliminates the ambiguity that exists in field theory. Regge behavior forces amplitudes to vanish in certain asymptotic regimes; addition of an entire function of the momenta to one-loop diagrams would inevitably spoil this property.

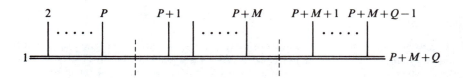

Figure 8.4. A general tree diagram with $P + M + Q$ ground-state particles factorized to give a tree with two arbitrary excited states and M ground states.

For example, the tree diagram of fig. 8.4 illustrates the interaction of $P + M + Q$ on-shell open-string states. It can be factorized as shown in the figure to obtain the amplitude for an arbitrary pair of 'excited' states to couple to M on-shell states. Ignoring the presence of unphysical states for the moment the one-loop amplitude is obtained by *sewing* the excited states together, *i.e.*, by inserting a propagator between the initial and the final excited states and summing over all possible states as well as integrating over their momenta. In the complete amplitude it is necessary to sum over loop diagrams with twists inserted in all possible ways in the internal propagators of the loop.

Just as in ordinary field theory, covariant string-theory formulas describe states of unphysical polarization circling in the loop. Care must be taken to somehow suppress their contribution. In early calculations of string loop diagrams, the propagation of unphysical states was avoided by inserting a rather complicated physical-state projection operator in the propagators. This ensured that the circulating particles corresponded only to physical states; the procedure was analogous to some early approaches to Yang–Mills theory. A more modern approach incorporates the Faddeev–Popov ghost modes in the calculations instead. This approach is far simpler, and is the approach that we will use in performing covariant calculations.

In the bosonic theory the inclusion of ghost modes is quite easy. The vertex operators, such as the tachyon vertex operator $e^{ik \cdot X}$, are constructed from X^μ only, without ghosts, where $X^\mu(\sigma, \tau)$ is the string coordinate defined in chapter 2. When ghosts are included in the formalism, these vertex operators are understood to include a unit operator in the ghost sector of the Fock space. The ghosts circulating around the loops can then cancel the contributions of unphysical states. This is their only role.

How can we be certain that the ghosts are really correctly canceling the contributions of the unphysical states? It is particularly important to address this question, since – pending a completely satisfactory derivation of loop amplitudes from a logically satisfying starting point – there is an element of guesswork in formulating the Feynman rules including the ghosts. To gain some insight into this important question, it is possible to do the calculations in light-cone gauge. In this case, there are no unphysical states propagating in the loop – neither states that violate the Virasoro conditions, nor null states, nor ghosts. All the states in the light-cone Fock space correspond to physical propagating degrees of freedom. The light-cone amplitudes are thus manifestly unitary – or at any rate, singularities that appear are due to physical intermediate states. It will be rather clear in our discussion that – at least for processes that are easily discussed in both formalisms – the light-cone approach gives the same answers that one obtains in the covariant treatment with ghosts. Ultimately, the rules involving Faddeev–Popov modes should be derived from a logically sound starting point, perhaps a gauge-invariant nonlinear field theory of strings.

A curious feature of string theories is that new singularities can arise due to divergences of sums over intermediate states. This feature already appeared in tree amplitudes, where we saw in chapters 1 and 7 that t-channel poles arise due to divergences in the sum over s-channel poles. In the case of loop diagrams even more remarkable things can happen. For instance, an open-string loop with suitable twists can actually give rise to closed-string poles. It was by trying to reconcile these singularities with unitarity that the significance of the critical dimension first became apparent; in the critical dimension, these singularities correspond to graviton poles, and (as we discussed in §1.5.6), they are the reason that a consistent string theory without gravity does not seem possible, at the quantum level.

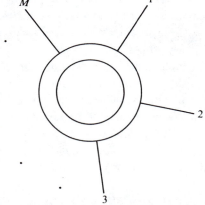

Figure 8.5. The planar loop diagram with M ground-state particles

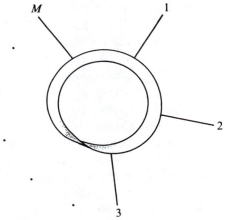

Figure 8.6. A nonorientable one-loop diagram with M external particles has a world sheet that is a Möbius strip.

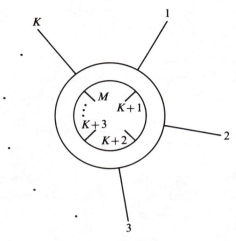

Figure 8.7. A nonplanar one-loop diagram in which K particles are attached to one boundary and $M - K$ to the other.

The simplest one-loop diagram in a theory of open strings corresponds to a process for which the world sheet is topologically an annulus or cylinder with M external states attached to one boundary as illustrated in fig. 8.5. (A world sheet with this topology is referred to as a *planar* diagram). The precise meaning of the parameters describing the annulus and the positions of the attached particles in this figure is explained later in this chapter. By including an odd number of twists in the world sheet it is possible to construct other one-loop diagrams associated with world sheets that are *nonorientable*, *i.e.*, Möbius strips having only one boundary (as in fig. 8.6). By using an even number of twists one can describe oriented surfaces in which particles are attached to both boundaries of the annulus as in fig. 8.7. These are called *nonplanar* diagrams. These various different contributions to the full one-loop open-string amplitude must be calculated separately, although much of the computation is similar for each of the diagrams. (In this respect, theories of oriented closed strings, which have only one diagram at each order, are a lot simpler.)

8.1.1 The Planar Diagrams

Let us consider bosonic open strings carrying group-theory quantum numbers of the type described in §6.1. Let n be the dimension of the fundamental representation of the gauge group – the representation of the charges that sit at the ends of the open string. Then the group-theory factor associated with the planar diagram (fig. 8.5) is

$$G_P = n\mathrm{tr}(\lambda_1\lambda_2\ldots\lambda_M), \qquad (8.1.1)$$

where the factor of n arises from the trace of the $n \times n$ unit matrix associated with the free boundary of the annulus. As in the case of tree diagrams, the matrices λ_r must be $n \times n$ matrices belong to the fundamental representation of the algebra of any of the allowed groups (*i.e.*, the classical groups $SO(n)$, $USp(n)$ and $U(n)$) if the states are at even mass levels. Hermitian matrices (denoted μ in §6.1) would be used for odd levels.

For simplicity and explicitness, we mostly consider processes in which the external states are either tachyons (an odd level) or massless vector particles (an even level), although essentially the same techniques can be used for arbitrary excited states. In either case the vertex for emitting an on-shell particle with momentum k_r at 'time' τ is denoted by $V(k_r, \tau)$, where

$$V(k_r, \tau) = e^{i\tau L_0}V(k_r, 0)e^{-i\tau L_0}. \qquad (8.1.2)$$

As in chapter 7, we frequently work with $x = e^{i\tau}$ and take x to be real, corresponding to a Wick-rotated time coordinate. In this case we write

$$V(k_r, x) = x^{L_0}V(k_r, 1)x^{-L_0}. \qquad (8.1.3)$$

We recall that, apart from the vertex (8.1.3), the main ingredient in the construction of tree diagrams in chapter 7 was the propagator, which for bosonic open strings was

$$\Delta = (L_0 - 1)^{-1}. \qquad (8.1.4)$$

To associate an amplitude with the diagram of fig. 8.5, we include a vertex (8.1.2) for each external line, and a propagator (8.1.4) for each internal line. The closed loop is represented by a trace in the Fock space of the internal lines. Putting things together in this way, the amplitude that we define is

$$A_P(1, 2, \ldots, M) = g^M G_P \int d^D p \, \mathrm{Tr}(\Delta V(k_1, 1)\Delta V(k_2, 1)$$
$$\ldots \Delta V(k_M, 1)). \qquad (8.1.5)$$

In the covariant formalism used here the trace runs over the infinite set of bosonic oscillator modes α_n^μ, $\mu = 0, \ldots, 25$, as well as the ghost oscillators

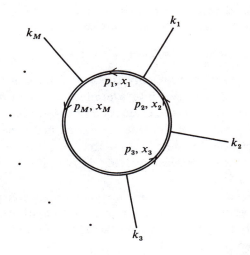

Figure 8.8. The kinematics for the calculation of the planar loop diagram.

b_n and c_n. In the light-cone gauge, the only modes entering in the trace would be the transverse oscillators α_n^i, $i = 1, \ldots, 24$. Just as in ordinary field theory, the poles of the propagators give rise to cuts associated with on-shell intermediate states.

The sequence of emitted particles $(1, 2, \ldots, M)$ corresponds to the order in which they are attached to the boundary in fig. 8.5. The cyclic property of the trace ensures that only the cyclic ordering matters. The full one-loop planar amplitude includes a sum over all cyclically inequivalent permutations of the external particles, each weighted with its own group-theory factor. The kinematics for this process is illustrated in fig. 8.8.

As in chapter 7, it is convenient to use the integral representation

$$\Delta = (L_0 - 1)^{-1} = \int_0^1 x^{L_0 - 2} dx \qquad (8.1.6)$$

for the open-string propagator. The vertex operator for emitting an on-shell tachyon of momentum k^μ (with $k^2 = 2$) is given by

$$V_0(k, 1) =: e^{ik \cdot X(1)} : . \qquad (8.1.7)$$

If the emitted particle is a massless vector boson, the vertex operator is given by

$$V(\zeta, k, 1) = \zeta \cdot \dot{X}(1) e^{ik \cdot X(1)}, \qquad (8.1.8)$$

where ζ^μ is the polarization vector of the particle and $\zeta \cdot k = k^2 = 0)$.

As explained in chapter 7, a convenient way to evaluate amplitudes with external vector particles is to use the vertex operator

$$V(k, \zeta, 1) = \exp\{\zeta \cdot \dot{X}(1) + ik \cdot X(1)\}, \qquad (8.1.9)$$

with the understanding that only the terms linear in the ζ's are to be kept. This vertex factorizes into a product of terms for each oscillator mode, which is helpful in the evaluation of the traces. (This will also be useful in the discussion of the heterotic string in the next chapter.) This vertex does not need to be normal ordered, since the only ordering factor that arises is an exponential involving ζ^2, which does not contribute to the terms linear in ζ^μ.

Let us now consider the one-loop planar diagram with M external tachyons. Inserting the integral representation for the propagators in (8.1.5) and using the fact that $x^{L_0} V(k_r, 1) = V(k_r, x) x^{L_0}$, the expression for the loop can be written as

$$A_P(1, 2, \ldots, M) = g^M G_P \int_0^1 \prod_{i=1}^M dx_i \int d^D p \mathrm{Tr}[V_0(k_1, x_1)$$

$$\times V_0(k_2, x_1 x_2) \ldots V_0(k_M, x_1 \ldots x_M) w^{L_0 - 2}]$$

$$= g^M G_P \int_0^1 \frac{dw}{w^2} \int_0^1 \prod_{r=1}^{M-1} \frac{d\rho_r}{\rho_r} \Theta(\rho_r - \rho_{r+1}) I(1, \ldots, M),$$

$$(8.1.10)$$

where

$$I(1, \ldots, M) = \int d^D p \mathrm{Tr}\left(V_0(k_1, \rho_1) \ldots V_0(k_M, \rho_M) w^{L_0}\right) \qquad (8.1.11)$$

and

$$\rho_r = x_1 \ldots x_r, \qquad (8.1.12)$$

$$w \equiv \rho_M = x_1 \ldots x_M. \qquad (8.1.13)$$

In writing (8.1.10), we have used the fact that the Jacobian for the transformation from the x_r variables that parametrize the individual propagators to the ρ_r variables is given by

$$\prod_{r=1}^M dx_r = dw \prod_{r=1}^{M-1} \frac{d\rho_r}{\rho_r}. \qquad (8.1.14)$$

The variables ρ_r are integrated on the interval $(w, 1)$ of the real axis of the complex ρ plane.

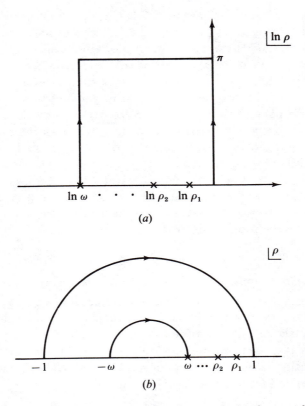

Figure 8.9. (a) The world sheet of the planar diagram depicted as a cylinder in the complex $\ln \rho$ plane. The lines marked by arrows are to be identified. The cylinder is of circumference $\ln w$ and length π and the particles are integrated around the boundary of the cylinder at $\mathrm{Im}(\ln \rho) = 0$ in a fixed sequence corresponding to a given Chan–Paton factor. (b) Diagram showing the world-sheet configuration corresponding to the integration variables ρ_r. The inner and outer arcs are to be identified with one another (in the sense indicated by the arrows) since points that differ by a factor of w are equivalent. Then the line segments on the positive and negative parts of the real axis become the boundaries of a cylinder or annulus.

Equations (8.1.10) and (8.1.11) can also be derived by the path-integral approach described in chapter 1. When it is done that way, one identifies the factor $I(1, \ldots, M)$ as the correlation function for a set of vertex operators on the boundary of a cylindrical world sheet. The cylinder can be pictured in terms of the variable $\ln \rho$ as depicted in fig. 8.9a. The external particles are attached to the boundary of the cylinder $\mathrm{Im}(\ln \rho_r) = 0$ at the points $-\mathrm{Re}(\ln \rho_r)$, which are integrated in sequence around the circumference of length $-\ln w$. The other boundary of the cylinder, with no attached particles, is at $\mathrm{Im}(\ln \rho) = \pi$ (i.e., negative ρ). By circulating the operator w^{L_0} around the trace in (8.1.11), using the cyclic property of the trace, it is evident that the integrand I is invariant under the simultaneous transformation of the ρ parameters $\rho \rightarrow w^n \rho$. This identifies a point ρ with equivalent points $w^n \rho$, where n is an arbitrary integer, corresponding to points that differ by an integer number of circuits around the cylinder. The domain of the complex ρ plane corresponding to the string world sheet in this parametrization therefore corresponds to the region shown in fig. 8.9b. The region is centered on the origin with inner radius equal to w and outer radius equal to 1. Points on the inner circle are identified radially with those on the outer one thereby forming the cylinder.

The integral over the loop momentum can be regarded as the zero-mode part of the trace in the expression for I. Each internal propagator in the loop includes a factor $x_r^{L_0}$ that contains the dependence on the momentum p_r carried by the propagator, where

$$p_r = p - k_1 - k_2 - \cdots - k_{r-1} = p + k_r + \cdots + k_M, \qquad (8.1.15)$$

and p is the loop momentum. This means that the complete dependence of the integrand of (8.1.11) on the loop momentum is given by $\prod_r x_r^{p_r^2/2}$. The integral of this term is a D-dimensional gaussian integral, which converges after one makes a Wick rotation that replaces the time-component p^0 by ip^0. (This is a standard step in the evaluation of Feynman diagrams.) The evaluation of the momentum integral is then straightforward to carry out by completing the square using the identity

$$\sum_{r=1}^{M} p_r^2 \ln x_r = \ln w \left(p + \sum_{r=1}^{M} k_r \frac{\ln \rho_r}{\ln w} \right)^2$$

$$- \sum_{r<s} k_r \cdot k_s \left\{ \ln \frac{\rho_s}{\rho_r} - \frac{\ln^2(\rho_s/\rho_r)}{\ln w} \right\}. \qquad (8.1.16)$$

The gaussian integral is then evaluated by shifting the loop momentum to $p' = p - \sum_r k_r \ln \rho_r / \ln w$. One obtains

$$\int d^D p \prod_{r=1}^{M} x_r^{p_r^2/2} = \left(\frac{-2\pi}{\ln w} \right)^{D/2} \prod_{1 \le r < s \le M} \left\{ c_{sr}^{-1/2} \exp \left(\frac{\ln^2 c_{sr}}{2 \ln w} \right) \right\}^{k_r \cdot k_s}. \qquad (8.1.17)$$

In this expression

$$c_{sr} \equiv \rho_s/\rho_r = x_{r+1} \ldots x_s. \qquad (8.1.18)$$

To evaluate the traces over the nonzero bosonic modes α_n^μ in (8.1.11), it is convenient to use methods based on the coherent-state representation of harmonic oscillators introduced in appendix 7.A. Recall that a coherent state is defined by $|z\rangle = \exp(za^\dagger)|0\rangle$, where z is an arbitrary complex number. The identity operator can be represented in terms of coherent states by

$$I = \frac{1}{\pi} \int d^2z |z\rangle e^{-|z|^2} \langle z|. \tag{8.1.19}$$

To prove this one notes that the projection of a coherent state onto a number basis state is $\langle m|z\rangle = z^m/\sqrt{m!}$. Therefore, performing the z integral one can show that $\langle m| I |n\rangle = \delta_{m,n}$, as required. It follows from this that the trace of an operator K is given by

$$\mathrm{Tr}(K) = \frac{1}{\pi} \int d^2z \, e^{-|z|^2} \langle z|K|z\rangle. \tag{8.1.20}$$

Substituting tachyon vertex operators in (8.1.11), the trace reduces to the product of traces over the individual harmonic oscillators so that

$$\mathrm{Tr}\left(V_0'(k_1,\rho_1)\ldots V_0'(k_M,\rho_M)w^{L_0'}\right) = \prod_{n=1}^\infty \prod_{\mu=0}^{D-1} T_n^\mu, \tag{8.1.21}$$

where a prime indicates that the zero-mode factor has been removed from the vertex operator. The ghost modes will be considered separately. Using (8.1.20), and the rule $x^{a^\dagger a}|z\rangle = |xz\rangle$, one obtains

$$
\begin{aligned}
T_n^\mu &= \int \frac{d^2z}{\pi} e^{-|z|^2} \langle z| \prod_{r=1}^M \exp\left[\frac{k_{r\mu}\alpha_{-n}^\mu}{n}\rho_r^n\right] \exp\left[-\frac{k_{r\mu}\alpha_n^\mu}{n}\rho_r^{-n}\right]|w^n z\rangle \\
&= \int \frac{d^2z}{\pi} e^{-(1-w^n)|z|^2} \exp\left[-\sum_{r<s} k_{r\mu}k_s^\mu \frac{1}{n}(\rho_s/\rho_r)^n\right] \\
&\quad \times \exp\left[\sum_{r=1}^M k_{r\mu}\left(\frac{1}{\sqrt{n}}\rho_r^n \bar{z} - \frac{1}{\sqrt{n}}(w/\rho_r)^n z\right)\right],
\end{aligned}
\tag{8.1.22}
$$

where the second expression is obtained by evaluating the matrix element by moving all the annihilation operators to the right and creation operators to the left and using the coherent-state formulas given in the appendix 7.A.

The complex gaussian integration in (8.1.22) can be evaluated by making use of the formula

$$\frac{1}{\pi} \int d^2z \, e^{-c|z|^2} e^{(az+b\bar{z})} = \frac{1}{c} e^{ab/c}, \tag{8.1.23}$$

where a, b and c are real variables. The result is that the product over all nonzero modes is given by

$$\prod_{n=1}^{\infty}\prod_{\mu=0}^{D-1} T_n^{\mu} = \prod_{n=1}^{\infty}(1-w^n)^{-D} \exp\left(-\sum_{r<s} k_r \cdot k_s \frac{c_{sr}^n + (w/c_{sr})^n - 2w^n}{n(1-w^n)}\right),$$

$$(8.1.24)$$

where momentum conservation $\sum k_r^{\mu} = 0$ has been used so that

$$\sum_{r<s} k_r \cdot k_s = -\frac{1}{2}\sum_{r=1}^{M} k_r^2 = -M. \qquad (8.1.25)$$

There is an additional contribution to the factor I that arises from the trace over the ghost and antighost modes. They do not appear in the formula for the tachyon vertex operator (which has the correct conformal dimension, $J = 1$, without any ghost contributions) but they do contribute to the propagator. In fact, we interpret the 'L_0' in the propagator to be the total L_0 of matter and ghosts, $L_0 = L_0^{(\alpha)} + L_0^{gh}$, where

$$L_0^{gh} = \sum_{n=1}^{\infty} n(c_{-n}b_n + b_{-n}c_n). \qquad (8.1.26)$$

Of course, in chapter 7 we did not include the ghost contribution to L_0 in defining the propagator. The inclusion of the ghost contribution to L_0 would hardly have mattered in chapter 7, since we considered initial and final states without ghosts or antighosts (these are precisely the states that are annihilated by L_0^{gh}), and we considered vertex operators that did not couple to the ghosts. The fact that it was not necessary to include ghosts to discuss tree diagrams should not seem surprising; it is likewise unnecessary in ordinary Yang–Mills theory to include the Faddeev–Popov ghosts in defining tree amplitudes. At the one-loop level, even when the external vertex operators are trivial unit operators in the ghost Fock space, ghosts can circulate around the loop and make an important contribution.

Including the L_0^{gh} contribution to L_0 in defining the propagator $\Delta = (L_0 - 1)^{-1}$ means that the required ghost factor is given by Tr $w^{L_0^{gh}}$. Here a slight amount of guesswork is required to define properly what one means by this 'trace'. Let us consider first the contribution of the c_1 and b_1 oscillators and their adjoints c_{-1} and b_{-1}. The identity operator in this space is

$$I_1 = |0,0\rangle \langle 0,0| + |1,0\rangle \langle 1,0| + |0,1\rangle \langle 0,1| + |1,1\rangle \langle 1,1|, \qquad (8.1.27)$$

where the labels refer to eigenvalues of $c_{-1}b_1$ and $b_{-1}c_1$, respectively. This suggests that

$$\mathrm{tr}(w^{c_{-1}b_1+b_{-1}c_1}) = 1 - w - w + w^2 = (1-w)^2. \qquad (8.1.28)$$

What requires some care are the minus signs on the right hand side of (8.1.28). It is natural to include these minus signs because the states $|1,0\rangle$ and $|0,1\rangle$ are fermionic. Thus, for example, $\mathrm{tr}(|1,0\rangle\langle 1,0|) = -\langle 1,0|1,0\rangle = -1$. In a similar fashion the c_n and b_n modes give a contribution $(1-w^n)^2$ for $n = 1,2,\ldots$. Thus, altogether, the contribution to the trace of ghost modes c_n, b_n with $n \neq 0$ is

$$\mathrm{Tr}\, w^{L_0^{gh}} = \prod_{n=1}^{\infty}(1-w^n)^2 = [f(w)]^2. \qquad (8.1.29)$$

The partition function $f(w)$ was introduced in §2.3.5. In terms of world-sheet path integrals, the minus signs in (8.1.28) correspond to path integrals on a cylinder with periodic ghost and anti-ghost boundary conditions in the time direction.

What about ghost zero modes? As we learned in §3.2.1, the ghost zero modes c_0 and b_0 are realized in a two-state Hilbert space with states $|\uparrow\rangle$ and $|\downarrow\rangle$ of ghost number $\pm 1/2$. These states are respectively bosonic and fermionic, so including their contribution as above would give a factor $1 - 1 = 0$. This cannot be right, so we will assume that the correct prescription is to suppress the ghost zero modes in defining the loop amplitudes.[*] Suppressing the ghost and antighost zero modes is certainly compatible with Lorentz invariance, and we will see that it gives the same answer that one gets in the manifestly unitary light-cone gauge.

Combining the contributions of the zero and the nonzero modes in (8.1.17) and (8.1.24) and the ghost contribution in (8.1.27) gives the result

$$I(1,\ldots,M) = [f(w)]^{2-D}\left(\frac{-2\pi}{\ln w}\right)^{D/2}\exp\left\{\sum_{r<s}k_r \cdot k_s \ln \psi_{rs}\right\}. \qquad (8.1.30)$$

Notice that the ghosts have turned the $-D$ in (8.1.24) into $-(D-2)$; they cancel the contributions of two of the D coordinates. The functions

---

[*] This ansatz has recently been justified, at least for open strings, by considerations involving gauge fixing of a gauge-invariant string field theory, but this is a subject that we will not consider in this book.

ψ_{rs} appearing in (8.1.30) are given by $\psi_{rs} \equiv \psi(c_{sr}, w)$, where

$$\ln \psi(c, w) = -\tfrac{1}{2}\ln c + \frac{\ln^2 c}{2\ln w}$$
$$- \sum_{m=1}^{\infty} \frac{c^m + (w/c)^m - 2w^m}{m(1 - w^m)} \cdot \qquad (8.1.31)$$

Using the identity

$$\sum_{m=1}^{\infty} \frac{1}{m} \frac{x^m}{1 - y^m} = -\sum_{n=0}^{\infty} \ln(1 - xy^n), \qquad (8.1.32)$$

$\psi(c, w)$ can be recast in the form

$$\psi(c, w) = \frac{1 - c}{\sqrt{c}} \exp\left(\frac{\ln^2 c}{2\ln w}\right) \prod_{n=1}^{\infty} \frac{(1 - w^n c)(1 - w^n/c)}{(1 - w^n)^2}. \qquad (8.1.33)$$

Various properties of ψ, including its relation to Jacobi θ functions, are discussed in appendix 8.A.

The quantity $\ln \psi_{rs}$ is proportional to the correlation function between two X's, $\langle X^\mu(\rho_r) X^\nu(\rho_s) \rangle$, which is also the Green function for the Laplace equation evaluated between the points ρ_r and ρ_s in the complex ρ plane. It would enter directly in this form in the world-sheet path integral approach that was sketched in chapter 1.

This function has a logarithmic singularity as $c_{sr} \to 1$, which corresponds to $\rho_r \to \rho_s$. This is just the logarithmic short distance singularity of the free boson propagator in two dimensions. The Green function satisfies boundary conditions appropriate to the periodicity properties of the loop diagram,

$$\psi(cw, w) = -\psi(c, w), \qquad (8.1.34)$$

which implies that any point ρ is identified with the point $w\rho$. The fact that ψ is invariant under this transformation is shown in appendix 8.A. This makes contact with the form of the scattering amplitudes embodied in the functional approaches to string theory.

The expression for the one-loop planar amplitude can now be assembled from (8.1.10) and (8.1.30), giving

$$A_P(1, 2\ldots, M) = g^M G_P \int_0^1 \frac{dw}{w^2} \int_0^1 \left(\prod_{r=1}^{M-1} \frac{d\rho_r}{\rho_r} \Theta(\rho_r - \rho_{r+1})\right)$$
$$\times [f(w)]^{2-D} \left(\frac{-2\pi}{\ln w}\right)^{D/2} \prod_{r<s} (\psi_{rs})^{k_r \cdot k_s}. \qquad (8.1.35)$$

Before discussing this result, let us quickly note how the same answer would arise in light-cone gauge, at least if the external particles all have zero k^+ so that the light-cone tachyon vertex operators $\exp\{ik \cdot X\}$ are simple. (Recall that X^- is a complicated operator in light-cone gauge.) In light-cone gauge there are no ghost modes, but $X^\mu(\sigma)$ has only $D - 2$ instead of D independent components (for nonzero modes), so the power of $f(w)$ in (8.1.24) is $-(D - 2)$. This precisely reproduces the effects of the ghosts. The similarity of the calculations in the two formulations is a feature of all the diagrams considered in this chapter.

The amplitude (8.1.35) possesses all the singularities in the variables

$$ s_{IJ} = - \left(\sum_{r=I}^{J} k_r \right)^2 \tag{8.1.36} $$

required by perturbative unitarity. They arise from various corners and edges of the integration region. For example, poles in the variables s_{IJ} arise from the limit in which the points corresponding to the particles labeled $I, \ldots, J$ come together on the world sheet, *i.e.*, $\rho_I \sim \rho_{I+1} \sim \cdots \sim \rho_J$, which is the limit in which $x_r \to 1$ for $r = I + 1, \ldots, J$. In this limit the ψ_{rs} vanish like $1 - c_{sr}$ for $I \leq r, s \leq J$. In order to exhibit the leading singularity it is helpful to define new variables η_r $(r = I + 1, \ldots, J)$ by $\eta_r = \rho_I - \rho_r$ and then to scale the η_r variables by a common factor of ϵ so that the singularity arises from the $\epsilon \to 0$ limit of the integration range. The most singular term in the ϵ integral is of the form

$$ \int_0^{} d\epsilon \, \epsilon^{J-I-1} \prod_{I \leq r < s \leq J} \epsilon^{k_r \cdot k_s} = \int_0^{} d\epsilon \, \epsilon^{-\frac{1}{2}s_{IJ} - 2}, \tag{8.1.37} $$

where use has been made of the fact that

$$ \sum_{I \leq r < s \leq J} k_r \cdot k_s = -\tfrac{1}{2}s_{IJ} - (J + 1 - I), \tag{8.1.38} $$

which follows from (8.1.36) and the tachyon mass shell $k_r^2 = 2$. The leading pole is the tachyon at $s_{IJ} = -2$. The expansion of the rest of the integrand as a series in ϵ gives poles corresponding to states at the various higher mass levels.

The amplitude (8.1.35) also contains the usual branch points in the invariant energy variables s_{IJ} associated with the thresholds for producing pairs of on-shell intermediate physical states. These *normal threshold* singularities emerge from the limit in which $x_I \to 0$ and $x_{J+1} \to 0$ $(J > I)$.

The singularities that arise include ones that would be labeled vertex corrections, mass corrections, and tadpole diagrams in a conventional point-particle field theory. In string theories all these kinds of Feynman diagrams correspond to different corners of the integration region of a single string diagram.

Figure 8.10. The insertion of the open-string loop on an external tachyon leg gives rise to an infinity due to an internal propagator evaluated on shell. This is an effect that is eliminated by a *finite* shift in the tachyon mass.

The self-energy correction to a propagator, for example, comes from the corner of the integration region in which the clusters of external particles on each side of the self-energy operator are close together in parameter space. A special case is when there is just one emitted particle on one side of the self-energy operator. This diagram contributes a radiative correction to the mass of the external particle. If one only considers external particles that are on mass shell (which is all that we have shown how to do), this leads to an infinite expression as depicted in fig. 8.10. These are relatively trivial divergences that do not really bear on the finiteness of the theory. They reflect the fact that the masses of the particles are shifted by radiative corrections. Nevertheless, learning to subtract them and extract the correct, finite remainders is an important step in learning to formulate the correct, finite Feynman rules; this is a subject that has been investigated only recently, and which is not considered in this book. For external massless vector states there is no divergence of this type, since gauge invariance protects a massless vector particle from getting a mass. This implies that its self-energy diagram vanishes on the mass shell and provides a zero to cancel the pole coming from the propagator in fig. 8.10.

In view of the form of the expression for $f(w)$, it is clear that the integrand in (8.1.35) can diverge at the corner of the integration at which

$w \rightarrow 1$, and so this region requires special consideration. This is the corner of the integral in which w^{L_0} does not vanish as L_0 becomes large. Since $L_0 = \frac{1}{2}p^2 + N$ is large for states of high mass or high Euclidean momentum, this can be regarded as the string analog of the ultraviolet region. It has, however, an alternative and more incisive interpretation.

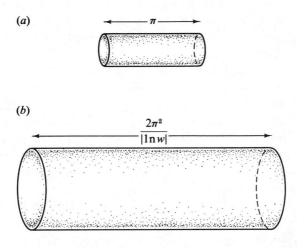

Figure 8.11. A cylinder of length π and circumference $\ln w$ is sketched in (a); it is conformally equivalent, as in (b), to a cylinder of circumference 2π and length $-2\pi^2/\ln w$.

Going back to (8.1.5) and (8.1.35), apart from factors involving the vertex operators, the formula for the amplitude contains a factor

$$A_0 = \int\limits_0^1 dw\, \text{Tr}\; w^{L_0-2}. \tag{8.1.39}$$

As we will demonstrate, the additional factors in (8.1.35) do not affect the essence of the discussion. The factor $w^{L_0} = \exp\{L_0 \ln w\}$ is an operator that propagates a string through a proper time $-\ln w$ (which is positive, since $w \leq 1$). Propagation of an open string through proper time $-\ln w$ can be represented by a path integral on a strip of width π and length $-\ln w$. We must also consider the trace in (8.1.39); this instructs us to identify the initial and final string states (before and after the propagation through imaginary time $-\ln w$). This turns the strip into a cylinder. This is why (8.1.39) can be represented by a path integral on a cylinder in the complex $\ln \rho$ plane of length π and circumference $\ln w$, as depicted in fig. 8.9a and again in fig. 8.11a.

The integration variable w plays a very natural role; it is the only invariant associated with the conformal structure of a cylinder. (See §3.3 for a discussion of these issues, albeit mainly in the case of closed strings.) The one-loop diagram naturally includes an integration over this parameter, and it is this integral that we have singled out in (8.1.39). For $w \sim 1$, the radius of the cylinder in fig. 8.9a is very small, so the open string is propagating for a very small real time; this is then what one might describe as the short-distance limit.

The alternative view of this divergence can be seen by using conformal invariance (which is valid only in the critical dimension!) to rescale the dimensions of the cylinder in fig. 8.11a by any desired amount. In particular, we can take this cylinder to have a fixed circumference 2π and a corresponding length $-2\pi^2 / \ln w$ (these conventions correspond to those of the more detailed analysis that follows). This is sketched in fig. 8.11b. Figure 8.11b can be viewed sideways as a diagram with a *closed* string propagating sideways for proper time $-2\pi^2 / \ln w$. It is now the length of the cylinder rather than its circumference that is to be integrated over. Now $w \rightarrow 1$ no longer appears as an ultraviolet limit but an 'infrared' limit of long-time propagation. Of course, this ability to look at things 'sideways' and view an ultraviolet effect as an infrared effect has no analog in ordinary field theory; it is one way of understanding the absence of ultraviolet difficulties in string theory. The reason that it is far preferable to understand the $w \rightarrow 1$ region as an infrared region rather than an ultraviolet region is that infrared difficulties, when present, have physical reasons, such as contributions of massless particles (or tachyons) whose role has not been properly taken into account; infrared divergences are not symptoms of an inconsistency of a theory, but on the contrary are often signals of interesting physics such as symmetry breaking or quark confinement. Ultraviolet divergences, on the other hand, if not renormalizable, signify the inconsistency of a theory. Also, the fact that the 'ultraviolet' divergence for w near one can be interpreted as an infrared effect has the following fundamental consequence. It means that this divergence must be absent in a theory in which physical principles such as supersymmetry enable us to predict the absence of infrared divergences. This will be the key to understanding infinity cancellations in chapter 10.

To continue our heuristic treatment of the region near $w = 1$, let us consider a closed string propagating for proper time $-\ln q \sim -2\pi^2 / \ln w$ between initial and final states $|i\rangle$ and $|f\rangle$, which we take to have zero space-time momentum. The amplitude for this process is given by an integral over the proper time

$$A_0 \sim \int_0 \frac{dq}{q} U, \qquad (8.1.40)$$

where U is the evolution operator that carries the initial state into the final one. This is given by

$$U = \langle f | e^{(L_0 + \tilde{L}_0 - 2)(2\pi^2 / \ln w)} | i \rangle, \qquad (8.1.41)$$

where $L_0 + \tilde{L}_0 - 2$ is the inverse closed-string propagator (and $L_0 = \tilde{L}_0$ for physical external states). In our problem, the role of the states $|f\rangle$ and $|i\rangle$ is played by the open boundaries in fig. 8.11, one of which has vertex operators attached and one not. $|f\rangle$ and $|i\rangle$ are states that are complicated to describe concretely (except by the characterization of them just given); in particular, it would be awkward (though possible) to describe them in Fock space. For our purposes, all we need to know is that $|i\rangle$ and $|f\rangle$ are states of zero momentum that are not orthogonal to the Fock-space vacuum $|0\rangle$. This being so, upon setting $\tilde{L}_0 = L_0$, (8.1.41) behaves for w very near 1 (which means very large proper time) as

$$U \sim \langle f | 0 \rangle \langle 0 | e^{4(L_0 - 1)\pi^2 / \ln w} | 0 \rangle \langle 0 | i \rangle, \qquad (8.1.42)$$

since all the other states have larger L_0 and hence give contributions that decrease exponentially relative to the tachyon contribution. As the zero-momentum tachyon has $L_0 = 0$, (8.1.42) behaves for $w \to 1$ as $\exp\{-4\pi^2 / \ln w\}$, so the integral in (8.1.40) above should diverge as

$$A_0 \sim \int_0 \frac{dq}{q} e^{-4\pi^2 / \ln w} \sim \int_0 \frac{dq}{q^3}. \qquad (8.1.43)$$

After the tachyon contribution of (8.1.43), the next most singular contribution to (8.1.41) would be the dilaton contribution. As the dilaton has $L_0 = 1$, its contribution to (8.1.41) would be proportional to

$$A_0' \sim \int_0 \frac{dq}{q}. \qquad (8.1.44)$$

Let us now see how to obtain these results by a precise mathematical analysis of (8.1.35). The region near $w = 1$ is conveniently analyzed by changing from the ρ_r variable to the variables z_r by

$$\ln z_r = 2i\pi\nu_r, \tag{8.1.45}$$

where

$$\nu_r = \frac{\ln \rho_r}{\ln w}, \tag{8.1.46}$$

(so that $\nu_M = 1$). The interval $(w, 1)$ of the real ρ axis is mapped onto the outer boundary of an annulus of unit radius in the z plane while the interval $(-w, -1)$ is mapped into the inner boundary of radius

$$q \equiv \exp i\pi\tau = \exp(2\pi^2/\ln w), \tag{8.1.47}$$

which is integrated between 0 and 1. This makes contact with fig. 8.5. The cylinder in fig. 8.11b is a picture of the world sheet in the $\ln z$ plane. The circumference of the cylinder is 2π, corresponding to the angular change in a circuit of the boundary of the annulus. The length of the cylinder is

$$-i\pi\tau = -\frac{2\pi^2}{\ln w} = -\ln q. \tag{8.1.48}$$

Since the point $w = 1$ is mapped to the point $q = 0$ the potential divergence under consideration is associated in the q parametrization with the hole in the center of the annulus shrinking to a point.[*] The variables z_r are integrated independently of q around the outer boundary of the annulus, except that $z_M = 1$. The others are required to keep their cyclic ordering on the boundary. Introducing this change of variables into the expression for $A_P(1, 2, \ldots, M)$ in (8.1.35) requires the Jacobian

$$\frac{dw}{w} \prod_{r=1}^{M-1} \frac{d\rho_r}{\rho_r} \Theta(\rho_r - \rho_{r+1})$$

$$= \frac{1}{2\pi^2}(-\ln w)^{M+1} \frac{dq}{q} \prod_{r=1}^{M-1} \Theta(\nu_{r+1} - \nu_r) d\nu_r. \tag{8.1.49}$$

The function ψ_{rs} transforms simply under this change of variables. This can be seen by writing it in terms of the Jacobi theta function $\theta_1(\nu|\tau)$.

---

[*] The conformal equivalence of the annulus to the cylinder considered earlier makes it clear that the space-time configuration of the world sheet does not necessarily have a vanishingly small hole in the limit that the hole in the annulus becomes tiny.

(Jacobi theta functions are described in appendix 8.A.) Using the properties listed in appendix 8.A, we see that

$$\psi(\rho, w) = -2\pi i \exp(i\pi\nu'^2/\tau') \frac{\theta_1(\nu'|\tau')}{\theta_1'(0 \,|\tau')}, \tag{8.1.50}$$

where

$$\nu' = \frac{\ln \rho}{2\pi i} = -\frac{\nu}{\tau} \tag{8.1.51}$$

and

$$\tau' = \frac{\ln w}{2\pi i} = -\frac{1}{\tau}. \tag{8.1.52}$$

The behavior of the ψ_{rs}'s under the transformation of variables defined by (8.1.46) and (8.1.48) can be deduced from the behavior of the function θ_1 under modular transformations (see appendix 8.A). The result is

$$
\begin{aligned}
\psi(\rho, w) &= \frac{2\pi i}{\tau} \frac{\theta_1(\nu|\tau)}{\theta_1'(0|\tau)} \\[2mm]
&= -\frac{1}{\tau} \exp(-i\pi\nu^2/\tau)\psi(z, q^2) \\[2mm]
&= -\frac{2\pi}{\ln q} \sin \pi\nu \prod_1^\infty \frac{1 - 2q^{2n}\cos 2\pi\nu + q^{4n}}{(1 - q^{2n})^2}.
\end{aligned}
\tag{8.1.53}
$$

The function $f(w) = \left(\theta_1'(0|\tau')/2\pi w^{1/8}\right)^{1/3}$ also transforms simply under the change of variables in (8.1.47) as shown in (8.A.25). The combination that enters into the expression for the loop amplitudes is

$$\frac{1}{w}\, [f(w)\,]^{-24} = \left(\frac{-\pi}{\ln q}\right)^{12} \frac{1}{q^2}\, [f(q^2)]^{-24}. \tag{8.1.54}$$

Substituting the transformed expressions into (8.1.35) results in an alternative expression for the amplitude in which the behavior near $w = 1$ or $q = 0$ is easy to analyze, since the functions $\ln \psi(z, q^2)$ and $f(q^2)$ have well-defined series expansions in powers of q^2 around $q = 0$. Substituting (8.1.49), (8.1.53) and (8.1.54) into (8.1.35) the amplitude for the scattering of M tachyons becomes (in D dimensions)

$$A_P(1,2,\ldots,M) = G_P g^M \pi^{M-1} \int_0^1 \prod_1^{M-1} \Theta(\nu_{r+1} - \nu_r) d\nu_r$$

$$\times \int_0^1 \frac{dq}{q} w^{(D-26)/24} q^{(2-D)/12} \left[f(q^2) \right]^{2-D}$$

$$\times \prod_{r<s} \left[\sin \pi \nu_{rs} \prod_{n=1}^{\infty} \frac{1 - 2q^{2n} \cos 2\pi \nu_{rs} + q^{4n}}{(1 - q^{2n})^2} \right]^{k_r \cdot k_s},$$

$$(8.1.55)$$

where $\nu_{rs} = \nu_s - \nu_r$. The first thing to note is that for the value $D = 26$, and only for this value, the integrand has a series expansion of the form $\int_0 dq(f_0 q^{-3} + f_1 q^{-2} + \ldots)$, where $f_0, f_1, \ldots$ are functions of the external momenta and the ν_r's. This is a power series in q because the powers of w and $\ln q$ have cancelled out of the integrand of (8.1.55). The powers of w only cancel for $D = 26$. The powers of $\ln q$, which cancel for all D originate as follows: the transformation of $f(w)$ in (8.1.54) contributes $1 - D/2$ powers, the ψ_{rs}'s in (8.1.53) contribute M powers, the Jacobian of (8.1.49) contributes $-M - 1$ powers and the loop momentum integration of (8.1.17) contributes $D/2$ powers. The singularities of the amplitude at $q = 0$ are associated with the vanishing of the inner radius of the annulus in fig. 8.5.

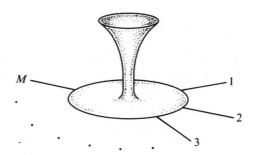

Figure 8.12. The divergences of the open-string one-loop diagram are illustrated by drawing the world-sheet configuration in which a long and thin closed string disappears into the vacuum via a disk.

Our preliminary discussion above obtained the same $\int_0^1 f_0 dq/q^3$ behavior via tachyon exchange in what we might describe (with tongue in cheek) as the 'crossed channel'. The nonleading divergence $\int_0^1 f_1 dq/q$ is likewise associated with the emission of the massless closed-string scalar state (the *dilaton*) at zero momentum. While the long cylinder of fig. 8.11b is the correct world-sheet picture, to understand what is going on in space-time we must recall that the tachyon or dilaton, like any mass eigenstate, is quasi-pointlike, with a size of order $\hbar$.[*] Hence in space-time the divergence comes from world sheets in which the long cylindrical world sheet is mapped to a narrow tube in space-time. Thus, the space-time picture is that of fig. 8.12. In supersymmetric theories, there is no tachyon; divergences associated with massless dilatons will be calculated and analyzed in the next two chapters.

Amplitudes with external massless vector states can be considered by the generalization of the above analysis in which the vertices are taken to be the vector emission vertices defined by (8.1.9). The calculation involves a trace over nonzero modes of the form of I, as defined in (8.1.11), but with vector emission vertices. The loop momentum integral gives a generalization of (8.1.17),

$$
\int d^D p \prod_{r=1}^{M} x_r^{p_r^2/2} e^{\zeta_r \cdot p_r} = \left(\frac{-2\pi}{\ln w}\right)^{D/2} \prod_{1 \le r < s \le M} \left\{ c_{sr}^{-1/2} \exp\left(\frac{\ln^2 c_{sr}}{2 \ln w}\right) \right\}^{k_r \cdot k_s}
$$

$$
\times \exp \sum_{1 \le r < s \le M} \left\{ (k_r \cdot \zeta_s - k_s \cdot \zeta_r)\left(\frac{\ln c_{sr}}{\ln w} - \tfrac{1}{2}\right) - \zeta_r \cdot \zeta_s \frac{1}{\ln w} \right\},
$$

$$(8.1.56)$$

and the product of the matrix elements of the trace over the nonzero modes generalizes the result in (8.1.24) to give

$$
\prod_{n=1}^{\infty} \prod_{\mu=1}^{D-1} T_n^{\mu} = \prod_{n=1}^{\infty} (1 - w^n)^{-D} \exp\left(-\sum_{r<s} k_r \cdot k_s \frac{c_{sr}^n + (w/c_{sr})^n - 2w^n}{n(1 - w^n)}\right)
$$

$$
\times \exp\left(\sum_{r<s}(\zeta_r \cdot k_s - \zeta_s \cdot k_r)\frac{c_{sr}^n - (w/c_{sr})^n}{(1 - w^n)}\right)
$$

$$
\times \exp\left(\sum_{r<s}\zeta_r \cdot \zeta_s n \frac{c_{sr}^n + (w/c_{sr})^n}{(1 - w^n)}\right).
$$

$$(8.1.57)$$

Combining the zero and the nonzero modes gives the expression for I

---

[*] Recall the discussion in §2.2.3.

with external massless vector states

$$I(1,\ldots,M) = [f(w)]^{2-D}\left(\frac{-2\pi}{\ln w}\right)^{D/2} \exp[\sum_{r<s}(k_r \cdot k_s \ln \psi_{rs}$$
$$+ (k_r \cdot \zeta_s - k_s \cdot \zeta_r)\eta_{rs} + \zeta_r \cdot \zeta_s \Omega_{rs})], \quad (8.1.58)$$

where terms involving ζ_r^2 in the exponent have been dropped, since only the terms linear in each ζ_r contribute to the amplitude. The function ψ_{rs} is defined as before while the other functions appearing in this expression are defined by $\eta_{rs} \equiv \eta(c_{sr}, w)$ with

$$\eta(c, w) = -\frac{1}{2} + \left(\frac{\ln c}{\ln w}\right) - \frac{c}{1-c} + \sum_{n=1}^{\infty}\left(\frac{w^n/c}{1-w^n/c} - \frac{cw^n}{1-cw^n}\right), \quad (8.1.59)$$

and $\Omega_{rs} \equiv \Omega(c_{sr}, w)$ with

$$\Omega(c, w) = -\frac{1}{\ln w} + \frac{c}{(1-c)^2} + \sum_{n=1}^{\infty}\left(\frac{w^n/c}{(1-w^n/c)^2} + \frac{cw^n}{(1-cw^n)^2}\right).$$
$$(8.1.60)$$

The function η_{rs} is proportional to $\langle \dot{X}^i(\rho_r)X^j(\rho_s)\rangle$, while Ω_{rs} is proportional to $\langle \dot{X}^i(\rho_r)\dot{X}^j(\rho_s)\rangle$. These correlation functions are related to ψ, given in (8.1.33), by

$$\eta(c, w) = c\frac{\partial}{\partial c}\ln\psi(c, w) \quad (8.1.61)$$

and

$$\Omega(c, w) = -c\frac{\partial}{\partial c}\eta(c, w). \quad (8.1.62)$$

The expression for the amplitude is given by

$$A_P(1, 2, \ldots, M) = g^M G_P \int_0^1 \frac{dw}{w^2} \int_0^1 \left(\prod_{r=1}^{M-1} \frac{d\rho_r}{\rho_r}\Theta(\rho_r - \rho_{r+1})\right) [f(w)]^{-24}$$
$$\times \left(\frac{-2\pi}{\ln w}\right)^{D/2} \prod_{r<s}(\psi_{rs})^{k_r \cdot k_s} \exp\left\{(k_r \cdot \zeta_s - k_s \cdot \zeta_r)\eta_{rs} + \zeta_r \cdot \zeta_s\Omega_{rs}\right\},$$
$$(8.1.63)$$

where it is understood that only the terms of first power in ζ_r are to be kept.

The transformations of the functions η_{rs} and Ω_{rs} under the change to the variables ν_r and τ (or, equivalently, z_r and q) are given in appendix 8.A. Using these it is straightforward to transform the expression for the amplitude to the coordinates of the annulus and verify that it has the same kind of divergence at the endpoint $q = 0$ as was found for the amplitude with external tachyons. This is not surprising, since the same expression for the scattering of M massless vector states could have been obtained by factorizing the amplitude with $2M$ tachyons on the massless vector poles in the channels formed by pairs of tachyons. It also follows, of course, from our heuristic treatment of the divergence, which did not depend on the particular choice of external states.

8.1.2 The Nonorientable Diagrams

Figure 8.6 shows an example of a Möbius strip diagram with a nonorientable world sheet. The corresponding amplitude contains the group-theory factor

$$G_N = \eta \text{tr} \left(\lambda^1 \lambda^2 \ldots \lambda^M \right). \tag{8.1.64}$$

The factor η, the group-dependent part of the twist operator (see §7.1.6 for a description of this operator), is $\eta = +1$ for the group $USp(2n)$, $\eta = -1$ for $SO(n)$ and $\eta = 0$ for $U(n)$. The group-theory factor for the planar amplitude in (8.1.1) contained a factor of $\text{tr}(1) = n$ that is not present for the Möbius strip. It was associated with the inner boundary of fig. 8.5, which has no counterpart in the present case, since a Möbius strip only has one boundary.

The Möbius strip diagram can be constructed by sewing a tree diagram in the manner described above but including a twist operator Ω in the appropriate place. In order to obtain the complete expression for the amplitude with a given order of particles around the boundary of the world sheet, *i.e.*, with a group-theory factor containing the λ matrices in a specific cyclic ordering, it is necessary to add together contributions obtained from diagrams with different distributions of twists on the internal propagators. Any odd number of twists must be allowed, since they all give Möbius strip diagrams. The case of no twists gives the planar diagram of the previous section and any positive even number of twists gives an orientable nonplanar diagram of the type considered in the next section.

The contribution to the amplitude from the particular diagram with a twist operator in the propagator between particles 1 and M is given by

$$
\begin{aligned}
A_N(1,2,\ldots,M) = g^M G_N \int d^D p \; \mathrm{Tr}\big(\Omega \Delta V_0(k_1,1) \\
\times \Delta V_0(k_2,1)\ldots \Delta V_0(k_M,1)\big)
\end{aligned}
\tag{8.1.65}
$$

where, as explained in §7.1.6, the twist operator is given by $\Omega = (-1)^{N-1}$ and N is the number operator made of bosonic (and ghost) modes. Equation (8.1.65) differs from (8.1.5) only by the addition of the twist operator and the replacement of G_P by G_N. Evaluating the oscillator traces as before, it is clear that one effect of the twist is to change the sign of the variable $w = x_1 x_2 \ldots x_M$ in (8.1.24) but not in the zero-mode momentum integral for which (8.1.17) is still appropriate. As a result the expression $\psi(c_{sr}, w)$ in (8.1.33) is replaced by $\psi^N(c_{sr}, w)$ defined by

$$
\begin{aligned}
\psi^N(c,w) = \frac{1-c}{\sqrt{c}} \exp\left(\frac{\ln^2 c}{2\ln w}\right) \\
\times \prod_{n=1}^{\infty} \frac{(1-(-w)^n c)\,(1-(-w)^n/c)}{(1-(-w)^n)^2}.
\end{aligned}
\tag{8.1.66}
$$

This can be recast in a form analogous to (8.1.50)

$$
\psi^N(\rho,w) = -2\pi i \exp\left(\frac{i\pi \nu'^2}{\tau'}\right) \frac{\theta_1(\nu'|\tau'+1/2)}{\theta_1'(0|\tau'+1/2)},
\tag{8.1.67}
$$

where the variables ν' and τ' are defined in (8.1.51) and (8.1.52). The factor of $\Omega = (-1)^N$ in (8.1.65) also results in the replacement $f(w) \to f(-w)$, both for the α oscillators and for the ghost oscillators, since N includes both kinds of modes.

The other contributions to the nonorientable one-loop amplitude with the same group-theory factor are obtained from loops with an odd number of Ω's in which the particles are arranged in the same sequence around the boundary of the Möbius strip. These diagrams contribute different pieces of the complete region of integration over the ρ_r variables in a rather nontrivial manner. For example, the diagrams that contribute to the amplitude with four external states are shown in fig. 8.13. Four of these have a single twisted propagator (including the diagram described above) while the other four diagrams each have three twisted propagators.

The world sheet for the Möbius strip is mapped into the semiannulus in the upper half ρ plane shown in fig. 8.14. In this case any point ρ

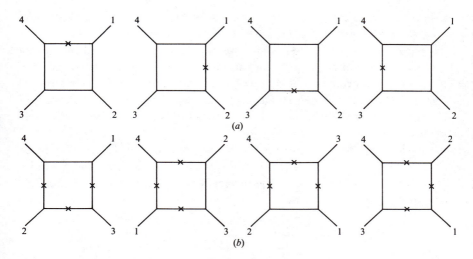

Figure 8.13. The contribution of the nonorientable loop amplitude of fig. 8.6 with four external particles has a world sheet that is a Möbius strip, where the ordering of the particles corresponds to the group-theory factor $\text{tr}\,(\lambda_1\lambda_2\lambda_3\lambda_4)$. (a) The four diagrams with a single twisted propagator. (b) The four diagrams with three twisted propagators.

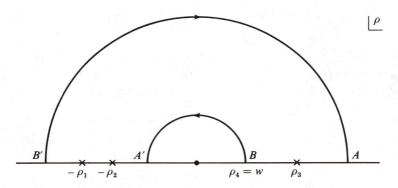

Figure 8.14. The representation of the Möbius strip as a semiannulus in the ρ plane. A point ρ is to be identified with a point $-w\overline{\rho}$ so that the arcs of the semiannulus are identified with a reversal of orientation as indicated by the arrows. The points A and A' are identified as are the points B and B'. The boundary of the Möbius strip is traced out along the real axis from A to B and from B' to A'.

is identified with $-w\overline{\rho}$ so that the inner and outer boundaries of the semiannulus are identified after a reversal of orientation as indicated by the arrows. The single boundary of the Möbius strip is mapped into the segments of the real axis (A, B) and (B', A'). The various diagrams in fig. 8.13 give contributions to the complete amplitude in which the integrands all have the same form but the range of the ρ_r variables is different. In the first of these diagrams the ρ_r variables are all integrated in the fixed sequence (1,2,3,4) in the interval $(w, 1)$ (between A and B). In the other diagrams at least one of the ρ_r variables is multiplied by -1. The result is that the contributions piece together so that in the sum of all the diagrams the variables are integrated in sequence along the whole boundary $(1, w)$ and $(-1, -w)$ $((A, B)$ and $(B', A'))$.

In order to study the divergence of this contribution to the amplitude from the limit $w = 1$ it is again useful to change to the variables ν_r and τ ((8.1.46) and (8.1.48)). Using the properties of ψ^N given in appendix 8.A, it can be recast in the form

$$
\psi^N(c, w) = \frac{4\pi^2}{\ln q} \frac{\theta_1\left(-\nu/2 \mid \tau/2 - 1/2\right)}{\theta_1'\left(0 \mid \tau/2 - 1/2\right)}
$$

$$
= -\frac{4\pi}{\ln q} \sin \frac{\pi\nu}{2} \prod_{1}^{\infty} \frac{1 - 2(-\sqrt{q})^n \cos \pi\nu + q^n}{\left(1 - (-\sqrt{q})^n\right)^2}.
$$

(8.1.68)

Also the partition function can be recast using the formula

$$
\frac{1}{w}[f(-w)]^{-24} = \left(\frac{-2\pi}{\ln q}\right)^{12} q^{-1/2}[f(-\sqrt{q})]^{-24},
$$

(8.1.69)

which is the counterpart of (8.1.54) for the nonorientable diagram.

The expression for the sum of all the terms contributing to the amplitude from a given cyclic ordering of the particles is (after setting $D = 26$)

$$
A_N(1, 2, \ldots, M) = G_N g^M \pi^{M-1} 2^{12-M} \int_0^2 \prod_1^{M-1} \Theta(\nu_{r+1} - \nu_r) d\nu_r \int_0^1 \frac{dq}{q^{3/2}}
$$

$$
\times [f(-\sqrt{q})]^{-24} \prod_{r<s} \left(\sin \frac{\pi\nu_{sr}}{2} \prod_1^{\infty} \frac{1 - 2(-\sqrt{q})^n \cos \pi\nu_{sr} + q^n}{\left(1 - (-\sqrt{q})^n\right)^2} \right)^{k_r \cdot k_s},
$$

(8.1.70)

where $\nu_{sr} = \nu_s - \nu_r$. Again, the powers of $\ln q$ cancel for $D = 26$. The range of the angular variables $2\pi\nu_r$ is now 0 to 4π, since it requires two revolutions to traverse the boundary of a Möbius strip.

The nonorientable loop expression is also divergent at the endpoint $q = 0$. To understand the divergence requires a discussion of some simple facts about the Möbius strip. The Möbius strip is easily drawn and visualized in the configuration shown in fig. 8.15a in which the boundary is not

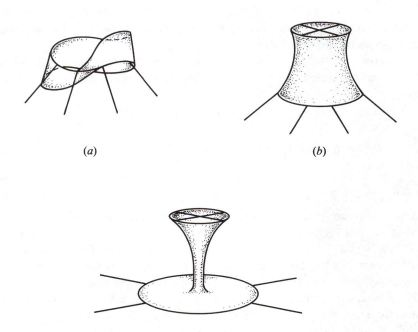

(a)   (b)

Figure 8.15. A Möbius strip is easily sketched, as in (a), in a configuration in which not all of the boundary is visible. As the boundary is a single ordinary circle, one can distort the Möbius strip to a configuration that is a simple cylinder near the boundary; this cylinder then ends at the 'cross-cap' that is symbolically indicated in (b). As in (c), the cylinder could extend for an extremely long distance before terminating in the cross-cap.

all visible and, in fact, is somewhat tangled. Drawing this configuration distracts attention from the fact that the boundary of the Möbius strip is actually a single ordinary circle. Near the boundary, the Möbius strip is just a cylinder. This cylinder, of course, does not end at a second boundary; it closes in on itself in a configuration rather difficult to draw or visualize, which is sometimes called a 'cross-cap'. This is sketched in fig. 8.15b. What would happen if one glues together the boundary component in fig. 8.15b? One would obtain a closed but not orientable world sheet – the real projective plane RP^2. This surface occurs as a world sheet of unoriented closed strings and is discussed in §8.3.1 below.

The behavior of the nonorientable open-string diagram for $w \to 1$ corresponds, as in fig. 8.15c, to a configuration in which a closed string propagates for a long proper time between the open boundary and the cross-cap. The tachyon divergence has nothing to do with details of the 'initial' and 'final' states in the 'crossed channel', but depends only on the tachyon (and dilaton) mass; this is why the unorientable diagram has the same divergence as the orientable one.

The form of the integrand of the nonorientable loop is closely related to that of the planar loop. In particular, the function ψ^N is related to ψ by

$$\psi^N(z, q^2) = 2\psi\left(z^{1/2}, -q^{1/2}\right). \tag{8.1.71}$$

Therefore by making the change of variables

$$q' = q^{1/4}, \qquad z' = z^{1/2} \tag{8.1.72}$$

in (8.1.70) the formulas for the planar and nonorientable loops can be made to look quite similar. (Because of (8.1.45), $z \rightarrow z^{1/2}$ corresponds to $\nu \rightarrow \nu/2$.) The divergences are then exhibited in the same form as in the case of the planar diagram. This fact is important in considering the possibility that there may be divergence cancellations in the sum of the planar and nonorientable diagrams. We will not pursue this for the bosonic theory, where the issue is obscured by the presence of a leading $1/q'^3$ divergence, associated with the closed-string tachyon, in addition to the $1/q'$ divergence associated with the emission of the dilaton at zero momentum. When we consider type I superstring theories in the next two chapters, we will show that there are no $1/q^3$ divergences, and it will then be more meaningful to focus on the cancellation of $1/q$ divergences.

8.1.3 Nonplanar Loop Diagrams

The third and final class of one-loop open-string diagrams to consider are those whose world sheet is topologically an annulus with external particles attached to both the inner and the outer boundaries (as in fig. 8.7). If we allow K external particles to be attached to the outer boundary and $(M - K)$ to the inner one, the diagram includes the group-theory factor

$$G_T = \text{tr}(\lambda^1 \lambda^2 \ldots \lambda^K) \, \text{tr}(\lambda^{K+1} \lambda^{K+2} \ldots \lambda^M), \tag{8.1.73}$$

where one trace factor is associated with each boundary. The planar diagram studied in §8.1.1 corresponds to the degenerate case $K = 0$.

In sewing trees to make the loop amplitude there are a number of terms with even numbers of twists that contribute to the same cyclic orderings

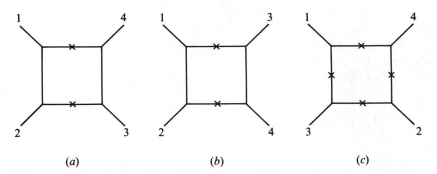

Figure 8.16. Contributions to the nonplanar four-particle amplitude with a group-theory factor tr $(\lambda_1\lambda_2)$ tr $(\lambda_3\lambda_4)$. (*a*) The two diagrams with an even number of twists. (*b*) The diagram with four twists.

of particles around each of the two boundaries. For example, the contributions to the term in the four-particle amplitude with group-theory factor tr $(\lambda_1\lambda_2)$ tr $(\lambda_3\lambda_4)$ are shown in fig. 8.16. These terms differ only in the relative ordering of pairs of particles on different boundaries. Just as in the case of the nonorientable diagrams, these different configurations give amplitudes that are integrals of the same integrand over different integration regions. In this case the world sheet for the amplitude is again mapped into the semiannular region in fig. 8.9 describing a cylindrical world sheet. The two intervals of the real axis between w and 1 and between $-w$ and -1 are the two boundaries of the cylinder on each of which two of the external states are attached. A careful analysis shows that the diagrams of fig. 8.16 combine to cover the entire region $w \leq \rho_1, \rho_2 \leq 1$, $-1 \leq \rho_3, \rho_4 \leq -w$ exactly once.

A convenient example to consider is the diagram with two twists shown in fig. 8.7. It is given (for external on-shell tachyon states) by

$$
\begin{aligned}
A_T(1,2,&\ldots,K;K+1,\ldots,M) \\
&= g^M G_T \int d^D p \operatorname{Tr}(\Omega \Delta V_0(k_1,1)\Delta \ldots \\
&\quad \times V_0(k_K,1)\Omega \Delta V_0(k_{K+1},1)\ldots \Delta V_0(k_M,1)).
\end{aligned} \tag{8.1.74}
$$

Note that particles $1,\ldots,K$ are on one boundary and $K+1,\ldots,M$ are on the other. In evaluating the trace in this expression the factors of Ω can be eliminated by commuting one of the Ω's through the intervening vertices and propagators associated with one of the boundaries – $1,\ldots,K$, say – until it is adjacent to the other Ω and drops out because $\Omega^2 = 1$. The only consequence of this is that the variables ρ_r in (8.1.11) are replaced by $-\rho_r$, i.e., $-\nu_r/\tau = \ln \rho_r/2\pi i$ is replaced by $-\nu_r/\tau + 1/2$ when $r = 1, \ldots, K$. The variables c_{sr} in (8.1.18) get a sign reversal only when r and s belong to different boundaries. As a result the factors $\psi(c_{sr}, w)$ remain unaltered when r and s are on the same boundary, while for contractions between particles on different boundaries the ψ_{rs}'s are replaced by ψ_{rs}^T defined by

$$
\psi^T(c, w) = \frac{1+c}{\sqrt{c}} \exp\left(\frac{\ln^2 c}{2 \ln w}\right) \prod_{n=1}^{\infty} \frac{(1 + w^n c)(1 + w^n/c)}{(1 - w^n)^2}
$$

$$
= 2\pi \exp\left(\frac{\ln^2 c}{2 \ln w}\right) \frac{\theta_2(\nu'|\tau')}{\theta_1'(0|\tau')}.
$$

(8.1.75)

Note that the $c \to -c$ replacement has only been made for the factors arising from the oscillator traces. The contribution from the momentum integral in (8.1.17) is unchanged from the case of the planar diagram.

In the example of the four-particle amplitude all the contributions have ρ_1 and ρ_2 multiplied by -1 and correspond to particles attached to the boundary between -1 and $-w$. In the first of the diagrams of fig. 8.16 $\rho_4 = w \le \rho_3 \le \rho_2 \le \rho_1 \le 1$. The sum of the other contributions in this figure fill in the rest of the integration region $-1 \le -\rho_1, -\rho_2 \le -w$ and $w = \rho_4 \le \rho_3 \le 1$, which corresponds to particles 1 and 2 attached to one of the boundaries and particles 3 and 4 to the other. The general result for M particles is that the two sets of ρ_r variables associated with each of the boundaries are integrated independently over one cycle around the boundary keeping the ordering of the particles on each of the boundaries fixed and holding one of them, say ρ_M, fixed at w.

The geometrical significance of the integration parameters again becomes clear after the transformation to the variables ν_r and τ. To exhibit it, we first make use of the transformation properties of the ψ^T function given in appendix 8.A, where it is shown that

$$
\psi^T = -\left(\frac{2\pi^2}{\ln w}\right) \frac{\theta_4(\nu|\tau)}{\theta_1'(0|\tau)}
$$

$$
= -\left(\frac{\pi}{\ln q}\right) q^{-1/4} \prod_{n=1}^{\infty} \frac{1 - 2q^{2n-1} \cos 2\pi\nu + q^{4n-2}}{(1 - q^{2n})^2}.
$$

(8.1.76)

The contributions to the nonplanar amplitude coming from the various distributions of even numbers of twists on the internal propagators are integrated over a range of the variables ν_r that is determined by the range of the ρ_r variables. When these contributions are pieced together, to give the total contribution associated with the group-theory factor in (8.1.72), the result is that the ν_r variables are integrated over one cycle of the outer and inner boundaries of the annulus independently, keeping the cyclic order of the particles fixed on each boundary separately (and with ν_M fixed at 1). Denoting this integration region by R, the overall contribution of the nonplanar diagram to the amplitude that multiplies the factor (8.1.72) for $D = 26$ is

$$
A_T = G_T \frac{1}{\pi} \int_R \prod_{r=1}^{M-1} d\nu_r \int_0^1 \frac{dq}{q^3} \left(\frac{-2\pi^2}{\ln q} \right)^M [f(q^2)]^{-24}
$$
$$
\times \prod_{1 \le r < s \le K} (\psi_{rs})^{k_r \cdot k_s} \prod_{K+1 \le r < s \le M} (\psi_{rs})^{k_r \cdot k_s} \prod_{\substack{1 \le r \le K \\ K+1 \le s \le M}} (\psi_{rs}^T)^{k_r \cdot k_s}.
$$

(8.1.77)

Since the formulas for ψ and ψ^T both contain a factor of $(\ln q)^{-1}$ the total $\ln q$ contribution from these factors is

$$
\prod_{r < s} (-\ln q)^{-k_r \cdot k_s} = (-\ln q)^M,
$$

(8.1.78)

and thus the $\ln q$ factors cancel as in the previous loop calculations. (This would not be the case for $D \ne 26$.) A new feature is introduced in this case by the factor $q^{-1/4}$ in the formula (8.1.76) for ψ^T. It results in a term

$$
\prod_{\substack{1 \le r \le K \\ K+1 \le s \le M}} q^{-k_r \cdot k_s / 4} = q^{-s/4},
$$

(8.1.79)

where

$$
s = - \left(\sum_{r=1}^K k_r \right)^2,
$$

(8.1.80)

and we have used the momentum conservation condition $\sum_{r=1}^M k_r = 0$. Physically, s is the square of the invariant energy associated with the channel $1, \ldots, K \to K+1, \ldots, M$ in which the incoming particles all belong to one boundary and the outgoing particles all belong to the other one.

Since the integrand of (8.1.77) contains a factor of $q^{-s/4-3}$, and the rest of the integrand is an analytic function of q^2 in the vicinity of $q = 0$, the integral does not have a divergence associated with $q \to 0$ for $s < -8$. When the integral is analytically continued in s to -8 and beyond, the onset of divergences in the q integration implies that there are singularities in s at positions given by $s/4 = -2,\ 0,\ 2,\dots.$. For $D \neq 26$, the integrand contains a nonvanishing power of $\ln q$, which causes these singularities to become branch points. This is a disaster for unitarity, since it implies the existence of contributions to the imaginary part of the loop amplitude, due to the discontinuities across the associated branch cuts, that do not correspond to phase-space integrals of products of tree amplitudes.

In previous chapters, we have seen many reasons for preferring $D = 26$ over $D < 26$ in the bosonic string theory. Only in $D = 26$ do different approaches to quantization (covariant or light-cone, for example) give the same answer. Precisely in 26 dimensions, the conformal anomaly cancels, the light-cone quantization gives a covariant answer, and the BRST operator is nilpotent. Nevertheless, a skeptic might insist that the old covariant quantization gives a Lorentz-invariant and ghost-free spectrum for any dimension less than 26. Why cannot this be the basis for a satisfactory interacting theory? What we have just seen is, perhaps, the decisive answer to this. For $D < 26$, one can indeed formulate a free theory, and even one that is consistent at tree level, but unitarity is lost at the one-loop level.[*]

When $D = 26$ the powers of $\ln q$ cancel and the singularities at $s/4 = -2,\ 0,\ 2,\dots$ are simple poles. In this case the only contributions to the imaginary part of the amplitude are δ functions at the pole positions. They can be reconciled with unitarity if the poles correspond to physical particle states of the theory. (This is how Lovelace first discovered the necessity of $D = 26$.) The fact that G_T contains separate trace factors for the initial and final states of the s channel implies that this channel is purely a singlet of the gauge group. Thus any particle poles in this channel are group-theory singlets. These states are just the closed-string states that are expected to arise, as can be seen by distorting the world surface of fig. 8.7. Also, the pole positions agree with the mass values previously established for closed bosonic strings.

The dependence on s in the integrand is similar to that anticipated in the discussion of the interpretation of the planar-loop divergence. In that case we argued that the divergences could be interpreted as arising from the closed-string states disappearing into the vacuum at zero momen-

---

tum, which is roughly the same as considering the nonplanar amplitude in the special case $K = 0$. By studying the factorization properties of the residues of the closed-string poles in the amplitude A_T, it is possible to verify that the spectrum of intermediate closed-string states is exactly the spectrum of the closed-string theory. (The technical details are quite complicated; we do not present them here.)

In addition to the closed-string poles there are open-string poles in the s channel with singlet quantum numbers. These poles arise from the part of the integral in which one of the two sets of ν_r variables (say, those with $r = 1, \ldots, K$) associated with all the particles on a boundary approach a common value independent of the value of q. The amplitude also has double open-string poles in the s channel that arise from the limit in which the coordinates ν_r of the other group of s-channel particles $(r = K + 1, \ldots, M)$ also approach a common value at the same time.

There is a coupling between the open-string poles and the closed-string poles, which arises in the simultaneous limit in which the parameter $q \to 0$ and the ν_r's for all the particles on a boundary come together. This coupling gives rise to a mixing of the particles of the closed-string sector with group-singlet states of the open-string sector. In particular, for the gauge group $U(n) = SU(n) \times U(1)$ there is a massless open-string vector state associated with the $U(1)$ factor, which can mix by this mechanism with the massless antisymmetric tensor state of the closed-string sector. When this happens they combine to give a massive antisymmetric tensor state by a kind of dynamical Higgs mechanism. However, as has already been pointed out, the group $U(n)$ is not consistent with supersymmetry in superstring theory. For the other allowed gauge groups, $SO(n)$ and $USp(n)$, the even mass levels do not contain singlets (they are just the adjoint representation). The odd levels do contain singlets, but none of them are degenerate with closed-string states, so there is no strong mixing of levels (at least in perturbation theory).

8.2 Closed-String One-Loop Amplitudes

We now consider one-loop scattering amplitudes with external closed strings. If there is an open string propagating around the loop, as in fig. 8.17a, the amplitude can be constructed exactly as in the analysis of open-string amplitudes in the previous section; one merely uses closed-string vertex operators instead of open-string vertex operators for the external lines. We therefore concentrate on the case shown in fig. 8.17b in which the internal and external lines are all closed strings. This case

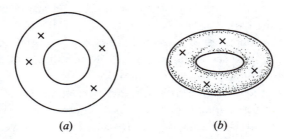

(a)   (b)

Figure 8.17. Scattering of external closed strings in a one-loop diagram in which the internal lines are (a) open strings or (b) closed strings.

presents very significant new features.

In the oriented bosonic closed-string theory a Feynman diagram is represented by a world sheet that is a closed orientable surface of genus l (*i.e.*, has l handles), where l is the number of loops. Since the topology is unique for each l, there is only a single string diagram to consider at any order in the perturbation expansion. For the unoriented closed-string theory, unitarity requires also including nonorientable surfaces (such as the Klein bottle). In either case there is no group-theory factor. The only diagram considered in this section is the oriented genus $l = 1$ surface, namely the torus. The amplitudes can be constructed by sewing together tree diagrams, just as we did for open-string loops. Even though there is a unique topology, it is still necessary in the operator formalism to add contributions from all possible orderings of the external particles in order to obtain the complete integration region. This fact already emerged in the analysis of closed-string tree amplitudes in §7.2.1. It reflects the fact that there is no natural ordering for external closed-string insertions on a world sheet.

8.2.1 The Torus

We now consider the torus amplitude for coupling M closed-string on-shell states. Arguments similar to those we have given for the open-string case would give

$$A_C(1, 2, \ldots, M) = \kappa^M \int d^D p \, \text{Tr} \left(\Delta V(k_1, 1) \Delta V(k_2, 1) \ldots V(k_M, 1) \right)$$

(8.2.1)

for a particular cyclic ordering. In this expression Δ is the closed-string propagator and V is a closed-string vertex operator, as described in §7.2.1. In contrast to the open-string case, (8.2.1) requires some modification.

First of all, it is necessary to sum over the ordering of the external states. Second, (8.2.1) overcounts the correct result by an infinite factor in a way that may at first sight be surprising.

Using world-sheet path integrals, as discussed in §1.4, the torus amplitude would be expressed in terms of integrals over a world sheet of toroidal topology, including an integral over the possible conformal structures of the torus. As we learned in §3.3, the conformal structures on the torus are given by a single complex parameter subject to rather subtle discrete equivalences under the action of the modular group $SL(2, Z)$. This actually coincides with the parameter τ introduced in §8.1, as will become apparent. We will see that analysis of (8.2.1) gives rise to the integration over the parameter τ but overlooks the discrete $SL(2, Z)$ equivalences. In the present formalism, we have to divide out the $SL(2, Z)$ action by hand. In chapter 11, we will develop a formalism in which the $SL(2, Z)$ equivalences arise automatically, but this formalism has other drawbacks, notably lack of manifest Lorentz covariance. A completely satisfactory formalism that would properly incorporate all physical principles has not yet been developed and would undoubtedly be a major advance. We will have much to say below about the role of the modular group, but first a mathematical analysis of (8.2.1) is necessary.

The vertex operators in (8.2.1) represent any physical closed-string states. Simple examples are the tachyon and the massless states discussed in §7.2.1. To keep things as simple as possible, we concentrate on amplitudes with external tachyons only. They illustrate most of the essential issues. In this case the amplitude can be written as

$$
A_C(1, 2, \ldots, M) = \left(\frac{\kappa}{4\pi}\right)^M \int \frac{d^2 w}{|w|^4}
$$

$$
\times \int \prod_{r=1}^{M-1} \frac{d^2 \rho_r}{|\rho_r|^2} \Theta\left(|\rho_r| - |\rho_{r+1}|\right) I_C(1, 2, \ldots, M),
$$

$$(8.2.2)$$

where

$$
I_C(1, 2, \ldots, M) = \int d^D p \ \mathrm{Tr}\left(V(k_1, \rho_1) \cdots V(k_M, \rho_M) w^{L_0} \overline{w}^{\tilde{L}_0}\right). \quad (8.2.3)
$$

In this expression the closed-string vertices are defined as in (6.61)

$$
V(k_r, \rho_r) \equiv V(k_r, \rho_r, \overline{\rho}_r) = V_R(\tfrac{1}{2} k_r, \rho_r) V_L(\tfrac{1}{2} k_r, \overline{\rho}_r), \quad (8.2.4)
$$

Here the complex parameters ρ_r and w are defined in terms of the z_r, which parametrize the propagators, by equations similar to (8.1.12) and

(8.1.13)

$$\rho_r = z_1 \ldots z_r,$$ (8.2.5)

$$\rho_M = w = z_1 \ldots z_M.$$ (8.2.6)

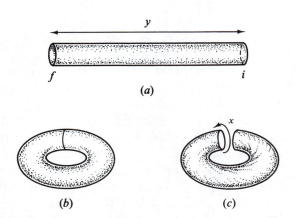

(a)

(b) (c)

Figure 8.18. A string propagating for imaginary time y between initial and final states $|i\rangle$ and $|f\rangle$ is indicated in (a). In (b) the ends of the cylinder are joined to make a closed torus corresponding to $\mathrm{Tr}\, e^{-yH}$, while in (c), a twist by an angle x is included before joining the ends of the cylinder.

The basic world-sheet structure of the torus can be made manifest by simple manipulations of (8.2.3). Let H and P be the 'Hamiltonian' and 'momentum' of the closed string, viewed as a $(1+1)$-dimensional quantum field theory. They are related to L_0 and $\tilde{L}_0$ by

$$L_0 = \frac{1}{2}(H + P), \quad \tilde{L}_0 = \frac{1}{2}(H - P).$$ (8.2.7)

Thus the factor $w^{L_0} \overline{w}^{\tilde{L}_0}$ in (8.2.3) becomes

$$(w\overline{w})^{H/2}(w/\overline{w})^{P/2}.$$ (8.2.8)

If we let

$$w\overline{w} = e^{-2y}, \quad w/\overline{w} = e^{2ix}$$ (8.2.9)

then ignoring the complicated vertex operator insertions, (8.2.3) would involve a factor of

$$\mathrm{Tr}\left(e^{-yH} e^{ixP}\right).$$ (8.2.10)

Now, $\exp\{-yH\}$ is the operator that propagates a closed string through imaginary time y. A matrix element $\langle f | \exp\{-yH\} | i \rangle$ could be represented as a path integral on a cylinder of circumference π and length y with boundary conditions at the end defined by the states $|i\rangle$ and $|f\rangle$, as in fig. 8.18a. The trace $\mathrm{Tr}\exp\{-yH\} = \sum_i \langle i|\exp\{-yH\}|i\rangle$ can be represented as a path integral on a torus by gluing the ends of the cylinder together as in fig. 8.18b. In (8.2.10), we have an additional factor of $\exp\{ixP\}$ in the trace. This is the operator that rotates the closed string by an angle $x = 2\sigma$, so the trace in (8.2.10) refers to a torus that is made by gluing together the ends of the open cylinder in fig. 8.18a with a relative twist corresponding to that angle. This is sketched in fig. 8.18c. It is in this way that the operator formalism reproduces integration over string world sheets. The complex variable $\tau = (x + iy)/2\pi$ is the unique invariant that determines the conformal structure of the torus; this parameter already entered in §3.3. Actually, τ is not quite an invariant of the complex structure, but must be subjected to certain discrete equivalences, as we have seen in §3.3. These have to be included by hand as a modification of (8.2.2), as is discussed in §8.2.2 below.

In the above, we have suppressed the vertex factors in (8.2.3). They of course are essential; they correspond to the possibility of inserting operators on the surface at positions depending on the values of the integration variables ρ_r. It is in this way that the operator formalism makes contact with world-sheet integrals over all possible world sheets and all possible positions of operator insertions.

In the present context the zero-mode momentum enters in a very similar manner as for the open-string loops (the main difference being some extra factors of $1/4$ in the coefficient of p^2 in L_0 and $\tilde{L}_0$). The zero-mode loop momentum integration gives a factor very similar to the one in (8.1.17)

$$\int d^D p \prod_{r=1}^{M} |z_r|^{p_r^2/4}$$

$$= \left(\frac{-4\pi}{\ln|w|}\right)^{D/2} \prod_{1 \leq r < s \leq M} \left\{ |c_{sr}|^{-1/2} \exp\left(\frac{\ln^2|c_{sr}|}{2\ln|w|}\right) \right\}^{k_r \cdot k_s/2}, \qquad (8.2.11)$$

where the variables $c_{sr} = \rho_s/\rho_r$ are now defined in terms of the complex variables ρ_r as in (8.1.18).

The evaluation of the oscillator traces also involves algebra similar to that of the open-string loops. In this case there is a factor of T_n^μ given by (8.1.24) for each of the left-moving nonzero modes multiplied by the complex conjugate of the same function for the corresponding right-moving mode. Combining these factors with the zero-mode factor in (8.2.11) gives the result (setting $D = 26$)

$$I_C(1, \ldots, M) = |f(w)|^{-48} \left(\frac{-4\pi}{\ln|w|} \right)^{13} \exp\left\{ \sum_{r<s} \tfrac{1}{2} k_r \cdot k_s \ln \chi_{rs} \right\}, \quad (8.2.12)$$

where the functions $\chi_{rs} \equiv \chi(c_{sr}, w)$, analogous to the ψ_{rs}'s, are given by

$$\chi(c, w) = \exp\left(\frac{\ln^2|c|}{2\ln|w|} \right) \left| \frac{(1-c)}{c^{1/2}} \prod_{m=1}^{\infty} \frac{(1 - w^m c)(1 - w^m/c)}{(1 - w^m)^2} \right|, \quad (8.2.13)$$

and $\ln \chi_{rs}$ is proportional to the correlation function $\langle \dot{X}^i(\rho_r) X^j(\rho_s) \rangle$ on the torus.

In order to study this expression in terms of variables that reveal the torus structure, it is useful to express χ in terms of the Jacobi theta function θ_1. To do this we define the variables

$$\nu_r = \frac{\ln \rho_r}{2\pi i}, \qquad \tau = \nu_M = \frac{\ln w}{2\pi i}, \quad (8.2.14)$$

in analogy with ν' and ρ' in (8.1.51) and (8.1.52) (so that $\operatorname{Im}\nu_r = -\ln|\rho_r|/2\pi$ and $\operatorname{Im}\tau = -\ln|w|/2\pi$). In terms of these $\chi_{rs} = \chi(\nu_{sr}, \tau)$, where

$$\chi(\nu, \tau) = 2\pi \exp\left\{ \frac{-\pi(\operatorname{Im}\nu)^2}{\operatorname{Im}\tau} \right\} \left| \frac{\theta_1(\nu|\tau)}{\theta_1'(0|\tau)} \right|, \quad (8.2.15)$$

and $\nu_{sr} = \nu_s - \nu_r$. The change of integration variables gives a Jacobian defined by

$$\frac{d^2 w}{|w|^2} \prod_{r=1}^{M-1} \frac{d^2 \rho_r}{|\rho_r|^2} = (4\pi^2)^M d^2\tau \prod_{r=1}^{M-1} d^2\nu_r. \quad (8.2.16)$$

By substituting (8.2.12) into (8.2.2) and changing integration variables one obtains the formula

$$A_C(1, 2, \ldots, M) = \left(\frac{\kappa}{4\pi} \right)^M \int \frac{d^2 w}{|w|^4} \int \prod_{r=1}^{M-1} \frac{d^2 \rho_r}{|\rho_r|^2} |f(w)|^{-48}$$

$$\times \left(\frac{-4\pi}{\ln|w|} \right)^{13} \prod_{r<s} (\chi_{rs})^{k_r \cdot k_s/2}$$

$$(8.2.17)$$

$$= (\pi\kappa)^M \int d^2\tau \int \prod_{r=1}^{M-1} d^2\nu_r \left(\frac{2}{\operatorname{Im}\tau} \right)^{13}$$

$$\times e^{4\pi \operatorname{Im}\tau} |f(w)|^{-48} \prod_{r<s} (\chi_{rs})^{k_r \cdot k_s/2},$$

where the ρ_r integration region is $0 < |w| = |\rho_M| \le |\rho_{M-1}| \le \ldots |\rho_1| \le 1$.

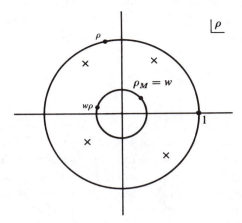

Figure 8.19. The mapping of the closed-string loop to the surface of a torus is represented by the annular region. The external particles, indicated by crosses, are integrated over the annular region with the boundary condition that identifies any point ρ on the outer boundary with the point $w\rho$ on the inner one.

The complete amplitude involves summing expressions of this type over all possible permutations of the external particles. The result is the same expression but with the ρ_r's integrated over the region $|w| \le |\rho_r| \le 1$ with no other constraint. This region is the annulus shown in fig. 8.19.

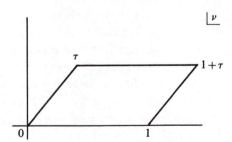

Figure 8.20. A representation of the integration region of the ν_r variables on the torus. With $\tau = (x + iy)/2\pi$, the torus is made by gluing together the sides of a parallelogram of height y and shift parameter x.

The integration variables of the closed-string loop amplitude have a simple geometrical interpretation in terms of a torus. Once we understand it, the correct choice for the integration region will become clear. The basic picture that will emerge is one in which the parameter τ characterizes conformally inequivalent geometries of a torus and the parameters ν_r

describes the position of the rth particle on the surface of the torus. We have already explained qualitatively why this happens. To see how this comes out of the formulas let us note first that

$$\chi(\nu+1,\tau) = \chi(\nu+\tau,\tau) = \chi(\nu,\tau), \qquad (8.2.18)$$

as is shown in appendix 8.A. Thus for fixed τ the function $\chi(\nu,\tau)$ is a doubly-periodic function of ν with periods 1 and τ. This means that the complex ν plane can be divided into parallelograms displaced from one another by integer multiples of 1 and τ, each one giving a copy of the same function, as shown in fig. 8.20. Therefore the function χ can be regarded as living on a single one of these parallelograms with opposite edges identified, in other words a torus. With $\tau = (x+iy)/2\pi$, this torus is made, as in fig. 8.20, by displacing a closed string by a height y with a shift of magnitude x, corresponding precisely to tr $\exp\{-yH+ixP\}$ as discussed earlier.

In the case of the genus zero surface there was an $SL(2,C)$ symmetry of the infinite plane that needed to be taken into account. So we may naturally ask whether there is an analog in the present case. The answer is clearly yes, as one can see by referring to fig. 8.20. The position of the center of the parallelogram is an arbitrary choice, its shape and orientation are important in order that it describe a given torus. In terms of the ν_r variables, this means that they can all be shifted by a common complex constant. In particular one of them may be fixed at a particular value and not integrated. In fact, the operator calculation gave the particular choice $\nu_M = \tau$. The effect of this $[U(1)]^2$ symmetry is actually easier to deal with than the $SL(2,C)$ symmetry of the tree amplitude. In the genus 1 case it does not really matter whether the coordinate ν_M is integrated over the torus or mapped to a particular point using the symmetry, because the group in question now has finite volume, namely the area of a parallelogram (Im τ), which could be compensated by including an extra factor of (Im $\tau)^{-1}$.

8.2.2 Modular Invariance

Our expression for the toroidal contribution to the scattering amplitude contains an integral over the complex parameter τ, which determines the conformal structure of the torus. We know from §3.3 that although a choice of τ uniquely specifies the conformal structure of the torus, the converse is not quite true. If

$$\begin{pmatrix} a & b \\ c & d \end{pmatrix}$$ (8.2.19)

is a 2×2 matrix with integer entries and determinant one, then τ and

$$\tau' = \frac{a\tau + b}{c\tau + d}$$ (8.2.20)

determine tori with the same complex structure. The τ integration should be restricted to a 'fundamental region' that includes each possible conformal structure on the torus precisely once. This does not arise automatically from the formalism that we are using here for loop diagrams, and we have to include the necessary modification by hand.

While a self-contained explanation of how the group $SL(2, Z)$ enters was already presented in §3.3, we pause here for an alternative and illuminating way of thinking about this. The Nambu–Goto action is invariant under reparametrizations of the string world sheet. The obvious reparametrizations are the infinitesimal ones

$$\sigma^\alpha \to \sigma^\alpha + \xi^\alpha(\sigma^\beta)$$ (8.2.21)

as well as reparametrizations that can be reached by exponentiating infinitesimal ones. The infinitesimal reparametrizations generate a group that is connected, of course, like the group generated by any set of infinitesimal reparametrizations. In our gauge fixing of the Nambu–Goto action, we only considered infinitesimal reparametrizations. This was actually correct at the tree level. According to classical theorems, the diffeomorphism groups of the tree level world surfaces (disk or sphere for open or closed strings) are connected and are generated by the infinitesimal reparametrizations.

At genus one and higher, the situation is quite different; the diffeomorphism group is not connected, and there are reparametrizations of the string world sheet that are not taken into account in the Faddeev–Popov gauge fixing. These are the modular transformations, as we will see.

Notice that in the assertion that the diffeomorphism group of the torus is not connected, the torus enters only as a topological entity; a choice of metric or conformal structure on the torus is not relevant. Our task is to pick one standard torus and discuss its diffeomorphism group. Consider building a torus by imposing an equivalence relation in the $\sigma_1 - \sigma_2$ plane (in this section we refer to the world-sheet coordinates as σ_1, σ_2, since

σ, τ would invite confusion with the traditional name τ for the modular parameter). In the preceding subsection we parametrized the torus by a complex variable ν with periods 1 and τ. The complex parameter ν can be replaced by a pair of real parameters σ_1 and σ_2, each of period one by defining

$$\nu = \sigma_1 + \sigma_2 \tau. \tag{8.2.22}$$

The equivalence relation that enters in defining the torus can then be expressed as

$$(\sigma_1, \sigma_2) \approx (\sigma_1 + n, \sigma_2 + m) \tag{8.2.23}$$

for arbitrary integers n and m.

Our claim is now that this torus admits diffeomorphisms that are not continuously connected to the identity. They are the transformations

$$(\sigma_1, \sigma_2) \rightarrow (a\sigma_1 + b\sigma_2, c\sigma_1 + d\sigma_2), \tag{8.2.24}$$

where $a, b, c,$ and d define an element of $SL(2, Z)$. Indeed, the transformation (8.2.24) is compatible with the equivalence relation (8.2.23) (points equivalent under (8.2.23) are mapped to equivalent points) only if $a, b, c,$ and d are integers. Furthermore, (8.2.24) is an invertible, one-to-one transformation of the torus to itself if and only if $ad - bc = 1$, the inverse of (8.2.24) being in this case

$$(\sigma_1, \sigma_2) \rightarrow (d\sigma_1 - b\sigma_2, -c\sigma_1 + a\sigma_2). \tag{8.2.25}$$

Thus, the group $SL(2, Z)$ can be realized as a group of reparametrizations of the torus. The transformation (8.2.24) can be reached continuously from the identity by exponentiating infinitesimal reparametrizations only if $a = d = 1, b = c = 0$. An arbitrary diffeomorphism of the torus can be put in the form (8.2.24) (for some choices of $a, b, c,$ and d) by multiplying by a diffeomorphism that can be reached continuously from the identity. Thus, the modular group $SL(2, Z)$ (and its multiloop analogs) is precisely the part of the reparametrization invariance of the Nambu–Goto action that is not taken into account in the conventional Faddeev–Popov gauge fixing.

Some particular examples of modular transformations are particularly easy to visualize. The transformation

$$\begin{pmatrix} 0 & 1 \\ -1 & 0 \end{pmatrix} \tag{8.2.26}$$

exchanges σ_1 and σ_2, with a change of sign. Its action on the modulus τ

of the conformal structure of the surface is just

$$\tau \to -1/\tau. \tag{8.2.27}$$

The transformation

$$\begin{pmatrix} 1 & 1 \\ 0 & 1 \end{pmatrix} \tag{8.2.28}$$

brings about a σ_2 dependent shift in σ_1. Its action on the modular parameter is

$$\tau \to \tau + 1 \tag{8.2.29}$$

With some effort it can be shown that (8.2.27) and (8.2.29) generate the whole modular group, so that in discussions of modular invariance it is for some purposes adequate to verify proper behavior under these transformations.

Although the Nambu–Goto action is formally invariant under infinitesimal reparametrizations, such invariance actually only holds quantum mechanically in 26 dimensions, where the anomaly cancels. Likewise, although $SL(2, Z)$ invariance on a toroidal world sheet is formally a consequence of the underlying reparametrization invariance, we must ask whether this invariance actually holds in the quantum theory or is spoiled by a 'global anomaly', *i.e.*, a quantum-mechanical breakdown of modular invariance. We will see below that $SL(2, Z)$ invariance is crucial for the absence of ultraviolet divergences in the theory, so an anomaly that spoils $SL(2, Z)$ would be disastrous. Furthermore, since $SL(2, Z)$ invariance is an integral part of the geometrical interpretation of the propagating string world sheet, an anomaly would be a disaster for the geometrical interpretation of string theory.

We now show explicitly that (in 26 dimensions!) the integrand in the toroidal diagram is indeed invariant under modular transformations, and therefore there is no one-loop anomaly in the $SL(2, Z)$ invariance. The amplitude (8.2.17) can be written in the form

$$A_C(1, 2, \ldots, M) = \int \frac{d^2\tau}{(\text{Im } \tau)^2} C(\tau) F(\tau), \tag{8.2.30}$$

where the measure $(\text{Im } \tau)^{-2} d^2\tau$ and each of the factors C and F will turn out to be separately invariant under modular transformations. The invariance of the measure follows from the fact that under $\tau \to (a\tau +$

b)/(cτ + d) we have

$$d^2\tau \rightarrow |c\tau + d|^{-4} d^2\tau \qquad (8.2.31)$$

and

$$\text{Im } \tau \rightarrow |c\tau + d|^{-2} \text{Im } \tau, \qquad (8.2.32)$$

These facts are easily established. The fact that the function

$$F(\tau) = 4(\kappa\pi)^M \text{Im}\tau \int \left(\prod_1^{M-1} d^2\nu_r \right) \prod_{r<s} (\chi_{rs})^{k_r \cdot k_s/2} \qquad (8.2.33)$$

is invariant under the $SL(2, Z)$ transformations follows from the fact that χ is a modular function of unit weight (as shown in appendix 8.A) together with the kinematic relation for on-shell closed-string tachyon states

$$\sum_{1 \leq r < s \leq M} k_r \cdot k_s = -4M. \qquad (8.2.34)$$

To verify the invariance of $F(\tau)$ it is necessary to transform the ν_r variables to $\nu_r' = \nu_r/(c\tau+d)$ at the same time as τ is transformed to $(a\tau+b)/(c\tau+d)$. The complex variable ν_r' is integrated over the parallelogram with corners at the points 0, $(d\tau' - b)$, $(a - b + d\tau' - c\tau')$ and $(a - c\tau')$, which has area Im τ'. The invariance of the transformed functions under $\nu' \rightarrow \nu' + 1$ and $\nu' \rightarrow \nu' + \tau'$ allows the ν' integration variables to be shifted so that they are integrated over the fundamental parallelogram bordered by the lines $(0, 1)$ and $(0, \tau')$ so the transformed integrand is integrated over the same region defined in terms of τ' as the untransformed integral was in terms of τ.

The invariance of the function $C(\tau)$, defined by

$$C(\tau) = (\tfrac{1}{2}\text{Im } \tau)^{-12} e^{4\pi\text{Im } \tau} |f(w)|^{-48}, \qquad (8.2.35)$$

is verified by using the transformation properties of $|f(w)|$ obtained from (8.1.54) (using the identifications $\ln w = 2i\pi\tau$ and $\ln q^2 = \ln w' = 2i\pi\tau'$), which gives

$$e^{4\pi\text{Im } \tau} |f(w)|^{-48} = |\tau'|^{-24} e^{4\pi\text{Im } \tau'} |f(w')|^{-48}. \qquad (8.2.36)$$

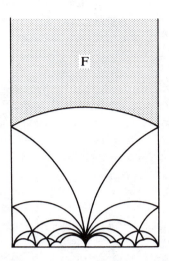

Figure 8.21. The τ plane. The curves separate fundamental regions of the modular group. The region marked F does not touch the Im $\tau = 0$ axis and is taken to be our 'standard' region.

8.2.3 The Integration Region

We have described the conformal structure of the one-loop diagram in terms of a complex variable τ of positive imaginary part. The upper half of the τ plane is mapped into itself by $SL(2, Z)$ in a rather complicated way. A region R of the upper half plane with the property that every point in the upper half plane can be mapped into R by a modular transformation in a unique way is called a fundamental region for the modular group. There is no unique way to split the upper half plane into fundamental regions, but a conventional and convenient choice is in fig. 8.21. One fundamental region in fig. 8.21, which we choose to view as the 'standard' one, is the region $-\frac{1}{2} \leq \mathrm{Re}\,\tau \leq \frac{1}{2}$, Im $\tau > 0$, $|\tau|^2 > 1$, which is marked F in fig. 8.21.

The expression (8.2.17) for the loop amplitude derived by sewing the tree diagram together involves integrating τ over a strip in the upper half plane, Im $\tau \geq 0$ and $-\frac{1}{2} \leq \mathrm{Re}\,\tau \leq \frac{1}{2}$. However, this corresponds to including an infinite number of fundamental regions that are mapped into each other by the action of the modular group. Integrating τ over the whole strip would definitely give an infinite and nonunitary amplitude (see fig. 8.3b), since the whole integral, including its discontinuity, would be counted an infinite number of times. This means that the expression given in (8.2.17) is incorrect. In order to avoid this overcounting the integration over τ must be truncated to just one region of the modular group, for example the region marked F in the figure.

In a very rough sense, one might think of Im τ as the imaginary time that our string has propagated in the loop. In this sense, the region of small Im τ might be viewed as an 'ultraviolet' region of small time propagation, while large Im τ is an 'infrared' region. The restriction to the fundamental region in this description eliminates the possibility of 'ultraviolet' divergences, while leaving the possibility of infrared divergences due to massless particles and tachyons.

The restriction on the integration region described above has to be implemented by hand in the construction described here, but it must emerge naturally in any consistent treatment based on a field theory of strings (such as the light-cone field theory), since such a field theory is guaranteed to be unitary. The restricted region of the τ integration has no simple interpretation in terms of the original propagators that make up the loop amplitude. The z_r parameters of the integral representation of the internal propagators in any individual contribution to the amplitude are integrated over limits that are coupled to each other when τ is restricted to F, which means that the loop amplitude is not equal to the sum of component loops of particles that make up the string spectrum.

The normal threshold singularities of the one-loop scattering amplitude required by unitarity can be exhibited as arising from endpoints of the integrations over the parameters ν_r and τ in much the same manner as in the case of open strings. Poles in any given channel arise from the region in which the variables associated with the particles in that channel approach each other on the surface of the torus, *i.e.*, the limit in which some of the variables $\nu_{sr} \to 0$. When $\nu_{sr} \sim 0$ the function χ_{rs} is particularly simple, being given by

$$\chi_{rs} \sim 2\pi|\nu_{rs}|. \tag{8.2.37}$$

The region of parameter space in which clusters of particles approach each other correspond to space-time embeddings of the world sheet in which clusters of particles are separated from the other particles by a long thin tube. The normal threshold singularities arise from limits in which w is close to zero (or Im $\tau \to \infty$) due to the vanishing of the z_r variables associated with internal propagators. This corresponds to a space-time picture in which the world sheet develops a pair of long thin internal tubes that separate clusters of the external states.

8.2.4 Analysis of Divergences

There are several possible limits in which divergences can arise. One is a divergence associated with the limit in which all the ν_r's except one approach each other. This corresponds to isolating the loop on an external tachyon leg, so that there is a singularity from a tachyon propagator evaluated on shell (fig. 8.10). This kind of divergence, which also arose for open-string loops, again can be interpreted as a renormalization of the mass of the tachyon. Although there are technical problems in calculating the appropriately renormalized theory in the absence of an off-shell formulation, this is no more of a conceptual problem than the analogous problem in conventional point field theory.

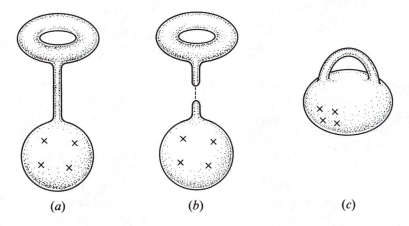

$$(a) \qquad\qquad (b) \qquad\qquad (c)$$

Figure 8.22. A string world sheet with a long neck connecting a tree diagram to a torus is sketched in (a). The integral over the length of the neck diverges due to propagation of the tachyon; the residue of the infinity has the interpretation indicated in (b). The world sheet of (a) can be conformally mapped into the picture sketched in (c) in which the world sheet has no long neck, and the external states are inserted very close together.

The really significant divergences arise from the region in which all the particles approach each other on the torus, *i.e.*, $\nu_{rs} \to 0$ for all r, s. Let us first discuss this region heuristically, as we did in the open-string case. Consider, as in fig. 8.22a, a string world sheet in which a sphere (tree diagram) with external vertex operators is connected by a long tube to a torus on which no vertex operators are inserted. The integration over the length of the neck gives rise to precisely the same $\int_0 dq/q^3$ divergence that we met in open-string diagrams. The coefficient of the divergence factorizes as in fig. 8.22b. On the other hand, by a conformal rescaling of the metric, the configuration of fig. 8.22a is equivalent to the configuration that we are discussing in which the particles are close together on the world sheet so that $\nu_{rs} \to 0$ for all r, s (as in fig. 8.22c).

Indeed, consider the two-dimensional metric

$$ds^2 = dr^2 + a(r)d\phi^2 \tag{8.2.38}$$

with $a(r)$ being a smooth function that behaves as r^2 for small r. The choice

$$a(r) = \sin^2 r, \qquad 0 \le r \le \pi \tag{8.2.39}$$

is the metric of a sphere. On the other hand, $a = $ constant would be the metric of a cylinder. By combining these, as in

$$\begin{aligned} a(r) &= \sin^2 r, && \text{if } 0 \le r \le \pi - \delta; \\ a(r) &= \sin^2(\pi - \delta), && \text{if } \pi - \delta \le r \le R \end{aligned} \tag{8.2.40}$$

we get (for small δ and large R) a sphere with a long thin tube coming out. Now, let us consider the conformal rescaling of the metric that converts (8.2.38) into the flat metric of the plane. We try to write

$$dr^2 + a(r)d\phi^2 = b(r)(dr'(r)^2 + r'(r)^2 d\phi^2) \tag{8.2.41}$$

where $b(r)$ and $r'(r)$ are functions that must be determined. Dropping the factor $b(r)$ on the right-hand side of (8.2.41) by a conformal rescaling of the metric, we would then arrive at the standard flat metric of the plane $ds^2 = dr'^2 + r'^2 d\phi^2$. It is easy to see that (8.2.41) requires that $a(dr'/dr)^2 = (r')^2$, which integrates to give

$$r' = \exp(\int_{r_0}^{r} dy/\sqrt{a(y)}), \tag{8.2.42}$$

where r_0 is an arbitrary constant of integration. This is a general formula; let us apply it to the case in which the metric is of the form (8.2.40). We want to consider in (8.2.40) the limit of very large R in which there is a very long neck in fig. 8.22a. In order to compare with the formulas for the loop amplitude, it is most useful to arrange that the part of the world sheet described by the metric of (8.2.40) is mapped to a region whose size is independent of R. This requires choosing $r_0 \sim R$ in (8.2.42), which in turn means that the region of $r \le \pi - \delta$ in (8.2.40) is mapped in the r'

variable to $r' \leq \exp(-R/\sin(\pi - \delta))$. Thus we have shown that the region in which the vertex operators are inserted nearby on the world sheet, or in other words the region $\nu_{rs} \to 0$ for all r, s, is conformally equivalent to the region with the infrared divergence suggested in fig. 8.22a.

To examine this region in a more precise mathematical way, it is convenient to define variables η_r by

$$\epsilon \eta_r = \nu_r - \nu_M \tag{8.2.43}$$

for $r = 1, 2, \ldots, M - 2$, where the real variable ϵ is $|\eta_{M-1}|$, i.e.,

$$\eta_{M-1} \equiv \epsilon\, e^{i\phi} = \nu_{M-1} - \nu_M. \tag{8.2.44}$$

The Jacobian for this transformation of variables is

$$\prod_{r=1}^{M-1} d^2\nu_r = i\epsilon^{2M-3} d\epsilon\, d\phi \prod_{r=1}^{M-2} d^2\eta_r. \tag{8.2.45}$$

Making use of (8.2.37) we can expand the integrand of the amplitude in (8.2.34) in a power series in ϵ and obtain a leading divergent contribution of the form

$$\int_0^1 \frac{d\epsilon}{\epsilon^3} d\phi \left(\prod_{r=1}^{M-2} d^2\eta_r \right) \prod_{1 \leq r < s \leq M-1} |\eta_r - \eta_s|^{k_r \cdot k_s/2} \int_F \frac{d^2\tau}{(\mathrm{Im}\,\tau)} C(\tau). \tag{8.2.46}$$

Here we see the expected $d\epsilon/\epsilon^3$ divergence, just as for open strings. The η_r integration is an M-point function on the sphere, while the τ integration is a vacuum-to-vacuum amplitude on the torus (an integral over the fundamental region of the modular group with no vertex operators inserted on the torus).

The next term in the expansion of the integrand gives a divergence of the form $\int d\epsilon/\epsilon$ corresponding to the propagation of a massless dilaton, rather than a tachyon, down the long neck of fig. 8.22a. The coefficient of this divergence,

$$\int_F d^2\tau (\mathrm{Im}\,\tau)^{-14} e^{4\pi \mathrm{Im}\,\tau} |f(e^{2\pi i\tau})|^{-48}, \tag{8.2.47}$$

should be proportional to the coupling of a dilaton to a toroidal world sheet, i.e., to the dilaton one-loop expectation value. This can be seen explicitly by comparing to the formula obtained directly from the one-loop closed-string amplitude with a single external on-shell dilaton. The vertex for emitting an on-shell dilaton at zero momentum is given by

$$V_D = \dot{X}^\mu_R(1)\dot{X}_{L\mu}(1), \tag{8.2.48}$$

so that the dilaton one-loop expectation value (in which $k = 0$) can be written as

$$\frac{\kappa}{4\pi} \int d^{26}p \int \frac{dz d\bar{z}}{(z\bar{z})^2} \text{Tr} \left(z^{L_0} \bar{z}^{\bar{L}_0} \dot{X}^\mu_R(z) \dot{X}_{L\mu}(\bar{z}) \right). \tag{8.2.49}$$

Writing $z = \exp\{2i\pi\tau\}$ this expression can easily be seen to be proportional to (8.2.47) with a finite constant of proportionality if the z integral is assumed to be restricted to a single fundamental region.

This restriction of the τ integration to a single fundamental region is required for agreement with (8.2.47), although it is not determined directly from the one-dilaton amplitude, since its normalization is not determined directly by unitarity. Since the dependence on the loop momentum enters the integrand in (8.2.49) via the gaussian factors $|z|^{p^2/4} \sim \exp\{-\pi \text{Im} \tau p^2/2\}$, the region of large loop momentum is suppressed by the restriction of the τ integration to the fundamental region, which does not include the region $\text{Im} \tau \sim 0$ (*i.e.*, $|z| \sim 1$). This is an explicit example of how the ultraviolet divergences usually associated with field-theory loop integrals are eliminated by the modular properties of closed-string theory.

We saw in §3.4.6 that Weyl invariance of string theory implies that the dilaton expectation vanishes at the level of tree diagrams. However, this is evidently not true at the one-loop level. In fact, not only is the coupling nonzero, but it is infinite since the integral in (8.2.47) is itself divergent at the endpoint $\text{Im} \tau \to \infty$. This divergence comes from the leading term in the expansion of the integrand of (8.2.49) in a power series in $z\bar{z}$. This term is associated with the tachyon state circulating around the loop and has nothing to do with the ultraviolet properties of the expression. The divergence can be viewed in space-time as originating from configurations in which the world sheet has the form of a torus of very large circumference.

8.2.5 The Cosmological Constant

In general relativity it is possible to add the term $\Lambda g^{\mu\nu}$ to the matter energy–momentum tensor, where Λ is the cosmological constant and $g^{\mu\nu}$ is the space-time metric tensor. This corresponds to a term in the Einstein–Hilbert action (in D dimensions) of the form

$$-\int d^D x \sqrt{g} \Lambda, \tag{8.2.50}$$

where $g = \det g_{\mu\nu}$. In the classical theory a nonzero value for Λ is equivalent to an extra energy density that is not generated by the matter in the universe. Astronomical data shows that the four-dimensional cosmological constant is fantastically small, less than about 10^{-120} in units of (Planck mass)^4. Unfortunately, even if it is set equal to zero in the classical theory the change of the vacuum energy due to quantum fluctuations generically results in a huge cosmological constant in conventional field theories. In supersymmetric theories these vacuum fluctuations vanish and there are no quantum corrections to the cosmological constant. However, supersymmetry is certainly broken in nature, at least at the scale associated with weak symmetry breaking, so that the observed vanishing of Λ remains an essential feature to be explained. Traditionally, particle physicists have ignored this problem in the absence of a sensible quantum theory of gravity but it is obviously an essential feature of physics that must be confronted by superstring theory.

At the tree level in string theory, the cosmological constant vanishes. This is shown by the fact that one does not run into any trouble if one attempts to quantize the free string propagating in flat Minkowski space. By contrast, if one attempts to consider propagation of a free string in a world with nonzero cosmological constant (de Sitter or anti-de Sitter space, for instance), one runs into a failure of conformal invariance along the lines of the discussion in §3.4. Granted that the cosmological constant vanishes in string theory at the tree level, we now want to determine what happens at the one-loop level.

To begin with, recall the calculation of the quantum corrections to Λ in ordinary field theory with point particles. The expression for the cosmological constant is given by the sum of the vacuum (or zero-point) energy of each species of particle. These quantum corrections to Λ can be determined from the term in the quantum effective action (*i.e.*, in the free energy) of the form of (8.2.50). To lowest order in perturbation theory this depends only on the kinetic terms in the field-theory action. Integrating over the field degrees of freedom in the partition function gives the expression for Λ arising from one particular species of particle with mass m,

$$\Lambda = \pm\frac{1}{2}\ln\det(p^2 + m^2) = \pm\frac{1}{2}\int \frac{d^D p}{(2\pi)^D}\ln(p^2 + m^2), \tag{8.2.51}$$

where the + sign arises if the particle is a boson and the − sign if it is a fermion. This expression is the most trivial kind of loop amplitude in which there are no interaction vertices, and it is also very obviously divergent in the absence of a regulator. In order to discuss the string calculation, it is useful to rewrite Λ by using

$$
\ln a = \lim_{\epsilon \to 0} \left[\int_{\epsilon}^{\infty} \frac{dt}{t} e^{-at} - \ln \epsilon \right].
\tag{8.2.52}
$$

Ignoring the divergent constant $\ln \epsilon$, the expression for Λ in (8.2.51) can be written as

$$
\Lambda = \mp \frac{1}{2} \int_{0}^{\infty} \frac{dt}{t} \int \frac{d^D p}{(2\pi)^D} e^{-t(p^2 + m^2)},
\tag{8.2.53}
$$

(where the − sign refers to bosons and the + sign to fermions). The fact that the integral diverges at the $t = 0$ limit is a symptom of ultraviolet divergences that need regularizing in ordinary field theories. However, as we will see, it causes no problem in string theory.

A plausible *ansatz* for the form of the cosmological constant in string theory is that it is the sum of the vacuum energies from each of the constituent states of the string. This is easy to evaluate, since the trace of the string propagator is the sum of propagators of the constituent states. The closed-string propagator introduced in §7.2.1 can be written as

$$
\Delta = \frac{\alpha'}{2\pi} \int_{|z| \leq 1} \frac{d^2 z}{|z|^2} z^{L_0 - 1} \bar{z}^{\tilde{L}_0 - 1}.
\tag{8.2.54}
$$

In comparing this with the propagator of a point particle, we can write

$$
z = e^{2i\pi\tau} \equiv e^{2i\pi(\operatorname{Re}\tau + i\operatorname{Im}\tau)},
\tag{8.2.55}
$$

so that $d^2 z / |z|^2 = 4\pi^2 d^2 \tau$. The integral over $\operatorname{Re}\tau$ projects onto the physical sector of states satisfying $L_0 = \tilde{L}_0$ while the $\operatorname{Im}\tau$ integration corresponds to the proper-time integration in (8.2.53), the relation of τ and t being $t = \pi\alpha' \operatorname{Im}\tau$. The sum of the contributions to the cosmological constant coming from the component fields of the string can therefore be expressed as

$$\Lambda = -\frac{1}{2} \int \frac{d^{26}p}{(2\pi)^{26}} \int d^2\tau \frac{1}{\operatorname{Im}\tau} e^{\pi(4-\alpha'p^2)\operatorname{Im}\tau} \operatorname{tr}\left(z^N \bar{z}^{\tilde{N}}\right)$$

$$= -\frac{1}{2} \left(\frac{1}{4\pi^2\alpha'}\right)^{13} \int d^2\tau \, (\operatorname{Im}\tau)^{-14} e^{4\pi\operatorname{Im}\tau} \left|f\left(e^{2i\pi\tau}\right)\right|^{-48}.$$

(8.2.56)

If we took the integration region in (8.2.56) to be the semi-infinite strip consisting of $0 < \operatorname{Im}\tau < \infty, -1/2 \leq \operatorname{Re}\tau \leq 1/2$, then (8.2.56) would correspond rather closely to (8.2.53). In fact, however, this is the wrong choice. The integrand of (8.2.56) is easily seen to be modular invariant by the same reasoning as we used above for the case of the dilaton expectation value. The τ integration region should therefore be restricted to cover a single fundamental region of the modular group shown in fig. 8.21. The restriction to a single fundamental region does not correspond to any simple operation that one might imagine in field theory. Roughly speaking, the exclusion of the region near $\operatorname{Im}\tau = 0$ ($|z| \sim 1$) is tantamount to cutting off the ultraviolet contributions to the vacuum energy of the component string states. The cosmological constant is still infrared divergent in the bosonic string theory due to the tachyon contribution, but the presence of the ultraviolet cut off is a significant departure from the expression obtained in point-particle theories.

The expression in (8.2.56) is proportional to the dilaton expectation value, (8.2.47). This relationship between the cosmological constant and the dilaton expectation value is the simplest example of a general phenomenon. The addition of a zero-momentum dilaton to an arbitrary process gives a result that is proportional to the derivative of the amplitude without the soft dilaton with respect to the string tension. This is closely related to a scaling behavior discussed in §13.2.

8.2.6 Amplitudes with Closed-String Massless States

We now briefly consider one-loop diagrams with external massless closed string states. A massless states with polarization tensor $\zeta_r^{\mu\nu}$ has a vertex operator $V_{\mu\nu}$ given by a product of left- and right-moving factors $V_{\mu R}\tilde{V}_{\nu L}$, where $V_{\mu R}$ and $\tilde{V}_{\nu L}$ correspond to emission vertices for massless open-string vectors. It is convenient to take the polarization tensors to be factorized in the form $\zeta_r^\mu \bar\zeta_r^\nu$. This is no loss of generality, since an arbitrary tensor can be written as the sum of products of vectors $\zeta_r^{\mu\nu} = \sum_n \zeta_{rn}^\mu \bar\zeta_{rn}^\nu$, so the full amplitude can be expressed as the sum of terms involving factorizing polarization tensors. The condition that the external states are physical requires $\zeta_r \cdot k_r = 0 = \overline{\zeta}_r \cdot k_r$ and $k_r^2 = 0$. Vertices $V(\zeta\overline{\zeta}, k)$ that factorize into a product of contributions from the left- and right-moving nonzero modes may be defined by $V_R(\zeta, \frac{1}{2}k, z)V_L(\overline{\zeta}, \frac{1}{2}k, \overline{z})$, where

$$V_R(\zeta, \tfrac{1}{2}k, z) = \exp\left(ik \cdot X_R(z) + \zeta \cdot \dot{X}_R(z)/z\right) \qquad (8.2.57)$$

and similarly for V_L, as explained in §7.1.4. In evaluating an amplitude we again only keep terms linear in the ζ's and the $\overline{\zeta}$'s. As in the case of open-string loops, the expression for the amplitude applies in either the covariant approach (with suitable ghost modes contributing to the propagators) or the light-cone approach (with the external momenta restricted to have vanishing + components and $\zeta_r^+ = \overline{\zeta}_r^+ = 0$). Since the z and $\overline{z}$ integrations involve the left- and right-moving modes independently, the evaluation of the traces over the nonzero modes in (8.2.1) factorizes into the product of a trace over each kind of mode. Each of these factors has a structure similar to the trace that arose in the open-string loops. The zero-mode momentum integral couples the left- and right-moving degrees of freedom, and so there are factors involving products of ζ and $\overline{\zeta}$.

8.3   Other Diagrams for Unoriented Strings

In theories with open strings or in the closed-string theory with unoriented strings there are additional diagrams that contribute to the same order of the perturbation expansion as the diagrams considered so far. These fall into two categories – 'higher-order' tree diagrams and additional loop diagrams. By a tree diagram we mean a diagram whose only singularities are one particle poles. By a higher-order tree diagram we mean a tree diagram that enters the perturbation expansion in a higher order than the order in which ordinary tree diagrams arise. The existence of higher-order tree diagrams is a peculiar feature of theories containing unoriented open and closed strings.

8.3.1 Higher-Order Tree Diagrams

In chapter 7 we studied string diagrams corresponding to a world sheet that is topologically a disk. With external open strings, this diagram is the leading contribution to the scattering amplitude, and is a perfectly ordinary tree diagram that can be conveniently described in the operator formalism. When there are two or more external open-string states and any number of external closed-string states, the amplitude can be investigated as in chapter 7. One picks any two external open strings, calls them the 'initial' and 'final' states $|i\rangle$ and $\langle f|$, and formulates the tree amplitude as

$$A = \langle f|V\Delta V\Delta V \dots V|i\rangle, \qquad (8.3.1)$$

where Δ is an open-string propagator and the V's are suitable open- or closed-string vertex operators.

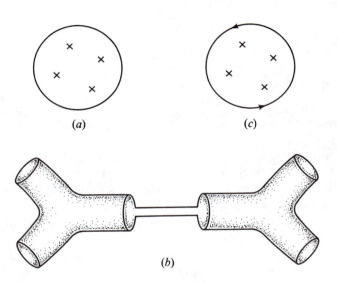

(a)

(c)

(b)

Figure 8.23. (*a*) A world sheet that is topologically a disk, with the external states being closed strings only. (*b*) Representation of the diagram as a closed-string tree diagram with intermediate open-string states. (*c*) With a change in the boundary conditions so that diametrically opposite points on the disk are identified (as indicated by the arrows) it becomes a projective plane. This gives a contribution to the purely nonorientable closed-string sector that modifies the tree contribution.

What happens if instead we consider, as in fig. 8.23*a*, a disk with external *closed* strings only? (Similar issues would arise in a diagram with precisely one external open string and any number of external closed strings.) This is definitely a tree diagram, since if it is 'cut' to reveal intermediate states, one finds only one-particle intermediate states. Depending on how the world sheet is cut these may either be open or closed-string states, as can be seen from fig. 8.23*b*. As a contribution to *closed*-string scattering, the disk is a *higher-order* tree diagram, since it has an extra factor of $\kappa \sim g^2$ compared to the Riemann sphere, which also contributes to scattering of closed strings. Like other diagrams considered in this chapter, it has a potential divergence due to tachyon emission into the vacuum.

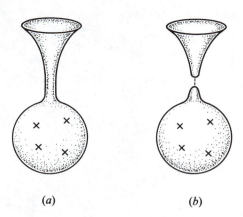

(a)   (b)

Figure 8.24. The dangerous configuration in the disk diagram with external closed strings is sketched in (a); the infrared divergence, if any, would factorize as in (b).

This is associated with the region of parameter space in which the radius of the disk becomes infinite, which is conformally equivalent to the region in which all the external particles come together on a disk of finite radius as shown in fig. 8.24a. The space-time configuration of the world sheet in this dangerous region is illustrated in fig. 8.24b. The residue of the infinity should be a closed-string tree amplitude on the sphere times a constant involving the coupling of a single closed string to the disk.

It would be quite clumsy to use the operator formalism to evaluate this diagram. The only reasonably simple way to evaluate this diagram is the world-sheet path integral approach that was developed in the introduction (§1.4.4).

The world sheet under discussion can be mapped to a variety of convenient configurations by various conformal transformations. For example, the mapping

$$z = -i\frac{\rho + r}{\rho - r} \tag{8.3.2}$$

maps the disk of radius r in the ρ plane into the upper half z plane. It proves more convenient to invert the disk by letting $\rho \to 1/\rho$ and to regard the world sheet as the plane with a disk of radius $q = 1/r$ cut out.

The problem of evaluating the amplitude for closed strings coupling to this world sheet is only a slight variation on the problem in §1.4.4 in which the world sheet was the Riemann sphere. As in that case, the amplitude can be written as

$$A(1, 2, \ldots, M) \sim n\kappa^{M-1} \int_{|\rho_r| \geq q} \prod_{r=1}^{M} d^2\rho_r \prod_{r,s}' \exp\left\{ \frac{1}{4} k_r \cdot k_s N(\rho_r; \rho_s, q^2) \right\},$$

$$(8.3.3)$$

where N is the propagator and the factor of n comes from the group-theory trace associated with the boundary of the disk. In the present case the propagator, $N(\rho_r; \rho_s, q^2)$, is the Green function for the two-dimensional Laplace equation on the punctured plane, satisfying

$$\partial_\rho^2 N(\rho; \rho', q^2) = 2\pi\delta^2(\rho - \rho') \tag{8.3.4}$$

together with Neumann boundary conditions at the edge of the disk, *i.e.*, vanishing of the normal derivative at the boundary, $\partial N(\rho, \rho'; q^2)/\partial|\rho| = 0$ at $|\rho| = q$. The expression for the propagator satisfying these conditions is easily obtained by the method of images, which gives

$$N(\rho; \rho', q^2) = \ln|\rho - \rho'| + \ln\left|\frac{q^2}{\rho\overline{\rho'}} - 1\right|. \tag{8.3.5}$$

The symbol $\prod_{r,s}'$ indicates that the infinite normal ordering terms are omitted from (8.3.3). This means that in the factor with $r = s$ the divergent piece of $N(\rho_r; \rho_r, q^2)$ (given by the first term in (8.3.5)) is omitted so that

$$N(\rho_r; \rho_r, q^2) = \ln\left|\frac{q^2}{\rho_r\overline{\rho}_r} - 1\right|. \tag{8.3.6}$$

(In fact, the divergent constant cancels against an infinite infrared constant that has been dropped from (8.3.5) in much the same manner as in the tree diagrams discussed in §1.4.4.)

The expression for the integrand is invariant under the group of transformations that maps the punctured plane onto itself. These are the $SU(1, 1)$ transformations

$$\rho_r \rightarrow \frac{\alpha\rho_r + q\beta}{\overline{\beta}\rho_r/q + \overline{\alpha}}, \tag{8.3.7}$$

where α and β are complex numbers satisfying $\overline{\alpha}\alpha - \overline{\beta}\beta = 1$, giving a total of three real parameters. To check invariance under these transformations it is useful to note that the various factors in the integrand transform as follows

$$d^2\rho \to \frac{d^2\rho}{|\bar{\alpha} + \bar{\beta}\rho/q|^4},$$

(8.3.8)

$$\rho_r - \rho_s \to \frac{\rho_r - \rho_s}{(\bar{\beta}\rho_r/q + \bar{\alpha})(\bar{\beta}\rho_s/q + \bar{\alpha})}$$

(8.3.9)

and

$$(q^2/\rho_r) - \bar{\rho}_s \to \frac{(q^2/\rho_r) - \bar{\rho}_s}{(\alpha + \beta q/\rho_r)(\alpha + \beta\bar{\rho}_s/q)}.$$

(8.3.10)

In order to avoid infinite overcounting, three of the integration variables can be fixed arbitrarily, just as in the treatment of the invariance of the upper half plane under $SL(2, R)$ transformations in the discussion of the open-string tree amplitudes in §1.5.2.

The integrand of the amplitude is also invariant under scale transformations

$$\rho_r \to e^\phi \rho_r,$$

(8.3.11)

provided the radius of the hole is changed so that

$$q \to e^\phi q.$$

(8.3.12)

This is evident from the fact that holes of any radius are all conformally equivalent to the upper half plane. The amplitude (8.3.3) can therefore be written in the form

$$A(1, 2, \ldots, M) \sim n\kappa^{M-1} \int_0^1 \frac{dq}{q} \int_{|\rho_r| \geq q} \left(\prod_{r=1}^M d^2\rho_r |q^2 - \bar{\rho}_r \rho_r|^2 \right) \frac{1}{V_4}$$

$$\delta^2(\rho_A - z_1)\delta^2(\rho_B - z_2) \prod_{1 \leq r < s \leq M} \left(|\rho_r - \rho_s| |q^2 - \rho_r\bar{\rho}_s| \right)^{k_r \cdot k_s/2}.$$

(8.3.13)

In writing this we have reinstated the scale invariance of the amplitude by integrating over the radius, q. This is compensated for by dividing out by the volume of the combined group of $SU(1, 1)$ and scale transformations. The δ functions fix the 'gauge' by setting any two of the complex ρ variables, labelled A and B, to arbitrary values, z_1 and z_2 (which fixes the four arbitrary gauge parameters). The volume factor, V_4, is given by

$$V_4 = \frac{q^2}{|\rho_A - \rho_B|^2 |q^2 - \rho_A \overline{\rho}_B|^2} \tag{8.3.14}$$

so that

$$\frac{\delta^2(\rho_A - z_1)\delta^2(\rho_B - z_2)}{V_4} \tag{8.3.15}$$

is invariant under $SU(1,1)$ and the scale transformations for any choice of A and B and of z_1 and z_2. For the special choice of parametrization in which $\rho_A = \rho_1 = \infty$ and $\rho_B = \rho_2 = 1$ the amplitude is given by

$$A(1, 2, \ldots, M) \sim n\kappa^{M-1} \int_0^1 \frac{dq}{q^3} |1 - q^2|^2$$

$$\times \int_{|\rho_r| \geq q} \left(\prod_{r=3}^M d^2\rho_r |q^2 - \overline{\rho}_r \rho_r|^2 |\rho_r|^{k_1 \cdot k_r / 2} \right) \tag{8.3.16}$$

$$\times \prod_{2 \leq r < s \leq M} \left(|\rho_r - \rho_s| |q^2 - \rho_r \overline{\rho}_s| \right)^{k_r \cdot k_s / 2}.$$

The closed-string double poles in this amplitude arise from the end-points where ρ_r's come together on the world sheet. There are also poles corresponding to open-string states that come from the region in which the ρ_r's approach the boundary of the disk. The overall constant can, in principle, be found from the unitarity condition that the diagram should factorize in the configuration shown in fig. 8.23b with the correct residues for the poles. As expected, the integral over the radius diverges like $\int_0^1 dq q^{-3} f(q^2)$, which is the same kind of divergence that arose in the case of the planar loop in §8.1.1. The leading divergent term is attributable to the tachyon state in the closed-string propagator that is coupling to the vacuum, as in the case of the open-string loop. The nonleading divergence, of the form $\int dq q^{-1} f'(0)$, is again an infrared effect due to the dilaton state in the closed-string propagator.

8.3.2 The Real Projective Plane

In addition to the disk diagram of the preceding subsection, there is another contribution to the closed-string amplitude at the same order in the coupling constant. It is the real projective plane, RP^2, which is a nonorientable surface. Let us first describe this manifold. (It was already briefly discussed in §8.1.2.) The ordinary two sphere S^2 has a natural embedding in three-dimensional real Euclidean space; it is the set of all x_k, $k = 1, 2, 3$ that obey

$$\sum x_k^2 = 1. \qquad (8.3.17)$$

To make RP^2 we impose a discrete equivalence relation

$$x_k \approx -x_k. \qquad (8.3.18)$$

While S^2 is simply connected, RP^2 has Z_2 (the discrete group with elements ± 1) for its fundamental group; because of the inversion in (8.3.18), RP^2 is unorientable. Just like S^2, RP^2 is a tree diagram in the sense that if it is 'cut', one obtains only one-particle intermediate states. However, just as in the case of the amplitude for closed strings coupling to a disk, the amplitude for closed strings coupling to RP^2 has double poles. It is therefore of higher order than the usual tree amplitudes defined on S^2.

There are a variety of ways to construct the scattering amplitude corresponding to a world sheet that is RP^2. It can be represented as a closed-string tree diagram with an insertion in a propagator. The requisite insertion can be described in light-cone gauge, for example, as a self-interaction of a single closed string (as will become clear in chapter 11). It is easier, however, to construct amplitudes associated with RP^2 by the world-sheet path integral method used in §8.3.1 to discuss the disk with external closed strings.

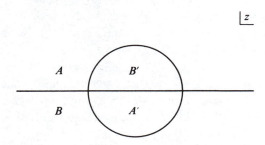

Figure 8.25. The fundamental region of the transformation $z \to -q^2/\bar{z}$ may be taken to be the upper half plane (A and B' in the figure) or the interior of the unit disk (A' and B' in the figure). The regions A and A' are mapped into each other by the transformation as are B and B'.

The definition of RP^2 shows that it is roughly 'half' of S^2. S^2, after removing a 'point at infinity', can be stereographically mapped to the entire complex z plane. The propagator of the free Bose field $X^\mu(\dot z)$ in the complex plane is – as we know from chapters 1 and 7 –

$$\langle X^\mu(z)\,X^\nu(z')\rangle = -\tfrac{1}{2}\eta^{\mu\nu}\ln|z-z'|. \qquad (8.3.19)$$

While the complex plane (plus a point at infinity) represents S^2, to obtain RP^2 we must impose the discrete equivalence relation (8.3.18). Allowing for an arbitrary rescaling of the z variable this relation can be written as

$$\bar z \approx -q^2/z. \qquad (8.3.20)$$

As long as we do not worry about the 'point at infinity', one 'fundamental region' of this transformation is the upper half plane. An equally good 'fundamental' region is the complex plane with a hole of radius q cut out (as shown in fig. 8.25). This is clear from the fact that the upper semicircle maps into the portion of the lower half plane that excludes the lower semicircle so that the punctured plane (the plane with the hole) is equivalent to the full upper half plane. Although this picture of a punctured plane looks similar to the disk diagram considered in the §8.3.1 in the case of RP^2, the boundary of the hole is treated differently, since diametrically opposite points are identified by virtue of (8.3.20). The disk has a boundary (or is noncompact if the boundary points are deleted), but RP^2 is a compact manifold without boundary. Using the Green function (8.3.19) in the upper half plane, and the characterization (8.3.20) of RP^2, we can deduce the Green function of the X^μ field on RP^2:

$$G(z;z',q^2) = -2\langle X^\mu(z)X^\nu(z')\rangle$$

$$(8.3.21)$$

$$= \eta^{\mu\nu}\left(\ln|z-z'| + \ln\left|\frac{q^2}{z\bar z'} + 1\right|\right).$$

The second term is essentially the contribution of an image charge required by the fact that the Green function must be invariant under (8.3.20). This expression applies whether the fundamental region is chosen to be the upper half plane or the punctured plane with a hole of radius q.

The contribution of this diagram to the amplitude is obtained in the same manner as the disk diagram of the §8.3.1. The Green function $N(\rho;\rho',q^2)$ (given in (8.3.5)) is replaced by the Green function

$$G(\rho;\rho',q^2) = N(\rho;\rho',-q^2) \qquad (8.3.22)$$

(given in (8.3.21)), which satisfies $G(\rho;\rho',q^2) = G(-\rho;\rho',q^2)$ at $|\rho| = q$ thereby identifying diametrically opposite points on the circumference.

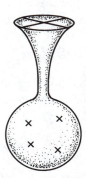

Figure 8.26. The projective-plane amplitude has a divergence of the now familiar type, with a tachyon propagating between a sphere with external lines attached and a projective plane without such external lines.

This relation between N and G is reminiscent of the way in which the Green function for the Möbius strip diagram ψ^N is related to that of the planar loop diagram ψ by replacing q^2 by $-q'^2$ (8.1.71). The expression for the projective-plane contribution to the amplitude therefore has the same form as (8.3.16) but with q^2 replaced by $-q^2$ and without the over-all factor of n, since the projective plane has no boundary. In particular, the projective-plane amplitude has the familiar $\int dq/q^3$ divergence, corresponding in this case to the configuration sketched in fig. 8.26. The similarity to the divergent contribution from closed-string insertions on the disk suggests that under suitable conditions these might cancel. This is a matter to which we will return in chapter 10.

8.3.3 Other Loop Diagrams

In the closed, oriented string theory the torus is the only possible diagram at order κ^M. However, in any theory in which the strings are unoriented it is necessary to include closed-string diagrams with nonorientable closed world sheets. At one-loop order this corresponds to the surface known as the Klein bottle. This is seen by constructing the type I closed-string loop diagram by sewing the tree diagrams and including a projection operator $\frac{1}{2}(1 + T)$ in the internal propagators in the loop, where T is the operator that interchanges the left- and right-moving oscillator spaces. This projection operator ensures that the only states circulating around the loop belong to the subspace that is symmetric under the interchange of the oscillators α_n^μ and $\tilde{\alpha}_n^\mu$, and therefore that the string is unoriented. This is analogous to the action of the twist operator in the open-string case. The result is a sum of many terms with interchange operators, T, distributed in all possible positions. In those contributions with an even number of T's the trace in the integrand of the amplitude factorizes into a product of two factors for each mode number in much the same way as in the orientable theory. The effect of an odd number of T factors in the operator inside the trace is to couple the left- and right-moving spaces in such a manner that the trace over oscillators of a given mode number goes around the loop twice in much the same way as the group-theory trace in the Möbius strip does. These terms describe a world sheet that is topologically a Klein bottle. Klein-bottle amplitudes have never been worked out in the literature, so little is known about their behavior. Until this is done, they remain a potential source of trouble for type I superstring theory.

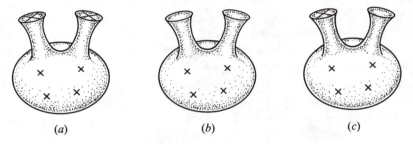

(a)   (b)   (c)

Figure 8.27. Representation of loop diagrams with external closed strings. (a) The Klein bottle. (b) The annulus. (c) The Möbius strip.

In terminology to be introduced in chapter 11, the Klein bottle can be represented as a tree diagram in which there are two successive instantaneous interactions of the type mentioned in §8.3.2 (see fig. 8.27a). Another obvious diagram to consider at the same order in the coupling constant is the diagram shown in fig. 8.27c, which involves two successive open-string insertions in a closed-string propagator. This is a process whose world sheet is topologically equivalent to an annulus with the closed strings attached to the interior of the surface. Finally, there is a diagram in which an insertion of an open-string propagator in a closed-string propagator is followed by an instantaneous self-interaction as shown in fig. 8.27c. This has a world sheet that is a Möbius strip and is also of order κ^M. Together with the torus, this exhausts the one-loop diagrams with external closed strings.

The sum of the three diagrams in fig. 8.27 combine to form the iteration of the insertion of a hole in the world sheet and attaching a cross-cap to the world sheet. These considerations are important in contemplating whether there might be cancellations between the divergences associated with the various diagrams.

8.4 Summary

In this chapter we have considered the one-loop amplitudes of the bosonic theory that are the lowest-order quantum corrections to the classical string field theory. We have emphasized operator methods for constructing these amplitudes in this chapter, but they can also be constructed by functional methods. (This will be described in the context of the light-cone gauge in chapter 11.)

The open-string one-loop diagrams are of three types. Planar diagrams are associated with a world sheet that can be represented in parameter space by an annulus with particles attached to one boundary. They are divergent at the endpoint of the integration over the variable w or q, which parametrizes the conformal structure of the annulus. This divergence corresponds to space-time configurations of the world sheet in which a disk with particles attached to the boundary is joined to one with no particles attached by a long thin tube that describes a zero-momentum closed-string dilaton state. Nonorientable diagrams, associated with world sheets that are Möbius strips, are divergent for similar reasons with the tube coupling to the vacuum via a cross-cap instead of a disk. The nonplanar orientable diagrams are not divergent. They contain closed-string bound states that arise as singularities in the channel with vacuum quantum numbers. It is only in the critical dimension ($D = 26$ in this case) that these singularities are poles in accord with unitarity.

In the case of the orientable closed-string theory, things are much simpler. The only one-loop diagram is associated with a world sheet that is topologically a torus. This can only be consistently defined in the critical dimension, since it is only then that the loop integrand is modular invariant. Even so, the amplitude is divergent due to singularities that arise at the endpoint at which all the external particles come together on the torus. Again, this divergence can be interpreted as an infrared divergence associated with the presence of tachyons and massless scalars in the theory.

There are various other kinds of quantum corrections to amplitudes in nonorientable theories with external closed strings, which correspond to adding handles or cutting windows out of the world sheet. Cutting a window out of the orientable world sheet of a closed-string tree diagram results in a quantum correction that is not a loop diagram but rather represents a mixing of open-string and closed-string states. There is a related contribution from the closed-string tree diagram with a nonorientable world sheet that is a projective plane. The one-loop diagrams with external closed-string states in the nonorientable theory can be classified according to the topology of their world sheets as a Klein bottle, an annulus and a Möbius strip in addition to the torus.

Appendix 8.A   Jacobi Θ Functions

The four Jacobi θ_k functions ($k = 1, \ldots, 4$) satisfy the heat equation

$$\frac{i}{\pi} \frac{\partial^2 \theta_k(\nu|\tau)}{\partial \nu^2} + 4\frac{\partial \theta_k(\nu|\tau)}{\partial \tau} = 0. \tag{8.A.1}$$

They are defined in terms of sums or in terms of products by

$$\theta_1(\nu|\tau) = i \sum_{n=-\infty}^{\infty} (-1)^n q^{(n-1/2)^2} e^{i\pi(2n-1)\nu}$$

$$= 2f(q^2)q^{1/4} \sin \pi\nu \prod_{n=1}^{\infty} \left(1 - 2q^{2n} \cos 2\pi\nu + q^{4n}\right), \tag{8.A.2}$$

$$\theta_2(\nu|\tau) = \theta_1(\nu + \tfrac{1}{2}|\tau) = \sum_{n=-\infty}^{\infty} q^{(n-1/2)^2} e^{i\pi(2n-1)\nu}$$

$$= 2f(q^2)q^{1/4} \cos \pi\nu \prod_{n=1}^{\infty} \left(1 + 2q^{2n} \cos 2\pi\nu + q^{4n}\right), \tag{8.A.3}$$

$$\theta_3(\nu|\tau) = e^{i\pi(\nu+\tau/4)}\theta_1(\nu + \tfrac{1}{2} + \tfrac{1}{2}\tau|\tau) = \sum_{n=-\infty}^{\infty} q^{n^2} e^{i\pi 2n\nu}$$

$$= f(q^2) \prod_{n=1}^{\infty} \left(1 + 2q^{2n-1} \cos 2\pi\nu + q^{4n-2}\right), \tag{8.A.4}$$

$$\theta_4(\nu|\tau) = -ie^{i\pi(\nu+\tau/4)}\theta_1(\nu + \tfrac{1}{2}\tau|\tau) = \sum_{n=-\infty}^{\infty} (-1)^n q^{n^2} e^{i\pi 2n\nu}$$

$$= f(q^2) \prod_{n=1}^{\infty} \left(1 - 2q^{2n-1} \cos 2\pi\nu + q^{4n-2}\right), \tag{8.A.5}$$

where

$$q = e^{i\pi\tau} \tag{8.A.6}$$

and

$$f(q^2) = \prod_{n=1}^{\infty} \left(1 - q^{2n}\right)$$

$$= \left(\theta_1'(0|\tau)/2\pi q^{1/4}\right)^{1/3}. \tag{8.A.7}$$

The notation $\theta'(0|\tau)$ conventionally represents a derivative with respect to the first variable (ν),

$$\theta_1'(0|\tau) \equiv \frac{\partial \theta_1(\nu|\tau)}{\partial \nu}\Big|_{\nu=0}, \tag{8.A.8}$$

The equality of the infinite series and infinite products in (8.A.2) – (8.A.5) is the content of the Jacobi triple product formula, which was presented in §3.2.4.

The function θ_1 satisfies the important periodicity relations

$$\theta_1(\nu + 1|\tau) = -\theta_1(\nu|\tau)$$

$$\theta_1(\nu + \tau|\tau) = -e^{-i\pi\tau - 2i\pi\nu}\theta_1(\nu|\tau), \tag{8.A.9}$$

with similar relations for the other θ_r functions that follow from their defining formulas.

The function ψ that arises in the planar loop amplitudes is conveniently expressed in terms of the θ_1 function by the formula

$$\psi(\rho, w) = \frac{1 - \rho}{\sqrt{\rho}} \exp\left(\frac{\ln^2 \rho}{2 \ln w}\right) \prod_{n=1}^{\infty} \frac{(1 - w^n \rho)(1 - w^n/\rho)}{(1 - w^n)^2} \tag{8.A.10}$$

$$= -2\pi i e^{-i\pi\nu^2/\tau} \frac{\theta_1(-\nu/\tau \mid -1/\tau)}{\theta_1'(0 \mid -1/\tau)}$$

where

$$\nu \equiv \frac{\ln z}{2i\pi} = \frac{\ln \rho}{\ln w}, \qquad \tau \equiv \frac{\ln q}{i\pi} = -\frac{2\pi i}{\ln w}, \tag{8.A.11}$$

so that

$$\frac{\ln \rho}{2i\pi} = -\frac{\nu}{\tau}, \qquad \frac{\ln w}{2i\pi} = -\frac{1}{\tau}. \tag{8.A.12}$$

Under the transformation $\rho \to w\rho$ ($\nu \to \nu + \tau$) the function $\psi(\rho, w)$ satisfies the relation

$$\psi(w\rho, w) = \psi(\rho, w), \tag{8.A.13}$$

which follows from (8.A.9).

Likewise, the function ψ^T is given by the formula

$$\psi^T(\rho, w) = \frac{1+\rho}{\sqrt{\rho}} \exp\left(\frac{\ln^2 \rho}{2\ln w}\right) \prod_{n=1}^{\infty} \frac{(1+w^n \rho)(1+w^n/\rho)}{(1-w^n)^2}$$

$$= 2\pi e^{-i\pi\nu^2/\tau} \frac{\theta_1(1/2 - \nu/\tau \mid -1/\tau)}{\theta_1'(0 \mid -1/\tau)},$$

(8.A.14)

and satisfies

$$\psi^T(w\rho, w) = \psi^T(\rho, w). \tag{8.A.15}$$

Similarly,

$$\psi^N(\rho, w) = \frac{1-\rho}{\sqrt{\rho}} \exp\left(\frac{\ln^2 \rho}{2\ln w}\right) \prod_{n=1}^{\infty} \frac{(1-(-w)^n \rho)(1-(-w)^n/\rho)}{(1-(-w)^n)^2}$$

$$= -2\pi i e^{-i\pi\nu^2/\tau} \frac{\theta_1(-\nu/\tau \mid 1/2 - 1/\tau)}{\theta_1'(0 \mid 1/2 - 1/\tau)},$$

(8.A.16)

which satisfies

$$\psi^N(w\rho, w) = \psi^N(\rho, w). \tag{8.A.17}$$

The θ_k functions have simple properties under modular transformations. The transformation $\tau \to -1/\tau$, for example, is easy to analyze by making use of a Poisson summation formula. The Poisson formula states

$$\sum_{n=-\infty}^{\infty} e^{-\pi n^2 A + 2n\pi A s} = \frac{1}{\sqrt{A}} e^{\pi A s^2} \sum_{m=-\infty}^{\infty} e^{-\pi A^{-1} m^2 - 2i\pi m s}. \tag{8.A.18}$$

This formula can be proved by using $\sum_m e^{2\pi i r m} = \sum_n \delta(r - n)$

$$\sum_{n=-\infty}^{\infty} e^{-\pi n^2 A + 2n\pi A s} = \sum_{m=-\infty}^{\infty} \int_{-\infty}^{\infty} dr \, e^{2i\pi r m} e^{-\pi r^2 A + 2r\pi A s}. \tag{8.A.19}$$

Upon carrying out the (Gaussian) integral over r on the right-hand side of (8.A.19), one obtains (8.A.18). Equation (8.A.18) will be generalized in appendix 9.B. The modular transformation law of θ_1, defined by the series in (8.A.2), can be deduced from (8.A.18) by making the identifications

$A = i\tau$ and $s = -\frac{1}{2} + \frac{1}{2\tau} + \frac{\nu}{\tau}$. We get

$$\theta_1 \left(-\frac{\nu}{\tau} \middle| -\frac{1}{\tau} \right) = \eta(\tau)^{1/2} \exp\left(\frac{i\pi\nu^2}{\tau} \right) \theta_1(\nu|\tau), \tag{8.A.20}$$

where

$$\eta = e^{i\pi/4}. \tag{8.A.21}$$

Furthermore the transformation $\tau \to \tau + 1$ leads rather simply to

$$\theta_1(\nu|\tau + 1) = \eta\theta(\nu|\tau). \tag{8.A.22}$$

The effect of an arbitrary modular transformation

$$\tau \to \frac{a\tau + b}{c\tau + d} \tag{8.A.23}$$

can be deduced by combining the two special transformations in (8.A.20) and (8.A.22) in arbitrary amounts, giving

$$\theta_1 \left(\frac{\nu}{c\tau + d} \middle| \frac{a\tau + b}{c\tau + d} \right) = \eta'(c\tau + d)^{1/2} \exp\left(\frac{i\pi c\nu^2}{c\tau + d} \right) \theta_1(\nu|\tau). \tag{8.A.24}$$

The coefficient η' in this equation is an eighth-root of unity; it is a subtle number-theoretic problem to determine which eighth root occurs for given a, b, c, d. We can deduce from (8.A.24) the transformation rule

$$\theta_1' \left(0 \middle| \frac{a\tau + b}{c\tau + d} \right) = \eta(c\tau + d)^{3/2}\theta_1'(0|\tau). \tag{8.A.25}$$

The other θ_r functions satisfy similar relations that can be deduced by substituting their expressions in terms of θ_1 into (8.A.24).

By setting $b = -c = 1$ and $a = d = 0$, the transformation properties of $f(q^2)$ follow from (8.A.7) and (8.A.25)

$$f(w) = \left[\frac{\theta_1'(0| - 1/\tau)}{2\pi w^{1/8}} \right]^{1/3} = \left(\frac{-\ln q}{\pi} \right)^{1/2} w^{-1/24} q^{1/12} f(q^2). \tag{8.A.26}$$

Similarly, the partition function relevant to the nonorientable loop calculation transforms according to the formula

$$f(-w) = \left(\frac{-\ln q}{2\pi} \right)^{1/2} w^{-1/24} q^{1/48} f(-\sqrt{q}). \tag{8.A.27}$$

The transformation properties of the various ψ functions can be obtained by applying formulas that follow from (8.A.24). Setting $a = 0$,

$b = 1$, $c = -1$ and $d = 0$ gives

$$\frac{\theta_1(-\nu/\tau \mid -1/\tau)}{\theta_1'(0 \mid -1/\tau)} = -1/\tau e^{i\pi\nu^2/\tau} \theta_1(\nu|\tau)/\theta_1'(0|\tau). \tag{8.A.28}$$

Choosing the same parameters and using the relationship between θ_1 and θ_2 in (8.A.3), one finds that

$$\frac{\theta_2(-\nu/\tau \mid -1/\tau)}{\theta_1'(0 \mid -1/\tau)} = \frac{i}{\tau} e^{i\pi\nu^2/\tau} \theta_4(\nu|\tau)/\theta_1'(0|\tau). \tag{8.A.29}$$

Finally, by using (8.A.24) with parameters $a = 1$, $b = 0$, $c = -2$ and $d = 1$

$$\frac{\theta_1(-\nu/\tau \mid 1/2 - 1/\tau)}{\theta_1'(0 \mid 1/2 - 1/\tau)} = \frac{2}{\tau} e^{i\pi\nu^2/\tau} \frac{\theta_1(-\nu/2 \mid \tau/4 - 1/2)}{\theta_1'(0 \mid \tau/4 - 1/2)}. \tag{8.A.30}$$

Using these formulas the expressions for the ψ functions can be written in a simple way in terms of the variables q and $z = \exp\{2i\pi\nu\}$ as

$$\psi(\rho, w) = -\frac{2\pi}{\ln q} \sin \pi\nu \prod_1^\infty \frac{1 - 2q^{2n} \cos 2\pi\nu + q^{4n}}{(1 - q^{2n})^2} \tag{8.A.31}$$

$$= -\frac{1}{\tau} e^{-i\pi\nu^2/\tau} \psi(z, q^2).$$

Similarly, the functions ψ^T and ψ^N can be written as

$$\psi^T(\rho, w) = -\frac{\pi}{\ln q} q^{-1/4} \prod_1^\infty \frac{1 - 2q^{2n-1} \cos 2\pi\nu + q^{4n-2}}{(1 - q^{2n})^2}, \tag{8.A.32}$$

and

$$\psi^N(\rho, w) = -\frac{4\pi}{\ln q} \sin \frac{\pi\nu}{2} \prod_1^\infty \frac{1 - 2(-\sqrt{q})^n \cos \pi\nu + q^n}{(1 - (-\sqrt{q})^n)^2}. \tag{8.A.33}$$

The other functions of relevance in the calculation of the open-string loops with external massless vector states are $\eta(w, \rho)$ and $\Omega(w, \rho)$ (defined in (8.1.59) and (8.1.60)). Their form under the change of variables ρ, w to z, q^2 (which corresponds to the changes $-\nu/\tau \to \nu$ and $-1/\tau \to \tau$) can be deduced by using (8.1.61) and (8.1.62), giving

$$\eta(\rho, w) = -\tau \eta'(z, q^2), \qquad (8.A.34)$$

and

$$\Omega(\rho, w) = \tau^2 \Omega'(z, q^2), \qquad (8.A.35)$$

The prime indicates that zero-mode terms are missing in the expressions on the right-hand sides of these equations, *i.e.*,

$$\eta'(\rho, w) = \eta(\rho, w) - \frac{\ln \rho}{\ln w}, \qquad (8.A.36)$$

$$\Omega'(\rho, w) = \Omega(\rho, w) + \frac{1}{\ln w}. \qquad (8.A.37)$$

The function χ in (8.2.13), which appears in the closed-string calculations, can be written in the form

$$\chi(z, w) \equiv \chi(\nu|\tau) = 2\pi e^{-\pi(\mathrm{Im}\,\nu)^2/\mathrm{Im}\,\tau} \left| \frac{\theta_1(\nu|\tau)}{\theta_1'(0|\tau)} \right|. \qquad (8.A.38)$$

The invariance of χ under the change of variable $\nu \to \nu + 1$ shown in (8.2.18) follows, since θ_1 simply changes sign (from (8.A.9)) under this transformation. The invariance of χ in (8.2.18) under $\nu \to \nu + \tau$ also follows, since the transformation of $|\theta_1|$ in (8.A.9) is balanced by the transformation of the exponential prefactor in (8.A.38).

It follows from (8.A.24) that χ transforms under (8.A.23) with weight 1, which means that

$$\chi\left(\frac{\nu}{c\tau + d}\bigg|\frac{a\tau + b}{c\tau + d}\right) = \frac{\chi(\nu|\tau)}{|c\tau + d|}. \qquad (8.A.39)$$

This completes our discussion of the formulas used in the one-loop calculations.

9. One-Loop Diagrams in Superstring Theory

In this chapter we discuss one-loop amplitudes in superstring theory. The operator methods used for the calculations of one-loop amplitudes with on-shell external states in chapter 8 will be applied to the calculation of the corresponding diagrams in superstring theories. The resulting amplitudes are expressed as integrals over world sheets that are closed and orientable in the case of the type II or heterotic theories. Type I theories are based on unoriented open and closed strings, and as a result their world sheets need not be orientable and can have boundaries. As in the case of the bosonic theory, the operator approach automatically gives the correct measure factors by elementary algebraic manipulations. In approaches that explicitly compute the sum over geometries, considerable care is required to correctly define and evaluate infinite determinants that give the measure. Nonetheless, such approaches have a number of advantages and certainly appear preferable for the study of multiloop amplitudes.

In the case of one-loop amplitudes in superstring theories, the light-cone gauge calculations based on the space-time supersymmetric formalism of chapter 5 and covariant calculations based on the formalism with world-sheet supersymmetry of chapter 4 are quite different, just as in the case of the calculation of the tree diagrams. For many purposes the light-cone method with manifest space-time supersymmetry is more economical, because there are fewer diagrams to calculate for a given process. For example, in theories with oriented closed strings (type II and heterotic) there is only one one-loop diagram, which corresponds to a world sheet that is topologically a torus. In the covariant formalism with world-sheet supersymmetry there are several contributions that must be calculated separately and then added together. These correspond to different fermion boundary conditions on the one-loop surface. The way in which these 'spin structures' piece together in this example and the generalization to surfaces of higher genus is one of the key issues in the consistency of superstring theory. It is considered in this chapter and in more detail in chapter 14.

9.1   Open-Superstring Amplitudes

We begin our investigation with the study of open-superstring loop amplitudes. These calculations are only directly relevant to type I superstring
theory, since it is the only one that contains open strings. The analysis is
also important, however, for the study of the type II and heterotic string
loop diagrams. In type II theories, left- and right-moving modes are described by essentially the same mathematics as open superstrings. Similarly, the open-superstring formulas are applicable to the right-moving
modes of the heterotic string theories.

In order to get a feeling for the behavior of these theories, it is very
instructive not only to formulate abstract formulas that describe general
cases, but to evaluate specific examples explicitly and explore some of their
salient properties. To do this in practice, it is convenient to concentrate
attention on amplitudes whose external states are ground states of the
theory, which means they are massless in the case of superstrings. There
is no fundamental obstacle to calculating processes with massive external
states as well, but the requisite vertex operators are more complicated.

In this chapter we also restrict our explicit calculations to examples
having four or fewer external particles. This does involve a rather significant restriction. When we consider, for example, M external vector
particles, each has a polarization vector ζ_i^μ and a momentum vector k_i^μ.
Only $M-1$ of the momenta k_i^μ are linearly independent. In general, amplitudes involve Lorentz-invariants formed from the polarization vectors and
momenta of external lines. When $M \geq 6$ the invariant factor can include
an antisymmetric ϵ tensor (with ten indices, since $D = 10$) contracted
into momenta and polarization vectors. For $M < 6$, no such terms are
possible, because an antisymmetric tensor can only give a nonvanishing
contraction with linearly independent vectors.

The ϵ tensor terms mentioned above are of particular interest, since
they can reveal potential anomalies of the quantum theory. The absence
of such anomalies is an important consistency condition for an acceptable
quantum theory. In chapter 10 we explicitly calculate the anomalies associated with certain hexagon string diagrams, but in this chapter we only
consider $M \leq 4$. When $M \leq 4$ we can use the light-cone gauge formalism
of chapter 5 without loss of generality. Since there are at most four linearly independent polarization vectors and three independent momenta
in a four point function, it is always possible to choose a Lorentz frame
in which the + light-cone component of all these vectors is zero. Then,
just as in the discussion of §7.4, amplitudes calculated in such a frame
uniquely determine the complete result in an arbitrary frame. By writing
a manifestly covariant formula for the amplitude that correctly describes

the result obtained in the particular frame, one achieves its extension to an arbitrary frame.

In §7.4.1 we derived vertex operators that describe the emission of massless superstring states in a frame with $k^+ = 0$. Also, in the light-cone gauge, polarization vectors satisfy $\zeta^+ = 0$, but just as in §7.4, we also set $\zeta^- = 0$. The results for massless boson and fermion emission vertices were

$$V_B(\zeta, k) = \zeta^i B^i e^{ik \cdot X} \tag{9.1.1}$$

$$V_F(u, k) = (u^a F^a + u^{\dot{a}} F^{\dot{a}}) e^{ik \cdot X}, \tag{9.1.2}$$

where

$$B^i = \dot{X}^i - \tfrac{1}{4} S \gamma^{ij} S k^j \tag{9.1.3}$$

$$F^a = (p^+/2)^{1/2} S^a \tag{9.1.4}$$

$$F^{\dot{a}} = (2p^+)^{-1/2} [(\gamma \cdot \dot{X} S)^{\dot{a}} + \tfrac{1}{3} : (\gamma^i S)^{\dot{a}} S \gamma^{ij} S : k^j]. \tag{9.1.5}$$

We can write open-superstring loop amplitudes in the same general form considered in chapter 8, namely

$$A = g^M \int d^{10}p \, \text{Tr}(\Delta V(1) \Delta V(2) \ldots \Delta V(M)), \tag{9.1.6}$$

where the vertex operators $V(r)$ can be either $V_B(\zeta, k)$ or $V_F(u, k)$ and the propagator is given, as before, by

$$\Delta^{-1} = L_0 = \tfrac{1}{2} p^2 + N \tag{9.1.7}$$

$$N = \sum_{n=1}^{\infty} (\alpha_{-n}^i \alpha_n^i + n S_{-n}^a S_n^a) \tag{9.1.8}$$

Also, we may associate twist operators

$$\Omega = -(-1)^N \tag{9.1.9}$$

with each of the propagators, depending on the topology of the diagram we wish to describe.

In the covariant calculations of chapter 8 it was necessary to include the contributions of Faddeev–Popov ghost coordinates in the trace. Even though they entered trivially in the vertices, they nonetheless gave a crucial factor of $[f(w)]^{-2}$ in the trace. In the physical light-cone gauge that we are now employing, there are no ghost coordinates, and the traces only refer to the modes of $X^i(\sigma, \tau)$ and $S^a(\sigma, \tau)$.

9.1.1 Amplitudes with $M < 4$ Massless External States

Using the expressions for the propagator and the ground-state emission vertices given above, superstring loop amplitudes can be expressed in terms of correlation functions in much the same way as bosonic loop amplitudes were in chapter 8, but now including traces over the anticommuting modes. This includes the trace over the fermionic zero modes S_0^a. As explained in appendix 9.A, this trace vanishes due to a supersymmetry trace identity, unless it includes at least eight S_0^a's. Since each ground-state boson emission vertex is at most bilinear in S_0^a, at least four external bosons are necessary to get a nonzero result for the zero-mode trace. This implies that amplitudes with zero, one, two or three external ground-state bosons vanish. This result does not depend on how many internal propagators are twisted and so it applies to the nonplanar and nonorientable loops as well as to the planar one. It also applies to the various closed-superstring theories, where there is a separate trace calculation for left-moving and right-moving modes.

A similar argument applies to processes with external ground-state fermions, as it obviously must, since these are related to the bosonic amplitudes by supersymmetry. The reasoning is slightly more subtle, however, since the fermion-emission vertex, V_F, contains a term cubic in S_0's as well as a term linear in S_0. Just counting powers of S_0 it would appear that there could be a nonvanishing loop with two external fermions and one external boson. This loop actually vanishes for reasons that are explained later in this chapter.

The vanishing of the one-loop diagram with no external particles is closely related to the one-loop vanishing of the vacuum energy, and hence the cosmological constant Λ. This connection between the vacuum energy and the cosmological constant was explained in chapter 8. The vanishing of the perturbative corrections to the cosmological constant is a feature of any theory with unbroken supersymmetry. If supersymmetry remains unbroken to all orders in the string perturbation theory, the amplitude with no external particles should vanish to all orders. If supersymmetry were to be broken in perturbation theory the resulting four-dimensional cosmological constant almost inevitably would be of order (Planck mass)^4, exceeding the experimental bound by a factor of about 10^{120}!

The vanishing of the open-string loop amplitude with one external open-string ground state is trivial, since there are no massless open-string states with vacuum quantum numbers. The vanishing of the loop amplitude with two massless external states implies the vanishing of the self-energies of the massless particles. This must be true to every order in perturbation theory. For the bosons it is an expected consequence of gauge invariance, while for the fermions it is a consequence of chirality. In each case there simply are not enough modes to describe a massive state.[*] Lastly, the vanishing of the three-particle amplitudes is quite reminiscent of nonrenormalization theorems for coupling constants which arise in supersymmetric field theories.

9.1.2 The Planar Diagrams

The amplitudes with four external ground states (any combination of fermions and bosons) are the first examples that give nonzero results. This is required by unitarity, of course, since they must possess imaginary parts that correspond to products of two-to-two tree amplitudes integrated over the phase space of the intermediate particles. Thus the fact that the tree amplitudes are nonzero implies that the loop amplitudes must also be nonzero. These calculations turn out to be relatively simple, because all the S oscillators in the vertex operators are used up in providing the eight S_0^a's needed to saturate the zero-mode trace. The calculations of the various kinds of open-string diagrams then become simple generalizations of the corresponding calculations in the bosonic string theory.

Consider first the four-particle one-loop diagram with external bosons only, so that every vertex is of the form given in (9.1.1). Since the trace over the S_0 modes vanishes unless there are at least eight powers of S_0 inside the trace, and each bosonic vertex operator can contribute at most two, it must do so. The only term with two of them is $-R_0^{ij}k^j$ in (9.1.3), where

$$R_0^{ij} = \tfrac{1}{4}S_0^a\gamma_{ab}^{ij}S_0^b. \tag{9.1.10}$$

Therefore once the S_0 trace is done, the vertex operators no longer con-

---

[*] A massless vector has eight polarizations, whereas a massive one requires nine. A massless Majorana–Weyl spinor has eight polarizations. Massive spinors cannot be Weyl, so the minimum number of components is 16. In each case there is no source, at least at any finite order of perturbation theory, of the extra modes required if these particles are to gain mass.

tribute terms involving the other S oscillators. The S_0 trace provides a factor in the amplitude

$$K = t_{i_1 j_1 i_2 j_2 i_3 j_3 i_4 j_4} \zeta_1^{i_1} \zeta_2^{i_2} \zeta_3^{i_3} \zeta_4^{i_4} k_1^{j_1} k_2^{j_2} k_3^{j_3} k_4^{j_4}, \qquad (9.1.11)$$

where the tensor t is defined to be

$$t^{i_1 j_1 i_2 j_2 i_3 j_3 i_4 j_4} = \text{Tr}\left(R_0^{i_1 j_1} R_0^{i_2 j_2} R_0^{i_3 j_3} R_0^{i_4 j_4}\right). \qquad (9.1.12)$$

The evaluation of this tensor is described in appendix 9.A. Remarkably, this factor K is exactly the same as the one found in §7.4.2 for the four-particle tree amplitudes. In the loop calculation the expression for the tensor t arises as a trace of eight S_0's, whereas the tree calculation was done in a quite different manner. Having discovered this result for bosons, it becomes plausible that supersymmetry ensures that the one-loop four-particle amplitudes involving fermions also have the same kinematic factors as the tree diagrams. In fact, this must be the case, because the various K factors given in §7.4.2 can all be related to one another by supersymmetry transformations.

Let us now consider the evaluation of the trace associated with the nonzero S_n^a modes. Since they drop out of the vertex operators after the S_0 trace is done, the propagator

$$\Delta = \int_0^1 x^{L_0 - 1} dx, \qquad (9.1.13)$$

defined in (9.1.7) and (9.1.8), is the only source of such modes in the evaluation of one-loop diagrams with four massless vectors. The contribution to the integrand of the loop amplitude from the trace over nonzero S_n^a modes is therefore given by

$$\text{Tr}\left(\prod_{a=1}^8 \prod_{n=1}^\infty w^{n S_{-n}^a S_n^a}\right) = \prod_{n=1}^\infty (1 - w^n)^8 = [f(w)]^8, \qquad (9.1.14)$$

(where $w = \prod_r x_r$). This trace is particularly easy to evaluate because each fermion level can only have occupation number 0 or 1, giving a factor of $1 - w^n$ for the nth level. The relative minus sign is due to the fact that an occupied level has an odd number of fermion modes. (The argument is the same as we gave for the ghost modes in §8.1.1.)

Having used up the prefactor B_i in each of the vertices (the factor that multiplies $\exp\{ik \cdot X\}$) in doing the zero-mode fermion trace, the trace over the bosonic modes has the same form as in §8.1.1, but with D set equal to ten, so that the result can simply be read off from there. In the light-cone approach, the trace over transverse oscillator modes gives a contribution

$$\text{Tr}\left(\prod_{i=1}^{8}\prod_{n=1}^{\infty} w^{\alpha^i_{-n}\alpha^i_n}\right) = [f(w)]^{-8}. \qquad (9.1.15)$$

The factors of $f(w)$ cancel between the bosonic and the fermionic traces. The complete result for the planar loop amplitude with a given cyclic ordering for the external massless states is

$$A_P(1,2,3,4) = g^4 G_P K \int \frac{dw}{w} \int \left(\prod_{r=1}^{3} \frac{d\rho_r}{\rho_r}\right)\left(\frac{-2\pi}{\ln w}\right)^5 \prod_{r<s} (\psi_{rs})^{k_r \cdot k_s}. \qquad (9.1.16)$$

Here the integration range is $w = \rho_4 \leq \rho_3 \leq \rho_2 \leq \rho_1 \leq 1$. The function $\ln \psi_{rs}$ is the Green function for the two-dimensional Laplace equation in the ρ plane (with Neumann boundary conditions) evaluated between points ρ_r and ρ_s on the boundary and was introduced in §8.1.1. K represents the kinematic factor given in (9.1.11), and G_P is the group-theory factor $\text{tr}(\lambda^1\lambda^2\lambda^3\lambda^4)$ introduced in chapter 6. The full planar loop amplitude is given by the sum over all noncyclic permutations of the external states.

The fact that the factors of $f(w)$ have canceled in (9.1.16) is a consequence of the one-to-one correspondence between α and S modes, which in turn is a consequence of space-time supersymmetry. It is explained later in this chapter how this result arises in the covariant RNS approach.

The planar loop amplitude of (9.1.16) is divergent due to the behavior of the integrand near $w = 1$. The analysis is almost identical to that of §8.1.1. To exhibit this divergence it is convenient to transform to the variables that express the amplitude as an integral over parameters defined on an annulus. These are the variables

$$z_r = e^{2i\pi\nu_r} \qquad \text{and} \qquad q = \exp\left(-\frac{2\pi^2}{\ln w}\right), \qquad (9.1.17)$$

where

$$\nu_r = \frac{\ln \rho_r}{\ln w}, \qquad (9.1.18)$$

introduced in §8.1.1. The result is an integral over the positions z_r of the particles around the outer boundary of the annulus and over the radius q

of the hole given by

$$A_P(1,2,3,4) = g^4 G_P K \int_0^1 \frac{dq}{q} \int \left(\prod_{r=1}^3 d\nu_r \right) \prod_{1 \leq r < s \leq 4} (\psi_{rs})^{k_r \cdot k_s}$$

$$= g^4 G_P K \int_0^1 \frac{dq}{q} F_P(q^2), \tag{9.1.19}$$

where (using an expression for ψ_{rs} in appendix 8.A)

$$F_P(q^2) = \int \left(\prod_{r=1}^3 d\nu_r \right) \prod_{1 \leq r < s \leq 4} \left[\sin \pi(\nu_s - \nu_r) \right.$$

$$\left. \times \prod_{n=1}^\infty \left(1 - 2q^{2n} \cos 2\pi(\nu_s - \nu_r) + q^{4n} \right) \right]^{k_r \cdot k_s}, \tag{9.1.20}$$

and $0 \leq \nu_1 \leq \nu_2 \leq \nu_3 \leq \nu_4 = 1$. (Other factors drop out using $\sum_{r<s} k_r \cdot k_s = 0$, valid for massless states.) $F_P(q^2)$ is an analytic function of q^2 at $q^2 = 0$ and therefore has a power series expansion in powers of q^2 starting with a constant term. In superstring theory the divergence at $q = 0$ is milder than in the bosonic theory (where the integral over q had the form $\int_0^1 dq f(q^2)/q^3$). We learned in chapter 8 that the divergence at $q = 0$ in the planar loop diagram can be interpreted as the result of the emission of a tachyon or a massless particle into the vacuum, *i.e.*, the exchange of either of these states in the 'crossed channel'. The supersymmetric theory with the GSO projection has no tachyon, so the most singular term comes from a massless dilaton propagating in the crossed channel, and we know from chapter 8 that $\int dq/q$ is the appropriate form of the divergence for such a process.

In the covariant approach the planar loop amplitude is obtained as the sum of a loop of fermions and one of bosons. Each of these has a divergence of the form $\int_0^1 dq f(q)/q^2$. This is less severe than in the bosonic string theory because the closed-string tachyon (before the GSO projection) is at $k^2 = 4$ instead of $k^2 = 8$. When the two terms are combined the result given here is reproduced. Thus the GSO conditions lead to cancellations of divergences – much more impressive than those of supersymmetric point-particle field theories.

The calculation of loops with external fermions involves the fermion vertex V_F, which has the form given in (9.1.2), (9.1.4) and (9.1.5). In performing the integration over the loop momentum there are now factors of p^+ (which always occur in integer powers, since there are always an even number of fermion vertices). The only terms that contribute to the amplitude are those for which the net power of p^+ in the integrand is zero. This is a consequence of the momentum integration, which for a term with a factor of $(p^+)^n$ in the integrand takes the form

$$\int dp^+ dp^- d^8\underline{p}(p^9 + ip^0)^n f(p^2) = \int_0^{2\pi} d\phi \int_0^\infty dr\, r^{n+1} e^{in\phi} \int d^8\underline{p} f(\underline{p}^2 + r^2),$$

(9.1.21)

where p^0 has been replaced by ip^0 (a Wick rotation) and polar coordinates have been introduced so that $p^9 + ip^0 = r \exp\{i\phi\}$. In all the examples that we consider the function $f(p^2)$ is just w^{p^2}. The integral (9.1.21) vanishes due to the ϕ integration unless $n = 0$. It is now evident why the loop diagrams with two ground-state fermions obey essentially the same counting rules as the ones with bosons. In order for the powers of p^+ to cancel in the integrand we must include an equal number of F^a and $F^{\dot{a}}$ factors in the integrand. Since an F^a can contribute at most one S_0 mode and an $F^{\dot{a}}$ can contribute at most three S_0 modes, the pair can contribute at most four S_0 modes. Therefore the one-loop three-point function with a boson and two fermions can only give six S_0's and vanishes just as in the purely bosonic case.

Four-point one-loop amplitudes with two or four external fermions can give nonzero results, of course. Again the S contributions are saturated by zero modes leaving trivial traces for the other S_n^a. Diagrams with two bosons and two fermions are of two types – one with adjacent fermions and one with the fermions separated by a boson. In each case there are two terms with no powers of p^+ that contribute to the integrand. Specifically, there are terms in which two powers of S_0 come from each V_B (without any accompanying powers of p^+), three powers from one of the V_F's (with a factor of $(p^+)^{-1/2}$) and the last power of S_0 from the other V_F (with a factor of $(p^+)^{1/2}$). Combining all the terms gives an expression for the full amplitude that is the same as the boson amplitude of (9.1.16) but with the Bose kinematic factor K replaced by the factor that multiplied the corresponding tree diagram with two external fermions and two bosons as in §7.4.2. The diagram with four external fermions likewise contributes the same result as that given in (9.1.16) but with the kinematic factor of §7.4.2 appropriate to that case. In summary, the amplitude given in (9.1.16) describes all one-loop planar diagrams provided one includes the appropriate kinematic factor K in each case.

In chapter 8 we noted the remarkable fact that the planar loop divergence has both an ultraviolet and an infrared interpretation. On the one

hand the calculation that gave it involved summing contributions of an infinite number of string states in the loop of arbitrarily high mass and spin
and integrating over ten-dimensional momenta. A field theorist would expect this to give a terrible ultraviolet divergence. Since the loop integral
contains w^{L_0}, with $L_0 = \frac{1}{2}p^2 + N$, these effects correspond to $L_0 \to \infty$
and hence are all converted to a singular behavior as $w \to 1$. In the
bosonic string theory they really did given a bad divergence $\sim dq/q^3$. In
terms of the w variable this is exponential because $q^{-1} \sim \exp[2\pi^2/(1-w)]$
for $w \to 1$. In the superstring case the lower space-time dimension and
boson–fermion cancellations (due to supersymmetry) have softened this
to dq/q, corresponding to dilaton emission into the vacuum, $i.e.$, in the
'crossed channel' of figs. 8.24a and 8.26.

By changing variables from w to q and studying the behavior of the integral near $q \to 0$ we are led to a completely different viewpoint concerning
the origin of the divergence. Now, instead of being due to ultraviolet
effects, it appears instead to be due to the emission of a massless scalar
('dilaton') of the supergravity multiplet, which then disappears into the
vacuum at zero momentum. This has consequences for the stability of the
vacuum, as the interpretation of the divergence now depends on a single
scalar state at zero momentum, $i.e.$, it is an infrared effect.

Adding a soft dilaton emission to an arbitrary process is equivalent to
taking a derivative with respect to the string tension. This is a very basic
property of string theory that is easy to see in the path integral approach.
The vertex for emitting a zero-momentum dilaton has the form[*]

$$V_D(k = 0) = \partial_\alpha X^i \partial^\alpha X^i. \tag{9.1.22}$$

This is proportional to the bosonic part of the superstring action, which is
multiplied by the string tension. In fact the dependence of the world-sheet
path integral on the string tension comes from a very simple factor

$$e^{-S} \sim \exp\left\{-\frac{T}{2} \int d^2\sigma \partial_\alpha X^i \partial^\alpha X^i\right\}. \tag{9.1.23}$$

From (9.1.22) and (9.1.23), it is evident that a derivative of the path
integral will bring about an insertion of a zero-momentum dilaton. We can
therefore interpret the fact that the divergent part of the loop amplitude
is proportional to the derivative with respect to the string tension of the

---

[*] This expression follows from the general structure of massless closed-string vertices given in §7.4.3, since the dilaton corresponds to the trace of the polarization
tensor ζ^{ii}.

tree amplitude as explicit verification that the divergence is due to an emitted soft dilaton making a transition into the vacuum. In this case, in contrast to the bosonic theory, there is no closed-string tachyon state, so the dilaton is the cause of the leading divergence.

9.1.3 Nonorientable Diagrams

In type I superstring theory, the strings are unoriented and therefore at one-loop level diagrams with an odd number of twisted propagators (for which the world sheet is a Möbius strip) must be included. As in the bosonic theory, the main difference from the planar loop is the replacement of w by $-w$ in the factors arising from the traces over oscillators with non-zero-mode numbers. In addition, there is a factor of $\eta = +1$, -1 or 0, depending on whether the group is $USp(2n)$, $SO(n)$ or $U(n)$, respectively. As discussed in §8.1.2, the complete nonorientable one-loop amplitude is pieced together from the contributions in which an odd number of twists are distributed on the internal propagators in all possible ways. The full contribution corresponding to a particular group-theory factor (such as $\text{tr}(\lambda_1 \lambda_2 \lambda_3 \lambda_4)$) comes from diagrams in which the particles are attached to the boundary of the Möbius strip in a given cyclic order. Only after adding all the diagrams and including the sum over noncyclic permutations of the external states is the full nonorientable amplitude obtained.

The trace over the fermionic S_n^a modes is again very simple when there are at most four external ground-state particles, since the zero-mode trace eliminates all the dependence on the nonzero fermion modes from the vertex. The one-, two- and three-particle amplitudes vanish by exactly the same reasoning as in the planar case. When there are four external states, the fermion zero-mode trace gives the same kinematic factor K as in the case of the planar amplitudes, since the twist operators do not contribute to the zero-mode dependences. Since this factor has total Bose or Fermi symmetry by itself, it does not depend on the ordering of the external particles. The effect of the twist operator in the trace in the integrand of the amplitude is simply to change w to $-w$ in the non-zero-mode part of the trace as in §8.1.2. As a result, the trace over the nonzero S_n^a modes contributes a factor of $[f(-w)]^8$, which cancels a corresponding factor in the denominator due to the bosonic α_n^i traces. The resulting contribution to the four-particle one-loop amplitude with a given cyclic order of the external states is given by

$$A_N(1,2,3,4) = g^4 G_N K \int_0^1 \frac{dw}{w} \left(\frac{-2\pi}{\ln w}\right)^5$$

$$\times \int \left(\prod_{r=1}^3 \frac{d\rho_r}{\rho_r}\right) \prod_{1 \le r < s \le 4} \left(\psi_{rs}^N\right)^{k_r \cdot k_s}, \qquad (9.1.24)$$

for a suitably defined ρ integration, which is more easily described after making one more change of variables. The function ψ^N is defined and discussed in appendix 8.A. Its logarithm, $\ln \psi_{rs}^N$, is proportional to the correlation function $\langle X^i(\rho_r) X^j(\rho_s) \rangle$ for points ρ_r and ρ_s that lie on the boundary of a Möbius strip. The group-theory factor G_N was defined in §8.1.2.

Changing variables to ν_r and q yields the expression

$$A_N(1,2,3,4) = 16\pi^3 g^4 G_N K \int_0^1 \frac{dq}{q} \int_R \left(\prod_{r=1}^3 d\nu_r\right) \prod_{r<s} \left(\psi_{rs}^N\right)^{k_r \cdot k_s}$$

$$= 16\pi^3 g^4 G_N K \int_0^1 \frac{dq}{q} F_N(-q^{1/2}), \qquad (9.1.25)$$

where

$$F_N(-q^{1/2}) = \int_R \left(\prod_{r=1}^3 d\nu_r\right) \prod_{1 \le r < s \le 4} \left[\sin \frac{\pi}{2}(\nu_s - \nu_r)\right.$$

$$\left. \times \prod_{n=1}^\infty \left(1 - 2(-q^{1/2})^n \cos \pi(\nu_s - \nu_r) + q^n\right)\right]^{k_r \cdot k_s}. \qquad (9.1.26)$$

The region R, as in the bosonic theory, is defined by $0 \le \nu_1 \le \nu_2 \le \nu_3 \le \nu_4 = 2$. Since the function F_N is so similar to F_P given in (9.1.20), it is natural to to change variables as in §8.1.2 to $\nu_r' = \nu_r/2$ and $q' = q^{1/4}$. This makes it possible to combine the expressions for the planar and nonorientable amplitudes. Including the trivial Jacobian for this change of variables, one has

$$F_N(-q^{1/2}) = F_N(-q'^2) = 8F_P(q'^2), \qquad (9.1.27)$$

and hence

$$\int_0^1 \frac{dq}{q} F_N(-q^{1/2}) = 32 \int_0^1 \frac{dq}{q} F_P(q^2). \qquad (9.1.28)$$

This relationship between the planar and nonorientable loop amplitudes means that their divergences are also related. This will be pursued further in §10.4.2.

9.1.4 Orientable Nonplanar Diagrams

One-loop diagrams with an even positive number of twisted propagators describe a world sheet that is topologically an annulus with the external states attached to both boundaries. Again, as discussed in §8.1.3, the full amplitude is constructed by adding contributions from all possible distributions of twists on internal propagators and adding diagrams with noncyclic permutations of the external states on the respective boundaries. The algebra simplifies considerably when there are only four external states for the same reasons (concerning S-oscillator traces) as in the case of the planar and nonorientable diagrams that we have already considered. Since the zero-mode trace is not affected by twist operators, the same kinematic factors K are produced as before.

Consider a four-particle process with vacuum quantum numbers in the s channel (as usual, $s = -(k_1 + k_2)^2 = -2k_1 \cdot k_2$) that has the group-theory factor $\text{tr}(\lambda_1\lambda_2)\text{tr}(\lambda_3\lambda_4)$. In this case there are two kinds of contributions to the amplitude. One of these has two twisted propagators, as in fig. 8.16a, while the other has four (with the external states in different cyclic order), as shown in fig. 8.16b. Each of these diagrams is calculated by the same procedure of sewing tree diagrams that led to the expression for the nonplanar loop amplitude of the bosonic theory in terms of the trace of a product of vertices in §8.1.3. Having used up two powers of the S-oscillator zero modes from each vertex in forming the kinematic factor K, the remaining part of each vertex is again simply $\exp\{ik_r \cdot X(0)\}$. The trace still includes the other S-oscillator modes. As in the case of the planar and nonorientable diagrams, factors of the partition function arising from the trace over nonzero bosonic modes cancel against ones from the trace over nonzero fermionic modes. The expression for the amplitude then becomes

$$A_T(1,2,3,4) = g^4 G_T K \int \prod_0^3 \frac{d\rho_r}{\rho_r} \frac{dw}{w} \left(\frac{-2\pi}{\ln w}\right)^5 \prod_{r<s} \left(\psi_{rs} \text{ or } \psi_{rs}^T\right)^{k_r \cdot k_s}.$$

(9.1.29)

The correlation function between two points ρ_r and ρ_s on different boundaries of the world sheet is $\ln \psi_{rs}^T$, so that ψ_{rs}^T (defined in §8.1.3) is to be used when the two points are on different boundaries of the world sheet, while ψ_{rs} is used when they are on the same boundary (as was the case for the corresponding process in §8.1.3). The boundaries of the world sheet are mapped onto the regions $(w, 1)$ and $(-1, -w)$ in the ρ plane with particles #1 and 2 attached to the points $-\rho_1$ and $-\rho_2$ on the negative axis and particles #3 and 4 to the points ρ_3 and $\rho_4 = w$ on the positive axis. The complete integration region is obtained by adding together various contributions, as discussed in §8.1.3.

Changing variables to those given in (9.1.17) and (9.1.18), which are appropriate to integrating over an annulus, the total expression for the amplitude becomes

$$
A_T(1, 2, 3, 4) = 16\pi^3 g^4 G_T K \int\limits_0^1 \frac{dq}{q} \int\limits_0^1 d\nu_1 d\nu_2 d\nu_3
$$

$$
\times \left(\psi_{12}\psi_{34}\right)^{k_1 \cdot k_2} \left(\psi_{14}^T \psi_{23}^T\right)^{k_1 \cdot k_4} \left(\psi_{13}^T \psi_{24}^T\right)^{k_2 \cdot k_4}.
$$

(9.1.30)

The ν_r variables are integrated independently over one complete circuit of the inner and outer circumferences of the annulus of radius q, holding particle #4 fixed at $\nu_4 = 1$. This contribution to the amplitude is finite as in the case of the bosonic theory. As explained in §8.1.3, the behavior of the integrand near $q = 0$ results in a sequence of closed-string poles in the s channel. In the present case, they occur at $s/2 = -k_1 \cdot k_2 = 0, 4, 8, \ldots$, which are the positions of the poles corresponding to the closed-string states of type I superstrings. In agreement with the findings of chapter 5, there is no tachyon in the closed-string spectrum.

In terms of a space-time picture, the dominant configurations of the world sheet that contribute to the s-channel singularities are those in which there is a long narrow tube joining two disks with the external states joined to the boundaries of the disks (just as for the nonplanar loop in §8.1.3). These are precisely the configurations that contribute an infrared divergence when external vertex operators are attached to one boundary only (*i.e.*, the planar diagram). When vertex operators are attached to both boundaries, the momentum flowing between the two boundaries gives an infrared regulator on the would-be infrared divergence. The infrared divergence is recovered for $s \to 0$, and this gives the closed string pole.

The residue of the ground-state closed-string pole at $s = 0$ is proportional to $g^4 G_T K$. Using the definition of K in (9.1.11), this residue is evidently quartic in the external momenta. This is in accord with the expectation that it is proportional to the sums of the squares of the coupling between the particles of the supergravity multiplet and a pair of open-string ground states (particles in the Yang–Mills supermultiplet). This includes the coupling of the graviton to the Yang–Mills sector, which is just the same as in Einstein's theory. Also included is the coupling of the antisymmetric field $B^{\mu\nu}$ to the Yang–Mills multiplet. Its couplings include Chern–Simons terms, which will be explained in chapter 13. These terms will be shown there to play an important role in ensuring the absence of anomalies in the low-energy theory.

There are also open-string poles with singlet quantum numbers in the s channel just as in the bosonic theory. However, there are no massless vector states with singlet quantum numbers for the symmetry groups allowed in superstring theories. Therefore there is no mechanism that can give mass to the antisymmetric tensor field analogous to the one mentioned in §8.1.3 for the bosonic theory with $U(n)$ Chan–Paton symmetry. Such a mechanism would in fact violate supersymmetry.

9.2 Type II Theories

The perturbation expansion of orientable closed-string theories consists of a single Feynman diagram at each order. At one loop, the only contributions come from world sheets that are topologically equivalent to a torus. In this respect these theories are much simpler than the type I theory, which involves several other closed-string one-loop diagrams, as discussed in §8.3.3. The trace in the integrand of the closed-string torus diagram factorizes into a trace for the left movers and one for the right movers (apart from the zero-mode loop momentum, which couples the two sectors). As we have discussed in §7.2.2, closed-string vertices are given by products of open-string vertices for left movers and right movers, with each factor carrying half of the momentum of the emitted state. In particular, emissions of massless closed-string states (the supergravity multiplet) are given by products of massless open-string vertices. In the case of the torus diagram the trace over oscillator modes is taken for each sector separately, so that aside from factors arising from the momentum integration, the integrand is the absolute square of the open-string loop integrand. As a result, the torus diagram with $M < 4$ massless external particles vanishes for exactly the same reasons as in the case of the open-string loops. Eight fermion zero modes are required for a nonzero trace, both for the left-moving sector and the right-moving one. However, each bosonic vertex can only supply two of them. Similarly, a fermion-emission vertex operator can supply one or three, but the two types of terms must appear an equal number of times.

In the one-loop Klein bottle diagram of the nonorientable (type I) theory the situation is more complicated, because the trace does not factorize into separate ones for left movers and right movers, as explained in §8.3.3. As a result, since both factors in the vertices end up inside the same trace, the diagram need not vanish when there are two or more vertices.

In addition to the vanishing of the cosmological constant ($M = 0$), the mass corrections ($M = 2$) and the corrections to the coupling constant ($M = 3$), a most striking feature is the vanishing of the diagram with a single zero-momentum dilaton ($M = 1$), which is the one-loop correction to the dilaton expectation value. In the tree approximation the dilaton expectation value vanishes by virtue of two-dimensional Weyl invariance of the string action, as we have discussed in §3.4.2. However, loop corrections to the dilaton one-point function do not generally vanish. Indeed, in the bosonic string theory we have seen that the divergence of the one-loop amplitudes can be interpreted as being due to the emission of a dilaton, which makes a transition into the vacuum at zero momentum. In view of the vanishing of the dilaton one-loop expectation value in superstring theories it is tempting to conjecture that the loop divergence should also be absent. In order to check this expectation, let us turn now to the explicit calculation of the contribution of the torus to the four particle amplitude.

9.2.1 Finiteness of the Torus Amplitude

Since the torus diagrams with $M < 4$ massless external particles vanish, the first nontrivial check of finiteness occurs for the diagram with $M = 4$ massless particles. (Unitarity implies that this amplitude cannot vanish.) In the one-loop open-string calculations of §9.1, we saw that the trace over S zero modes completely used up the prefactors of the vertex operators in the case of four massless external states. The trace over the remaining nonzero S modes then became easy, given by

$$\text{Tr}\, w^{\sum_1^\infty nS^a_{-n}S^a_n} = [f(w)]^8 . \tag{9.2.1}$$

The corresponding closed-string calculation exhibits the same basic behavior. If we take the external polarization tensors to factorize (*i.e.*, $\zeta^{AB} = \zeta^A\tilde\zeta^B$, where $A, B, \ldots$ denote vector or spinor indices as appropriate to describe the emitted states) the trace over right-moving S zero modes results in a kinematic factor K of the type described in §9.1, while the trace over the other S modes gives the factor in (9.2.1). Similarly, the trace associated with the left-moving $\tilde S^a_0$ gives a kinematic factor $\tilde K$ and the other left-moving $\tilde S^a_n$ gives a factor $[f(\overline{w})]^8$. Altogether, therefore, there is a closed-string kinematic factor

$$K_{cl} = K\tilde K \tag{9.2.2}$$

and a contribution to the measure of $|f(w)|^{16}$ from the S-mode traces.

Since the four-particle open-string loop diagrams were shown to have the same kinematical factors as the corresponding tree amplitudes, it follows that the same is true for four-particle closed string amplitudes. Thus the kinematic factor K_{cl} is identical to the one constructed for four-particle open-string trees in §7.4.3. There it was written out in the form

$$\zeta_1^{AA'}\zeta_2^{BB'}\zeta_3^{CC'}\zeta_4^{DD'}K_{ABCD}(k/2)K_{A'B'C'D'}(k/2). \tag{9.2.3}$$

The type IIA and IIB amplitudes are distinguished by whether the spinors associated with the two factors have the same or opposite chirality.

The factor $|f(w)|^{16}$ introduced by the S traces is exactly canceled by partition-function factors arising from the bosonic modes. The partition function for the right-moving α_n^i modes is $[f(w)]^{-8}$ (as in the open-string case), while the left-moving modes $\tilde\alpha_n^i$ contribute the complex conjugate factor $[f(\overline{w})]^{-8}$.

The remainder of the calculation is identical to that described in §8.2.1 for the torus diagram of the bosonic string theory with external tachyons. There are only a few minor modifications. The fact that $D = 10$ instead of 26 has already been taken into account in the discussion of the powers of $f(w)$. The formulas have no other explicit D dependences. The other difference, of course, is that now the external states satisfy $k^2 = 0$ instead of $k^2 = 2$, which changes some factors. For example, factors in χ_{rs} that are independent of r and s now cancel in the product (see below), since $\sum_{r<s} k_r \cdot k_s = 0$.

By the same steps as were described in the earlier sections of this chapter and the previous one, the expression for the four-particle torus amplitude in either of the type II superstring theories is found to be given by

$$A_C(1,2,3,4) = \left(\frac{\kappa}{4\pi}\right)^4 \int \frac{d^2w}{|w|^2} \int \left(\prod_{r=1}^{3} \frac{d^2\rho_r}{|\rho_r|^2}\right) I(1,2,3,4), \tag{9.2.4}$$

where

$$I(1,2,3,4) = \int d^{10}p|w|^{p^2/4}\mathrm{Tr}[w^N\overline{w}^{\tilde N}V(1,\rho_1)$$
$$\times V(2,\rho_2)V(3,\rho_3)V(4,\rho_4)] \tag{9.2.5}$$
$$= K_{cl}\left(\frac{-4\pi}{\ln|w|}\right)^5 \prod_{r<s}(\chi_{rs})^{k_r\cdot k_s/2}.$$

The integration region is pieced together from the different orderings of the emitted states, in a way that is identical to the case of the bosonic closed-string calculation described in §8.2.1. Each of these configurations has the $|\rho_r|$ arranged in a given sequence. In the complete expression for the amplitude the ρ_r variables are integrated independently over the annulus $|w| \le |\rho_r| \le 1$. As in §8.2.3, we shall see that the w integral must be restricted to avoid an infinite overcounting of equivalent string world-sheet configurations. As in chapter 8, it is convenient to make the change of variables

$$\nu_r = \ln \rho_r / 2i\pi, \qquad \tau = \ln w / 2i\pi, \tag{9.2.6}$$

(so that $\nu_4 = \tau$). Then the amplitude can be rewritten as

$$A_C(1,2,3,4) = (\pi\kappa)^4 K_{cl} \int \frac{d^2\tau}{(\text{Im}\,\tau)^2} F_C(\tau), \tag{9.2.7}$$

where

$$F_C(\tau) = \frac{1}{(\text{Im}\,\tau)^3} \int \prod_{r=1}^{3} d^2\nu_r \prod_{r<s} (\chi_{rs})^{k_r \cdot k_s / 2}. \tag{9.2.8}$$

The ν_r integration can be restricted to cover the torus exactly once, since the integrand has the same invariance (under $\nu \to \nu + 1$ and $\nu \to \nu + \tau$) as in §8.2.1.

In order for the τ integration to include each conformally inequivalent torus geometry once and only once, it should be restricted to a single region, say the fundamental region F described in §8.2.3. However, this only makes sense if any choice of region gives the same result, i.e., if the integrand has modular invariance. Modular invariance was demonstrated for the torus diagram in the bosonic string theory in §8.2.2, and must now be re-examined for type II superstrings. A failure of modular invariance would represent a physical inconsistency.

Our goal is now to demonstrate that the integral (9.2.7) is invariant under modular transformations $\tau \to (a\tau + b)/(c\tau + d)$, where a, b, c, d are integers satisfying $ad - bc = 1$. Since the measure $d^2\tau/(\text{Im}\,\tau)^2$ is modular invariant, it suffices to demonstrate invariance of the function $F_C(\tau)$. As we explained in chapter 8 the limits of the ν integration are τ dependent. Thus to restore them to a period parallelogram it is necessary to accompany the transformation of τ with the mapping $\nu \to \nu/(c\tau + d)$ for each of the ν_r. Then we can use the transformation formula (in appendix 8.A),

$$\chi \left(\frac{\nu}{c\tau + d} \bigg| \frac{a\tau + b}{c\tau + d} \right) = \frac{\chi(\nu | \tau)}{|c\tau + d|}, \tag{9.2.9}$$

to deduce that

$$\prod_{r<s}(\chi_{rs})^{k_r \cdot k_s/2} \rightarrow |c\tau + d|^{-\sum_{r<s} k_r \cdot k_s/2} \prod_{r<s}(\chi_{rs})^{k_r \cdot k_s/2}. \qquad (9.2.10)$$

However, this factor is invariant, since $\sum_{r<s} k_r \cdot k_s = 0$ due to the mass-lessness of the external particles. The rest of the formula for $F_C(\tau)$ is also invariant, since the transformation of the measure, $\prod_{r=1}^{3} d^2\nu_r$, gives a factor of $|c\tau + d|^{-6}$, which cancels a corresponding factor from the transformation of $(\text{Im}\tau)^{-3}$. This means, as in the closed bosonic theory, that the τ integration can be restricted to a single fundamental region. By choosing this to be the region F in fig. 8.21, any possible divergences associated with $\tau = 0$ are avoided.

The expression for the amplitude has the expected poles and normal thresholds dictated by unitarity in the various channels. The poles in channels associated with clusters of external particles arise, as explained in §8.2.4, from the limits in which these particles come together on the surface of the torus. In terms of the space-time viewpoint this corresponds to configurations of the world sheet in which groups of particles are separated by a long thin tube. In the example we have been discussing, namely the superstring torus diagram with four massless particles, there are no mass zero poles in any channel. This is in accord with the vanishing of the vertex and self-energy diagrams for massless states, since these diagrams would be factors in the residues of any such poles. Most importantly, there is no divergence arising from the ν_r integrations. In the case of the bosonic string theory, the region in which all the ν_r's come together was found to be responsible for a divergence, so this is the region that it is particularly important to check. It is investigated as in §8.2.4, by making a change of variables to η_r , ϵ and ϕ defined by

$$\epsilon\eta_r = \nu_r - \nu_M, \qquad r = 1, 2, \ldots, M - 2$$
$$\epsilon e^{i\phi} = \nu_{M-1} - \nu_M. \qquad (9.2.11)$$

Using the same asymptotic estimates of the χ_{rs} factors as before (and $(\epsilon)^{\sum k_r \cdot k_s} = 1$), the total ϵ dependence near $\epsilon = 0$ is given by

$$\int_0^{} d\epsilon\, \epsilon^5 f(\epsilon^2) \int_F \frac{d^2\tau}{(\text{Im}\tau)^5}, \qquad (9.2.12)$$

which is convergent. Since the region of parameter space in which $\epsilon \rightarrow 0$ corresponds to a dilaton tadpole insertion, this is the expected result; a nonzero value of the tadpole amplitude would violate supersymmetry.

We have now seen that the torus diagram with four massless external particles is finite for type II superstring theory. This result surely extends to M-particle torus amplitudes. After all, if one of them were divergent, one could factorize on a massless pole in a three-particle channel to get a residue that factorizes into a product of a tree with $M - 2$ legs and a loop with four legs. Since the tree is certainly finite, the divergence would have to be present in the loop, contradicting the result we have just obtained.

It is also plausible that the finiteness result extends to multiloop amplitudes, although the argument is not yet rigorous. The one-loop calculation has demonstrated that there is no divergence associated with all the particle coordinates (ν_r) approaching a common value on the world sheet. This depends on the absence of a dilaton expectation value on a genus one surface, since the region in which all (ν_r) approach a common value is related by a conformal transformation to a dilaton coupling to the genus one surface. It very probably follows from space-time supersymmetry that the dilaton expectation value vanishes to all orders; attempts to prove this using the covariant fermion vertex operator described in §7.3.5 have been made. Assuming that such an argument is established in due course, what other potential problems are there in loop diagrams? One might expect trouble at singular limits of the geometry, corresponding to corners of the space of Teichmüller parameters that are the multiloop analog of τ. Such problems would again be related to dilaton couplings, and should be absent due to space-time supersymmetry, though a complete proof has not yet appeared.

9.2.2 Compactification on a Torus

We now explore the loop diagrams that arise in the simplest example of string compactification, namely compactification on a torus. The purpose of the discussion is in part to demonstrate explicitly how modular invariance survives compactification, in part to prepare for a later discussion of the low-energy limit of the theory in four dimensions (with this simple compactification) and in part to set the stage for our later discussion of a more realistic form of compactification, based on orbifolds. Thus, we consider a situation in which $(10 - D)$ spatial dimensions of a superstring theory are circles with periodic boundary conditions $(X^I \approx X^I + 2\pi R^I)$ so that the lattice formed by the conjugate momenta is the trivial cubic lattice given by $p^I = m^I/R^I$.

Compactification on a circle was described in §6.4.1. The extension to a direct product of $10 - D$ circles, giving the torus T^{10-D}, is completely straightforward. Since the torus has no curvature, the compactification affects only the zero modes of the expansion of the $X^I(\sigma, \tau)$ coordinates, where I labels the dimensions that have been compactified ($I = 1, \ldots, 10- D$). The calculations of the loop diagrams are therefore the same as before apart from modifications to the treatment of the loop momentum. In particular, the momentum component in the I direction only takes discrete values, $p^I = m^I/R^I$, where R^I is the radius and m^I is an arbitrary integer. Thus in defining the loop integral the momentum integrations must be replaced by sums for these directions. The analysis in §6.4.1 showed that a closed string is also characterized by winding numbers, which count the number of times it wraps around each of the circular dimensions. Thus, in defining the loop amplitude, it is also necessary to include a sum over all possible winding numbers.

Let us begin our study of the effects of introducing nontrivial space-time topology by considering open-string one-loop diagrams. The gaussian integral over each of the components of the loop momentum associated with each compact dimension previously gave a factor of

$$\int_{-\infty}^{\infty} w^{\alpha' p^2} \, dp = (-\pi/\alpha' \ln w)^{1/2}. \tag{9.2.13}$$

This must be replaced in the compactified theory by the sum

$$\frac{1}{R} \sum_{m=-\infty}^{\infty} w^{\alpha' m^2/R^2}. \tag{9.2.14}$$

The coefficient $1/R$ is the counterpart of dp in the integral, since it is the change in $p = m/R$ from one value of m to the next. The radius R is chosen, for convenience, to be the same in all compact dimensions. (More generally, the compact $(10 - D)$-dimensional space may be a torus associated with a less trivial lattice Λ in which case $\int d^{10-D}p$ is replaced by $\sum_{p \in \Lambda}$.) The ratio of the factors in (9.2.13) and (9.2.14), denoted by F_1, can be written as

$$F_1 = a \left(\frac{-\ln w}{\pi} \right)^{1/2} \theta_3 \left(0 \left| \frac{a^2 \ln w}{i\pi} \right. \right), \tag{9.2.15}$$

where

$$\theta_3(0|\tau) = \sum_{m=-\infty}^{\infty} e^{i\pi m^2 \tau} \tag{9.2.16}$$

and

$$a^2 = \frac{\alpha'}{R^2} \qquad (9.2.17)$$

is a dimensionless parameter. Given the loop amplitude for the uncompactified theory, the only modification arising from toroidal compactification of $10 - D$ dimensions is the appearance of an extra factor of $(F_1)^{10-D}$ in the integrand.

The behavior of θ_3 under the change of variables $\ln w \to \ln q = 2\pi^2/\ln w$ follows from that of θ_1 in appendix 8.A (or may be proved directly by use of the Poisson summation formula, which was described in appendix 8.A). This gives

$$F_1(a, q) = \theta_3 \left(0 \left| \frac{\ln q}{2\pi i a^2} \right. \right) = \sum_{n=-\infty}^{\infty} q^{n^2/2a^2}. \qquad (9.2.18)$$

It is fortunate that this factor F_1 does not introduce powers of $\ln q$ in the formula for the loop amplitude. Introducing $(F_1)^{10-D}$ into the planar and the nonorientable loop amplitudes does not alter their general features. In the region of the divergence ($q \to 0$) the modification factor $F_1(a, 0)$ equals one, so that the nature of the divergences is not modified. It is no longer true, however, that the integrands for the planar loop amplitude and nonorientable amplitude are related by (9.1.25), since $F_1(a, -q^{1/4}) \neq F_1(a, q)$.

Since the theory now contains closed-string states of nonzero winding numbers, there should also be a modification to the spectrum of closed-string states. This must be reflected in the structure of the nonplanar loop diagram. This can easily be checked, since the positions of the closed-string poles in the s channel are given by expanding the integrand in a power series in q^2. This series now has additional terms coming from $(F_1)^{10-D}$ so that new poles occur at positions given by

$$\alpha's = N + \frac{1}{2a^2} \sum_{I=1}^{10-D} (n^I)^2, \qquad (9.2.19)$$

where n^I are the integers arising from the expansion in (9.2.18). This is the same spectrum as that of the closed-string theory compactified on a torus (§6.4.1) with the n^I's identified with the winding numbers. (The intermediate states in this example all have zero Kaluza–Klein charges, since the external states have $p^I = 0$.)

Let us now consider the torus diagram for closed-string scattering in the presence of the toroidal compactification. In this case the strings can have arbitrary winding numbers and the mode expansion of $X^I(\sigma, \tau)$ $(I = D, \ldots, 9)$ is given by (see §6.4.1)

$$X^I(\tau, \sigma) = X^I_R(\tau - \sigma) + X^I_L(\tau + \sigma)$$

$$X^I_R(\tau - \sigma) = x^I_R + \left(\frac{m^I}{2R} - n^I R\right)(\tau - \sigma) + \frac{i}{2}\sum_{n \neq 0}\frac{1}{n}\alpha^I_n e^{-2in(\tau-\sigma)}$$

$$X^I_L(\tau + \sigma) = x^I_L + \left(\frac{m^I}{2R} + n^I R\right)(\tau + \sigma) + \frac{i}{2}\sum_{n \neq 0}\frac{1}{n}\tilde{\alpha}^I_n e^{-2in(\tau+\sigma)}.$$

$$(9.2.20)$$

Before compactification the momentum integral gave a factor of

$$\int_{-\infty}^{\infty} |w|^{\alpha' p^2/2} dp = (\alpha' \text{Im}\tau)^{-1/2} \tag{9.2.21}$$

for each direction, where $w = \exp(2\pi i \tau)$. This integral must now be replaced by a double sum over winding numbers n and the charges m

$$\frac{1}{R}\sum_{m,n=-\infty}^{\infty} w^{(am-n/a)^2/4}\overline{w}^{(am+n/a)^2/4}, \tag{9.2.22}$$

so that the integrand must be modified by a factor of $(F_2)^{10-D}$, where F_2 is the ratio of these factors. The exponents in this equation are the zero-mode parts of L_0 and $\tilde{L}_0$ deduced from (9.2.20), as given in §6.4.1. The ratio of (9.2.21) and (9.2.22) can be expressed as

$$F_2(a, \tau) = a(\text{Im}\tau)^{1/2}\sum_{mn}\exp\left[-2i\pi mn \text{Re}\tau - \pi(a^2 m^2 + n^2/a^2)\text{Im}\tau\right].$$

$$(9.2.23)$$

It is important that the factor $(F_2)^{10-D}$ should not spoil the modular invariance of the integrand of the loop amplitude. Since the entire group of modular transformations is generated by the two transformations $\tau \to \tau + 1$ and $\tau \to -1/\tau$, it is sufficient to check these two cases. Invariance of F_2 under the first of these is trivial. Proof of the invariance under the second transformation involves a generalization of the Poisson summation formula to a double sum. The proof, given in appendix 9.B, shows that

$$F_2(a, \tau) = F_2(a, -1/\tau), \tag{9.2.24}$$

as required. Thus, even after compactification, the torus diagram still has the modular invariance required for a consistent interpretation as a string theory. Note that it was crucial to include all the winding states to achieve this result.

In the example just described, modular invariance did not restrict the allowed radii R^I of the compact space. In the discussion of the heterotic string in §9.3.2 we shall see that modular invariance imposes stringent restrictions.

9.2.3 The Low-Energy Limit of One-Loop Amplitudes

The fact that at low-energy superstring theory reduces to supergravity coupled to super Yang–Mills is one of its major features. Explicit evidence for this has been given at the level of the tree diagrams of perturbation theory around flat ten-dimensional space-time (*i.e.*, for classical string field theory) in §7.4. The three-particle vertices given there were found to be exactly those of the low-energy field theory in the case of type 1 open strings. In the case of heterotic strings, we learned that there are also order α' corrections that vanish at low energy. In the case of four-particle amplitudes, we found that the low-energy field theory results are multiplied by a ratio of Γ functions that approaches one at low energy.

We can now consider the corresponding limit for the one-loop quantum corrections, which should reduce to the one-loop corrections to the appropriate supersymmetric field theory based on point particles. Clearly, if we just take the low-energy limit of the expressions for the loops in flat ten-dimensional space (which is equivalent to $\alpha' \to 0$ at the order that we are considering here), the result has the usual divergences of ordinary higher-dimensional field theory. An infinite answer is also obtained by eliminating the extra dimensions by compactifying them on a torus and letting the radii of the torus vanish ($a \to \infty$) before taking the low-energy limit. If this were possible it would yield a consistent string theory in fewer than ten dimensions. What prevents this is that an infinite number of closed-string states with nonzero winding numbers have masses proportional to the radius of the extra dimensions, so there is a condensation of such states at zero mass when the radius vanishes.

We would like to consider a coupled limit in which $10 - D$ dimensions become vanishingly small at the same time that the states of nonzero winding number become infinitely massive and decouple. This is achieved by the limit

$$R \to 0, \quad \text{and} \quad \alpha' a^2 = \frac{\alpha'^2}{R^2} \to 0. \qquad (9.2.25)$$

Although compactification on a general torus is not meant to be physically realistic, this method serves to illustrate the way in which the low-

energy limit of the quantum corrections to string theory correspond to the
loop corrections to point-particle theories in dimensions in which they are
not divergent. A single expression for a string loop amplitude gives rise,
in the low-energy limit, to the sum of all the Feynman diagrams with the
same topology. This means that if there are cancellations of divergences
between different Feynman diagrams these are automatically incorporated
in this procedure.

The low-energy limit of the open-string sector of type I theories gives
$N = 4$ Yang–Mills theory.[*] The one-loop amplitudes vanish when there
are fewer than three external ground states, which corresponds to the
absence of a coupling constant or mass renormalization of the low-energy
theory in any dimension. The open-string amplitude with four external
massless states has the same overall kinematic factor K for both tree and
one-loop amplitudes (and probably multiloop amplitudes as well). The
fact that the group-theory factors of the tree amplitudes and the planar
loop corrections are also proportional means that we can write the sum
of the tree diagrams and the planar loop corrections in the form

$$A(1, 2, 3, 4) = g_{10}^2 K \left\{ \frac{1}{st} \frac{\Gamma(1 - \alpha's)\Gamma(1 - \alpha't)}{\Gamma(1 - \alpha's - \alpha't)} + c_1 f^{(1)} + \ldots \right\}, \quad (9.2.26)$$

where the group-theory factor $\mathrm{tr}\,(\lambda_1 \lambda_2 \lambda_3 \lambda_4)$ has been suppressed, c_1 is an
overall dimensionless constant determined by unitarity and $f^{(1)}$ is given
by

$$f^{(1)} = \frac{g_{10}^2}{\alpha'} \int_0^1 \frac{dq}{q} [F_1(a, q)]^{10 - D} \int_0^1 \prod_{r=1}^3 (d\nu_r \theta(\nu_{r+1} - \nu_r))$$

$$\times \left[\frac{\psi_{12}\psi_{34}}{\psi_{13}\psi_{24}} \right]^{-\alpha's} \left[\frac{\psi_{23}\psi_{14}}{\psi_{13}\psi_{24}} \right]^{-\alpha't} . \quad (9.2.27)$$

The dots in (9.2.26) indicate the presence of other quantum corrections,
including those coming from nonorientable and multiloop diagrams. In the
low-energy limit the nonorientable loop turns out to give a contribution
of the same form as the planar loop. In the case of the nonplanar loop
there is no tree diagram with the same group-theory factor as the loop,
but the rest of the expression reduces in the low energy limit to the same

---

[*] Here we use the terminology $N = 4$ Yang–Mills for the theory in any dimension
D although it has four spinorial supercharges only in its four-dimensional form.
The analogous statement also applies to the terminology used below for $N = 8$
supergravity.

form as the planar loop. The analysis in both these cases is very similar to the planar loop and is not presented here.

The ten-dimensional Yang–Mills coupling constant g_{10} is dimensionally [length]^3, while the coupling g_D when $10 - D$ dimensions are compactified is defined by

$$g_D = g_{10} R^{(10-D)/2}. \tag{9.2.28}$$

In order to define a limit in D dimensions in which we are left with a one-loop Yang–Mills result with finite g_D but no gravitational effects, it is necessary that the D-dimensional gravitational coupling constant

$$\kappa_D = \kappa_{10} R^{(10-D)/2} \sim \frac{g_D^2}{\alpha'} R^{(D-10)/2} \tag{9.2.29}$$

vanishes. This is the case provided that $R^{(10-D)/2}/\alpha' \to 0$.

It is convenient to consider the special case of (9.2.25) in which $\alpha' \to 0$ with fixed a (in which case the vanishing of κ_D requires that $D < 6$), but the result holds for the general limit. The number of noncompact dimensions D is treated as a continuous parameter in the formulas.

In the limit $\alpha' \to 0$ with a fixed the powers of ψ_{rs} in (9.2.27) approach zero, and the leading behavior is controlled by the divergent endpoint $q \sim 1$ (*i.e.*, $w = 2\pi^2/\ln q \sim 0$) when $D < 6$. This uses the fact that F_1 has the asymptotic form

$$F_1(a, q) \sim a \left(-\frac{2\pi}{\ln q} \right)^{1/2}, \tag{9.2.30}$$

while

$$\psi_{rs}(c, w) \sim (1 - x)(1 - w/x) \exp\left(\frac{-\ln x \ln z(w/x)}{2 \ln w} \right). \tag{9.2.31}$$

Substituting these asymptotic expansions into $f^{(1)}$, after a certain amount of algebra the leading behavior of (9.2.27) is found to be given by

$$f^{(1)} \sim g_D^2 \pi^{\gamma-1} \Gamma(-\gamma) \int_0^1 \left(\prod_{i=1}^{4} d\eta_i \right) \delta(1 - \sum_i \eta_i)(\eta_1 \eta_3 s + \eta_2 \eta_4 t)^\gamma, \tag{9.2.32}$$

where the variables η_i are defined by

$$\eta_i = \nu_i - \nu_{i-1}, \tag{9.2.33}$$

and

$$\gamma = \tfrac{1}{2} D - 4. \tag{9.2.34}$$

This analysis can be used to obtain the expression for the one-loop

$N = 4$ super Yang–Mills amplitude in any dimension D. The expression (9.2.32) diverges for even dimensions $D \geq 8$, which is in agreement with calculations of the ultraviolet divergences of $N = 4$ Yang–Mills theory. When $D = 6$ the limit is finite, while the expression diverges at $D = 4$ due to infrared effects. The function $f^{(1)}$ diverges like $(D - 4)^{-2}$ near $D = 4$, which is the behavior typical of an infrared divergence in on-shell scattering of four Yang–Mills bosons in dimensional regularization.

In a similar fashion one can obtain one-loop amplitudes for $N = 8$ supergravity from low-energy limits of type II amplitudes. It is only in $D = 10$ dimensions that there is a distinction between the chiral and nonchiral theories so that the rather trivial low-energy compactification being discussed here does not distinguish between type IIA and type IIB theories.

The one-loop diagrams with fewer than four external ground states vanish again in this case so that the first nontrivial loop amplitude has four external massless states. In this case the sum of the tree diagrams and the quantum corrections can be written as

$$A(1,2,3,4) = \kappa_{10}^2 K_{cl} \left\{ -\frac{4}{stu} \frac{\Gamma(1 - \alpha's/4)\Gamma(1 - \alpha't/4)\Gamma(1 - \alpha'u/4)}{\Gamma(1 + \alpha's/4)\Gamma(1 + \alpha't/4)\Gamma(1 + \alpha'u/4)} \right.$$
$$\left. + d_1 g^{(1)} + \dots \right\},$$

$$(9.2.35)$$

where d_1 is another dimensionless constant determined by unitarity,

$$g^{(1)} = \frac{\kappa_{10}^2}{\alpha'} \int_F \frac{d^2\tau}{(\mathrm{Im}\tau)^2} [F_2(a,\tau)]^{10-D} F_C(\tau) \qquad (9.2.36)$$

and $F_C(\tau)$ is defined in (9.2.8). The dots once more indicate the presence of higher-order quantum corrections.

The low-energy limit can be considered again by taking $\alpha' \to 0$ with a and κ_D held fixed. As in the case of the open-string amplitude, the loop diverges for $D \geq 8$, while for $D < 8$ the region of integration that dominates (9.2.36) is the endpoint $\mathrm{Im}\tau \to \infty$, since in this limit

$$F_2(a,\tau) \sim a(\mathrm{Im}\tau)^{1/2}. \qquad (9.2.37)$$

Substituting the asymptotic form for the functions inside (9.2.36) gives (after some algebra) the asymptotic form of $g^{(1)}$ as $\alpha' \to 0$

$$g^{(1)} \sim \kappa_D^2 \Gamma(-\gamma) \int_0^1 d\rho_3 \int_0^{\rho_3} d\rho_2 \int_0^{\rho_2} d\rho_1 [s\rho_1\rho_2 + t\rho_2\rho_3$$

$$+ u\rho_3\rho_1 + t(\rho_1 - \rho_2)]^\gamma$$

$$+ \text{ terms that symmetrize } s, t \text{ and } u,$$

(9.2.38)

where $\gamma = D/2 - 4$ as before.

This asymptotic expression for $g^{(1)}$ can be shown to be proportional to the asymptotic expression for $f^{(1)}$, symmetrized in s, t and u, i.e.,

$$g^{(1)} \sim f^{(1)}(s,t) + f^{(1)}(t,u) + f^{(1)}(u,s). \qquad (9.2.39)$$

This means that $g^{(1)}$ is also finite when $D = 6$. It has a softer infrared divergence than $f^{(1)}$ as $D \to 4$, due to cancellations arising from the symmetrization. Specifically, $g^{(1)} \sim (D-4)^{-1}$, which is the characteristic infrared divergence of gravity in dimensional regularization. In conclusion, both $N = 4$ Yang–Mills and $N = 8$ supergravity are ultraviolet finite at one loop for $D < 8$ and infrared finite for $D > 4$. These results were obtained rather more easily from considering limits of string-theory amplitudes than would have been possible in the field theories themselves.

9.3   The Heterotic String Theory

In chapter 6 we learned that the internal gauge symmetry degrees of freedom of the heterotic string theories can be expressed either in a fermionic or a bosonic formulation. The tree amplitude calculations in §7.4.4 were based on the bosonic formulation, and that is again the formalism that we will use for the study of one-loop amplitudes in this section. Recall that in this approach the internal gauge symmetry is described by sixteen left-moving bosonic coordinates that parametrize the maximal torus of $E_8 \times E_8$ or spin$(32)/Z_2$.

The generators of $E_8 \times E_8$ or $SO(32)$ can be divided into 16 'neutral' ones belonging to the Cartan subalgebra and 480 'charged' ones with charges described by a 16-component vector K^I, which may also be regarded as Kaluza–Klein momenta associated with the 16-dimensional torus. In formulating vertex operators for states of the Yang–Mills supermultiplet, the required structure is somewhat different for the neutral and charged states. The construction was described in detail in §6.4.3 and applied to the evaluation of some tree amplitudes in §7.4.4. For neutral states belonging to the Cartan subalgebra of the adjoint representation or supergravity multiplet states (which are singlets of the gauge group) the internal momentum is $K^I = 0$. For the 480 charged states the vector K^I corresponds to a root vector of the gauge group and has $K \cdot K = 2$. The vertex operators associated with the charged states contain cocycle factors discussed in §6.4.4 and §6.4.5.

If we use the space-time supersymmetric light-cone gauge formalism of chapter 5 to describe the right-moving modes, as in the preceding sections, then we can achieve the same simplification for the heterotic superstrings as we found for the type II superstring theories. As was the case with the other superstring theories, amplitudes with fewer than four external ground states vanish due to the trace associated with the right-moving S_0^a modes. Also, as in the other theories, the four-particle amplitudes are especially simple, since vertex operators associated with the right-moving modes are effectively just those of the Veneziano model with no prefactors once the S_0^a trace has been evaluated.

9.3.1 The Torus with Four External Particles

Since the simplest nontrivial example of a heterotic string loop amplitude has four massless external particles, that is the case we consider. At one-loop order there is only one diagram, the torus diagram. As in the case of type II superstrings, the world sheet must be closed and orientable, and the torus is the unique possibility at one loop.

The kinematics is the same as that described in chapter 8 (see fig. 8.8). The space-time momenta of the emitted states are denoted K_r^I ($I = 1, 2, \ldots, 16$) and the charge of the rth propagator P_r^I is given by

$$P_r^I = P^I - \sum_{s=1}^{r-1} K_s^I. \qquad (9.3.1)$$

The loop momentum P^I is summed over all the sites of the appropriate lattice Λ.

The one-loop amplitudes are again given by 'sewing' tree diagrams, which gives for four external particles

$$A_H(1,2,3,4) = \left(\frac{\kappa}{4\pi}\right)^4 \int \frac{d^2w}{|w|^2} \int \prod_{r=1}^{3} \frac{d^2\rho_r}{|\rho_r|^2} I(1,2,3,4), \qquad (9.3.2)$$

as in (9.2.4). Now the correlation function $I(1, 2, 3, 4)$ is given by

$$I(1, 2, 3, 4) = \int d^{10}p|w|^{p^2/4} \sum_{P \in \Lambda} \overline{w}^{P^2/2} \text{Tr}[w^N \overline{w}^{\tilde{N}-1} V(1, \rho_1)$$
$$\times V(2, \rho_2)V(3, \rho_3)V(4, \rho_4)].$$
(9.3.3)

Note that we have included factors

$$w^{L_0}\overline{w}^{\tilde{L}_0-1} = |w|^{p^2/4}w^N\overline{w}^{\tilde{N}-1+P^2/2},$$
(9.3.4)

because the superstring sector has no normal ordering constant, whereas the bosonic sector has one of -1. The vertex operator $V(k_r, \rho_r)$ describes the emission of any of the ground-state particles of the heterotic string. Since the trace over the right-moving S_0^a modes uses up all the dependence on S modes in the vertices, the contribution of the right-moving S_n^a modes to the trace becomes $[f(w)]^8$, just as for the right-moving modes of the type II theories.

It is convenient to express the second-rank polarization tensors (including the spinors) of the supergravity multiplet as sums of products of the polarization vectors of the Yang–Mills supermultiplet (*i.e.*, to write $\zeta_r^{\mu\nu} = \sum_n \zeta_{rn}^\mu \overline{\zeta}_{rn}^\nu$, $\zeta_r^{a\mu} = \sum_n u_{rn}^a \overline{\zeta}_{rn}^\mu$, ...). In this way the overall kinematic factor arising from the trace over the fermion zero modes is the sum of factors of the same form as the factor K that arose in the open-string case (we shall only consider one of these terms and so we shall drop the label n).

The factors $\overline{\zeta}_{r\mu}\dot{X}_L^\mu(\overline{z}_r)$ ($\mu = 0, \ldots, 9$) associated with the vertices for the emission of gravitons can be dealt with in the same way as the vector-emission vertices in the bosonic open-string theory by using $\exp[\overline{\zeta}_r \cdot \dot{X}_L(\overline{z}_r)]$ and extracting the terms linear in $\overline{\zeta}_r$ at the end of the calculation. The vertices for the emission of the sixteen neutral gauge particles contain factors of $\overline{\xi}_r^I \dot{X}_L^I(\overline{z}_r)$ that can also be dealt with by exponentiation in the form $\exp[\overline{\xi}_r^I \dot{X}_L^I(\overline{z}_r)]$ and by extracting the terms linear in $\overline{\xi}_r$ at the end of the calculation. The factor $\overline{\xi}_r^I$ can be regarded as a 'polarization' vector in the extra sixteen dimensions that labels the neutral gauge particles. By including both types of factors simultaneously, we obtain formulas that describe emission of an arbitrary combination of gravitons and neutral gauge particles. Since the right movers are described supersymmetrically, we can also describe their fermionic supersymmetry partners by modifying the choice of K factors appropriately.

The integration over the ten-dimensional loop momentum p^μ is similar to that in the bosonic theory in §8.2.1. But now there is additional p

dependence in the integrand introduced by the zero mode of $\dot{X}_L^\mu$

$$\int d^{10}p \prod_{r=1}^{M} |z_r|^{p_r^2/4} e^{\zeta_r \cdot p_r/2} = \left(\frac{-4\pi}{\ln|w|}\right)^5 \prod_{r<s} \left\{ |c_{sr}|^{-1/2} \exp \frac{\ln^2 |c_{sr}|}{2\ln|w|} \right\}^{k_r \cdot k_s/2}$$

$$\times \exp \sum_{1\le r<s\le M} \left\{ (k_r \cdot \overline{\zeta}_s - k_s \cdot \overline{\zeta}_r) \left(\frac{\ln|c_{sr}|}{2\ln|w|} - \frac{1}{4} \right) - \overline{\zeta}_r \cdot \overline{\zeta}_s \frac{1}{2\ln|w|} \right\}.$$

$$(9.3.5)$$

This is the same formula as §8.1.1 with x_r replaced by $\sqrt{z_r}$ and ζ_r replaced by $\overline{\zeta}_r/2$.

The sum over the discrete values of the loop momenta P^I is evaluated in much the same way as for the continuous momenta, namely by completing a square. The algebra is the same as in §8.1.1, which gives

$$\sum_{P\in\Lambda} \prod_{r=1}^{M} z_r^{P_r^2/2} e^{\overline{\xi}_r \cdot P_r} = \mathcal{L} \prod_{r<s} \left\{ \overline{c}_{sr}^{-1/2} \exp \left(\frac{\ln^2 \overline{c}_{sr}}{2\ln\overline{w}} \right) \right\}^{K_r \cdot K_s}$$

$$\times \exp \left\{ \sum_{1\le r<s\le M} (K_r \cdot \overline{\xi}_s - K_s \overline{\xi}_r) \right.$$

$$\times \left. \left(\frac{\ln \overline{c}_{sr}}{\ln\overline{w}} - \frac{1}{2} \right) - \overline{\xi}_r \cdot \overline{\xi}_s \frac{1}{\ln\overline{w}} \right\}$$

$$= \mathcal{L} \prod_{r<s} \exp \left\{ K_r \cdot K_s \left(i\pi\overline{\nu}_{sr} - \frac{i\pi\overline{\nu}_{sr}^2}{\overline{\tau}} \right) \right.$$

$$\left. + (K_r \cdot \overline{\xi}_s - K_s \cdot \overline{\xi}_r) \left(\frac{\overline{\nu}_{sr}}{\overline{\tau}} - \frac{1}{2} \right) - i\frac{\overline{\xi}_r \cdot \overline{\xi}_s}{2\pi\overline{\tau}} \right\},$$

$$(9.3.6)$$

where, as usual, the complex conjugate of (9.2.6) gives

$$\overline{\nu}_{sr} = \overline{\nu}_s - \overline{\nu}_r, \qquad \overline{\nu}_r = -\ln \overline{p}_r/2\pi i \qquad (r = 1,\dots,M-1) \qquad (9.3.7)$$

and

$$\overline{\nu}_M \equiv \overline{\tau} = -\ln \overline{w}/2\pi i. \qquad (9.3.8)$$

The lattice sum is contained in the function $\mathcal{L}$, defined by

$$\mathcal{L} = \sum_{P\in\Lambda} e^{-i\pi\overline{\tau}(P+S)^2}, \qquad (9.3.9)$$

where

$$S^I = \sum_{r=1}^{4} \left(\frac{\ln \overline{p}_r}{\ln\overline{w}} K_r^I + \frac{\overline{\xi}_r^I}{\ln\overline{w}} \right) = \sum_{r=1}^{4} \left(\frac{\overline{\nu}_r}{\overline{\tau}} K_r^I + \frac{i\overline{\xi}_r^I}{2\pi\overline{\tau}} \right), \qquad (9.3.10)$$

where terms involving $\xi_r \cdot \xi_r$ have been dropped, since only the terms linear in ξ_r contribute to the amplitude.

The factors from nonzero modes in the trace contained in the function I are evaluated in a way that is very similar to the previous examples. The calculation may be divided into two factors so that

$$I = I_1 \times I_2, \tag{9.3.11}$$

where the factor I_1 comes from the (left- and right-moving) space-time modes α_n^i and $\tilde{\alpha}_n^i$ together with the trace over the fermion modes S_n^a, while the factor I_2 comes from the modes of the extra sixteen left-moving coordinates $\tilde{\alpha}_n^I$.

The fermionic trace contributes a factor $K\,[f(w)]^8$ to I_1. The zero-mode momentum factor (9.3.5) combines with the factors arising from the nonzero modes in I_1 that occur in the function

$$\text{Tr}\left\{ \exp \sum_1^M \left(ik_r \cdot X(\overline{z}_r) + \overline{\zeta}_r \cdot P_L(\overline{z}_r) \right) \right\}, \tag{9.3.12}$$

that forms the integrand of the loop diagram to give a factor similar to the bosonic one in the type II theories. The result has a form that is reminiscent of the open-string expression in §8.1.1:

$$\begin{aligned} I_1 =&\, K\,[f(\overline{w})]^{-8} \left(\frac{-4\pi}{\ln|w|} \right)^5 \prod_{r<s} (\chi_{rs})^{k_r \cdot k_s/2} \\ &\times \exp\left\{ \tfrac{1}{2}(k_r \cdot \overline{\zeta}_s - k_s \cdot \overline{\zeta}_r)\hat{\eta}_{rs} + \overline{\zeta}_r \cdot \overline{\zeta}_s \hat{\Omega}_{rs} \right\}, \end{aligned} \tag{9.3.13}$$

where $\chi_{rs} = \chi(\overline{c}_{sr}, \overline{w})$, with similar definitions for $\hat{\eta}_{rs}$ and $\hat{\Omega}_{rs}$. The functions $\hat{\eta}_{rs}$ and $\hat{\Omega}_{rs}$ are in fact proportional to the correlation functions $\langle \dot{X}^\mu(\overline{z}_r) X_\mu(\overline{z}_s)\rangle$ and $\langle \dot{X}^\mu(\overline{z}_r)\dot{X}_\mu(\overline{z}_s)\rangle$, respectively. They differ from the unhatted functions defined in chapter 8 only in their zero modes (due to the fact that the zero modes couple the left-moving and right-moving coordinates).

$$\begin{aligned} \hat{\eta}_{rs} \equiv \hat{\eta}(\overline{c}_{sr}, \overline{w}) &= \eta(\overline{c}_{sr}, \overline{w}) - \frac{\ln \overline{c}_{sr}}{\ln \overline{w}} + \frac{\ln |c_{sr}|}{\ln |w|} \\ &= \eta_{rs} - \frac{\overline{v}_{sr}}{\overline{\tau}} + \frac{\text{Im}\,\overline{v}_{sr}}{\text{Im}\,\overline{\tau}}, \end{aligned} \tag{9.3.14}$$

and

$$\begin{aligned} \hat{\Omega}_{rs} \equiv \hat{\Omega}(\overline{c}_{sr}, \overline{w}) &= \Omega(\overline{c}_{sr}, \overline{w}) + \frac{1}{\ln \overline{w}} - \frac{1}{2 \ln |w|} \\ &= \Omega_{rs} - \frac{1}{2i\pi\overline{\tau}} - \frac{1}{4\pi \text{Im}\,\overline{\tau}}. \end{aligned} \tag{9.3.15}$$

The factor I_2 associated with the extra sixteen dimensions is obtained by combining (9.3.6) with the factors that arise from the nonzero modes. To begin with there is the factor of $\bar{\epsilon}(K_1, K_2, K_3, K_4)$ that comes from the cocycle factors in the vertices with emitted charged Yang–Mills particles. Apart from the fact that the momentum P^I takes discrete values, the calculation is essentially the same as the open-string calculation, and gives

$$I_2 = \bar{\epsilon}[f(\overline{w})]^{-16} \mathcal{L} \prod_{r<s} (\psi_{rs})^{K_r \cdot K_s}$$

$$\times \exp\left[(K_r \cdot \bar{\xi}_s - K_s \cdot \bar{\xi}_r)\eta_{rs} + \bar{\xi}_r \cdot \bar{\xi}_s \Omega_{rs}\right]. \tag{9.3.16}$$

Writing the measure in terms of the variables ν_r and τ

$$\frac{1}{\overline{w}} \frac{d^2w}{|w|^2} \prod_{r=1}^{M} \frac{d^2\rho_r}{|\rho_r|^2} = \frac{(2\pi)^8}{\overline{w}} d^2\tau \prod_{r=1}^{M-1} d^2\nu_r, \tag{9.3.17}$$

all the massless four-particle one-loop amplitudes are contained in

$$A_H(1,2,3,4) = (\pi\kappa)^4 \bar{\epsilon} K \int_R d^2\tau \int \prod_r d^2\nu_r \left(\frac{-4\pi}{\ln|w|}\right)^5 \frac{1}{\overline{w}} [f(\overline{w})]^{-24}$$

$$\times \mathcal{L} \exp \sum_{r<s} \left(\tfrac{1}{2}(k_r \cdot \bar{\zeta}_s - k_s \cdot \bar{\zeta}_r)\hat{\eta}_{rs} + (K_r \cdot \bar{\xi}_s - K_s \cdot \bar{\xi}_r)\eta_{rs}\right.$$

$$\left. + \bar{\zeta}_r \cdot \bar{\zeta}_s \hat{\Omega}_{rs} + \bar{\xi}_r \cdot \bar{\xi}_s \Omega_{rs}\right) \prod_{r<s} (\chi_{rs})^{k_r \cdot k_s/2} (\psi_{rs})^{K_r \cdot K_s}.$$

$$\tag{9.3.18}$$

Any process involving external ground states can be obtained from this formula by choosing the rth state to have nonvanishing $\bar{\zeta}_r^\mu$ for a gravitational particle, nonvanishing $\bar{\xi}_r^I$ for a neutral particle in the Yang–Mills multiplet or $K_r \neq 0$ for a charged Yang–Mills particle and keeping only those terms that are linear in the nonvanishing $\bar{\zeta}_r$'s and $\bar{\xi}_r$'s, as explained earlier.

The integration region in (9.3.18) is the same as for the torus diagram of the other theories we have explained. The crucial question is whether F_H defined by the formula

$$A_H(1,2,3,4) = (\pi\kappa)^4 \bar{\epsilon} K \int \frac{d^2\tau}{(\mathrm{Im}\tau)^2} F_H(\tau), \tag{9.3.19}$$

is modular invariant. This is the issue to which we now turn.

9.3.2 Modular Invariance of the $E_8 \times E_8$ and $SO(32)$ Theories

Although there are many aspects of the demonstration of modular invariance that follow the same path as for the type II theories, the heterotic theory is crucially different due to the fact that it incorporates internal symmetry *via* the lattice Λ. This enters the expression for the amplitude (9.3.18) in the factor $\mathcal{L}$. Modular invariance imposes important restrictions on possible lattices.

We begin the discussion by showing that the region of integration over the ν_r variables can again be restricted to a fundamental parallelogram corresponding to the periods 1 and τ in the complex ν plane. To prove this we have to check the invariance under the transformations

$$\nu_r \to \nu_r + 1 \tag{9.3.20}$$

and

$$\nu_r \to \nu_r + \tau, \tag{9.3.21}$$

which correspond to $z_r \to z_r \exp\{2\pi i\}$ and $z_r \to w z_r$. The first of these is an obvious invariance of all the factors in the amplitude apart from those arising from the right-hand side of (9.3.6), since they are either functions of $|\rho_r|$ or have a series expansion in powers of ρ_r. However, the left-hand side of (9.3.6) is obviously invariant under $z_r \to z_r \exp\{2\pi i\}$, since P_r is a vector on the root lattice so that $P_r^2/2$ is an integer. The second of the transformations on the ν_r variables is a symmetry of all the functions in the integrand of the amplitude. For ψ_{rs} and χ_{rs} this follows from the properties listed in appendix 8.A, while for the other functions it is a consequence of their definitions in terms of derivatives of ψ_{rs}. The fact that $\mathcal{L}$ is invariant under (9.3.21) follows from its definition by shifting $\ln \overline{\rho}_r$ by $\ln \overline{w}$ in the exponent of (9.3.9), which is equivalent to shifting P in the exponent by a lattice vector K_r leaving the sum invariant. This means that the ν_r region of integration can indeed be restricted to a fundamental region in the ν plane so that it covers the torus exactly once.

Now we turn to the question of modular invariance, where something new will arise. In order to study the behavior of the amplitude under the global diffeomorphisms generated by modular transformations, it is sufficient to consider the two transformations

$$\tau \to \tau' = \tau + 1 \tag{9.3.22}$$

and

$$\tau \to \tau' = -\frac{1}{\tau}, \tag{9.3.23}$$

since they generate the entire modular group, as was described in §8.2.2.

The first of these corresponds to the change from w to $w \exp\{2\pi i\}$, which is a special case of (9.3.20) with $r = M$. In order to study the behavior under the second transformation, it is necessary, as before, to transform the ν_r variables simultaneously by

$$\nu \to \nu' = -\frac{\nu}{\tau} \tag{9.3.24}$$

so that the ν' integral covers the same domain expressed in terms of the transformed τ variable as the original domain covered by ν expressed in terms of τ.

The various functions that enter the expression for the amplitude have simple transformations under this change of variables. The transformation of $\psi(\overline{\nu}, \overline{\tau})$, given in appendix 8.A, can be written as

$$\psi(\overline{\nu}, \overline{\tau}) = \frac{1}{\overline{\tau}'} \exp\left(\frac{i\pi\overline{\nu}'^2}{\overline{\tau}'}\right) \psi(\overline{\nu}', \overline{\tau}'). \tag{9.3.25}$$

Similarly, from equations of that appendix

$$\eta(\overline{\nu}, \overline{\tau}) = \overline{\tau}' \left(\eta(\overline{\nu}', \overline{\tau}') - \frac{\overline{\nu}'}{\overline{\tau}'}\right) \tag{9.3.26}$$

and

$$\Omega(\overline{\nu}, \overline{\tau}) = \overline{\tau}'^2 \left(\Omega(\overline{\nu}', \overline{\tau}') + \frac{i}{2\pi\overline{\tau}'}\right). \tag{9.3.27}$$

Using (9.3.14) and (9.3.15) it is easy to check that the hatted functions have the simple transformation properties

$$\hat{\eta}(\overline{\nu}, \overline{\tau}) = \overline{\tau}' \hat{\eta}(\overline{\nu}', \overline{\tau}') \tag{9.3.28}$$

and

$$\hat{\Omega}(\overline{\nu}, \overline{\tau}) = \overline{\tau}'^2 \hat{\Omega}(\overline{\nu}', \overline{\tau}'). \tag{9.3.29}$$

All the factors considered so far are analogous to those that appeared in the type II theories. We now turn to the factor $\mathcal{L}$, defined in (9.3.9), which has no analog in the earlier models. In order for the amplitude to be modular invariant it is important that $\mathcal{L}$ transform simply. In appendix 9.B we show that this only happens if the lattice Λ is self-dual. In that case $\mathcal{L}$ transforms as

$$\mathcal{L} \equiv \mathcal{L}(\overline{\nu}_r, \overline{\tau}, \overline{\xi}_r) = (\overline{\tau}')^8 \mathcal{L}(\overline{\nu}'_r, \overline{\tau}', \overline{\tau}'\overline{\xi}_r) e^{-i\pi\overline{\tau}S^2}, \tag{9.3.30}$$

where we have used the fact that $\overline{\tau}S^I(\overline{\nu}_r, \overline{\tau}, \overline{\xi}_r) = S^I(\overline{\nu}'_r, \overline{\tau}', \overline{\tau}'\overline{\xi}_r)$, which follows from the definition of S^I in (9.3.10). Writing out $i\pi\overline{\tau}S^2$ explicitly gives

$$
-i\pi\bar{\tau}S^2 = \sum_{1\leq r<s\leq4} \left(iK_r \cdot K_s \frac{\pi\bar{\nu}_{sr}^2}{\bar{\tau}} + i\bar{\xi}_r \cdot \bar{\xi}_s \frac{1}{2\pi\bar{\tau}} \right.
$$
$$
\left. - (K_r \cdot \bar{\xi}_s - K_s \cdot \bar{\xi}_r)\frac{\bar{\nu}_{sr}}{\bar{\tau}} \right)
$$
$$
\tag{9.3.31}
$$
$$
= -\sum_{1\leq r<s\leq4} \left(iK_r \cdot K_s \frac{\pi\bar{\nu}'^2_{sr}}{\bar{\tau}'} + i\bar{\tau}'^2\bar{\xi}_r \cdot \bar{\xi}_s \frac{1}{2\pi\bar{\tau}'} \right.
$$
$$
\left. - \bar{\tau}'(K_r \cdot \bar{\xi}_s - K_s \cdot \bar{\xi}_r)\frac{\bar{\nu}'_{sr}}{\bar{\tau}'} \right).
$$

This expression contains just the pieces of the zero modes of the transformed functions $\ln\psi(\bar{\nu}'_{sr},\bar{\tau}')$, $\bar{\tau}'\eta(\bar{\nu}'_{sr},\bar{\tau}')$ and $\bar{\tau}'^2\Omega(\bar{\nu}'_{sr},\bar{\tau}')$ that are removed by the transformations in (9.3.25)–(9.3.27). So $\mathcal{L}$ combines with the other factors to give a combination that transforms simply. The fact that the transformation of $\mathcal{L}$ is only simple if the lattice is self-dual is of central importance in the theory. We have already seen that the only allowed lattices are those corresponding to groups with 'even' roots (*i.e.*, roots of length squared 2). We now find that consistency of the one-loop amplitudes imposes the further restriction of self-duality. As was discussed in §6.4.7, the only self-dual lattices of rank 16 are the lattices Γ_{16} and $\Gamma_8 + \Gamma_8$ associated with the groups spin(32) and $E_8 \times E_8$. The requirement of modular invariance is the first truly persuasive argument we have given for imposing the self-duality constraint. Considerations based on anomaly cancellations in §13.5.3 will select the same gauge groups for $D = 10$ theories with $N = 1$ supersymmetry.

The transformation properties of the factors in the integrand can now be summarized. First, there are factors involving ψ_{rs} associated with the M_C external massless states that have nonzero K_r's

$$
\prod_{1\leq r<s\leq4} (\psi(\bar{\nu}_{sr},\bar{\tau}))^{K_r\cdot K_s} = (\bar{\tau}')^{M_C} \prod_{1\leq r<s\leq4} (\psi(\bar{\nu}_{sr},\bar{\tau}))^{K_r\cdot K_s}, \tag{9.3.32}
$$

where the fact that $\sum_{1\leq r<s\leq4} K_r \cdot K_s = -\frac{1}{2}\sum_{r=1}^4 (K_r)^2 = -M_C$ has been used. There is also a factor of $\bar{\tau}'$ from each factor of η_{rs} and $(\bar{\tau}')^2$ from each Ω_{rs}, so that M_U external uncharged Yang–Mills states (*i.e.*, those that have $K_r = 0$ and nonzero $\bar{\xi}_r$) give an overall factor of $(\bar{\tau}')^{M_U}$. Similarly, M_G external gravitational particles (with $K_r = 0$ and $\zeta_r \neq 0$) contribute a factor $(\bar{\tau}')^{M_G}$, as follows from the transformations of $\hat{\eta}_{rs}$ and $\hat{\Omega}_{rs}$. Thus, altogether, a factor of $\bar{\tau}'$ arises for each kind of particle giving $(\bar{\tau}')^4$ in the four-particle case.

The transformation of the function $f(\overline{w})$ can be expressed in the form

$$\frac{1}{\overline{w}}[f(\overline{w})]^{-24} = (\overline{\tau}')^{-12}\frac{1}{\overline{w}'}[f(\overline{w}')]^{-24}, \qquad (9.3.33)$$

(where $w' = \exp\{4\pi^2/\ln w\}$). The remaining terms in the integrand of (9.3.18) are

$$\frac{d^2\tau}{(\mathrm{Im}\tau)^2}\prod_{r=1}^{3}\frac{d^2\nu_r}{(\mathrm{Im}\tau)^3}\prod_{r<s}(\chi_{rs})^{k_r \cdot k_s/2}. \qquad (9.3.34)$$

Precisely this combination was shown to be invariant under modular transformations by the arguments following (9.2.5).

The factor of $(\overline{\tau}')^{-12}$ in (9.3.33) is exactly cancelled by the 8 powers of τ' from the transformation of $\mathcal{L}$ (9.3.30) and the $M_C + M_U + M_G = 4$ powers of τ' coming from the other functions, so that the integrand $F_H(\tau)$ is invariant under the transformation $\tau \to -1/\tau$.

The modular invariance of the integrand again allows the τ integral to be restricted to a fundamental region in the τ plane, which is conveniently chosen to be the region F in fig. 8.21. This region avoids potential singularities at $\tau = 0$ leading to a consistent expression for the loop amplitude. We should also check for divergences from the endpoints that gave divergences in the bosonic theory.

The obviously dangerous endpoint is the one in which the particles come together on the surface of the torus, the limit in which $\nu_r \sim \nu_s$. The analysis of this limit follows closely that of type II theories in §9.2.1 and involves substituting the change of variables in (9.2.11) into (9.3.18). For simplicity suppose that the external states have nonzero K_r so that there are no terms with $\overline{\xi}_r$ or $\overline{\zeta}_r$. The expression for S^I in (9.3.10) vanishes when all the ρ_r are equal so that $\mathcal{L}$ has no ν_r dependence. The only terms that change the analysis from that of the type II theories are the factors of $(\psi_{rs})^{K_r \cdot K_s}$ in (9.3.18). Since $\psi_{rs} \sim \nu_{rs}$ these contribute $(\epsilon)^{\sum_{r<s} K_r \cdot K_s} = \epsilon^{-4}$ (where we have used the fact that $K_r^2 = 2$). The net effect is that near the endpoint $\epsilon = 0$ the loop amplitude behaves as

$$\int_0 \epsilon d\epsilon \int_F \frac{d^2\tau}{(\mathrm{Im}\tau)^5}\frac{1}{\overline{w}}[f(\overline{w})]^{-24}\,\mathcal{L}, \qquad (9.3.35)$$

where $\mathcal{L}$ is given by (9.3.9) with $S^I = 0$. The τ integral apparently diverges in the limit $\mathrm{Im}\tau \to \infty$ ($\overline{w} \to 0$), where $\mathcal{L} \sim 1$ (the contribution of the term with $P = 0$ in (9.3.9)). However, the angular integral cancels the apparent singular behavior of $1/\overline{w}$ and the result is that the expression for the loop is finite. There is a straightforward generalization to the cases with external particles described by ξ_r or ζ_r.

The fact that the loop amplitudes for the heterotic string theory have modular-invariant integrands only for self-dual lattices is of great significance, since this restricts the possible groups to just $\text{spin}(32)/Z_2$ and $E_8 \times E_8$, as described in chapter 6. In §9.5.3 and §9.5.4 we will show that by giving up the assumption of space-time supersymmetry it is possible to formulate a modular-invariant and tachyon free heterotic string theory based on the gauge group $SO(16) \times SO(16)$ as well.

9.4   Calculations in the RNS Formalism

Up to this point, the calculations of this chapter have been based entirely on the light-cone gauge formalism of chapter 5. There are a number of reasons why it is useful to also consider calculations based on the RNS formalism of chapter 4 that utilizes fermionic coordinates that are space-time vectors rather than space-time spinors. For one thing, this is the only approach in which we know how to do manifestly covariant calculations (although the light-cone gauge can also be used in the RNS formalism). Also, not being restricted to special momenta, generalizations to diagrams with numerous emitted states or multiple loops seem to be more straightforward, though they are not explored in this book. Finally, studying the scattering amplitudes in the RNS formalism enables us to take a critical look at the extent to which the GSO projection is necessary. This is indeed the issue with which we begin the next subsection, before turning in the following subsection to the actual loop calculations.

9.4.1 Modular Invariance and the GSO Projection

The RNS model of closed strings has a massless spin 3/2 particle. For consistency, a massless spin 3/2 particle must be coupled to a conserved supersymmetry current (or decouple at low energies). Yet without the GSO projection, the RNS model lacks space-time supersymmetry (and the spin 3/2 particle does not decouple at low energies). It is therefore natural to suspect that the RNS model is inconsistent at the interacting level unless one imposes either the GSO projection to achieve space-time supersymmetry or some other modification to eliminate the massless spin 3/2 particle. We can now see how this comes about: the RNS model, without the GSO projection or a related projection that eliminates the massless spin 3/2 particle, lacks modular invariance at the one-loop level.

We first discuss the right-moving fermions of the RNS model. Recall

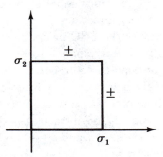

Figure 9.1. A path integral on a torus; the fermions may obey + or − boundary conditions in the σ_1 and σ_2 directions, respectively.

that on a closed string, $0 \le \sigma \le \pi$, the fermions may obey either periodic or antiperiodic boundary conditions. We refer to these as + and − or even and odd boundary conditions, respectively. The even and odd boundary conditions give rise, respectively, to fermions and bosons. If we carry out a path integral on a torus, then the boundary conditions must be specified in more detail, as in fig. 9.1. As in §8.2.2, we suppose that that the torus is parametrized by $\nu = \sigma_1 + \sigma_2 \tau$, so that the fundamental periods are $\sigma_1 \to \sigma_1 + 1$ and $\sigma_2 \to \sigma_2 + 1$. The boundary conditions may be separately periodic or antiperiodic in the σ_1 and σ_2 directions. There are thus four possibilities, which we denote as $(++), (+-), (-+)$, and $(--)$, with the first $\pm$ sign referring to the behavior in the σ_1 direction and the second referring to the behavior in the σ_2 direction. The four possible boundary conditions are often referred to as 'spin structures', this being a special case of a more general concept that will be investigated further in chapter 14.

In the absence of any projection, loop amplitudes contain a factor of

$$\mathrm{Tr}\, e^{-yH} \tag{9.4.1}$$

(where y is given by $\tau = (x+iy)/2\pi$ as in §8.2.1) for propagation through imaginary time y. It is essential now to remember that in the path integral formulation of quantum statistical mechanics, the partition function of fermions is computed using *antiperiodic* boundary conditions in the σ_2 direction. The trace in (9.4.1) is thus naturally represented by a path integral with antiperiodic $(-)$ boundary conditions in the σ_2 direction. If on the other hand we wish to calculate the quantity

$$\mathrm{Tr}\,(-1)^F\, e^{-yH}, \tag{9.4.2}$$

with $(-1)^F$ being the operator used in the GSO projection that counts the number of world-sheet fermions modulo two, then we must use $+$ boundary conditions in the σ_2 direction.

Therefore, in the absence of the GSO projection, the contribution of the NS sector to a loop amplitude corresponds to $(--)$ boundary conditions, while the contribution of the R sector corresponds to $(+-)$ boundary conditions. The combination of partition functions $(--)$ and $(+-)$ is not modular invariant. The modular transformation $(\sigma_1, \sigma_2) \rightarrow (\sigma_2, -\sigma_1)$ exchanges the role of σ_1 and σ_2, so it turns $(+-)$ into $(-+)$. To get a modular-invariant theory, the $(--)$ and $(+-)$ contributions of the RNS model must be supplemented by $(-+)$. But $(-+)$ is a partition function for NS states ($-$ boundary conditions in the σ_1 direction) with an insertion of $(-1)^F$ ($+$ boundary conditions in the σ_2 direction). Thus, in the NS sector the inclusion of both $(--)$ and $(-+)$ boundary conditions on the toroidal world sheet amounts to replacing (9.4.2) with

$$\text{Tr}\,(1 + (-1)^F)e^{-yH}. \tag{9.4.3}$$

This is the GSO projection – the projection on states even under $(-1)^F$.[*] Thus we have learned at the one-loop level that for the sake of unitarity – or at least modular invariance – it is necessary to make the GSO projection in the NS sector. At the one-loop level these considerations do not force us to make the GSO projection in the R sector, since the combination $(+-), (-+)$ and $(--)$ is modular invariant.

The reason that that combination is modular invariant is as follows. Modular transformations $(\sigma_1, \sigma_2) \rightarrow (a\sigma_1 + b\sigma_2, c\sigma_1 + d\sigma_2)$ change the boundary conditions and therefore permute $(++), (+-), (-+)$ and $(--)$. However, $(++)$ is modular invariant by itself, since boundary conditions that are completely periodic in both σ_1 and σ_2 remain so after taking linear combinations of σ_1 and σ_2. Modular transformations cannot mix $(+-), (-+)$ and $(--)$ with $(++)$ (the latter being modular invariant), so they must permute $(+-), (-+)$, and $(--)$ among themselves. Hence, it is possible to try to make a modular invariant theory with the inclusion only of $(+-), (-+)$ and $(--)$. This is a theory in which the GSO projection is made for bosons only. Including $(++)$ would mean making the GSO projection for fermions too, but since $(++)$ is modular invariant by itself,

---

[*] We noted in §4.3.3, where the GSO projection was introduced, that *a priori* the overall sign of the operator $(-1)^F$ is ill-defined. We now have a precise prescription for defining it, namely modular invariance. Given a choice for the normalization of the $(--)$ contribution to the one-loop diagram, a modular transformation gives a definite answer for $(-+)$ with a determined sign, which indeed corresponds to the sign that gives a supersymmetric theory with no tachyon.

modular invariance at one-loop level does not force us to make the GSO projection for fermions. It can actually be shown by a more intricate argument that the analog of modular invariance at the two-loop level requires the GSO projection for the R as well as NS sectors. We will not enter into this and content ourselves with having explained why the GSO projection is needed in the NS sector.[†]

If one does make the GSO projection in the R sector, that sector is described by the sum of $(+-)$ and $(++)$ path integrals, by analogy with our comments on the NS sector. Adding these all up, we have $(--), (-+), (+-)$ and $(++)$ contributions, which must all be included in the form of the theory with space-time supersymmetry. The supersymmetric form of the theory with both bosons and fermions and with the GSO projection can thus be described very succinctly: one sums over all possible boundary conditions. The multiloop generalization of this will be explored in chapter 14.

There is a subtlety that should be brought out here. It is rather clear that to achieve modular invariance, once we include the $(--)$ sector it is *necessary* to also include the $(-+)$ and $(+-)$ sectors (into which $(--)$ transforms under modular transformations) in order to achieve modular invariance. It is not clear that including $(-+)$ and $(+-)$ along with $(--)$ is *sufficient* for modular invariance. Indeed, consider a modular transformation that maps $(--)$ into itself. The general form of such a transformation is $(\sigma_1, \sigma_2) \rightarrow (a\sigma_1 + b\sigma_2, c\sigma_1 + d\sigma_2)$, where a and d are odd and b and c are even. Such a transformation leaves the $(--)$ *boundary conditions* invariant, but it is not obvious that it leaves the $(--)$ *path integral* invariant. There might be an anomaly. One of our main conclusions in the next subsection will be that in ten dimensions there is no such anomaly, at the one-loop level. This result has also been proved in higher orders, but the requisite analysis is rather complicated and is not presented in this book.

These considerations have several interesting generalizations. In stating the argument, we assumed that the theory is to contain both bosons and fermions (both NS and R states), and we used the action of the modular group on the R contribution to argue for the necessity of the GSO projection in the NS sector. It is possible to avoid this assumption. Modular transformations can be used to prove that the theory must contain both R and NS sectors.

[†] It is conceivable that in some more subtle way unitarity requires the GSO projection of the R sector even at one-loop level.

It is clearly impossible to have an interacting theory with R states only (states from the NS sector appear as poles in the scattering of two states from the R sector), so the interesting question to consider is whether one can have a consistent (modular-invariant) theory based on the NS sector alone. In such a theory, the boundary conditions on the torus would be $(--)$ only (or $(--)$ and $(-+)$ if we decide to make a GSO projection). While $(--)$ is invariant under the $\sigma_1 - \sigma_2$ interchange considered earlier, it is not invariant under other modular transformations. Indeed, the modular transformation $(\sigma_1, \sigma_2) \rightarrow (\sigma_1+\sigma_2, \sigma_2)$ converts $(--)$ into $(+-)$, proving that R states must be present if NS states are. We can then, of course, use the previous argument to prove that the GSO projection must be made at least in the NS sector.

Another interesting variant of this argument arises in the fermionic formulation of the $SO(32)$ and $E_8 \times E_8$ heterotic theories. In this case, one must decide what boundary conditions to use for internal fermions that carry group quantum numbers. Considerations analogous to the above (including the two-loop argument in the R sector for which the reader is referred to the references) show that to achieve modular invariance, it is necessary to include the analog of both R and NS sectors and to make the analog of the GSO projection. These steps were taken in §6.3 with only partial motivation.

In one respect the above discussion is quite satisfactory. It is puzzling that there should exist in the RNS model (without the GSO projection) a massless spin $3/2$ particle without supersymmetry. We have at least partly identified what is wrong with the theory in this form. Yet there is a logical gap in the discussion. Instead of modifying the theory with the GSO projection to achieve space-time supersymmetry, might it not be possible to modify the theory to eliminate the massless spin $3/2$ particle?

This is indeed possible, if one uses an option that was neglected in the above. In the foregoing we considered right-moving fermions in isolation. The RNS model has left-moving fermions as well, and the same is true for the heterotic theories in their fermionic formulation. For definiteness let us consider the type II theory. There are sixteen possible boundary conditions, which we might denote as $(\pm\pm; \pm\pm)$, with the first pair of signs referring to left movers and the second pair of signs to right movers. The supersymmetric form of the theory corresponds to including separate R and NS sectors and making separate GSO projections for both left and right movers. This is the only way to achieve modular invariance if the allowed choices of boundary conditions for left movers are independent of the allowed choices of boundary conditions for right movers. There is no necessity to make the latter assumption, however, and if we relax it then it is possible to find other ways to achieve modular invariance. The most obvious modular-invariant way to correlate the boundary conditions of left movers with the boundary conditions of right movers is to say that left movers and right movers should have the same boundary conditions. Thus, the collection of four sets of boundary conditions $(++;++), (+-;+-), (-+;-+)$, and $(--;--)$ is modular invariant. Whatever the action of a modular transformation on left movers, it does the same to right movers, and so permutes these four sets.

The procedure just described gives a theory that is not supersymmetric. In fact, the theory has no fermions at all, since left-moving R (or NS) states are coupled with right-moving R (or NS) states, making bosons in each case. Because of the absence of fermions (and the presence, as well, of a tachyon – we have not removed it from the spectrum, since we did not make a GSO projection!), the model just described probably has little practical application. It shares the faults of the Veneziano model, but probably also enjoys the same degree of consistency. Its significance is that it illustrates that there are in fact two ways to avoid the potential inconsistencies associated with the massless spin 3/2 particle of the RNS model: one can modify the theory to achieve space-time supersymmetry or eliminate the massless spin 3/2 particle.

There are a number of other unitary theories that can be constructed roughly along these lines. In each case, the GSO projection is replaced by some other projection that eliminates the massless spin 3/2 particle rather than achieving space-time supersymmetry. Most of these theories have tachyons, but one, with gauge group $SO(16) \times SO(16)$, does not. Instead of formulating this theory in the manner of the present discussion, we will formulate it in a different way in the following section.

9.4.2 The Loop Calculations

The calculation of superstring loop diagrams in the RNS formulation requires the addition of contributions from all the possible fermionic boundary conditions or 'spin structures'. In the operator approach this is achieved by adding fermionic loop diagrams built out of integrally-moded fermionic operators (R sector) and bosonic loop diagrams built out of half-integrally moded fermionic operators (NS sector). In addition, one must include in each case the supplementary $(-1)^F$ projections (GSO conditions) that ensure space-time supersymmetry. It is, of course, important to include the ghosts and antighosts associated both with the Fermi modes and the Bose modes in the covariant version of this approach.

Covariant boson emission vertices do not involve the ghost modes, which only enter via the propagators. Just as in the bosonic string theory the net effect of the ghost coordinates is to cancel two powers of the partition functions that arise from the bosonic and fermionic modes. With external fermions the situation is much more complicated, since the fermion-emission vertex involves the ghost modes in an important manner (see §7.3.5). Many of the basic calculations with external fermions are much simpler in the light-cone gauge.

The various kinds of loop diagrams with external massless vector particles are built from the vertices

$$V_R(k, \zeta) = \zeta \cdot \psi e^{ik \cdot X} \tag{9.4.4}$$

for emission from a fermionic string and

$$V_{NS}(k, \zeta) = \left(\zeta \cdot \dot{X} + k \cdot \psi \zeta \cdot \psi \right) e^{ik \cdot X} \tag{9.4.5}$$

for emission from a bosonic string. Including the GSO projection operators $\frac{1}{2}(1 + G)$ in open-string loops with circulating bosons or $\frac{1}{2}(1 + \overline{\Gamma})$ with circulating fermions leads to four contributions to each of the topologically distinct open-string amplitudes. (G and $\overline{\Gamma}$ were defined in §4.3.3.) Since the heterotic string is supersymmetric only in its right-moving coordinates the counting is the same in that case. (If the internal symmetry is included by means of left-moving fermion modes then the analog of the GSO projection operator combines different boundary conditions in that sector also.) The type II theory involves similar counting for the right movers and the left movers separately, leading to 16 terms for the general one-loop diagram.

Amplitudes consist of parity-conserving and parity-violating pieces. The parity-violating pieces arise from fermion loops containing an odd number of $\overline{\Gamma}$'s. These parity-violating fermionic loops must contain at least ten γ^μ matrices to give a nonzero result for the trace over the Dirac zero modes, since the presence of a γ_{11} kills it otherwise. This gives rise to a ten-dimensional ϵ tensor whose indices are contracted with linearly independent momenta and polarization vectors associated with the emitted particles. This means that at least six external vector particles are needed (since with five of them there would only be four independent momenta and five polarization vectors). It is precisely this parity-violating part of the hexagon diagram that can give rise to anomalies. That is the subject of the next chapter.

The vanishing of the cosmological constant for open superstrings (the $M = 0$ loop) was derived earlier from a supersymmetry trace identity (the S_0 trace). In the RNS formalism it arises as a cancellation between three nonzero contributions. Similar remarks apply to the vanishing of the cosmological constant in the type II and heterotic theories. This cancellation can now be seen as a consequence of Jacobi's relation (§4.3.3), which was earlier used to verify that there are an equal number of bosonic and fermionic states at each level of the spectrum of the theory. The vanishing of the two-particle and three-particle scattering amplitudes can similarly be derived by piecing together the different contributions.

9.5 Orbifolds and Twisted Strings

In this section we describe a generalization of the GSO projection that has many interesting applications. It can be used to give an exactly soluble model of string compactification, which is almost as simple as the compactification on a flat torus considered in §9.2.2 above. Also, it can be used to construct a very interesting new tachyon-free ten-dimensional string theory in which space-time supersymmetry is absent, or perhaps present at a deep level but spontaneously broken.

9.5.1 Generalization of the GSO Projection

Consider a closed-string theory with a discrete symmetry group F. For instance, if we compactify on a flat torus, as in §9.2.2, F may be simply a discrete symmetry group of the torus; this is the example that we will consider in the next subsection.

Propagating on the string world sheet are various fields, such as boson fields $X^\mu(\sigma_1, \sigma_2)$ and perhaps world-sheet Fermi fields. Let us call the generic field propagating on the world sheet $\phi(\sigma_1, \sigma_2)$. Consider a one-loop world-sheet path integral on a torus, which we can think of as defined by the $\sigma_1 - \sigma_2$ plane with the equivalence relations $\sigma_1 \approx \sigma_1 + 1$ and $\sigma_2 \approx \sigma_2 + 1$. What boundary conditions should we impose on $\phi(\sigma_1, \sigma_2)$? The obvious choice would be to require that

$$\phi(\sigma_1 + 1, \sigma_2) = \phi(\sigma_1, \sigma_2 + 1) = \phi(\sigma_1, \sigma_2). \qquad (9.5.1)$$

Because we assume the existence in the theory of a discrete symmetry group F, it is possible to contemplate a generalization of (9.5.1). We may take any two elements g and h of F and require that

$$\phi(\sigma_1 + 1, \sigma_2) = g\phi(\sigma_1, \sigma_2), \quad \phi(\sigma_1, \sigma_2 + 1) = h\phi(\sigma_1, \sigma_2). \qquad (9.5.2)$$

Actually, it follows from (9.5.2) that

$$gh\phi(\sigma_1, \sigma_2) = g\phi(\sigma_1, \sigma_2 + 1) = \phi(\sigma_1 + 1, \sigma_2 + 1)$$

$$= h\phi(\sigma_1 + 1, \sigma_2) = hg\phi(\sigma_1, \sigma_2).$$

(9.5.3)

The equality of the first and last expressions in (9.5.3) is an unreasonable requirement, which is likely to lead to inconsistencies, unless

$$gh = hg. \tag{9.5.4}$$

We assume that we should restrict ourselves to boundary conditions defined by pairs (g, h) of group elements that obey this condition.

If we choose to include a contribution with boundary conditions (g, h), what other contributions must be included to achieve modular invariance? Under a transformation $(\sigma_1, \sigma_2) \rightarrow (\sigma_1', \sigma_2') = (a\sigma_1 + b\sigma_2, c\sigma_1 + d\sigma_2)$, how do g and h transform? The shift $(\sigma_1', \sigma_2') \rightarrow (\sigma_1' + 1, \sigma_2')$ is equivalent to $(\sigma_1, \sigma_2) \rightarrow (\sigma_1 + d, \sigma_2 - c)$, so inspection of (9.5.2) reveals that under this transformation ϕ transforms into $g^d h^{-c}\phi$. Likewise, under $(\sigma_1', \sigma_2') \rightarrow (\sigma_1', \sigma_2' + 1)$, ϕ is transformed to $g^{-b} h^a \phi$. In sum, the modular transformation in question transforms (g, h) to

$$(g', h') = (g^d h^{-c}, g^{-b} h^a). \tag{9.5.5}$$

In particular, g' and h' commute if g and h do, so it is consistent to restrict our attention to commuting pairs (g, h).

Of course, we are entitled to restrict ourselves to the case $g = h = 1$, this being the theory that we started with in the first place. If, however, we wish to generalize the original theory, we must include contributions with $g, h \neq 1$. We cannot simply add to the original theory an arbitrarily selected contribution with boundary conditions (g, h), since this would not be modular invariant. If we wish to add some new contributions to the original path integral $(g = h = 1)$ we must add a modular-invariant combination. The general theory of what modular-invariant combinations are allowed is rather complicated, especially if one incorporates constraints that arise at the two-loop level, but there is one rather simple choice that meets the eye. One can sum over all boundary conditions (g, h), subject to the one constraint that $gh = hg$. The prescription of summing over all commuting pairs (g, h) is clearly a modular-invariant prescription, since we have seen that modular transformations permute pairs of this type. We focus on this choice, which is a kind of generalization of the GSO projection. More elaborate choices exist if F is complicated enough, but in the particular cases we consider below, the summation over all commuting pairs (g, h) is the only possibility.

Just as in our discussion of the GSO projection, choosing boundary conditions in a modular-invariant way does not guarantee that the resulting amplitudes are modular invariant. There may be quantum anomalies such that the quantum-mechanical path integrals on the torus do not possess the full modular symmetry that formally would appear to follow from the choices of boundary conditions. Such anomalies do not arise (in ten dimensions) in the case of the GSO projection, but in the more general situations that we will discuss, modular anomalies can arise, and avoiding them is an important restriction.

9.5.2 Strings on Orbifolds

In §9.2.2 above, we discussed the propagation of strings on $M^4 \times T^6$, with M^4 being four-dimensional Minkowski space and T^6 being a six-dimensional torus. This compactification is relatively simple to analyze mathematically, but far from realistic. It leaves unbroken $N = 4$ or $N = 8$ supersymmetry in four dimensions (depending on which superstring theory we start with); this is not a good starting point for a realistic four-dimensional theory. Simple toroidal compactification also gives rise to unrealistic gauge groups and matter representations in four dimensions. We now describe a generalization of this discussion that is almost as tractable mathematically and far more realistic.

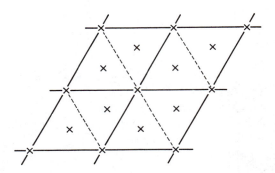

Figure 9.2. A triangular lattice in the complex z plane; it is used in constructing a torus with Z_3 symmetry. The 'fundamental region' in the construction of this torus consists of two triangles, and contains three fixed points of the Z_3 symmetry.

Rather than being abstract, let us consider a concrete example. To begin with, we consider a special torus T_0 made (fig. 9.2) by the following identifications of points in the complex z plane:

$$z \approx z + 1 \approx z + e^{2i\pi/3}.$$ (9.5.6)

The 'fundamental region' in the complex z plane consists of two triangles in the figure. What is special about this particular torus is that it admits the Z_3 symmetry generated by

$$\alpha_0 : z \to e^{2i\pi/3} z.$$ (9.5.7)

On the torus there are three points left invariant by this transformation. They are the points

$$z = \frac{k}{\sqrt{3}} e^{i\pi/6}$$ (9.5.8)

for $k = 0, 1, 2$. (In the complex plane, only $z = 0$ is invariant under (9.5.7), but the other points in (9.5.8) are shifted by lattice vectors, i.e., by integer multiples of 1 and $\exp\{2i\pi/3\}$. We only consider $k = 0, 1, 2$ since other values are related to these by (9.5.6).)

Now consider three complex variables z_i, $i = 1, 2, 3$. With the identifications $z_i \approx z_i + 1 \approx z_i + \exp\{2i\pi/3\}$, we obtain three tori T_i, $i = 1, 2, 3$ isomorphic to the above. The product $T = T_1 \times T_2 \times T_3$ is a torus of real dimension six (complex dimension three), and we wish to consider string propagation on $M^4 \times T$, or more precisely on a certain modification of this.

T admits the Z_3 symmetry

$$\alpha : z_i \to e^{2i\pi/3} z_i, \quad i = 1, 2, 3.$$ (9.5.9)

This symmetry has $3^3 = 27$ fixed points (points that are left invariant), these being points in which each z_i takes one of the values in (9.5.8). We denote the Z_3 group generated by α as F.

Now, we would like to supplement (9.5.6) with an additional equivalence relation, this being the statement that two points on T are considered equivalent if they are related by the F action. Thus, denoting the coordinates of a point on T as $\{z_i\}$, we impose the condition

$$\{z_i\} \approx \{e^{2i\pi/3} z_i\}.$$ (9.5.10)

We denote as Z the space T/F of equivalence classes of points on T subject to this equivalence relation. Z is not a manifold, because the equivalence relation (9.5.10) introduces conical singularities at the 27 fixed points of

α; this is a matter that will be further discussed in §16.3. Because of these conical singularities, Z is called an 'orbifold' rather than a manifold. Even though Z is not a manifold, string propagation on $M^4 \times Z$, which is what we now consider, seems to make perfectly good sense.

Consider first the propagation of a point particle on Z (or on $M^4 \times Z$, but it is Z that we really wish to understand at the moment). On T the point particle would be described by a wave function $\psi(z_i)$ belonging to a Hilbert space K_0. Working on $Z = T/F$ means that we restrict the wave function $\psi(z_i)$ to obey

$$\psi(e^{2i\pi/3} z_i) = \psi(z_i). \tag{9.5.11}$$

The Hilbert space K of states of a point particle propagating on Z is the subspace of K_0 consisting of wave functions that obey (9.5.11) or, in other words, the space of states that are invariant under α. The operator that projects onto states that are invariant under α is

$$P = (1 + \alpha + \alpha^2)/3. \tag{9.5.12}$$

P is the operator that projects the big Hilbert space K_0 onto the subspace K of states of a particle propagating on $Z = T/F$. If H is the Hamiltonian of the point particle, the partition function on T is Tr $\exp\{-\beta H\}$, but on $Z = T/F$ it is Tr $\exp\{-\beta H\}P$.

Following this logic for strings, the first step in discussing string propagation on Z (or actually $M^4 \times Z$) is to consider string propagation on T (the relevant Hilbert space was discussed in §9.2.2 above) and then project this onto the F-invariant subspace. Consider a one-loop diagram in which the string world sheet is as in the preceding subsection a parallelogram in the $\sigma - \tau$ plane with opposite sides identified. If the strings are propagating on T, we perform a path integral on this parallelogram with periodic boundary conditions. If, instead, the strings are propagating on $Z = T/F$, we must as in the point-particle case project the string wave functions onto F-invariant states. As in the GSO case, the projection onto states invariant under F can be made by summing over boundary conditions in the τ direction or in other words by including the boundary conditions $(1,1), (1,\alpha)$, and $(1,\alpha^2)$ in the language of the previous subsection.

In the point-particle case, this projection is the end of the story, but in the string case we must go on, because the sum over boundary conditions $(1,1), (1,\alpha)$, and $(1,\alpha^2)$ is not modular invariant. In fact it is clear from (9.5.5) that once we include the three combinations $(1,1), (1,\alpha)$, and $(1,\alpha^2)$ with equal weight, we are forced to include all nine combinations

(α^k, α^m), $k, m = 1, 2, 3$, with equal weight. The need to include these new contributions is somewhat analogous to the fact that modular invariance forces one to supplement the NS sector with the R sector, as we saw in §9.4.1 above.

There is an alternative language for explaining the need to include the new sectors. Consider a closed string on $M^4 \times Z$; we describe M^4 with standard Minkowski coordinates x^μ, $\mu = 0, \ldots, 3$, and we describe Z with the complex coordinates z_i, $i = 1, 2, 3$ introduced earlier. Ordinary closed-string boundary conditions would be

$$(x^\mu(\sigma + \pi), z_i(\sigma + \pi)) = (x^\mu(\sigma), z_i(\sigma)). \qquad (9.5.13)$$

On Z, however, the points $\{z_i\}$ and $\{\exp(2i\pi/3)z_i\}$ are equivalent, so instead of (9.5.13) we are entitled to consider closed strings that obey

$$(x^\mu(\sigma + \pi), z_i(\sigma + \pi)) = (x^\mu(\sigma), e^{2i\pi/3}z_i(\sigma)). \qquad (9.5.14)$$

This corresponds to boundary conditions twisted by α in the σ direction. This 'twisted sector' must still be projected onto states invariant under $F = Z_3$, which is again done by summing over boundary conditions $1, \alpha$, or α^2 in the τ direction. On the $\sigma - \tau$ torus the contribution of the twisted sector consists of terms with boundary conditions $(\alpha, 1), (\alpha, \alpha)$, and (α, α^2). In addition, we can replace $\exp\{2i\pi/3\}$ in (9.5.14) with $\exp\{4i\pi/3\}$, corresponding to strings that are twisted in the σ direction by α^2. Projecting this sector onto F-invariant states, the requisite contributions to world-sheet path integrals correspond to the inclusion of terms with boundary conditions $(\alpha^2, 1), (\alpha^2, \alpha)$, and (α^2, α^2). All in all, the untwisted states and the two types of twisted sectors, all projected onto F-invariant subspaces, are represented by summing over all nine possible boundary conditions on the torus.

There are a variety of reasons that string propagation on $M^4 \times Z$ is a far more realistic model of compactification than string propagation on $M^4 \times T$. First of all, compactification on $M^4 \times Z$ spontaneously breaks some of the supersymmetries. Of the underlying ten-dimensional super-symmetries, only those that are invariant under F remain as unbroken four-dimensional supersymmetries after compactifying on $M^4 \times Z$. If we start with $N = 1$ supersymmetry in ten dimensions, then compactification on $M^4 \times T$ leaves $N = 4$ supersymmetry in four dimensions. Of these four supersymmetries, only one is invariant under F, so compactifying on $M^4 \times Z$ breaks the potential $N = 4$ supersymmetry down to $N = 1$, which is far more reasonable as a starting point for phenomenology in four dimensions.[*]

Another desirable feature of compactification on an orbifold instead of an ordinary torus is that there are interesting possibilities for gauge symmetry breaking. Let G be the ten-dimensional gauge group. Let u be an element of G that obeys $u^3 = 1$, so that it generates a Z_3 subgroup of G. In the above orbifold construction, instead of requiring that states be invariant under the rotation α, we could require that they be invariant under the combined rotation and gauge transformation

$$\overline{\alpha} = \alpha \cdot u. \qquad (9.5.15)$$

The whole construction that we have carried out above can be repeated using $\overline{\alpha}$ and the group $\overline{F}$ that it generates instead of α and F. Formally the logic is the same, but the physics is somewhat different and even more interesting. One obtains a mechanism for spontaneous breaking of both supersymmetry and gauge symmetries. Supersymmetries that do not commute with α and gauge symmetries that do not commute with u are all spontaneously broken.

Consider, for example, the heterotic superstring theory with $E_8 \times E_8$ gauge group. We choose u to belong to the first E_8. As we know from appendix 6.A, E_8 contains a maximal subgroup $E_6 \times SU(3)$. The center of $SU(3)$ is Z_3, with a generator that we will call u. The subgroup of E_8 that commutes with u is $E_6 \times SU(3)$. The orbifold compactification with $\overline{\alpha} = \alpha \cdot u$ thus breaks E_8 to $E_6 \times SU(3)$ (while leaving unbroken the second E_8). This is a considerable improvement over the ordinary toroidal compactification, which leaves $E_8 \times E_8$ unbroken, because while E_8 is not viable as a four-dimensional grand-unified group, E_6 is. (E_8 has no complex representations, but the **27** of E_6 accommodates a single generation of quarks and leptons in a relatively economical way.)

---

[*] To see that it is $N = 1$ supersymmetry in four dimensions that survives, one reasons as follows. The ten-dimensional Lorentz group $SO(1,9)$ has a subgroup $SO(1,3) \times SO(6)$, where $SO(1,3)$ is the four dimensional Lorentz group and $SO(6) \approx SU(4)$ is the subgroup of $SO(1,9)$ that commutes with $SO(1,3)$. The four supersymmetries of $N = 4$ supersymmetry in four dimensions transform as the **4** of $SU(4)$. In the construction of the Z orbifold, the particular symmetry operator α was chosen to be one that belongs to an $SU(3)$ subgroup of $SO(6)$ or $SU(4)$. Precisely one element of the **4** of $SU(4)$ is invariant under $SU(3)$, and this is the one unbroken supersymmetry. If α were chosen instead to belong to $U(3)$ but not $SU(3)$, supersymmetry would be broken entirely in the reduction to four dimensions. Essentially the same considerations will play a major role in chapters 15 and 16.

The construction of the spectrum of the string propagating on $M^4 \times Z$ is not difficult, but will not be presented here. With the above choice of u, one finds 36 (!) chiral generations of quarks and leptons, transforming as the **27** of E_6. Nine of these come from the untwisted sector and twenty-seven from the twisted sector. It is quite attractive that the representation of E_6 that is desirable for phenomenology appears automatically, and that without putting them in by hand in any obvious way one gets out a multiplicity of fermion generations (in fact, too many).

If one restricts oneself to compactification on orbifolds that leave unbroken $N = 1$ supersymmetry in four dimensions, then the physics of orbifolds is rather similar to that of compactification on manifolds of $SU(3)$ holonomy, which we will discuss in detail in chapters 15 and 16. In fact, as we will see in §16.10, such orbifolds can be viewed as limiting cases of certain manifolds of $SU(3)$ holonomy. Whether orbifolds are 'just' limiting cases or are the most interesting case remains to be seen.

9.5.3 Twisted Strings in Ten Dimensions

In the above, we have considered orbifolds as an approach to compactification from ten dimensions to four dimensions. An orbifold-like construction can also be used to construct interesting ten-dimensional string theories, somewhat similar to the models considered in §9.4.1 above, but more interesting because one of them is tachyon free. The tachyon-free model arises as a twisted variant of the $E_8 \times E_8$ or $SO(32)$ heterotic string. It can be constructed with either of those two theories as the starting point; we consider the $E_8 \times E_8$ case.

In constructing an orbifold, one begins with a group F that acts as a symmetry group of an already known string theory. What are the possible symmetry groups of the ten-dimensional $E_8 \times E_8$ theory? There is the Lorentz symmetry group $SO(1,9)$ and the $E_8 \times E_8$ gauge group. If we wish to construct a Poincaré-invariant string theory in ten dimensions, then F must commute with the Poincaré group. The only element of $SO(1,9)$ with that property is the 2π rotation

$$x = \exp\left(2\pi J_{12}\right). \tag{9.5.16}$$

We could try to make an orbifold-like construction using the group $F_0 \approx Z_2$ generated by x. It is possible to generalize this, however. Let y be an element of $E_8 \times E_8$ with $y^2 = 1$, and let $z = xy$. Then $z^2 = 1$, so z generates a group isomorphic to Z_2, which we call F. We discuss the theory obtained by twisting the $E_8 \times E_8$ string theory by F – projecting it onto states invariant under F and then, as in the previous section, adding the 'twisted' sector to achieve modular invariance. The resulting theory lacks supersymmetry (or perhaps should be interpreted as a theory in which supersymmetry is present at an underlying level but is spontaneously broken), since the ten-dimensional supersymmetries are odd under z.

It turns out that there is precisely one choice of y for which the theory that we construct is modular invariant and tachyon-free. As we know from appendix 6.A, the E_8 Lie algebra contains $SO(16)$ as a maximal subalgebra. Under $SO(16)$ the adjoint representation of E_8 decomposes as $\mathbf{120} \oplus \mathbf{128}$. In the center of $SO(16)$ there is an element that takes the form

$$y_0 = -1. \tag{9.5.17}$$

in the fundamental representation. Under y_0, the $\mathbf{120}$ of $SO(16)$ is even, while y_0 can be chosen so that the $\mathbf{128}$ of $SO(16)$ is odd. In $E_8 \times E_8$, let $y_0^{(1)}$ and $y_0^{(2)}$ be, respectively, the elements of the first and second E_8's with the above-stated properties. In the construction we are considering, only one choice of y gives a tachyon-free theory; it is

$$y = y_0^{(1)} \times y_0^{(2)}. \tag{9.5.18}$$

Of course, $y = 1$ for any $E_8 \times E_8$ singlet.

To construct the twisted string theory, the first step is to project the $E_8 \times E_8$ heterotic superstring on the subspace $z = 1$; then to achieve modular invariance we must add a 'twisted sector' in which the boundary conditions are twisted by z in the σ direction, just as in the construction of orbifolds.

The projection on $z = 1$ in the ordinary or untwisted sector is easy to describe. With $z = xy$ and $x^2 = y^2 = 1$, $z = 1$ is equivalent to $x = y$. States of $x = 1$ or $x = -1$ are bosons and fermions, respectively, and the projection means that from the spectrum of the $E_8 \times E_8$ theory we must keep the bosons of $y = 1$ and the fermions of $y = -1$. The gravitino, for instance, has $x = -1, y = +1$, so it is removed by the projection, in agreement with our expectation that there cannot be a massless spin 3/2 particle in a theory that is not supersymmetric. The graviton has $x = y = 1$ so it is retained; this theory, like other closed-string theories, includes gravity. What is the gauge group of the twisted theory? Among the gauge bosons of the $E_8 \times E_8$ theory, the ones that have $y = 1$ are those that transform as

$$(\mathbf{120}, \mathbf{1}) \oplus (\mathbf{1}, \mathbf{120}) \tag{9.5.19}$$

under $SO(16) \times SO(16)$. This is the adjoint representation of $SO(16) \times SO(16)$, so (as we do not get any additional massless bosons from the twisted sector) the unbroken gauge group of the twisted theory is $SO(16) \times SO(16)$. As for massless fermions, the states in the adjoint representation of $E_8 \times E_8$ that have $y = -1$ are precisely those that transform as

$$(\mathbf{128}, \mathbf{1}) \oplus (\mathbf{1}, \mathbf{128}), \qquad (9.5.20)$$

so these are the massless fermions that we obtain from the untwisted sector.

Analysis of the twisted sector requires somewhat more care. At this point it is worthwhile to make a definite choice of which construction of the theory we wish to consider. For left-moving modes, we use the fermionic description of $E_8 \times E_8$ in terms of two groups of left-moving fermions λ^A and $\tilde{\lambda}^B$, with $A, B = 1, \ldots, 16$. For right-moving modes, we use the manifestly supersymmetric light-cone gauge formalism described in chapter 5. Thus, there are eight right-moving fermions S^a transforming in one of the two spinor representations of $SO(8)$, the $\mathbf{8_s}$, which we describe as positive chirality. The S^a obey periodic boundary conditions

$$S^a(\sigma + \pi) = S^a(\sigma) \qquad (9.5.21)$$

in the supersymmetric form of the theory developed in chapter 5. The ground-state energy of the right-moving modes (including X^μ as well as S^a) is zero. The ground state has a degeneracy that arises by quantizing the zero modes of S^a and consists of sixteen states transforming as $\mathbf{8_v} \oplus \mathbf{8_c}$ of $SO(8)$. In particular, the massless fermions have negative chirality, opposite to the S^a. Of course, when we project onto states of $z = +1$, the surviving $SO(16) \times SO(16)$ fermions are those that transform as (9.5.20).

Now we wish to construct the twisted sector. In this sector right-moving modes have boundary conditions in the σ direction twisted by x, while left-moving modes have boundary conditions in the σ direction twisted by y. For the right-moving modes, it is relatively easy to figure out what this means. The modes that are odd under the 2π rotation x are the S^a, so (9.5.21) is replaced by

$$S^a(\sigma + \pi) = -S^a(\sigma). \qquad (9.5.22)$$

Thus, in the twisted sector the 'supersymmetric' fermions S^a are subject to boundary conditions analogous to those in the NS sector of the RNS formulation. Therefore, we can borrow the analysis from chapter 4. The ground state $|\Omega\rangle$ of the right movers is nondegenerate, and has $L_0 = -1/2$. Since the ground state of the left movers will turn out to be at $\tilde{L}_0 = 0$, the right-moving ground state is eliminated when we impose the condition

$\tilde{L}_0 = L_0$. As for massless states of the right movers, these are built just as in the analogous discussion in chapter 4 by acting on $|\Omega\rangle$ with an S^a mode of energy $+1/2$:

$$|\Omega^a\rangle = S^a_{-1/2}|\Omega\rangle. \tag{9.5.23}$$

Notice that $|\Omega^a\rangle$ has the same chirality as the S^a and thus opposite chirality from the massless fermion states found in the untwisted sector. There are no massless bosons in the twisted sector.

Now we must discuss the left-moving modes in the twisted sector. The degrees of freedom that are odd under y are the $SO(16) \times SO(16)$ fermions λ^A and $\tilde{\lambda}^B$, so the 'twisting' means that we must reverse the boundary conditions obeyed by the λ^A and $\tilde{\lambda}^B$. Here, however, we find ourselves in a quandary. The fermionic construction of the $E_8 \times E_8$ model described in §6.3.2 already required us to sum over all possible boundary conditions in the σ direction of the modes λ^A and $\tilde{\lambda}^A$, the possibilities being

$$(PP), (PA), (AP) \text{ and } (AA). \tag{9.5.24}$$

Here, for instance, (PP) is a sector in which the λ^A and $\tilde{\lambda}^A$ are all periodic in σ. Since we are already summing over all possible boundary conditions in σ before 'twisting' by y, it appears at first sight that the twisting operation can merely permute the terms in (9.5.24) and give us nothing new. This would be disastrous, since the left-moving sector without twisting contains a massless vector $(\tilde{\alpha}_{-1}|\Omega\rangle)$; if this is combined with the right-moving states in (9.5.23), the theory possesses a massless spin 3/2 particle despite the absence of space-time supersymmetry.

Before leaping to this conclusion, let us think more carefully about the GSO-like projection that is used in obtaining $E_8 \times E_8$ symmetry in the fermionic description. We defined in §6.3.2 operators $(-1)^{F_1}$ and $(-1)^{F_2}$ that obeyed

$$(-1)^{F_1}\lambda^A = -\lambda^A(-1)^{F_1}, \quad (-1)^{F_1}\tilde{\lambda}^B = +\tilde{\lambda}^B(-1)^{F_1}$$

$$(-1)^{F_2}\lambda^A = +\lambda^A(-1)^{F_2}, \quad (-1)^{F_2}\tilde{\lambda}^B = -\tilde{\lambda}^B(-1)^{F_2}. \tag{9.5.25}$$

We noted in §6.3.2 that these properties defined the operators $(-1)^{F_1}$ and $(-1)^{F_2}$ only up to sign. The GSO-like projection on states of $(-1)^{F_1} = (-1)^{F_2} = 1$ gave $E_8 \times E_8$ symmetry in the supersymmetric theory investigated in §6.3.2 if the signs are right. For instance, the left-moving ground state $|\Omega\rangle$ in the PP sector must have $(-1)^{F_1} = (-1)^{F_2} = 1$. We now know from our investigations of modular invariance of the $E_8 \times E_8$

theory in §9.3.2 above that in the supersymmetric form of the theory the sign choices that lead to $E_8 \times E_8$ symmetry are also the ones that give modular invariance.

Now we can state what it means to 'twist' by a transformation y that at first sight appears to merely permute the four sectors in (9.5.24). The answer turns out to be that the twisted sector is one in which one must make the opposite projection; thus, in the twisted PP sector, for example, the left-moving ground state $|\Omega\rangle$ has $(-1)^{F_1} = (-1)^{F_2} = -1$. Under this assumption, the lowest-lying states of $(-1)^{F_1} = (-1)^{F_2} = +1$ are

$$|\Omega_{ab}\rangle = \lambda^A_{-1/2} \tilde{\lambda}^B_{-1/2} |\Omega\rangle. \tag{9.5.26}$$

We know from chapter 6 that the left-moving ground state has $\tilde{L}_0 = -1$, so the states $|\Omega_{ab}\rangle$ have $\tilde{L}_0 = 0$. Coupled with the right-moving states in (9.5.23), this gives massless fermions of positive chirality that transform as $(\mathbf{16}, \mathbf{16})$ under $SO(16) \times SO(16)$. With the opposite GSO-like projection from chapter 6, the left-moving twisted $(AP), (PA)$ and (AA) sectors are easily seen not to contribute any massless particles. So combining our results from the twisted and untwisted sectors, the massless fermions (apart from some $SO(16) \times SO(16)$ singlets) transform as

$$(\mathbf{128}, \mathbf{1})_- \oplus (\mathbf{1}, \mathbf{128})_- \oplus (\mathbf{16}, \mathbf{16})_+ , \tag{9.5.27}$$

where the subscript denotes the ten-dimensional chirality.

To justify the assertion that the left-moving twisted sector is one in which we are to make the opposite GSO projection from usual, we write down the modular-invariant partition function of this theory. It is proportional to

$$\text{tr} \left(\exp(i\tau L_0 + i\bar{\tau}\tilde{L}_0) \right) \sim \frac{1}{\theta_1'^4(\tau)} \left\{ \frac{\theta_2^4(\tau)}{\theta_2^8(\bar{\tau})} + \frac{\theta_4^4(\tau)}{\theta_4^8(\bar{\tau})} - \frac{\theta_3^4(\tau)}{\theta_3^8(\bar{\tau})} \right\}. \tag{9.5.28}$$

The energetic reader should be able to derive this formula and verify modular invariance. The first term on the right-hand side of (9.5.28) represents the contribution of the untwisted sector projected on $z = +1$. The last two terms correspond to the twisted sector. It is really the analysis of the modular-invariant formula (9.5.28), that justifies our previous interpretation of the left-moving twisted sector.

9.5.4 Alternative View Of The $SO(16) \times SO(16)$ Theory

We will now briefly describe how the theory just constructed could be obtained if we describe the right-moving modes in the RNS formalism, while still using (as above) the fermionic description of $E_8 \times E_8$.

The construction in §6.3 of the supersymmetric $E_8 \times E_8$ theory involved projection operators $(-1)^{F_1}$ and $(-1)^{F_2}$ for each of the four left-moving sectors AA, AP, PA and PP. In the RNS framework, one also has a GSO projection operator $(-1)^{F_R}$ for right-movers. Defining the overall signs of these operators is somewhat delicate; the supersymmetric and $E_8 \times E_8$ invariant form of the theory involves choosing the signs so that the massless gauge and gravitational supermultiplets are all even.

Describing right-movers in the RNS formalism, how can we recover the $SO(16) \times SO(16)$ theory which was just constructed? It all boils down to suitable changes in the signs of the various GSO-like operators. The requisite changes in sign vary from sector to sector and are

$$
\begin{array}{cc}
\underline{\text{in sector}} & \underline{\text{change}} \\
(\text{R}; A, A) \text{ and } (\text{NS}; P, P) & (-1)^{F_1} \text{ and } (-1)^{F_2} \\
(\text{R}; P, A) \text{ and } (\text{NS}; A, P) & (-1)^{F_R} \text{ and } (-1)^{F_1} \\
(\text{R}; A, P) \text{ and } (\text{NS}; P, A) & (-1)^{F_R} \text{ and } (-1)^{F_2}.
\end{array}
\tag{9.5.29}
$$

This can be shown to be the only set of sign changes that gives a modular invariant and tachyon-free theory. Let us examine the spectrum of state that results. In the R sector (space-time fermions) we have

$$
\tfrac{1}{8}(\text{mass})^2 = N_{\text{R}} = \tilde{N} - a,
\tag{9.5.30}
$$

whereas in the NS sector (space-time bosons) we have

$$
\tfrac{1}{8}(\text{mass})^2 = N_{\text{NS}} - \tfrac{1}{2} = \tilde{N} - a.
\tag{9.5.31}
$$

Here, as in the $E_8 \times E_8$ case, $a = 1$ for the AA sector, $a = 0$ for the AP or PA sector, and $a = -1$ for the PP sector. Let us first look for tachyons. Since the N's are never negative, a tachyon would require $a = 1$ and $\tilde{N} = \tfrac{1}{2}$. However this implies that $N_{\text{NS}} = 0$ in the AA sector. This is not possible because $(\text{NS}; A, A)$ is one of the sectors we have not changed. Just as in the $E_8 \times E_8$ theory, N_{NS} has eigenvalues $1/2, 3/2, \ldots$ in this sector. Thus there are no tachyons.

Next let us investigate the massless states. Using the rules given above the only sectors that can give rise to them are $(\text{R}; A, A)$ with $N_{\text{R}} = 0$ and $\tilde{N} = 1$, $(\text{R}; P, A)$ and $(\text{R}; A, P)$ with $N_{\text{R}} = \tilde{N} = 0$, $(\text{NS}; A, A)$ with $N_{\text{NS}} = 1/2$ and $\tilde{N} = 1$. The only massless bosons come from the last sector and have quantum numbers

$$
(\mathbf{8_v}; \mathbf{1}, \mathbf{1})_R \times [(\mathbf{8_v}; \mathbf{1}, \mathbf{1}) \oplus (\mathbf{1}; \mathbf{120}, \mathbf{1}) \oplus (\mathbf{1}; \mathbf{1}, \mathbf{120})]_L.
\tag{9.5.32}
$$

The notation here is that the first label refers to the transverse $SO(8)$ group, while the next two refer to the two $SO(16)$ factors. The subscripts R and L refer to right-moving and left-moving modes, respectively. The product $\mathbf{8_v} \times \mathbf{8_v} = \mathbf{35_v} \oplus \mathbf{28} \oplus \mathbf{1}$ gives the graviton, antisymmetric tensor, and dilaton, all singlets of the gauge group. The other terms give $(\mathbf{8_v}; \mathbf{120}, \mathbf{1}) \oplus (\mathbf{8_v}; \mathbf{1}, \mathbf{120})$, which are the Yang–Mills gauge fields for $SO(16) \times SO(16)$. Note that in the $E_8 \times E_8$ theory there were additional massless vectors arising from the (NS, P, A) and (NS, A, P) sectors that gave the $(\mathbf{8_v}; \mathbf{128}, \mathbf{1}) \oplus (\mathbf{8_v}; \mathbf{1}, \mathbf{128})$ required to complete the adjoint representation of $E_8 \times E_8$. The change in assignment of $(-1)^F$ for the NS and A pieces of these sectors now prevents them from contributing massless modes. This sector now starts with the states satisfying $N_{\text{NS}} = 1$ and $\tilde{N} = \frac{1}{2}$, corresponding to $(\text{mass})^2 = 4$

The massless fermions of the $SO(16) \times SO(16)$ theory are of two types. First of all consider $(\text{R}; A, A)$ states with $N_{\text{R}} = 0$ and $\tilde{N} = 1$. The Fock space description of the left-moving modes must take the form $\psi^A_{-1/2} \tilde{\psi}^B_{-1/2} |0\rangle$ to satisfy the rules $(-1)^{F_1} = (-1)^{F_2} = -1$ and $\tilde{N} = 1$. Thus tensoring with an $\mathbf{8_s}$ spinor for the right-movers, the quantum numbers are altogether $(\mathbf{8_s}; \mathbf{16}, \mathbf{16})$. Additional spinors are obtained from the (R, P, A) and (R, A, P) sectors with $N = \tilde{N} = 0$. In this case the reversal of $(-1)^F$ for the R states implies that they must be spinors, $\mathbf{8_c}$, of the opposite chirality. The P ground states are 128-component spinors and the A ground states with $(-1)^F = 1$ are singlets. Thus altogether we obtain massless spinors $(\mathbf{8_c}; \mathbf{128}, \mathbf{1}) \oplus (\mathbf{8_c}; \mathbf{1}, \mathbf{128})$. This completes the description of the massless spectrum.

The first excited states are only a 'half step' up from the ground state – in other words they have $(\text{mass})^2 = 4$, half the value of the first excited levels of the other two heterotic theories. As we have already mentioned, these states are given by the (NS, A, P) and (NS, P, A) sectors with $N_{\text{NS}} = 1$ and $\tilde{N} = \frac{1}{2}$. The $N_{\text{NS}} = 1$ states are given by $\alpha^i_{-1} |0\rangle$ and $b^i_{-1/2} b^j_{-1/2} |0\rangle$ corresponding to $\mathbf{8_v}$ and $\mathbf{28}$ representations of $SO(8)$. These combine to give the 36-dimensional antisymmetric tensor of the massive little group $SO(9)$. Thus the quantum numbers in this case are $(\mathbf{36}; \mathbf{128}, \mathbf{16}) + (\mathbf{36}; \mathbf{16}, \mathbf{128})$.

The $SO(32)$, $E_8 \times E_8$ and $SO(16) \times SO(16)$ theories are certainly intimately related; they can be described in terms of the same degrees of freedom, with different boundary conditions. Many physicists have suggested that these three theories may actually all be different vacuum states in one underlying theory. It is very possible that the same may be true for the type IIA and IIB theories.

We have described just a few of the modular invariant ten dimensional string theories in which supersymmetry is broken by modifications of boundary conditions. (Apart from the tachyon-free $SO(16) \times SO(16)$

theory, a simple construction of a tachyonic but modular invariant theory was given at the end of §9.4.1.) There are many variants of these constructions which lead to modular-invariant ten dimensional string theories with tachyons. In a number of cases, the tachyons are charged under the gauge group.

In a theory with tachyons, one may be tempted to interpret the tachyon as an instability which may lead to compactification down to a stable and tachyon-free state. In the Veneziano model, this seems unlikely for the following reason. Consider the Veneziano model formulated on $M^n \times K$, where K is a compact manifold of dimension $26 - n$. Consider a tachyon whose momentum p^μ only has components tangent to M^n. Its vertex operator $\exp\{ip \cdot X\}$ is decoupled from the compact dimensions and will have the standard form, leading to the standard value of the tachyon mass, regardless of the choice of K. On the other hand, in the ten dimensional modular invariant theories with charged tachyons, the tachyon vertex operator is quite different and is in no way decoupled from the process of compactification. In those models, the tachyon mass can definitely be affected by compactification.

9.6 Summary

We have described the one-loop quantum corrections to superstring theories using the light-cone formalism with manifest space-time supersymmetry. As in the bosonic theory, one-loop open-string diagrams are divergent due to the emission of a zero-momentum closed-string state into the vacuum. Since there are no closed-string tachyons in superstring theories this divergence can be entirely associated with the massless dilaton. Possible cancellation of the divergences in the planar diagram and the nonorientable diagram will be explored in the next chapter.

The one-loop amplitudes of type II theories are modular invariant and finite. The condition that the loop integrand should be modular invariant restricts the possible heterotic string theories. The only supersymmetric possibilities are those associated with the self-dual lattices $\Gamma_8 + \Gamma_8$ or Γ_{16}. We have explored the role of the GSO projection in achieving modular invariance, as well as some alternatives to it that eliminate the massless spin 3/2 particle rather than giving space-time supersymmetry. The resulting insights have also led to an interesting approach to compactification of higher dimensions, namely 'orbifolds', as well as the construction of a heterotic string theory whose gauge group is $SO(16) \times SO(16)$. The latter is a tachyon-free theory without space-time supersymmetry.

Appendix 9.A　　Traces of Fermionic Zero Modes

In this appendix we assemble a number of properties needed for calculations involving traces over fermionic zero modes.

The zero-mode operators S_0^a span the 16-dimensional space $|i\rangle$ and $|\dot{a}\rangle$, where the states are normalized so that $\langle i|j\rangle = \delta_{ij}$ and $\langle \dot{a}|\dot{b}\rangle = \delta_{\dot{a}\dot{b}}$. The identity operator in the S_0 space is therefore given by

$$I = |i\rangle\langle i| + |\dot{a}\rangle\langle\dot{a}|, \tag{9.A.1}$$

which leads to the expression for the S_0 trace of any operator A

$$\mathrm{tr}_{S_0}(A) = \langle i|A|i\rangle - \langle\dot{a}|A|\dot{a}\rangle, \tag{9.A.2}$$

where the $-$ sign arises because the fermionic states $|\dot{a}\rangle$ and $\langle\dot{a}|$ anticommute.

The action of the S_0's on these states was given in §7.4.1 in terms of matrices $\gamma_{a\dot{a}}^i$ introduced in appendix 5.A as

$$S_0^a|\dot{a}\rangle = \frac{1}{\sqrt{2}}\gamma_{a\dot{a}}^i|i\rangle \tag{9.A.3}$$

and

$$S_0^a|i\rangle = \frac{1}{\sqrt{2}}\gamma_{a\dot{a}}^i|\dot{a}\rangle. \tag{9.A.4}$$

We also saw in chapter 7 that there is only one independent tensor that can be made out of a pair of S_0's, namely

$$R_0^{ij} = \frac{1}{4}\gamma_{ab}^{ij}S_0^a S_0^b, \tag{9.A.5}$$

which is the operator that changes the helicity of a state without changing its occupation number. The action of R_0^{ij} on the states is easily deduced by iterating (9.A.3) and (9.A.4), giving

$$R_0^{ij}|k\rangle = \delta^{jk}|i\rangle - \delta^{ik}|j\rangle, \tag{9.A.6}$$

and

$$R_0^{ij}|\dot{a}\rangle = -\frac{1}{2}\gamma_{\dot{a}\dot{b}}^{ij}|\dot{b}\rangle. \tag{9.A.7}$$

The R_0^{ij} satisfy the angular-momentum algebra

$$[R_0^{ij}, R_0^{kl}] = \delta^{jk}R_0^{il} + \delta^{il}R_0^{jk} - \delta^{ik}R_0^{jl} - \delta^{jl}R_0^{ik}. \tag{9.A.8}$$

The formulas for products of R_0's and S_0's acting on the basis states can be obtained by repeated application of these formulas. For example, the product of two R_0's gives

$$R_0^{ij} R_0^{kl} |m\rangle = \delta^{mk} \delta^{il} |j\rangle - \delta^{mk} \delta^{jl} |i\rangle - \delta^{ml} \delta^{ik} |j\rangle + \delta^{ml} \delta^{jk} |i\rangle, \qquad (9.A.9)$$

and

$$R_0^{ij} R_0^{kl} |\dot{a}\rangle = \frac{1}{4} (\gamma^{kl} \gamma^{ij})^{\dot{a}\dot{b}} |\dot{b}\rangle. \qquad (9.A.10)$$

In the loop calculations we need to evaluate the trace of products of S_0's. Evidently there has to be an even number of S_0's in the trace to get a nonzero result. Since any bilinear made out of S_0's can be written in terms of products of R_0's it is sufficient to consider traces of powers of such products.

We begin with the trace of the identity operator, which is easily seen to vanish by substituting $A = I$ in (9.A.2),

$$\mathrm{tr}_{S_0}(I) = \delta^{ii} - \delta^{\dot{a}\dot{a}} = 8 - 8 = 0. \qquad (9.A.11)$$

The trace of R_0^{ij} vanishes in a rather trivial way, since

$$\langle k | R_0^{ij} | k \rangle = 0 = \langle \dot{a} | R_0^{ij} | \dot{a} \rangle. \qquad (9.A.12)$$

The boson contribution to $\mathrm{tr}_{S_0}(R_0^{ij} R_0^{kl})$ is calculated from (9.A.9) to be $2 \left(\delta^{jk} \delta^{il} - \delta^{ik} \delta^{jl} \right)$. This cancels against the trace in the fermion sector obtained by multiplying (9.A.10) by $\langle \dot{a} |$, so that

$$\mathrm{tr}_{S_0}(R_0^{ij} R_0^{kl}) = 0. \qquad (9.A.13)$$

In fact, the trace vanishes unless all eight S_0^a's are present. In other words, of all 2^8 operators 1, S_0^a, $S_0^{a_1} S_0^{a_2}$, ..., $S_0^{a_1} S_0^{a_2} \cdots S_0^{a_8}$, only the last one has a nonvanishing trace. To prove this, form 'raising' and 'lowering' operators

$$B_1 = (S_0^1 + i S_0^2)/\sqrt{2}$$
$$B_1^\dagger = (S_0^1 - i S_0^2)/\sqrt{2} \qquad (9.A.14)$$

with $\{B_1^\dagger, B_1\} = 1$ and similarly for the three remaining pairs of S_0's. Then in terms of eigenstates $|0\rangle$ and $|1\rangle$ of the number operator $B_1^\dagger B_1 =$

$iS_0^1 S_0^2$, the traces in the S_0^1–S_0^2 sector are

$$\text{tr}(1) = \langle 0|0\rangle - \langle 1|1\rangle = 1 - 1 = 0$$

$$\text{tr}(B_1^\dagger B_1) = \langle 0|\, B_1^\dagger B_1 \,|0\rangle - \langle 1|\, B_1^\dagger B_1 \,|1\rangle = 0 - 1 = -1. \tag{9.A.15}$$

In terms of S_0^1 and S_0^2 this reads $\text{tr}(1) = 0$ and $\text{tr}(S_0^1 S_0^2) = i$. Generalizing to the case of eight S_0's, we learn that

$$\text{tr}(S_0^{a_1} S_0^{a_2} \cdots S_0^{a_8}) = \epsilon^{a_1 a_2 \dots a_8}. \tag{9.A.16}$$

The first nonvanishing trace occurs when there are four R_0's (*i.e.*, eight powers of S_0),

$$t^{ijklmnpq} = \text{tr}_{S_0}\left(R_0^{ij} R_0^{kl} R_0^{mn} R_0^{pq} \right). \tag{9.A.17}$$

The general structure of this tensor can be deduced from (9.A.5) and (9.A.16). Evidently it must have antisymmetry in each pair of indices ij, kl, mn and pq, and symmetry under the interchange of any pair of these index pairs. The latter symmetry is a consequence of the algebra, (9.A.8), and the fact that the trace of three R_0's vanishes. The full tensor is given by

$$
\begin{aligned}
t^{ijklmnpq} = -\tfrac{1}{2}\epsilon^{ijklmnpq} \\
-\tfrac{1}{2}\Big\{ (\delta^{ik}\delta^{jl} - \delta^{il}\delta^{jk})(\delta^{mp}\delta^{nq} - \delta^{mq}\delta^{np}) \\
+ (\delta^{km}\delta^{ln} - \delta^{kn}\delta^{lm})(\delta^{pi}\delta^{qj} - \delta^{pj}\delta^{qi}) \\
+ (\delta^{im}\delta^{jn} - \delta^{in}\delta^{jm})(\delta^{kp}\delta^{lq} - \delta^{kq}\delta^{lp}) \Big\} \\
+ \tfrac{1}{2}\Big\{ \delta^{jk}\delta^{lm}\delta^{np}\delta^{qi} + \delta^{jm}\delta^{nk}\delta^{lp}\delta^{qi} + \delta^{jm}\delta^{np}\delta^{qk}\delta^{li} \\
+ \text{45 terms obtained by antisymmetrizing} \\
\text{on each pair of indices}\Big\}.
\end{aligned}
\tag{9.A.18}
$$

The ϵ tensor and the two terms inside curly brackets are the three independent tensor structures allowed by the symmetries. It is therefore possible to determine their coefficients by considering three special cases.

The tensor t arises very naturally from the S_0 trace in four-particle one-loop calculations, but it is not nearly so obvious that it should also appear in the corresponding tree-diagram calculations of chapter 7. In fact, it is the same and the kinematic factor K that arose in the bosonic tree-diagram calculation in §7.4.2 can be written as

$$K = t^{ijklmnpq} k_1^i \zeta_1^j k_2^k \zeta_2^l k_3^m \zeta_3^n k_4^p \zeta_4^q. \tag{9.A.19}$$

The same correspondence is also valid for the kinematic factors that describe external fermions, as well.

Appendix 9.B   Modular Invariance of the Functions F_2 and $\mathcal{L}$

In this appendix we prove the modular invariance of the functions F_2 and $\mathcal{L}$. Since the modular group is generated by $\tau \to -1/\tau$ and $\tau \to \tau+1$, and the latter is manifest in each case, it suffices to study the transformations of the functions F_2 and $\mathcal{L}$ under the change of variables $\tau \to -1/\tau$.

We start by deducing a more general formula. Consider the quantity $\mathcal{F}(x)$, a function of a vector $x \equiv (x_1, x_2, \dots, x_p)$ defined by

$$\mathcal{F}(x) = \sum_{n \in Z^p} \exp[-\pi(n+x) \cdot A \cdot (n+x)], \tag{9.B.1}$$

where A is a positive definite $p \times p$ matrix. The components of the vector n are summed over all integer values. The function $\mathcal{F}(x)$ is periodic in x with integer period for each of its components, so that $\mathcal{F}(x) = \mathcal{F}(x + e_i)$, where e_i denotes a vector with the ith component equal to 1 and the rest zero. $\mathcal{F}(x)$ can therefore be expressed as a Fourier series

$$\mathcal{F}(x) = \sum_{m \in Z^p} e^{2i\pi m \cdot x} \mathcal{G}(m), \tag{9.B.2}$$

with the inverse transform given by

$$\mathcal{G}(m) = \int_0^1 d^p x \, e^{-2i\pi m \cdot x} \mathcal{F}(x). \tag{9.B.3}$$

Inserting the definition of $\mathcal{F}$ into the right-hand side of this equation gives

$$
\begin{aligned}
\mathcal{G}(m) &= \sum_{n \in Z^p} \int_0^1 d^p x \exp[-\pi(n+x) \cdot A \cdot (n+x) - 2\pi i m \cdot x] \\
&= \int_{-\infty}^{\infty} d^p x \exp[-\pi x \cdot A \cdot x - 2\pi i m \cdot x] \\
&= \int_{-\infty}^{\infty} d^p x \exp[-\pi(x + im \cdot A^{-1}) \cdot A \cdot (x + iA^{-1} \cdot m) \\
&\qquad\qquad - \pi m \cdot A^{-1} \cdot m] \\
&= (\det A)^{-1/2} \exp[-\pi m \cdot A^{-1} \cdot m],
\end{aligned}
$$

(9.B.4)

where the sum over the integer components of n has been evaluated by extending the range of the integrals over the components of x to the whole of R^p and the resulting gaussian integral has been evaluated. Substituting this formula back into (9.B.2) gives,

$$
\begin{aligned}
\mathcal{F}(x) &= \sum_{n \in Z^p} \exp[-\pi(n+x) \cdot A \cdot (n+x)] \\
&= (\det A)^{-1/2} \sum_{m \in Z^p} \exp[-\pi m \cdot A^{-1} \cdot m + 2\pi i m \cdot x],
\end{aligned}
$$

(9.B.5)

which is the basic identity that we are seeking.

As a first application of this transformation consider the function that arises in the toroidal compactification of type II theories,

$$
F_2(a, \tau) = a(\text{Im}\tau)^{1/2} \sum_{n_1 n_2} \exp[-2\pi i n_1 n_2 \text{Re}\tau - \pi(a^2 n_1^2 + n_2^2/a^2)\text{Im}\tau].
$$

(9.B.6)

This is a special case of (9.B.1) with $p = 2$ and $x = 0$. The matrix A, given by

$$
A = \begin{pmatrix} a^2 \text{Im}\tau & i\,\text{Re}\tau \\ i\,\text{Re}\tau & a^{-2}\text{Im}\tau \end{pmatrix},
$$

(9.B.7)

satisfies the identities

$$
\det A = |\tau|^2, \qquad A^{-1}(\tau) = A(-1/\tau).
$$

(9.B.8)

The latter follows from

$$\text{Im}(1/\tau) = -\text{Im}(\tau/|\tau|^2) \tag{9.B.9}$$

and

$$\text{Re}(1/\tau) = \text{Re}(\tau/|\tau|^2). \tag{9.B.10}$$

Substituting into (9.B.5) gives the relation

$$F_2(a,\tau) = a(\text{Im}\tau)^{1/2} \sum_n \exp(-\pi n \cdot A \cdot n)$$

$$= a \left(\frac{\text{Im}\tau}{|\tau|^2} \right)^{1/2} \sum_m \exp(-\pi m \cdot A^{-1} \cdot m). \tag{9.B.11}$$

It follows from this that $F_2(a,\tau) = F_2(a,-1/\tau)$, as required.

Let us now consider $\mathcal{L}(\bar{\nu},\bar{\tau},\zeta)$ defined by

$$\mathcal{L} = \sum_{P\in\Lambda} e^{-i\pi\bar{\tau}(P+S)^2}, \tag{9.B.12}$$

and

$$S^I = \sum_{r=1}^4 \left(\frac{\ln \bar{\rho}_r}{\ln \bar{w}} K_r^I + \frac{\bar{\xi}_r^I}{\ln \bar{w}} \right) = \sum_{r=1}^4 \left(\frac{\bar{\nu}_r}{\bar{\tau}} K_r^I + \frac{i\bar{\xi}_r^I}{2\pi\bar{\tau}} \right). \tag{9.B.13}$$

In this case the application of (9.B.5) requires the identifications $p = 16$, $A_{ij} = i\bar{\tau}g_{ij}$, where $g_{ij} = \sum_{I=1}^{16} e_i^I e_j^I$ is the metric on the lattice Λ and $x_i = \sum_{I=1}^{16} S^I e_i^I$. Substituting in the general formula gives

$$\mathcal{L} = \sum_{P\in\Lambda} e^{-i\pi\bar{\tau}(P+S)^2} = \frac{(\bar{\tau})^{-8}}{\sqrt{\det g}} \sum_{M\in\Lambda^*} e^{i\pi(M+\bar{\tau}S)^2/\bar{\tau} - i\pi\bar{\tau}S^2}. \tag{9.B.14}$$

This transformation is only simple when the lattice is self-dual, in which case $\sqrt{\det g} = 1$ and $\Lambda^* = \Lambda$.

Notice that in the special case that the external particles have $K_r = 0$ and $\zeta_r^I = 0$, $\mathcal{L}$ is just the Θ function for the lattice (described in §6.4.7). If $\det g = 1$, the transformation formula becomes

$$\sum_{P\in\Lambda} q^{P^2} = \left(-\frac{\pi}{\ln q} \right)^8 \sum_{P\in\Lambda} e^{\pi^2 P^2/\ln q}, \tag{9.B.15}$$

where $q = \exp\{-i\pi\bar{\tau}\}$.

10. The Gauge Anomaly in
Type I Superstring Theory

The supersymmetric Yang–Mills multiplet in ten dimensions is parity violating; the massless fermions are spinors of $SO(1,9)$ of one chirality or the other, but not both. In the context of string theory, this means that the massless fermion states of the open superstring have parity-violating gauge couplings. In fact, parity violation appears because there are two possible choices for the GSO projection in the Ramond sector, and one or the other must be chosen. Because of the role of the open superstring in constructing closed-string theories, most of the closed-string theories are likewise parity violating. For the type I and type IIB theories, the use of one GSO projection or the other introduces parity violation for closed strings just as it does for open strings. Parity violation is avoided in the type IIA theory, because one GSO projection is made for right-moving modes on the world sheet while the opposite GSO projection is made for left movers, and the overall system is invariant under simultaneous reflections or parity transformations of the world sheet and space-time. It is not invariant under separate world-sheet or space-time reflections. The heterotic theories are parity violating in the space-time sense (and on the world sheet), because a parity-violating right-moving multiplet is coupled to a parity-conserving left-moving multiplet.

Parity-violating gauge couplings of massless fermions are likewise conspicuous in four-dimensional physics; they are a basic part of the structure of the $SU(2) \times U(1)$ model of electroweak interactions. The neutrino has negative helicity and weak hypercharge $Y = -1$; its antiparticle, the antineutrino, has positive helicity and weak hypercharge $Y = +1$.

While the gauge couplings of light fermions in four dimensions are parity violating, their gravitational couplings are parity conserving. According to the four-dimensional CPT theorem, for every four-dimensional fermion state with one helicity, there is an antiparticle with the same mass and opposite helicity. The gravitational interactions of a fermion depend only on mass and helicity, so the overall gravitational couplings of four-dimensional fermions conserve parity. This is not so in ten dimensions; the ten-dimensional CPT operation leaves invariant the chirality of a massless fermion. Thus, the gravitational couplings of a supersymmetric Yang–Mills multiplet in ten dimensions, as well as the gauge couplings, violate parity. Compactification from ten to four dimensions automatically eliminates parity violation from low-energy gravitational couplings, but as we will see in detail in chapter 14, under suitable conditions we are left with parity-violating gauge couplings in four dimensions. On the other hand, it seems to be impossible to generate parity-violating gauge couplings in four dimensions starting with a multidimensional theory in which the gauge couplings conserve parity.[*] The parity violation that is almost forced on us in ten dimensions because the supersymmetric Yang–Mills multiplet violates parity is thus a crucial ingredient in trying to describe nature via string theory.

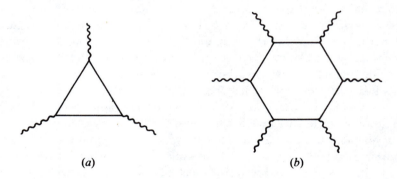

Figure 10.1. In four-dimensional gauge theories with parity-violating gauge couplings, an anomaly can arise in the triangle diagram with external gauge bosons, depicted in (a). The analogous issue in ten dimensions involves a hexagon diagram with external gauge bosons or gravitons, depicted in (b).

At the same time, parity-violating gauge couplings can lead to inconsistencies. Four-dimensional gauge theories with parity-violating gauge couplings, for example, are inconsistent unless the Adler–Bell–Jackiw triangle anomaly cancels. The most familiar potential anomaly involves the triangle anomaly with three external gauge bosons, depicted in fig. 10.1a; an anomaly can also arise in the triangle with an external gauge boson

---

[*] Parity can indeed be spontaneously broken in the process of compactification, but the resulting parity violation does not show up in four-dimensional *gauge couplings*. This will be clear from our analysis in chapter 14 of the fermion quantum numbers that arise in four dimensions after compactification.

and two external gravitons, although this case is less familiar. The cancellation of these anomalies is an important ingredient in the consistency of the standard electroweak theory. The analogous question in ten dimensions concerns the hexagon diagram shown in fig. 10.1*b* with external gauge bosons and gravitons.

If one investigates the hexagon diagram in ten-dimensional supersymmetric fields theories, one immediately observes that there is no possible anomaly in the type IIA supergravity theory (it conserves parity), while the anomalies in the type IIB supergravity theory can be seen to cancel in a striking fashion upon summing the contributions of particles of various spin. Study of ten-dimensional theories with $N = 1$ supergravity seems to reveal that they are all afflicted with hexagon anomalies; for instance, the hexagon diagram with only external gauge bosons receives a possibly anomalous contribution from gluinos, and this contribution is nonzero for every nonabelian gauge group. These points will be further elucidated in chapter 13. Here, after a brief discussion of anomalies in $D = 10$ super Yang–Mills theory, we study the analogous issues in string theory. In fact, we study the hexagon diagram with external gauge bosons in type I superstring theory. A surprising result will emerge: the anomaly cancels if and only if the gauge group is $SO(32)$. An analysis of the mechanism behind this result in chapter 13 will show that it can be interpreted in field theoretic language in terms of the existence of additional anomalous diagrams beyond that of fig. 10.1*b*. This interpretation suggests that the other hexagon anomalies also cancel for the type I theory with gauge group $SO(32)$, although the relevant calculations have not been performed. It also indicates the possible existence of an anomaly-free ten-dimensional superstring theory with gauge group $E_8 \times E_8$, and indeed this clue played a role in stimulating the discovery of the heterotic superstring theory, which was described in chapter 6.

10.1 Introduction to Anomalies

A general discussion of the meaning and structure of anomalies in ordinary point-particle field theories is deferred to §13.2. Here we settle for some brief introductory remarks to set the stage for the string calculation.

10.1.1 Anomalies in Point-Particle Field Theory

An anomaly is the breakdown of a classical conservation law due to quantum-mechanical loop corrections. Anomalies are important in four-dimensional quantum field theory for a variety of reasons. Their interpretation and significance is quite different depending on whether the anomalous symmetry is global or local. Certain global symmetries need to be broken for phenomenological reasons, a job that may in some cases be performed by anomalies. Local symmetries, on the other hand, must not be broken by anomalies, since this leads to fatal inconsistencies.

An example of a global symmetry whose breaking is desirable is the classical scale invariance of QCD with massless quarks. This needs to be broken (by an anomaly in the trace of the energy–momentum tensor) in order that hadrons can emerge as bound states with nonzero masses. Another example of a welcome breaking of a global symmetry is also provided by QCD with n massless quarks. Classically there is a global $U_L(n) \times U_R(n)$ symmetry corresponding to independent rotations of the left-handed and right-handed components of the quarks. This symmetry needs to be broken to a diagonal $SU(n)$ subgroup to account for the approximate vectorial $SU(n)$ symmetry of the observed hadrons.

When a gauge symmetry is broken by quantum-mechanical anomalies the situation is quite different. Gauge invariance is crucial for a consistent interpretation of a quantum theory with gauge fields. In particular, it is responsible for decoupling longitudinally polarized gauge fields. Such longitudinal states must decouple from physical processes if the theory is to be unitary.

Theories in which the gauge couplings violate parity are sometimes called chiral gauge theories. A chiral gauge theory is the same thing as a gauge theory in which the left handed fermions transform differently under the gauge group from the way the right handed fermions transform. Let R and $\tilde{R}$ be the representations of the gauge group furnished by left and right handed fermions, respectively. In four dimensions, the antiparticles of left handed fermions are right handed and vice-versa. CPT symmetry therefore requires in four dimensions that $\tilde{R}$ should be the complex conjugate representation of R. Consequently, in four dimensions, chiral gauge theories, which are gauge theories in which R and $\tilde{R}$ are distinct representations, are the same as theories in which the representation R is complex (not equivalent to its complex conjugate). Thus gauge groups without complex representations (such as $SO(32)$ and $E_8 \times E_8$) cannot give a chiral theory.

In higher dimensions things can be very different. In odd dimensions there are no Weyl spinors, no left–right asymmetry (parity violation) and no anomalies. Such dimensions are certainly not promising for phenomenology. In even dimensions, there are important differences between $4k$ and $4k + 2$ dimensions. In $4k$ dimensions for $k > 1$, the situation is for our present purposes similar to four dimensions.

The case in which something really new happens is that of $4k + 2$ dimensions, and this is the case of interest to superstring theory, since the world sheet has dimension two and space-time has dimension ten! In $4k + 2$ dimensions the antiparticle of a left-handed particle is again left-handed, and it is therefore possible to construct theories containing fermions of one chirality only. This is why the gravitational interactions can violate parity in $4k + 2$ dimensions. Furthermore, fermions of given chirality in $4k + 2$ dimensions must belong to a real representation of the gauge group. (If we start with a complex representation Q, then the antiparticles belong to $\overline{Q}$, so altogether we have the real representation $Q \oplus \overline{Q}$.) Gauge anomalies potentially arise whenever the representations R and $\tilde{R}$ of left- and right-handed fermions transform differently under the gauge group, so that gauge couplings violate parity. In four dimensions, as we have already noted, R is distinct from $\tilde{R}$ if and only if it is a complex representation. In $4k + 2$, R and $\tilde{R}$ are always real, as we have observed above, but need not be equal; when R and $\tilde{R}$ are distinct representations, the gauge couplings violate parity, and gauge anomalies may arise.

There are various ways of calculating anomalies (*i.e.*, the anomalous divergences of classically conserved currents) in ordinary field theory. The effective action $\Gamma(A)$ for a Yang–Mills field A_μ coupled to a chiral spinor ψ is defined by

$$\exp[i\Gamma(A)] \sim \int D\psi \, D\overline{\psi} \exp\left\{ - \int dx \overline{\psi} \left(\frac{1 + \Gamma_{11}}{2} \right) \Gamma \cdot D\psi \right\}, \quad (10.1.1)$$

where D_μ is the gauge-covariant derivative and $(1 + \Gamma_{11})/2$ is the chirality projection operator. The generator of gauge transformations is the operator

$$D_\mu \frac{\delta}{\delta A_\mu} \qquad (10.1.2)$$

and accordingly the quantity

$$G = D_\mu \frac{\delta}{\delta A_\mu} \Gamma(A). \qquad (10.1.3)$$

measures the failure of the effective action to be gauge invariant. Actually,

$$J_\mu = \frac{\delta}{\delta A_\mu} \Gamma \qquad (10.1.4)$$

is the gauge current induced by the background field, so

$$G = D_\mu J^\mu. \qquad (10.1.5)$$

Thus, a failure of gauge invariance ($G \neq 0$) is the same thing as a breakdown of current conservation in the background field. It is possible to calculate the anomaly directly from the functional integral formulation provided that divergences are carefully regulated. In this approach the anomaly is found to originate from a phase in the fermion measure. Alternatively, the anomaly can be calculated directly from the Feynman diagrams.

In the case of gauge currents, the physically meaningful question is whether the longitudinal components of the associated gauge fields decouple or not. The most direct approach to studying the question is to consider diagrams whose external particles consist of a set of on-mass-shell gauge fields. Then one can take the polarization vector of one of them to be longitudinal with all the others transverse and physical and check whether the resulting expression vanishes or not. If it does for all possible S matrix elements, then the theory is anomaly-free. Careless arguments can easily give a vanishing result as a consequence of the classical symmetry. It is necessary to carefully regulate divergent integrals before evaluating the anomaly in order not to make subtle errors associated with unjustified shifts of integration variables. (This is analogous to the calculation of the anomaly in the Virasoro algebra, where careless manipulations of divergent sums also would give wrong answers.)

There is a wide range of methods for regulating divergences in quantum field theory, all of which can be used for calculating anomalies. They include Pauli–Villars, Fujikawa's method of suppressing fermion modes of high momentum in the functional integral, dimensional regularization (supersymmetric or not as appropriate), point-splitting methods and so forth. Anomalies arise in parity-violating theories, because in the case of parity-violating gauge or gravitational interactions, all of these regularizations spoil gauge invariance or general covariance; the violation of these symmetries does not necessarily disappear when the regulator is removed, and is known as the anomaly. The result for the anomaly calculated by any such method is the same up to irrelevant terms that can be absorbed in a local counterterm of the effective action. A true anomaly is one that cannot be canceled by adding such a counterterm.

10.1.2 The Gauge Anomaly in $D = 10$ Super Yang–Mills Theory

Anomalies occur in parity-violating loop amplitudes only, since parity-conserving amplitudes can be regularized in a way that respects gauge invariance and general covariance. Parity-violating amplitudes contain an ϵ tensor arising from the couplings of chiral particles in the loop. Rather than discussing general concepts (to which we return in chapter 13), let us illustrate them with a specific example that is of considerable interest in its own right, namely the Yang–Mills gauge anomaly in the $D = 10$ super Yang–Mills theory. This theory, which was described in appendix 4.A, is particularly relevant, since it is part of a low-energy approximation to the superstring theories that we study in the next section. By showing how the anomaly analysis works for this theory we can develop many of the tools that are required for the string calculation. It also puts us in a better position to identify the features that distinguish the two problems.

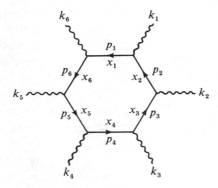

Figure 10.2. Illustration of the kinematics of a hexagon diagram.

The $D = 10$ super Yang–Mills theory contains vector gauge fields and Majorana–Weyl spinors, both in the adjoint representation of the gauge group. Consider a one-loop diagram with M external on-shell gauge particles and fermions for the internal lines, as shown in fig. 10.2. The formal expression for this diagram (up to a numerical coefficient that does not concern us) is

$$A_M \sim T \int d^{10}p \, \text{tr} \left[\frac{\Gamma \cdot p_1}{p_1^2} \Gamma \cdot \zeta_1 \frac{\Gamma \cdot p_2}{p_2^2} \cdots \Gamma \cdot \zeta_M \left(\frac{1 + \Gamma_{11}}{2} \right) \right]. \quad (10.1.6)$$

The group-theory factor is

$$T = \text{Tr}(t^{a_1} t^{a_2} \ldots t^{a_M}), \quad (10.1.7)$$

where a_i labels the charge of the ith emitted boson and the t's are in the adjoint representation of the gauge algebra, because that is the representation to which the fermions belong. The vector $\zeta_i^\mu(k_i)$ describes the polarization of the ith boson. Since it is required to be physical and on-shell,

$$k_i \cdot \zeta_i = k_i \cdot k_i = 0. \tag{10.1.8}$$

The momenta of internal fermion lines are given by

$$p_i = p - \sum_{j=1}^{i-1} k_j. \tag{10.1.9}$$

The chirality projection operator $(1 + \Gamma_{11})/2$ is introduced to describe the chirality of the circulating fermions. Such a projection operator accompanies each propagator, but they can be combined in a single term as shown. An ϵ term arises from the Dirac trace of the Γ_{11} term, so this is the only piece that is relevant to the anomaly analysis. The result of doing the trace with the Γ_{11} and carrying out the integral must be to give an ϵ tensor contracted into ten of the momenta and polarization vectors. However, only $M - 1$ of the momenta k_i are linearly independent (since $\sum k_i = 0$), and thus the first case in which there are enough linearly independent k_i and ζ_i to have a chance of giving a nonzero result is $M = 6.^*$ Therefore the hexagon is the simplest diagram that is potentially anomalous, and henceforth we restrict our attention to that case.

The Γ_{11} piece of integral A_6 is linearly divergent. (This is the characteristic behavior of loop amplitudes that give anomalies.) Therefore it needs to be regulated. In the Pauli–Villars method, for example, one subtracts an identical expression with the massless propagators $\Gamma \cdot p/p^2$ replaced by massive ones $(\Gamma \cdot p + m)/(p^2 + m^2)$.

To test for an anomaly one now chooses one of the polarization vectors (#6, say) to be longitudinal by setting $\zeta_6^\mu = k_6^\mu$. The regulated integral can then be evaluated explicitly, with the entire result coming from the regulator term. The final result for the anomaly is then obtained by letting the regulator mass $m \to \infty$. This gives a finite result for the anomaly

$$G \sim T \epsilon_{\mu_1 \ldots \mu_5 \nu_1 \ldots \nu_5} \zeta_1^{\mu_1} \cdots \zeta_5^{\mu_5} k_1^{\nu_1} \cdots k_5^{\nu_5}. \tag{10.1.10}$$

(The algebraic details are omitted, since they are step for step the same as the ones we present in §10.2.1 for the string calculation.)

The result described above shows that the $D = 10$ super Yang–Mills theory has a gauge anomaly for every choice of gauge group! Only in the abelian case, when the theory becomes free, can T vanish. Thus, by itself, super Yang-Mills theory in ten dimensions cannot be a consistent theory.

---

* The generalization to an arbitrary even number of dimensions is $M = 1 + D/2$.

The field-theory anomaly is proportional to the trace (10.1.7), with the matrices in the adjoint representation of the gauge group. This cannot vanish in all cases. The kinematical factor that this multiplies in (10.1.10) has total symmetry in the five physical external lines. Thus when we add the permutations required by Bose symmetry, five of the indices in the trace become symmetrized. In view of the cyclic symmetry of the trace, this implies that only the totally symmetrized trace contributes. To project out this term, it is convenient to define $t = \sum_a c_a t^a$, where the c_a are arbitrary constants. This represents an arbitrary element of the algebra. Only the symmetrized trace contributes to $\mathrm{Tr}(t^6)$, which therefore contains all the relevant group-theory information. Such expressions can be related to analogous ones based on other representations of the matrices. For example, suppose that λ^a is in the fundamental representation of an $SO(n)$ group and $\lambda = \sum c^a \lambda^a$. In this case t has $n(n-1)/2$ rows and columns, which can be labelled by a pair of antisymmetrized n-valued indices. λ, on the other hand, is an $n \times n$ matrix. They are related by

$$t_{ab,cd} = \tfrac{1}{2}\big(\lambda_{ac}\delta_{bd} - \lambda_{bc}\delta_{ad} - \lambda_{ad}\delta_{bc} + \lambda_{bd}\delta_{ac}\big). \tag{10.1.11}$$

Using the symbol 'Tr' to denote a trace in the adjoint representation and 'tr' to denote one in the fundamental representation it is easy to show that

$$\mathrm{Tr}t^6 = (n-32)\mathrm{tr}\lambda^6 + 15\mathrm{tr}\lambda^4\mathrm{tr}\lambda^2. \tag{10.1.12}$$

Similarly, for a $USp(n)$ group

$$\mathrm{Tr}t^6 = (n+32)\mathrm{tr}\lambda^6 + 15\mathrm{tr}\lambda^4\mathrm{tr}\lambda^2 \tag{10.1.13}$$

and for a $U(n)$ group

$$\mathrm{Tr}t^6 = n\mathrm{tr}\lambda^6 + 15\mathrm{tr}\lambda^4\mathrm{tr}\lambda^2. \tag{10.1.14}$$

(These matters are discussed further in §13.5.3.) Thus, among all simple classical groups there is one case in which $\mathrm{tr}\lambda^6$ can be eliminated, namely $SO(32)$. (There are also many examples that are products of simple classical groups.) We should emphasize, however, that the super Yang–Mills theory is anomalous even in all such cases, because a $\mathrm{tr}\lambda^4\mathrm{tr}\lambda^2$ term survives. In the next section we show that in the superstring extension new effects lead to a cancellation of these terms for $SO(32)$.

10.1.3 Anomalies in Superstring Theory

We saw in §7.1.5 that in the string theory there are an infinite number of gauge invariances associated with the decoupling of zero-norm states. The longitudinal mode of the massless vector state, $k \cdot \alpha_{-1}|0; k\rangle = L_{-1}|0; k\rangle$, is the most basic example. Decoupling of the zero-norm states is one aspect of the consistency and unitarity of the theory. There are a variety of ways to study anomalies in field theory; some of them do not generalize easily to string theory with our present understanding of this subject. For example, in field theory, one can inquire as to the gauge invariance of the fermion determinant in an external gauge field; one cannot at present ask the analogous question in string theory, since string theory is formulated as an on-shell theory. There are, however, ways of studying anomalies in field theory that do generalize to string theory. In field theory, gauge invariance entails the decoupling from the S matrix of the longitudinal mode of the vector meson, and this decoupling breaks down when anomalies are present. The question of anomalies can thus be posed in string theory by asking whether the longitudinal mode of the massless vector meson really decouples from the S matrix at the one-loop level. In string theory one would really like to verify the decoupling of all of the zero-norm states, but this question has been explicitly investigated only for the massless level, and this is the case we consider.

It turns out in this investigation to be rather useful to work in a covariant gauge. By completely specifying the gauge, as in the light-cone treatment, manifest Lorentz invariance is lost. In such a gauge a Lorentz transformation includes a compensating gauge transformation, adjusted so that the transformed theory is in the same gauge as the untransformed one. A breakdown of gauge invariance due to anomalies is therefore manifested by a violation of Lorentz invariance. No simple way of studying this is known, in part because the light-cone vertices take a simple form only if the momenta and polarizations have vanishing + components; but under this restriction, the parity-violating amplitudes vanish, and it is not possible to probe for anomalies. So we shall restrict our considerations to the covariant formulation of the theory.

Just as in field theory, it is possible in string theory to give formal proofs of the decoupling of the longitudinal mode of the vector meson from the loop diagram. In string theory, such a formal proof can be based on a canceled propagator argument, of the type sketched in §7.1.5. The canceled propagator argument must be used cautiously in loop amplitudes, however, since canceling a propagator might introduce divergent behavior in a way that compensates for the zero introduced by the canceled propagator to give a finite answer. This is what happens in the calculation of the hexagon anomaly in type I theories.

The analysis of the gauge invariance of the tree diagrams of the superstring theories is very similar to that of the bosonic theory discussed in §7.1.5. Consider, as an example, an open-string tree diagram with fermion end states and with emitted massless vector states. The rules for calculating diagrams involve the fermion propagator, $S = 1/F_0 = F_0/L_0$ and the emission vertex $W(\zeta, k, 1) = \zeta \cdot \psi(1) \exp\{ik \cdot X(1)\}$. When the polarization of the rth state is set equal to the momentum the vertex can be written in the form

$$W(k_r, k_r, 1) = k_r \cdot \psi(1)V_0(k_r, 1) = i\sqrt{2}[F_0, V_0(k_r, 1)]. \qquad (10.1.15)$$

As in the bosonic theory, the factors of F_0 in the commutator cancel against adjacent propagators and cause the amplitude to vanish. This argument does not involve the GSO projection operator – the theory is gauge-invariant without the need for space-time supersymmetry.

In the absence of the GSO projection, the above argument would still work if the fermion propagator were

$$\Delta^m = \frac{1}{(F_0 - im)} = \frac{F_0 + im}{L_0 + m^2} \qquad (10.1.16)$$

rather than $1/F_0$. Indeed, the vertex operator for a longitudinal vector meson can be written as

$$i\sqrt{2}[F_0 - im, V_0] \qquad (10.1.17)$$

so the canceled propagator argument still goes through. This is significant because, as in field theory, it is useful to regulate the one-loop diagram by subtracting a fermion loop with $m \neq 0$. The gauge-invariant regulator just stated[*] shows that in the absence of the GSO projection, there is no one-loop anomaly in gauge invariance. This is not surprising, since without the GSO projection, the theory is parity conserving. If, however, we carry out the GSO projection, all is different. In this case the vertices have factors of $(1 + \overline{\Gamma})$ in them, and the vertex of a longitudinal vector meson can no longer be written as a commutator with $F_0 - im$, since im commutes with $(1 + \overline{\Gamma})$, whereas F_0 converts $(1 + \overline{\Gamma})$ into $(1 - \overline{\Gamma})$. Just as in field theory, the factors of $\overline{\Gamma}$ make regularization impossible and introduce the possibility of an anomaly.

---

[*] This regulator violates other principles of string theory, which require the intercept $m = 0$, but it preserves the gauge invariance of the massless vector meson, which is all we need for the present discussion.

10.2   Analysis of Hexagon Diagrams

The hexagon diagrams responsible for the pure gauge anomaly in type I superstring theories are those that contain fermion loops and six external Yang–Mills particles. Since these particles are open-string modes, the diagrams divide into the three classes – planar, nonorientable and nonplanar – described in chapters 8 and 9. Associated with each diagram is a distinct Chan–Paton group-theory factor, introduced in accordance with the rules described in §6.1. This factor is made from traces of λ matrices belonging to the fundamental representation of the gauge algebra $SO(n)$, $USp(n)$ or $U(n)$. For planar diagrams this factor is

$$G_P = n\mathrm{tr}\,(\lambda_1\lambda_2\cdots\lambda_6)\,, \qquad (10.2.1)$$

where the cyclic sequence of the matrices inside the trace corresponds to the cyclic sequence in which the emitted particles are attached to the boundary of the diagram. As explained in previous chapters, this factor multiplies a dynamical expression (to be described) that has the same cyclic symmetry. The expressions that describe the other 59 cyclically inequivalent permutations must be added to form the full planar one-loop amplitude. The factor of n comes from the trace of the unit matrix, which is associated with the boundary of the annulus that has no particles attached.

The nonorientable hexagon diagram has a similar group-theory factor. In this case the world sheet (which is a Möbius strip) has only one boundary, to which all the particles are attached. As a result there is no factor of n. Just as in §8.1.2 and §9.1.3, the appropriate group-theory factor for the Möbius strip is

$$G_N = \eta\mathrm{tr}\,(\lambda_1\lambda_2\cdots\lambda_6)\,, \qquad (10.2.2)$$

where η is a factor due to twisting an odd number of lines (-1 for $SO(n)$, $+1$ for $USp(n)$ and 0 for $U(n)$), as explained in §6.1. In particular, this group-theory factor accompanies the diagram in which the external particles are arranged in the same cyclic order as in the trace, with a twist between the emission vertices for particles #6 and #1. Other diagrams with one, three or five twists also contribute terms with the same group-theory factor provided they correspond to arranging the emissions in the same cyclic order on the boundary of the Möbius strip. As in the loop calculations of earlier chapters, this division of the calculation into these pieces with different odd numbers of twists is somewhat artificial from the point of view of the picture of the world sheet in which the only distinctive feature is the cyclic order of the particles around the boundary of the Möbius strip. In the operator formalism that we are using the sum is required in order to obtain the correct integration region corresponding to a given cyclic ordering. As in the preceding chapters, one can save some labor by just calculating one of them and extending the integration region appropriately.

The final class of diagrams consists of the nonplanar diagrams with two or four twists, for which the group factor is a product of traces of the form

$$G_T = \text{tr}\,(\lambda_1 \lambda_2 \lambda_3 \lambda_4)\,\text{tr}\,(\lambda_5 \lambda_6)\,.\tag{10.2.3}$$

For the groups $SO(n)$ and $USp(2n)$ all the terms factorize into traces containing two λ_r's and four λ_r's only, since the trace of an odd number vanishes due to the antisymmetry of the λ's. For the $U(n)$ groups, diagrams with three particles attached to each boundary of the world sheet or those in which there is just one particle on one of the boundaries can also occur. Since $U(n)$ groups have already been shown to be inconsistent with supersymmetry, they do not need to be considered (although the anomaly is yet another argument for their inconsistency). Any particular partition and ordering of the λ_r's in the group factor, such as (10.2.3), corresponds to several diagrams with two twists or four twists. The total nonplanar amplitude is given by summing all inequivalent choices.

The division of the amplitude into three kinds of diagrams is very natural in type 1 superstring theory, since each type of factor is associated with a world sheet of a different topology. In the case of an ordinary field theory of point particles the Feynman rules do not distinguish a twisted from an untwisted propagator, so the contributions all have the same dynamical form. Since the massless fermions circulating around the loop in that case belong to the adjoint representation, it is more usual to express the anomaly in terms of a trace of matrices in the adjoint representation, $\text{Tr}(t_1 \cdots t_6)$, as we have done in §10.1.2. However, it is possible to re-express these traces in terms of ones based on matrices in the fundamental representation of the gauge group, as we indicated there. They can then be classified in the same manner as in the string theory. In the calculation of the next section we will discover that the crucial distinction between the field-theory calculation of §10.1.2 and the corresponding string-theory calculation is the cancellation of terms of the form G_T in the string case due to effects associated with the closed-string sector of the theory.

The expression for the complete one-loop amplitude is given by summing the terms with the three types of topologies. The nonplanar diagrams have a distinct group-theory structure and must give a vanishing anomaly by themselves. The planar and nonorientable diagrams have the

same group-theory structure and can therefore conspire to give canceling anomaly contributions. Each of the three types of terms is given by
the sum of four contributions corresponding to the four possible boundary conditions (spin structures) labeled $(--)$, $(-+)$, $(+-)$ and $(++)$ in
the notation of chapter 9. This combination of diagrams diverges like
$\int dq/q = 2\pi^2 \int dw/w \ln^2 w$ near $q = 0$ (or $w = 1$) for either the planar or
nonorientable diagrams. The fact that this is a milder divergence than in
the bosonic theory is due to a cancellation between $(--)$ and $(+-)$, the
contributions from the bosonic and fermionic loops with no $\overline{\Gamma}$ factors in
the trace. The contributions from diagrams with circulating fermions or
bosons that include a factor of $\overline{\Gamma}$ are individually divergent like $\int dq/q$.
Of the four types of terms (for each topology), the three $(--)$, $(+-)$, and
$(-+)$ give parity-conserving amplitudes, while the fourth one $(++)$ gives
a parity-violating one, since it contains a factor of Γ_{11} in the trace giving
rise to an ϵ tensor.

In §10.4.2 we will see that the the divergences in the parity-conserving
planar and nonorientable loop amplitudes cancel for the group $SO(32)$.
Although the light-cone gauge formalism is very useful for studying that
cancellation it is not suitable for analyzing parity-violating amplitudes
without much more difficulty. The potential anomalies that we now wish
to study come entirely from the parity-violating $(++)$ term, and that it
why it is much more convenient to use the covariant formalism of chapter 4
for this calculation.

10.2.1 The Planar Diagram Anomaly

The formal expression for the Feynman diagram that describes the parity-
violating part of the fermion loop with six external gauge particles is

$$A_P(1, 2, \cdots, 6) = g^6 G_P \int d^{10}p \, \mathrm{Tr} \left(\overline{\Gamma} SW(1)SW(2) \cdots SW(6) \right). \quad (10.2.4)$$

The fermion propagator is

$$S = F_0/L_0. \quad (10.2.5)$$

Also,

$$\overline{\Gamma} = \Gamma_{11}(-1)^{\sum d_{-n} \cdot d_n}, \quad (10.2.6)$$

so that $(1 + \overline{\Gamma})/2$ is the GSO projection operator of the fermion sector
(as in §4.3.3), and the on-shell emission vertices for the massless gauge

particles are

$$W(r) \equiv W(\zeta_r, k_r, 1) = \zeta_r \cdot \psi(1) e^{ik_r \cdot X(1)}. \tag{10.2.7}$$

The kinematics for the process is shown in fig. 10.2. Fermionic and bosonic ghost modes must be included in the definition of the propagators, but their effect on the amplitude is just to cancel two powers of the partition function, as in the case of the bosonic theory described in chapter 8. Equation (10.2.4) describes a term for which the fermionic partition function is $[f(w)]^8$. It cancels against the partition function of the bosonic modes, just as in the supersymmetric light-cone gauge calculations. In this case the bosonic and fermionic ghost contributions to the loop amplitude also cancel.

Since the amplitude in (10.2.4) is infinite it needs to be regulated. The anomaly is evaluated, as in §10.1.2, by replacing the polarization vector of one of the emitted gauge particles by its momentum. Because the diagram is divergent, in contrast to the case of the tree diagrams, the canceled propagator argument cannot be applied to argue that the amplitude is gauge invariant. This would miss the non-gauge-invariant behavior inevitably introduced by regulator terms.

The amplitude in (10.2.4) formally has cyclic symmetry in the six external lines. Whether this continues to be true after the divergence is canceled by a regulator term depends on the particular choice of the regulator. A choice that breaks the cyclic symmetry can be averaged over cyclic permutations to restore it if one wishes. Since at the present stage in the development of string theory the Feynman rules are introduced in a somewhat *ad hoc* manner so as to implement unitarity, rather than being derived from a fundamental action principle, it is not entirely clear what requirements, if any, a regularization method should satisfy. Two different methods for regularizing the divergences in the diagrams relevant to the calculation of anomalies have been studied. They both lead to the same conditions for the cancellation of anomalies, but they give somewhat different formulas for them when they are nonzero. One is the natural string-theory generalization of the Pauli–Villars method and the other is closely related to the Fujikawa method. Both are described in this chapter. The Pauli–Villars method is treated in detail in the text, and the second method is relegated to appendix 10.A.

As mentioned earlier, the Pauli–Villars method consists of subtracting from the divergent diagram a similar diagram with massive fermion propagators and then taking the limit $m \to \infty$ at the end of the calculation. The divergences can be arranged to cancel, since they are short-distance effects that do not depend on any finite masses. Since the regularized amplitudes are finite, naive manipulations, such as canceled propagator arguments, can be used with impunity. Therefore, since they give no anomaly for the original term without the mass parameter, the entire anomaly originates from the regulator term. As we will show, a finite nonzero anomaly is obtained in the limit $m \to \infty$.

In order to apply the Pauli–Villars method to superstring theory, we need to define a massive fermionic string propagator. Since F_0 is the string generalization of $i\Gamma \cdot p$, the natural choice is

$$S^m = \frac{1}{F_0 - im} = \frac{F_0 + im}{L_0 + m^2}. \tag{10.2.8}$$

The regulator term, $A_P^m(1, 2, \ldots, 6)$ is then obtained from (10.2.4) by replacing the propagators by S^m, defined by (10.2.8). In contrast to Pauli–Villars regulation of ordinary field theory the regulator mass m is never larger than all the masses of the physical particles in the string theory, since they have unbounded masses.

When we introduce integral representations for the factors $(L_0 + m^2)^{-1}$, as in previous string amplitude calculations, this results in the loop integrand containing an extra factor of w^{m^2}. Recall that the unregulated integral diverges at the $q = 0$ or $w = 1$ endpoint like

$$\int dq/q = 2\pi^2 \int dw/\ln^2 w. \tag{10.2.9}$$

Simply subtracting the term A_P^m changes the divergence to

$$\int dw(1 - w^{m^2})/\ln^2 w \sim m^2 \int dw/\ln w, \tag{10.2.10}$$

which still diverges at $w = 1$. In order to subtract the divergence completely it is necessary to take a combination of three terms, such as

$$A_P - 2A_P^m + A_P^{\sqrt{2}m}, \tag{10.2.11}$$

where the divergent pieces combine in the form

$$\int dw(1 - 2w^{m^2} + w^{2m^2})/\ln^2 w \sim m^4 \int dw, \tag{10.2.12}$$

which is finite at $w = 1$. Once it has been decided that the expression is finite and naive manipulations can be carried out, it does not matter much whether there are one or two regulator terms are used. (In fact, the expression for the anomalous part of A_P^m turns out to be the same as for $A_P^{\sqrt{2}m}$.) Bearing this point in mind, for simplicity of exposition we shall consider the combination

$$A_P^{reg} = A_P - A_P^m, \qquad (10.2.13)$$

as the regulated amplitude in what follows.

The regulated amplitudes do not satisfy the usual requirements for string amplitudes, since the introduction of the mass parameter m destroys conformal invariance. In particular, the massive propagators in the regulator term destroy the L_1 gauge conditions that previously ensured the decoupling of the longitudinal modes of the external gauge particles. This is exactly what we expect a regulator to do for an amplitude that has an anomaly.

The procedure for evaluating the anomaly arising from (10.2.13) (more correctly, from (10.2.11)) is to test for gauge invariance on each of the external legs by replacing the polarization vector for each line in turn by the corresponding momentum. The expression for A_P has cyclic symmetry in the external particles, since $\bar{\Gamma}$ can be moved around the trace by naive manipulations, which are valid for the regularized expression. However, A_P^m does not have cyclic symmetry, since $\bar{\Gamma}$ is next to vertex #6 and does not anticommute with the S^m. As a result the vertex for line #6 has a different status from the other five. We will discover that this particular regularization scheme enforces gauge invariance for all the external particles except #6, which must bear the entire brunt of the anomaly.

Let us begin by considering potential anomalies associated with lines #1–5. We can make use of the identity

$$W(k, k, 1) = k \cdot \psi(1)e^{ik \cdot X(1)} = [F_0, e^{ik \cdot X(1)}] \qquad (10.2.14)$$

to write A_P as the difference of two terms with canceled propagators on either side of the vertex under consideration. In the term A_P^m it is useful to write (10.2.14) as

$$W(k, k, 1) = [F_0 - im, V_0(k, 1)]. \qquad (10.2.15)$$

This also leads to the difference of two terms each of which contain canceled propagators adjacent to vertex under consideration. Choosing, for example, to study line #1, two of these four terms are given by

$$G_P \int d^{10}p \left\{ \text{Tr}[V_0(k_1)\frac{F_0}{L_0}\zeta_2 \cdot \psi(1)V_0(k_2)\dots\frac{F_0}{L_0}\zeta_6 \cdot \psi(1)V_0(k_6)\overline{\Gamma}] \right.$$

$$\left. -\text{Tr}[V_0(k_1)\frac{F_0 + im}{L_0 + m^2}\zeta_2 \cdot \psi(1)V_0(k_2)\dots\frac{F_0 + im}{L_0 + m^2}\zeta_6 \cdot \psi(1)V_0(k_6)\overline{\Gamma}] \right\},$$

$$(10.2.16)$$

while the other two terms have a canceled propagator between particles #1 and #2. The trace over the fermionic zero modes (the Dirac matrices) is very simple, because the Γ_{11} factor in $\overline{\Gamma}$ requires ten Γ matrices to contribute a nonzero result. These have to come from the zero modes of the five F_0's in the numerators of the propagators and the five $\psi(1)$'s in the vertices. This uses up all the fermionic modes in the vertices and in the numerators of the propagators. Using the fact that

$$\text{tr}\,(\Gamma_{11}\Gamma_{\mu_1}\dots\Gamma_{\mu_{10}}) = 32\epsilon_{\mu_1\dots\mu_{10}}, \qquad (10.2.17)$$

the resulting expression is proportional to

$$\int d^{10}p \; \epsilon_{\mu_1\dots\mu_5\nu_1\dots\nu_5} \zeta_2^{\mu_1}\dots\zeta_6^{\mu_5}(p-k_1)^{\nu_1}k_2^{\nu_2}\dots k_5^{\nu_5}$$

$$\times \left\{ \text{Tr}[V_0(k_1)\frac{1}{L_0}V_0(k_2)\dots\frac{1}{L_0}V_0(k_6)] \right.$$

$$\left. -\text{Tr}[V_0(1)\frac{1}{L_0 + m^2}V_0(2)\dots\frac{1}{L_0 + m^2}V_0(k_6)(-1)^{\sum d_{-n}\cdot d_n}] \right\}.$$

$$(10.2.18)$$

The factors of L_0 in the denominators in this expression contain the five momenta $p_2,\dots,p_6$, where $p_r = p - \sum_{s=1}^{r-1}k_s$. Therefore, by making a shift in the definition of the loop momentum from p to $p+k_1$, the formula no longer depends on k_1 or k_6. Thus each of the two terms must vanish, because there are not enough independent momenta in the integrand to saturate the indices of the ϵ tensor. The analysis for lines #2–5 works the same way.

Shifting the definition of the loop momentum is only a valid manipulation in the regulated expression. (In the unregulated expression for the anomaly it is certainly not valid to redefine the loop momentum. This is the classic mistake that leads to the conclusion that there is no anomaly.) The argument that the sum of these terms vanishes is correct but a little cavalier, since each of them involves the product of the vertices $V_0(k_1, 1)$ and $V_0(k_6, 1)$ at the same point on the world sheet (*i.e.*, without the intervening propagator). To treat this properly requires separating the two vertices infinitesimally and then considering the limit in which they co-incide. This calculation bears a close resemblance to the point-splitting method of calculating the anomaly described in appendix 10.A. The result is that each of the two terms in (10.2.18) gives the same finite answer so that their difference is indeed zero.

The calculation of the divergence on line #6 is rather different because of the $\overline{\Gamma}$ factor adjacent to the vertex. In this case it is convenient to re-express the relation in (10.2.18) (using the fact that $\{F_0, \overline{\Gamma}\} = 0$) as

$$k_6 \cdot \psi V_0(k_6)\overline{\Gamma} = \{F_0, V_0(k_6)\overline{\Gamma}\}$$

$$= \{F_0 - im, V_0(k_6)\overline{\Gamma}\} + 2im V_0(k_6)\overline{\Gamma}, \tag{10.2.19}$$

where the first relation is to be used in A_P, whereas the second relation is appropriate to A_P^m. By writing the relation in this manner there are again two terms with canceled propagators arising from both A_P and A_P^m. These vanish in pairs by the same argument as before leaving a nonzero term that contains the anomaly proportional to

$$W_P \sim G_{Pm} \int d^{10}p \operatorname{Tr}\left(\frac{F_0 + im}{L_0 + m^2}\zeta_1 \cdot \psi(1) V_0(k_1, 1)\ldots\right.$$

$$\left.\times \zeta_5 \cdot \psi(1) V_0(k_5, 1)\frac{F_0 + im}{L_0 + m^2}V_0(k_6, 1)\overline{\Gamma}\right). \tag{10.2.20}$$

The trace over the Dirac matrices again involves a Γ_{11}, which requires at least ten Γ matrices to give a nonzero result (10.2.17). There are actually eleven possible sources for these Γ matrices. There is a Γ matrix in the F_0's in each of the six propagators and in the $\psi(1)$ factors in each of the vertices for particles #1 to #5. This means that the one of the im terms in a propagator is left over after the Dirac trace is performed. (The other possible candidates for a left-over term are all linear in fermion oscillators d_n, but such terms vanish when the d_n traces are performed.) The result involves a linear combination of terms arising from the ways in which ten Γ matrices can be selected out of the eleven available. Five of the Γ matrices are contracted into the five remaining polarization tensors in each of the terms while the other five are contracted into five out of the six momenta in the propagators, $p_r = p - \sum_{i=1}^{r-1} k_r$, so that the Dirac trace results in a factor proportional to

$$im\ \epsilon_{\mu_1...\mu_5\nu_1...\nu_5} \zeta_1^{\mu_1} ... \zeta_5^{\mu_5} [(p-k_1)^{\nu_1} ... (p-k_1-k_2-k_3-k_4-k_5)^{\nu_5}$$
$$- p^{\nu_1}(p-k_1-k_2)^{\nu_2} ... (p-k_1-k_2-k_3-k_4-k_5)^{\nu_5} + \cdots$$
$$- p^{\nu_1}(p-k_1)^{\nu_2} ... (p-k_1-k_2-k_3-k_4)^{\nu_5}]$$

$$= -im\ \epsilon(\zeta, k),$$

(10.2.21)

where $\epsilon(\zeta, k)$ is proportional to the anomaly in the low-energy theory and is defined by

$$\epsilon(\zeta, k) = \epsilon_{\mu_1...\mu_5\nu_1...\nu_5} \zeta_1^{\mu_1} ... \zeta_5^{\mu_5} k_1^{\nu_1} ... k_5^{\nu_5}. \tag{10.2.22}$$

The resulting anomaly is

$$W_P \sim \lim_{m\to\infty} m^2 \epsilon(\zeta, k) G_P \int d^{10}p \text{Tr}\left(\frac{1}{L_0+m^2}V_0(k_1,1)\cdots\right.$$
$$\left. \times V_0(k_5,1)\frac{1}{L_0+m^2}V_0(k_6,1)\Gamma_d\right), \tag{10.2.23}$$

where $\Gamma_d = (-1)^{\sum d-n\cdot d_n}$ The calculation has now reduced to a form that is very similar to the calculations in chapter 8 of loop amplitudes in the bosonic theory with the difference that the momenta in (10.2.23) satisfy $k_r^2 = 0$ and $D = 10$.

The evaluation of the integral proceeds by replacing each propagator by its integral representation $(L_0+m^2)^{-1} = \int_0^1 x_r^{L_0+m^2-1}dx_r$. The manipulations described in §8.1.1 now relate the expression for the anomaly to correlation functions between vertex operators. They are evaluated at points ρ_r (where $\rho_r \equiv x_1...x_r$) on an annulus, which is built as in chapters 8 and 9 by identifying any real point ρ with the points $w^n\rho$; here $w \equiv \rho_6$ and n is an arbitrary integer. Using the relations in §8.1.1, we see that

$$W_P \sim \lim_{m\to\infty} m^2 \epsilon(\zeta, k) G_P \int_0^1 \frac{dw}{w} \int_R \left(\prod_{r=1}^5 \frac{d\rho_r}{\rho_r}\right) w^{m^2} I(1,\ldots,6), \tag{10.2.24}$$

where

$$I(1,\ldots,6) \equiv \int d^{10}p w^{p^2/2} \text{Tr}\left(w^N V_0(k_1,\rho_1)\ldots V_0(k_6,\rho_6)\Gamma_d\right)$$
$$= \left(\frac{-2\pi}{\ln w}\right)^5 \prod_{r<s} [\psi(c_{sr},w)]^{k_r\cdot k_s}, \tag{10.2.25}$$

and the region R is given by $1 \geq \rho_1 \geq \ldots \rho_5 \geq w$. The function $\ln\psi(c_{sr}, w)$

(where $c_{sr} = \rho_s/\rho_r$), defined in §8.1.1, is the correlation function between $X(\rho_r)$ and $X(\rho_s)$. In this case there is no factor of $[f(w)]^{2-D}$, since the trace over the Fermi modes (and associated ghosts) gives a factor of

$$\text{Tr}(-w^n)^{\sum d_{-n} \cdot d_n} = [f(w)]^8, \qquad (10.2.26)$$

which cancels the partition functions arising from the trace over α modes and their ghosts.

In view of (10.2.24), it is evident that in the limit $m \to \infty$ the integral is dominated by the region $w \sim 1$. This is in accord with the expectation that any nonzero anomaly arises from a conspiracy between the infinity of the amplitude at $w = 1$ and the zero due to the canceled propagator. In order to exhibit the region around $w = 1$ it is easiest to transform the integration variables to z_r and q defined by

$$z_r = \exp\left(\frac{2\pi i \ln \rho_r}{\ln w}\right) \quad \text{and} \quad q = \exp\left(\frac{2\pi^2}{\ln w}\right), \qquad (10.2.27)$$

as in §8.1.1. Then the dominant region is given by $q \sim 0$, the region in which the hole in the annulus approaches zero radius. Under this transformation $\psi(c_{sr}, w)$ transforms into $\psi(z_s/z_r, q^2)$ according to

$$\psi(c_{sr}, w) = -\frac{i\pi}{\ln q} \exp\left[-\frac{(\ln z_s/z_r)^2}{4 \ln q}\right] \psi(z_s/z_r, q^2), \qquad (10.2.28)$$

as explained in appendix 8.A. Combining this with the momentum integral (see §8.1.1)

$$\int d^{10}p \prod_{r=1}^{6} x_r^{p_r^2/2} = \left(\frac{-2\pi}{\ln w}\right)^5 \prod_{1 \le r < s \le 6} \left\{c_{sr}^{-1/2} \exp\left(\frac{\ln^2 c_{sr}}{2 \ln w}\right)\right\}^{k_r \cdot k_s}, \qquad (10.2.29)$$

I can be re-expressed in terms of z_r and q by

$$I = \left(\frac{-\ln q}{\pi}\right)^5 \prod_{r<s} (\psi(z_s/z_r, q^2))^{k_r \cdot k_s} \exp - \left(k_r \cdot k_s \frac{\ln^2(z_s/z_r)}{2 \ln q^2}\right)$$

$$= \left(\frac{-\ln q}{\pi}\right)^5 \langle p = 0 | \text{Tr}\left(q^{2N} V_0(k_1, z_1) \dots V_0(k_6, z_6) \Gamma_d\right) | p = 0 \rangle, \qquad (10.2.30)$$

which is the correlation function between the vertices on the boundary of an annulus or cylinder. The symbol $|p = 0\rangle$ implies evaluation at zero momentum. The trace applies to all oscillator modes. By recasting the formula in the latter form given in (10.2.30), we have succeeded in exhibiting all the q dependence in a particularly useful way for the subsequent analysis.

As in §8.1.1, the integration measure in terms of the variables on the annulus is given by

$$\int_0^1 \prod_1^6 \frac{dx_r}{x_r} = \int_0^1 \frac{dw}{w} \int_0^1 \left(\prod_1^5 \frac{d\rho_r}{\rho_r} \Theta(\rho_r - \rho_{r+1}) \right)$$

$$= \int_0^1 \frac{dw}{w} (\ln w)^5 \int_0^1 \left(\prod_{r=1}^5 d\nu_r \Theta(\nu_{r+1} - \nu_r) \right),$$

(10.2.31)

where $\nu_r \equiv \ln z_r / 2i\pi = \ln \rho_r / \ln w$ (with $\nu_6 = 1$). Substituting this change of variables and the expression for I into (10.2.24) gives the expression for the anomaly

$$W_P \sim \lim_{m \to \infty} \epsilon(\zeta, k) m^2 G_P \int_0^1 \left(\prod_{r=1}^5 d\nu_r \Theta(\nu_{r+1} - \nu_r) \right) \int_0^1 dw\, w^{m^2 - 1}$$

$$\times \langle p = 0 | \mathrm{Tr} \left(q^{2N} V_0(k_1, z_1) \dots V_0(k_6, z_6) \Gamma_d \right) | p = 0 \rangle.$$

(10.2.32)

Evidently, the only contribution to the trace that can survive the $m \to \infty$ limit is the term with $N = 0$, since q vanishes exponentially fast as $w \to 1$ and N has discrete (integer) eigenvalues. This can be demonstrated explicitly by considering the w integral

$$\int_0^1 w^{m^2 - 1} e^{4\pi^2 N / \ln w}\, dw = \frac{1}{m^2} \int_0^\infty e^{-\lambda - 4\pi^2 m^2 N / \lambda}\, d\lambda,$$

(10.2.33)

which is equal to $1/m^2$ when $N = 0$, whereas terms with nonzero N vanish exponentially as $m \to \infty$. As a result the expression for the anomaly arising from the planar loop diagram is given by

$$W_P \sim \epsilon(\zeta, k) G_P \int_0^1 \left(\prod_{r=1}^5 d\nu_r \Theta(\nu_{r+1} - \nu_r) \right) \langle 0 | V_0(k_1, z_1) \dots V_0(k_6, z_6) | 0 \rangle$$

$$= \epsilon(\zeta, k) G_P \int_0^1 \left(\prod_{r=1}^5 d\nu_r \Theta(\nu_{r+1} - \nu_r) \right) \prod_{r<s} (z_s - z_r)^{k_r \cdot k_s},$$

(10.2.34)

where $z_r = \exp(2\pi i \nu_r)$. The last step is very similar to the evaluation of a bosonic tree diagram. It uses the correlation function derived for tree amplitudes in appendix 7.A. The total contribution to the hexagon gauge anomaly from planar diagrams is obtained by summing over permutations of the external lines.

The anomaly has the same form as the anomaly in the super Yang–Mills field theory of §10.1.2 multiplied by a factor that resembles a tree diagram, but both comparisons are imprecise. The super Yang–Mills gauge anomaly in §10.1.2 involves a trace of matrices in the adjoint representation of the gauge algebra, whereas the matrices in the trace here are in the fundamental representation of a classical algebra. In the low-energy limit, $k_r \ll \sqrt{T}$ (where the string tension has been set to $T = 1/\pi$) the anomaly reduces to the expression that arises from the massless hexagon loop in supersymmetric Yang–Mills theory, aside from the important change in the group-theory factor. The relationship between the group-theory factors can be understood as follows. The field-theory trace $\mathrm{Tr}(t_1 t_2 \ldots t_6)$ (with matrices in the adjoint representation) can be re-expressed as a superposition of terms of the form $\mathrm{tr}(\lambda_1 \lambda_2 \ldots \lambda_6)$ and $\mathrm{tr}(\lambda_1 \ldots \lambda_4)\mathrm{tr}(\lambda_5 \lambda_6)$ (including other permutations) for any classical group, as explained in §10.1.2. The particular coefficients that arise in this transcription depend on the group. When this transcription is made the terms with six λ matrices inside a single trace precisely correspond to the low-energy limit of the string-theory anomaly arising from the planar and nonorientable diagrams summed over all permutations of the lines. One might expect a similar relation between the terms with a product of two traces and the anomalies associated with nonplanar string-theory diagrams. However, as we explain later in this chapter, this is where new string-theory effects give cancellations that persist even in the low-energy limit.

The dynamical factor that multiplies the trace differs from a tree amplitude in two respects. First of all, it has double poles in channels formed by sets of adjacent particles. For example, there are double poles in the channel formed by particles 1 to k, which come from the integration endpoint where $z_1 \sim z_2 \sim \ldots \sim z_k$ and simultaneously $z_{k+1} \sim \ldots \sim z_6 = 1$. Furthermore, there are no massless poles in any channel, which is not surprising, since a nonzero residue at such poles would correspond to an anomaly in a diagram with less than six external massless particles. It is noteworthy that the string-theory anomaly contains poles with nonzero masses in various channels. By factorizing on these poles it is possible to derive expressions for anomalous diagrams with fewer than six external particles, some of which are massive. These are examples of anomalies in gauge invariances associated with massive modes mentioned at the beginning of §10.1.3.

In appendix 10.A the calculation of the anomaly is repeated using a regulator that cuts off the propagators with factors $\exp\{-\eta_r L_0\}$, where the limit $\eta_r \to 0$ is taken at the end of the calculation. This procedure is equivalent to cutting off the integration region of the vertex operator correlation function so that the values of the $\ln \rho_r$ can never coincide. The minimum value of $\ln \rho_r - \ln \rho_{r+1}$ is $-\eta_{r+1}$. When a single propagator is regulated in this manner the result for the anomaly has the same form as the result obtained by Pauli–Villars regularization. If the values of all the η_r are taken to be equal, so that the regularization treats the particles symmetrically, the result is different. The functional form of the integrand is the same, but variables ν_r parametrizing the particles on the boundary of a disk, are integrated over a restricted region (the combination R_1 and R_2 described in appendix 10.A.)

10.2.2 The Anomaly in the Nonorientable Diagram

The anomaly arising from the nonorientable diagrams is evaluated by combining the methods of §8.1.2 for nonorientable bosonic loop diagrams with those of the preceding subsection. The contributions to a given group-theory factor come from a sum of several diagrams with an odd number of twisted propagators. The treatment of the zero-mode Dirac trace is identical to that in the planar loop diagram, so that the expression for the anomaly again boils down to a structure very similar to the corresponding bosonic loop diagram. This gives an expression for the nonorientable contribution to the anomaly similar to (10.2.32), but with q^2 replaced by $-\sqrt{q}$ and z_r replaced by $z_r^{1/2}$ as in (7.61),

$$W_N \sim \lim_{m\to\infty} \epsilon(\zeta,k)m^2 G_N \int_0^2 \left(\prod_{r=1}^5 d\nu_r \Theta(\nu_{r+1} - \nu_r) \right) \int_0^1 dw w^{m^2-1}$$

$$\times \langle p = 0| \mathrm{Tr}\left((-1)^N q^{N/2} V_0(k_1, \sqrt{z_1}) \ldots V_0(k_6, \sqrt{z_6}) \Gamma_d \right) |p = 0\rangle,$$

$$(10.2.35)$$

where the normalization corresponds to that of (10.2.32). In this case the integrand is similar to the correlation of six V_0's on the boundary of a Möbius strip so that the angular variables $\nu_r = \ln z_r/2\pi i$ are integrated in sequence between zero and two. After taking the limit $m \to \infty$ and performing the w integral the expression can be written as

$$W_N \sim \epsilon(\zeta, k) G_N \int_0^2 \left(\prod_{r=1}^5 d\nu_r \Theta(\nu_{r+1} - \nu_r) \right) \tag{10.2.36}$$

$$\times \langle 0|V_0(k_1, \sqrt{z_1}) \dots V_0(k_6, \sqrt{z_6})|0 \rangle = \eta \frac{32}{n} W_P,$$

where we have changed variables from $\sqrt{z_r}$ to z_r (*i.e.*, from $\nu_r/2$ to ν_r, which gives a factor of 32) and used the fact that $G_N = \eta G_P/n$. As a result the total anomaly with the group-theory factor $\text{tr}(\lambda_1 \dots \lambda_6)$ is given by

$$W = W_P + W_N, \tag{10.2.37}$$

which vanishes for the gauge group $SO(32)$ and for no other group.

The relative factor of $\eta 32/n$ in (10.2.36) has a simple interpretation in the low-energy effective super Yang–Mills field-theory calculation in §10.1.2. In that case the total anomaly was proportional to $\text{Tr} t^6$, which decomposed into traces in the fundamental representation, as in (10.2.17) for $SO(32)$, for example. The term $n \text{tr} \lambda^6$ in that equation comes from the low energy limit of the planar hexagon diagram. The low-energy limit of any of the nonorientable diagrams gives the term $\text{tr} \lambda^6$. The number of such diagrams is equal to the number of ways in which an odd number of twists can be distributed on the internal legs of a hexagon diagram, which is 32. Each one of these contributes a dynamical factor of $\epsilon(\zeta, k)$ to the anomaly and, after Bose symmetrization, this leads to the overall factor of 32. The last term in (10.2.17), $15 \text{tr} \lambda^4 \text{tr} \lambda^2$, has the group-theory structure of a nonplanar diagram.

10.2.3 Absence of Anomalies in Nonplanar Diagrams

We will now analyze the nonplanar hexagon diagram and show that it gives no anomaly. This should not be too surprising, since nonplanar diagrams are not divergent, and the canceled propagator argument ought to be valid. We now verify this in detail.

The calculation of these diagrams, which have an even number of twisted propagators, involves the same zero-mode trace over Dirac matrices as the other loop diagrams. The expression for the anomaly again resembles the expression for the bosonic-string nonplanar loop in §8.1.3. The major difference from the earlier calculations is that the factor of q^{2N} in (10.2.32) is replaced by $q^{2N-s/4}$ (where s is the invariant energy-squared of the singlet channel), which vanishes in the limit $q \to 0$ for all values of N. More precisely, it vanishes for $s < 0$ and is defined to vanish for all other values by analytic continuation.

The fact that the contribution to the anomaly from the nonplanar diagram vanishes might appear to contradict the calculation of the anomaly in the super Yang–Mills field theory. The field-theory anomaly always contains terms of the form $\text{tr}(\lambda_1 \ldots \lambda_4)\text{tr}(\lambda_5\lambda_6)$, even in the case of the group $SO(32)$ when $\text{tr}(\lambda_1 \ldots \lambda_6)$ terms are absent. The resolution is that the nonplanar loop diagram contains massless closed-string bound states in addition to the states of the open-string sector. The presence of the closed-string poles in the s channel was discussed in §8.1.3, where it was seen that the factor $q^{-s/4}$ is responsible for these poles. This means that the low-energy limit of the theory has extra contributions in which the massless closed-string poles play a critical role. These poles occur in just those singlet channels in which the field-theory result is nonvanishing, so it is possible for them to give the required cancellation. In chapter 13, we will explore this phenomenon systematically from the low-energy point of view, so here we only make a few remarks.

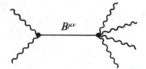

Figure 10.3. A tree diagram with the exchange of the $B^{\mu\nu}$ field between a vertex with two gauge particles and one with four gauge particles. This tree diagram arises as a piece of the low-energy limit of the nonplanar string hexagon loop diagram and is anomalous in just the manner that cancels the usual low-energy hexagon anomaly.

The exchange of closed strings in the 5–6 channel arises from configurations in which the world sheet has a long narrow tube joined at one end to a disk with two particles attached to its boundary and at the other end to a disk with four particles attached to its boundary. At low energy the massless closed-string particles survive and so, in addition to the expected anomalous hexagon diagram, there is a *tree* diagram in which the massless particles of the closed-string sector are exchanged between a pair of gauge particles at one vertex and four gauge particles at the other. This tree diagram is anomalous if the particle exchanged in the tree is the massless antisymmetric tensor field $B_{\mu\nu}$, since it is then possible to get an ϵ tensor contracted into ten external vector indices. The vertex coupling the $B_{\mu\nu}$ field to two gauge particles has the form $\partial_{[\rho}B_{\mu\nu]}\text{tr}\left(A^\rho F^{\mu\nu}\right)$, as we will see in §13.5.3; here $F^{\mu\nu}$ is the Yang–Mills field defined in terms of the vector potential A^μ and [...] indicates antisymmetrization of indices. The vertex that couples the $B_{\mu\nu}$ field to four gauge particles will be seen in §13.5.3 to have the form $\epsilon_{\mu_1...\mu_{10}} B^{\mu_1\mu_2} F^{\mu_3\mu_4} \ldots F^{\mu_9\mu_{10}}$. The tree diagram with $B_{\mu\nu}$ exchange (fig. 10.3) consists of several terms obtained by contracting out the $B_{\mu\nu}$ fields in each of these vertices with an intermediate propagator containing a factor $1/(k_5 + k_6)^2$. There is one term in the tree amplitude due to this diagram that has an anomaly with a coefficient that is exactly the value needed to cancel the contribution due to the hexagon loop diagram. The appearance of anomalies in tree diagrams may be an unfamiliar phenomenon. As we will explain in detail in chapter 13, this reflects the fact that the coupling of B to two gauge bosons is gauge invariant only if B transforms nontrivially under Yang–Mills gauge transformations, while the coupling of B to four gauge bosons is gauge invariant only if B does not transform at all; the 'interference' term between these two couplings, which is the diagram of fig. 10.3 thus violates gauge invariance regardless of what gauge transformation law is assumed for B.

10.3 Other One-Loop Anomalies in Superstring Theory

We have seen that amplitudes with external gauge particles must vanish when we substitute $\zeta_\mu = k_\mu$ for one of the polarization vectors as an expression of on-shell gauge invariance or current conservation. In similar fashion, if the polarization tensor of an external graviton is taken to be $\zeta_{\mu\nu} = k_\mu\zeta_\nu + k_\nu\zeta_\mu$, the amplitude should vanish to reflect the decoupling of longitudinally polarized gravitons required by energy–momentum conservation. If this does not happen, one speaks of a 'gravitational anomaly'.

As we will discuss further in §13.3.2, hexagon diagrams with external gravitons can give rise to gravitational anomalies. In the type I theory these diagrams consist of several distinct topological types described in §8.3.3. The diagram in which the world sheet is a torus is also the only closed-string loop diagram in the type II theories. Since this loop diagram is finite it is entirely plausible that the canceled propagator argument can be used to prove that the diagram is not anomalous. This is in accord with the result of the anomaly calculation in the low-energy theory (described in §13.5.2), which shows that the hexagon anomalies precisely cancel for the combination of massless states in the type II theory. type I closed strings can also couple to a Klein bottle, an annulus and a Möbius strip at this order in κ. Therefore, if all anomalies cancel for the $SO(32)$ theory, as is suggested by the low energy analysis of chapter 13, then the gravitational anomalies associated with these diagrams should cancel for $SO(32)$, though not for other groups.

In addition to anomalies with external gauge bosons or gravitons only, 'mixed anomalies' can arise in diagrams with both external gravitons and Yang–Mills particles. The one-loop diagrams of this type (for type I superstrings) are the annulus and the Möbius strip. The analysis of the low-energy theory in chapter 13 suggests that these anomaly contributions should also cancel in the $SO(32)$ case.

The analysis above has been presented for type I superstrings. However, as we show in §13.5, the low-energy analysis suggests that the anomalies cancel in the type IIB and heterotic theories; indeed, the low-energy analysis of anomalies stimulated the discovery of the heterotic theories. It is plausible that the loop diagram with six external particles is finite in these theories and that the canceled propagator argument leads to the vanishing of all the anomalies. Since the finiteness of the loop diagram is linked to modular invariance, there seems to be a clear connection between the absence of ten-dimensional gauge and gravitational anomalies and modular invariance, which corresponds to invariance under two-dimensional global diffeomorphisms of the toroidal world sheet.

10.4   Cancellation of Divergences for $SO(32)$

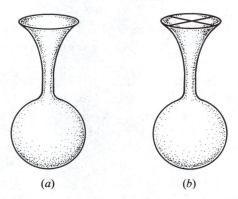

(a)   (b)

Figure 10.4. The divergences in the planar and nonorientable diagrams arise from dilaton tadpoles on a disk or on RP^2, as sketched in (a) and (b), respectively.

We recall that in chapter 9 we found divergences of the form $\int_0 dq/q$ in both the planar and nonorientable loop diagrams. In each case, these divergences represent massless dilaton tadpoles, as in fig. 10.4. The divergence in the planar loop is proportional to the dilaton tadpole computed on a disk, while the divergence in the nonorientable loop is proportional to the dilaton tadpole computed on RP^2. We now wish to determine under what conditions, if any, these divergences cancel.

10.4.1 Dilaton Tadpoles and Loop Divergences

Let us first think briefly about the meaning of dilaton tadpoles, in the context of the limiting low-energy field theory. Let ϕ be the dilaton field, and let $V(\phi)$ be the potential energy for the dilaton. The tadpole is just

$$\frac{\partial V}{\partial \phi}, \qquad (10.4.1)$$

and its vanishing is one of the equations of motion of the theory.

Now, in four dimensions, there are quite a few supersymmetric field theories, and it is possible to include a scalar potential $V(\phi)$ in these theories. There are not very many supersymmetric field theories in ten dimensions, and in attempting to construct such theories one learns that it is impossible to add a potential $V(\phi)$ to ten-dimensional supergravity without violating supersymmetry explicitly. Thus, in ten dimensions the dilaton tadpole must vanish by virtue of supersymmetry – though an analogous statement would not be valid in four-dimensional supergravity.

Since the infinities of the type I theory reflect dilaton tadpoles, and the dilaton tadpoles violate supersymmetry, the infinities must cancel when supersymmetry is valid. The type I theory is definitely not supersymmetric (at the one-loop level) for gauge groups other than $SO(32)$, since supersymmetry implies general covariance, and we will see in chapter 13 that for groups other than $SO(32)$ there are gravitational anomalies. However, it seems (combining our above results with those in chapter 13), that all anomalies cancel in the case of $SO(32)$, so that supersymmetry is presumably valid in that case. In this case, the dilaton tadpoles must cancel for $SO(32)$, and the one-loop infinities must cancel also. This is the question that we now investigate here.

The reader may wonder about the following. If we believe that the dilaton tadpole cancels between the disk and RP^2 in the case of $SO(32)$, why do we not investigate this directly, rather than probing the question indirectly by studying one-loop infinities? In fact, the dilaton tadpole on the disk and on RP^2 represents the expectation value of the dilaton vertex operator (at zero momentum) on the respective world sheets. The calculation on either world sheet separately is not difficult, up to normalization, and is easily seen to give a nonzero answer in each case. Unfortunately, the respective normalization factors are all-important in this problem, and are difficult to determine, since they depend on finite factors that are left over after the Faddeev–Popov gauge fixing in the two cases. The most straightforward way to determine the correct normalization of the dilaton tadpoles on the disk and on RP^2 seems to be to obtain these tadpoles as the coefficients of the divergences in the orientable and unorientable loop diagrams, whose normalizations are known from unitarity.

We now turn to the mathematical analysis. The expressions for the planar and the nonorientable loop diagrams in §9.1.2 and §9.1.3 diverge at the $q \to 0$ (or $q' \to 0$) endpoints. The residue of either of these divergences is proportional to

$$F_P(0) = \int_0^1 \left(\prod_{r=1}^3 \Theta\left(\nu_{r+1} - \nu_r\right) d\nu_r \right) \prod_{r<s} (\sin \pi(\nu_s - \nu_r))^{2\alpha' k_r \cdot k_s}, \quad (10.4.2)$$

with $\nu_4 = 1$. (In the analysis of this section it is useful to include explicit α' dependences.) By making a change of variables from ν_3 to

$$x = \frac{\sin \pi(\nu_2 - \nu_1) \sin \pi \nu_3}{\sin \pi(\nu_3 - \nu_1) \sin \pi \nu_2} \quad (10.4.3)$$

with Jacobian

$$\frac{\partial \nu_3}{\partial x} = \frac{1}{\pi} \text{Im}\left(\frac{1}{(x-y)}\right) \quad (10.4.4)$$

and $y = \exp\{i\pi\nu_1\} \sin \pi(\nu_2 - \nu_1)/\sin \pi\nu_2$, it is straightforward to show that

$$F_P(0) = \frac{1}{\pi} \int_0^1 dx \int_0^1 d\nu_2 \int_0^{\nu_2} d\nu_1 x^{-\alpha's}(1-x)^{-\alpha't} \text{Im}\left(\frac{1}{x-y}\right). \quad (10.4.5)$$

This can be evaluated by using the relation

$$\int_0^{\nu_2} \frac{d\nu_1}{x-y} = \int_0^{\nu_2} \frac{d\nu_1 \left(e^{2i\pi\nu_2} - 1\right)}{e^{2i\pi\nu_1} + (x-1)e^{2i\pi\nu_2} - x}$$

$$= \frac{1 - e^{2i\pi\nu_2}}{(x-1)e^{2i\pi\nu_2} - x} \left(\frac{1}{2i\pi} \ln\left(\frac{x}{x-1}\right) - \nu_2\right), \quad (10.4.6)$$

which is easily integrated over ν_2, giving

$$F_P(0) = -\frac{1}{2\pi^2} \int_0^1 \left(\frac{\ln x}{1-x} + \frac{\ln(1-x)}{x}\right) x^{-\alpha's}(1-x)^{-\alpha't} dx. \quad (10.4.7)$$

Now compare this result with the expression for the tree diagram with the same ordering of the external particles given in §7.4.2, which can be written (with α' reinstated) as

$$A_{tree} = -2g^2(\alpha')^2 K G_P \frac{\Gamma(-\alpha's)\Gamma(-\alpha't)}{\Gamma(1 - \alpha's - \alpha't)}$$

$$= 2g^2 K G_P \frac{\alpha'}{t} \int_0^1 x^{-\alpha's-1}(1 - x)^{-\alpha't} dx. \tag{10.4.8}$$

It follows from this that the planar loop divergence is proportional to

$$K g^4 G_P F(0) \sim \frac{g^2}{\alpha'} \frac{\partial}{\partial \alpha'} A_{tree}. \tag{10.4.9}$$

10.4.2 Divergence Cancellations

The nonorientable loop amplitude has exactly the same form as the planar one with the variable q^2 replaced by $-q^2$ and with a different coefficient. Hence, for a special gauge group the divergences of these two amplitudes may cancel. Since $\eta G_P = n G_N$ for the group $SO(32)$ ($n = 32$ and $\eta = -1$), the two contributions combine to give

$$A_P + A_N = 8\pi^3 g^4 K G_P \int_{-1}^1 \frac{d\lambda}{\lambda} F(\lambda), \tag{10.4.10}$$

where $\lambda = q^2$. This expression is nonsingular if we assume a regulator that treats the point at $\lambda = 0$ by a symmetric cutoff – in other words, if the integral is defined by a principal-value prescription.

   This argument for the finiteness of the $SO(32)$ theory given above is not entirely rigorous. The strength of the individual infinities depends on the choice of variables. If we had redefined the q variable in the expression for the nonorientable loop amplitude in another way, the formal cancellations of infinities would not take place. The choice of the q' variable used appears to be the natural one, because it leads to expressions for the integrands of the planar and the nonorientable loops that combine into an extended integration region. The real issue to resolve is whether there is a gauge-invariant way of regulating divergent loop amplitudes that justifies this particular choice. The fact that gauge anomalies also cancel at one loop for this group (as discussed in this chapter and chapter 13) is further evidence for the special nature of the group $SO(32)$ in type I theories. It also supports the belief that the infinity cancellation argument could be made rigorous.

There is no conceptual difficulty in considering amplitudes with more than four external states, but explicit calculations are algebraically tedious, since in the general case there are complicated contributions from the nonzero fermion modes in the prefactors of the vertices. In fact, for loops with $M > 4$ external bosonic lines it is probably more convenient to use the covariant formalism based on ψ^μ fields, even though bosonic and fermionic loops must be separately calculated and combined.

The relation of the infinities to dilaton tadpoles means that the cancellation seen in the four-particle amplitudes must also be a property of one-loop amplitudes with any number of external states. The pictures of the world sheets in figs. 8.12 and 8.15c show that the divergence cancellation can be viewed as a cancellation between the contributions from a disk (for the planar diagram) and from the real projective plane (for the nonorientable diagram), which couple to the world sheet by a long narrow tube. These configurations can not depend on how many external states are attached to the world sheet and hence must be independent of the number of external particles. This argument should even apply for multiloop diagrams. If the type I $SO(32)$ superstring theory has multiloop divergences, they must have some completely different origin.

In addition to the loop diagrams described here, which are all of order g^4, there are other diagrams that contribute to the possible divergences in the theory at low order. These were described in §8.3.3 for the case of the bosonic theory. For example, the disk diagram with external closed-string states (fig. 8.23a) has a divergence that would have to cancel with the divergence from the diagram that is described by the projective plane. Similarly, the divergences associated with the Klein bottle and those associated with the closed-string processes in which the external states couple to an annulus and to a Möbius strip must all cancel. Figure 8.27 shows that the latter sum of divergences can be formally expressed as an iteration of the combination of the disk and the projective plane. Since the local behavior on the world sheet that is responsible for these divergences is identical to that of the loop diagrams we have described in detail, it seems inevitable that the infinity cancellation would work in exactly the same way for this pair of diagrams.

10.5   Summary

In this chapter we have described the calculation of the one-loop contributions to the gauge anomaly in type I superstring theory. They were studied by considering hexagon diagrams in which the external states are massless on-shell gauge particles. The cancellation of such anomalies is required for the consistency of the theory. The string theory calculation generalizes the analogous calculation in Yang–Mills field theory, which was also briefly described. The planar and nonorientable hexagon diagrams with chiral fermions circulating around the loop give rise to gauge anomalies, each having the same form. The requirement that these anomalies cancel uniquely determines the gauge group to be $SO(32)$. Nonplanar diagrams are separately free from anomalies regardless of the gauge group. This fact is closely linked to the fact that nonplanar diagrams contain the bound states of the gravitational sector. In order to understand this cancellation, we noted that at low energy the anomaly in the nonplanar hexagon diagram includes the contribution of a massless antisymmetric tensor exchange. We will see in chapter 13 that this exchange can contribute an anomaly that exactly cancels the fermion loop anomaly with the corresponding group-theory factor.

Appendix 10.A   An Alternative Regulator

This appendix describes a method of regulating hexagon string diagrams that can be interpreted as a 'point-splitting' of the vertices on the world sheet. This is achieved by modifying the propagator carrying momentum p_r by a factor of $\exp\{-\eta_r L_0\}$, where η_r is a small positive parameter that goes to zero at the end of the calculation. This cutoff has the effect of suppressing states inside the propagators with high value of $L_0 = N + p^2/2$. It therefore bears a close resemblance to the method used by Fujikawa of regulating the fermion determinant by suppressing high-momentum modes with a gaussian momentum cutoff. In the string context it also suppresses states of high mass. The modified propagator has the form

$$S_{\eta_r} = e^{-\eta_r L_0} \frac{F_0}{L_0} = F_0 \int_0^{y_r} x_r^{L_0-1} dx_r, \qquad (10.\text{A}.1)$$

where $y_r = \exp\{-\eta_r\}$, which means that ρ_r/ρ_{r-1} is always less than $\exp\{-\eta_r\}$ so that the points ρ_r and ρ_{r-1} cannot coincide.

This regularization is symmetric in all the external particles if all six η_r's are chosen equal. This choice eliminates all poles from the amplitude, since in any of the channels formed by the external particles the corresponding set of ρ variables can never approach each other on the world sheet. An alternative possibility that leaves most of the poles would be to take all the η_r's to be zero except the one corresponding to one of the canceled propagators adjacent to the line whose anomaly is being considered.

To start with, let us consider general values of η_r. Since all the external lines are equivalent with this regulator, consider taking the divergence on line #6. As before, the anomaly only arises from the parity-violating amplitude in (10.2.4). By using the relation in (10.2.14), the regulated anomaly is given as the difference of two terms with canceled propagators. For the case of a planar diagram this can be written as

$$W_P \sim \lim_{\{\eta_r\}\to 0} G_P \int d^{10}p \left(\mathrm{Tr}[e^{-\eta_1} W(\zeta_1, k_1, 1) \frac{F_0 e^{-\eta_2 L_0}}{L_0} \cdots V_0(k_6, 1)\overline{\Gamma}] \right.$$
$$\left. - \mathrm{Tr}[\frac{F_0 e^{-\eta_1 L_0}}{L_0} W(\zeta_1, k_1, 1) \ldots W(\zeta_5, k_5, 1) e^{-\eta_6 L_0} V_0(k_6, 1)\overline{\Gamma}] \right).$$

$$(10.A.2)$$

Substituting the integral representation for each of the propagators gives

$$W_P \sim G_P(W_P^1 + W_P^2), \tag{10.A.3}$$

where

$$W_P^1 = \lim_{\{\eta_r\}\to 0} \int d^{10}p \int_0^{y_1} \frac{dx_1}{x_1} \cdots \int_0^{y_5} \frac{dx_5}{x_5} \mathrm{Tr}[w^{L_0} W(k_1, \rho_1) F_0 \cdots$$
$$\times W(k_5, \rho_5) F_0 W(k_6, \rho_6)\overline{\Gamma}], \tag{10.A.4}$$

and

$$W_P^2 = - \lim_{\{\eta_r\}\to 0} \int d^{10}p \int_0^{y_2} \frac{dx_2}{x_2} \cdots \int_0^{y_6} \frac{dx_6}{x_6} \mathrm{Tr}[w^{L_0} F_0 W(k_1, \rho_1) \cdots$$
$$\times W(k_5, \rho_5) F_0 W(k_6, \rho_6)\overline{\Gamma}]. \tag{10.A.5}$$

In the term W_P^1, $x_6 \equiv y_6$, while in the term W_P^2, $x_1 \equiv y_1$. In both W_P^1 and W_P^2, $w = x_1 x_2 \ldots x_6$ and $\rho_r = x_1 \ldots x_r$. Note, in particular, that in W_P^1

$$\ln w = \ln \rho_5 - \eta_6, \tag{10.A.6}$$

while in W_P^2

$$\nu_1 = \frac{\ln \rho_1}{\ln w} = \frac{-\eta_1}{\ln w}. \tag{10.A.7}$$

The Dirac trace is evaluated in the same manner as the steps leading to (10.2.16), giving terms that are similar in form to the first term of

(10.2.16). W_P^1 contains the integral

$$\int d^{10}p \epsilon_{\mu_1\ldots\mu_5\nu_1\ldots\nu_5} \zeta_1^{\mu_1} \ldots \zeta_5^{\mu_5} p^{\nu_1} k_1^{\nu_2} \ldots k_4^{\nu_4}$$
$$\times \text{Tr}\left(w^{L_0} V_0(k_1,\rho_1)\ldots V_0(k_6,\rho_6)\Gamma_d\right), \tag{10.A.8}$$

while W_P^2 contains

$$\int d^{10}p \; \epsilon_{\mu_1\ldots\mu_5\nu_1\ldots\nu_5} \zeta_1^{\mu_1} \ldots \zeta_5^{\mu_5} (p-k_1)^{\nu_1} k_2^{\nu_2} \ldots k_5^{\nu_5}$$
$$\times \text{Tr}\left(w^{L_0} V_0(k_1,\rho_1)\ldots V_0(k_6,\rho_6)\Gamma_d\right). \tag{10.A.9}$$

In the context of (10.2.16), integrals of this type were combined with the Pauli–Villars regulator terms so that the $\eta_r \to 0$ limit could be taken and a naive shift of the loop momentum led to the vanishing of each term. Here, it is important to consider the loop integral before the limit is taken. In both the terms the p integration involves the factors

$$\frac{1}{2}\sum_{r=1}^{6} p_r^2 \ln x_r = \frac{1}{2}p^2 \ln w + p_\mu \sum_{r=1}^{6} k_r^\mu \ln \rho_r + \frac{1}{2}\sum_{r<s} k_r \cdot k_s \ln(\rho_s/\rho_r). \tag{10.A.10}$$

This means that the factor of p^ν in (10.A.8) and (10.A.9) can be replaced by a term linear in $\partial/\partial p^\nu$, i.e.,

$$p^\nu \to \frac{1}{\ln w} \frac{\partial}{\partial p_\nu} - \sum_{r=1}^{6} k_r^\nu \frac{\ln \rho_r}{\ln w}. \tag{10.A.11}$$

After integrating by parts, only the second term contributes to either (10.A.8) or (10.A.9). Using momentum conservation to eliminate k_6^ν, the terms that survive contraction with the ϵ tensor in either case have the form

$$\epsilon(\zeta,k)\left(\sum_{r=1}^{6} \eta_r/\ln w\right)\int d^{10}p \text{Tr}\left(w^{L_0} V_0(\rho_1)\ldots V_0(\rho_6)\overline{\Gamma}\right). \tag{10.A.12}$$

In the $\eta_r \to 0$ limit this expression is dominated by the region of integration near $w = 1$.

The term W_P^1 can now be written as

$$W_P^1 \sim \epsilon(\zeta, k) \int_0^{\eta_1} \frac{dx_1}{x_1} \ldots \int_0^{\eta_5} \frac{dx_5}{x_5} \left(\sum_{r=1}^6 \eta_r / \ln w \right) \tag{10.A.13}$$

$$\times \int d^{10}p \, \mathrm{Tr} \left(w^{L_0} V_0(\rho_1) \ldots V_0(\rho_6) \Gamma_d \right).$$

In order to isolate the region near $w = 1$, it is convenient to change variables once again to $\nu_r = \ln \rho_r / \ln w$ ($r = 1, \ldots, 5$) and $q = \exp(2\pi^2 / \ln w)$. Note that in this case q is not independent of ν_5, by virtue of (10.A.6), which implies

$$\nu_5 = \frac{\ln \rho_5}{\ln w} = \frac{\ln \rho_5}{\ln \rho_5 - \eta_6}. \tag{10.A.14}$$

so that

$$q = \exp[-2\pi^2 (1 - \nu_5) / \eta_6]. \tag{10.A.15}$$

This vanishes exponentially as $\eta_6 \to 0$ when $1 - \nu_5$ is finite. The change of variables gives

$$\frac{d\rho_r}{\rho_r} = \ln w \, d\nu_r, \tag{10.A.16}$$

for $r = 1, 2, 3, 4$ while for $r = 5$

$$\frac{d\rho_5}{\rho_5} = \frac{1}{\eta_6} \ln^2 w \, d\nu_5, \tag{10.A.17}$$

so that the integration measure in (10.A.13) is given by

$$\prod_{r=1}^5 \frac{d\rho_r}{\rho_r} = \frac{1}{\eta_6} \ln^6 w \prod_{r=1}^5 d\nu_r. \tag{10.A.18}$$

The trace in (10.A.13) is expressed in terms of the transformed variables by (10.2.30). Notice, in particular, that all the powers of $\ln w$ cancel out and the result is proportional to $\sum_{r=1}^6 \eta_r / \eta_6$. Since q vanishes exponentially fast in the limit $\eta_r \to 0$, only the term with $N = 0$ contributes to the result, as before, giving

$$W_P^1 \sim \epsilon(\zeta, k) \frac{\sum_{r=1}^6 \eta_r}{\eta_6} \int_{R_1} \prod_{r=1}^5 d\nu_r \langle 0 | V_0(z_1) \ldots V_0(z_6) | 0 \rangle. \tag{10.A.19}$$

The integration region R_1 is discussed below. The expression for W_P^2 is the same as W_P^1 but with η_1, given by (10.A.7), replacing η_6 in the denominator. The integration region R_2 is also different.

A symmetric regulator choice sets all the η_r's equal to a common value η before taking the limit $\eta \to 0$. In this case the region of integration in W_P^1 is given in terms of the ρ_r variables by

$$\rho_6 = \rho_5 - \eta \leq \rho_4 - 2\eta \ldots \leq \rho_1 - 5\eta \leq -6\eta. \qquad (10.A.20)$$

Using $\nu_r = \ln \rho_r / \ln w$ and (10.A.14), it follows that R_1 is given by

$$1 + \nu_{r-1} \leq \nu_5 + \nu_r \qquad \text{for} \quad r = 1, \ldots 6, \qquad (10.A.21)$$

where $\nu_6 = 1$ and $\nu_0 \equiv 0$. Similarly, the region R_2 appropriate to the expression for W_P^2, using (10.A.7), is given by

$$\nu_1 + \nu_{i-1} \leq \nu_i. \qquad (10.A.22)$$

The two regions R_1 and R_2 are nonoverlapping, since $\nu_1 + \nu_5 \geq 1$ in R_1, whereas $\nu_1 + \nu_5 \leq 1$ in R_2. The integrand is the same as obtained by the Pauli–Villars method, but the total region formed by adding W_P^1 and W_P^2 is smaller than the one that arose in the Pauli–Villars method. The two methods give exactly the same criteria for anomaly cancellations, however.

The result is especially simple if the η_r's with $r \leq 5$ are set to zero before η_6 is. Then the procedure is not symmetric in its treatment of the external states. In that case $W_P^2 = 0$ and the result for W_P^1 is identical to the complete anomaly obtained by the Pauli–Villars method in (10.2.34) (which also treated the states asymmetrically). Similarly, setting all the η_r's to zero before η_1 gives $W_P^1 = 0$, and then the total anomaly comes from W_P^2, which is equal to W_P in (10.2.34). Each of these special examples is equivalent to regulating just one of the propagators adjacent to vertex #6, the vertex with the longitudinal state attached.

11. Functional Methods in the Light-Cone Gauge

Perturbative calculations based on the operator formalism described in chapters 7 – 10 can be used when at least one of the strings at each interaction vertex is a physical on-shell state. That approach is effective for calculations of tree and one-loop amplitudes but is not applicable for multiloop amplitudes, since these necessarily involve at least one vertex coupling three internal (and therefore off-shell) strings.

In principle, the systematic derivation of rules for calculating arbitrary string-theory diagrams should follow from a second-quantized string field theory. This has been developed in the light-cone gauge for both bosonic strings and superstrings. A major effort is currently underway to develop covariant gauge-invariant action principles, as well. Such a formulation might provide a deeper understanding of the geometric significance of string theories.

We will not develop string field theory in this book, but in this chapter we do develop a 'first-quantized' approach to the Feynman rules of light-cone string field theory. Apart from exhibiting at least some of the ingredients of string field theory, this approach has several other virtues. The equivalence of the first-quantized Feynman rules to world-sheet path integrals – the only approach to multiloop amplitudes that we have mentioned hitherto – is important in understanding the unitarity of the latter. Also, while we have extensively discussed conformal invariance and conformal mappings in previous chapters, the explicit and detailed applications of these concepts in this chapter should shed a new light on them. Most of the explicit calculations in this chapter will be restricted to the bosonic theory. In the last section we will describe the extension of these ideas to the space-time supersymmetric light-cone formalism.

11.1 The String Path Integral

Just as with ordinary point-particle theories, it is possible to anticipate the structure of the rules for calculating diagrams implied by the field theory within the framework of the first-quantized theory. The basic idea is to describe string scattering amplitudes by an extension of the

Feynman path-integral approach to point-particle quantum mechanics. In this approach a string-theory scattering amplitude is described as a sum over all connected world surfaces joining the incoming and outgoing strings weighted by $\exp iS$. The surface may have handles attached or holes cut out (in the case of theories with open strings), corresponding to the loop corrections of the second-quantized field theory.

11.1.1 The Analog Model

Let us begin by sketching the goal of this chapter. In §1.4 we described a general ansatz expressing string scattering amplitudes in terms of path integrals on Riemann surfaces. The integrals can be reduced (by Gaussian integration) to integrals over a finite number of parameters (moduli of the surface and positions at which vertex operators are inserted). According to this ansatz, for instance, the amplitude for the scattering of M ground-state tachyons (in the bosonic string theory) at an arbitrary order in perturbation theory (*i.e.*, with a world sheet that has an arbitrary number of handles attached or holes cut out) is given by

$$A(1, 2, \ldots, M) = \int d\mu(z_1, z_2, \ldots, z_M, \gamma)$$

$$\times \exp\left(-\frac{1}{2\pi T} \sum_{r > s} p_r \cdot p_s N(z_r; z_s, \gamma)\right). \qquad (11.1.1)$$

Here the z_i are the positions of vertex operators, and γ are moduli of the surface. The measure $d\mu$ is, according to our analysis in chapters 1 and 3, to be determined by calculating the functional determinants of the string coordinates X^μ and the conformal ghosts b, c. Also, $N(z_r; z_s, \gamma)$ is the Green function of the Laplace equation on the world sheet with Neumann boundary conditions (the normal derivative of N vanishes on the world-sheet boundary, if any).[*] For external open strings the z_i are points on the boundary of the world sheet, while for closed strings the external particles are attached to the interior of the world sheet. In its historical origins, (11.1.1) (originally with an imprecise prescription for determining the measure $d\mu$) was known as the analog model.

Equation (11.1.1) is a beautiful formula, which summarizes much of what we currently know about string theory. However, there are a few peculiar features. It is not evident why (11.1.1) obeys unitarity. It is not

---

[*] This boundary condition expresses the result of §2.1.3, namely that the σ derivative of $X^\mu(\sigma, \tau)$ should vanish at the ends $\sigma = 0, \pi$ of the string.

obvious what the connection is, if any, of (11.1.1) with standard Feynman-diagram prescriptions for obtaining scattering amplitudes.

Our goal in this chapter is to shed some light on these questions by deriving (11.1.1) from a standard quantum mechanical formalism, involving the use of a light-cone gauge Hamiltonian. The techniques in question were pioneered by Mandelstam. They provide a well-defined quantum-mechanical framework for perturbative calculations. Rules for calculating perturbation-theory diagrams in this gauge can be derived from a hermitian quantum-mechanical Hamiltonian, so unitarity is guaranteed.

This method treats the string as a quantum-mechanical system with the interactions between strings incorporated in a manner that ensures unitarity. The integration variables in (11.1.1) (the positions of vertex operator insertions and the moduli of the surface) are identified with physical parameters such as the times of interaction and the fraction of the total p^+ momentum carried by intermediate string states. A drawback of the method is certainly that Lorentz covariance is not manifest; proving it will be one of our aims. Another drawback is that the origin of gauge invariance and general covariance (or their stringy generalizations) is not elucidated; a proper elucidation of these concepts remains a task for the physics of tomorrow.

For most of this chapter, we will discuss the bosonic string theory, sketching analogous results for supersymmetric strings in §11.7.

11.1.2 The Free String Propagator

In order to illustrate some of the subtle issues involved in the functional formalism it is useful to consider the free string propagator in some detail. Following Feynman's treatment of quantum mechanics the functional integral for a free propagator defining the amplitude for a string in an arbitrary state $|X_1(\sigma)\rangle$ at τ_1 to evolve into an arbitrary state $|X_2(\sigma)\rangle$, at τ_2 is given by the functional integral

$$G(X_1(\sigma), \tau_1; X_2(\sigma), \tau_2) = \langle X_2| \int \mathcal{D}X^i(\sigma, \tau) \exp(i \int_{\tau_1}^{\tau_2} d\tau \int d\sigma \mathcal{L})|X_1\rangle,$$

(11.1.2)

where $X^i(\sigma, \tau_A) \equiv X^i_A(\sigma)$, $|X_A\rangle \equiv |X_A(\sigma)\rangle$ $(A = 1, 2)$ and $\int \mathcal{D}X^i(\sigma, \tau)$ denotes the integral over all surfaces joining the initial and final string configurations. $\mathcal{L}$ is the light-cone gauge Lagrangian density of the bosonic string theory; we are familiar with it from chapter 2. It is often convenient for calculational purposes to let $\tau \to -i\tau$ so that the parameters can be regarded as forming a single complex variable (as was done in chapters 1 and 3)

$$\rho = \tau + i\sigma. \tag{11.1.3}$$

This is the convention in the rest of this chapter. We also scale the σ parameter so that it spans the range

$$0 \le \sigma \le 2\pi|p^+|, \tag{11.1.4}$$

with a corresponding scaling of the τ parameter so that

$$X^+ = \tau/2\pi T, \tag{11.1.5}$$

(which means replacing $2|p^+|\sigma$ by the new σ and $2|p^+|\tau$ by the new τ). As before, the string tension is often set to $T = 1/\pi$ for convenience (*i.e.*, $\alpha' = 1/2$) and p^+ is taken to be positive for incoming strings. With these conventions

$$\exp\left(i\int d\tau d\sigma \mathcal{L}\right) \to \exp\left(-\frac{1}{2\pi}\int d\tau \int_0^{2\pi|p^+|} d\sigma \partial_\alpha X^i \partial^\alpha X^i\right). \tag{11.1.6}$$

The states are normalized so that

$$\langle X_2|X_1\rangle \equiv \int DX_1^i G(X_1(\sigma), \tau; X_2(\sigma), \tau) = 1. \tag{11.1.7}$$

Recall that physical closed-string states do not depend on the origin of the sigma coordinate. Therefore a physical closed-string state, $|X(\sigma)\rangle_{\text{phys}}$, has a normalization that is expressed in terms of unconstrained states by projecting onto the subspace of states that satisfy the constraint

$$|X(\sigma)\rangle_{\text{phys}} = \frac{1}{2\pi|p^+|}\int_0^{2\pi|p^+|} d\sigma_0 |X(\sigma + \sigma_0)\rangle. \tag{11.1.8}$$

In this expression the state with $X(\sigma)$ shifted to $X(\sigma + \sigma_0)$ is generated from the unshifted state by

$$|X(\sigma + \sigma_0)\rangle = \exp\left(\frac{i\sigma_0}{|p^+|}\int_0^{2\pi|p^+|} d\sigma X'^i(\sigma)\frac{\partial}{\partial X^i(\sigma)}\right)|X(\sigma)\rangle \tag{11.1.9}$$

$$= \exp\left(i(L_0 - \tilde{L}_0)\frac{\sigma_0}{|p^+|}\right)|X(\sigma)\rangle.$$

The effect of the σ_0 integral in (11.1.8) is to impose the usual subsidiary constraint $L_0 = \tilde{L}_0$ on the physical space of closed-string states. As a result of (11.1.8) the normalization of a physical state is seen to satisfy

$$\text{phys}\langle \mathbf{X}|\mathbf{X}\rangle_{\text{phys}} = \frac{1}{2\pi|p^+|} \int\limits_{0}^{2\pi|p^+|} d\sigma_0 \langle \mathbf{X}(\sigma)|\mathbf{X}(\sigma+\sigma_0)\rangle. \qquad (11.1.10)$$

In performing closed-string calculations the external states are taken to be unprojected and the answer is integrated over the σ variables for each of the external strings. The result is therefore a factor of $1/\sqrt{2\pi p_r^+}$ in the normalization of each external closed-string state of momentum p_r. Furthermore, in the calculation of a scattering amplitude the integration over σ for each external state introduces an additional factor of $2\pi p_r^+$ for each closed string giving a net factor of $\sqrt{2\pi p_r^+}$, which must be kept track of in obtaining the expression for the covariant closed-string scattering amplitudes.

The functional integral has to be defined in a manner that introduces a regulator to take care of the divergences associated with the short-wavelength modes. These infinities can then be absorbed into the definition of the coupling constant g and the phase of the scattering amplitude. To be more explicit consider the propagator in the form

$$G(\mathbf{X}_1(\sigma), \tau_1; \mathbf{X}_2(\sigma), \tau_2) = \langle \mathbf{X}_2| \exp[-P^- X^+]|\mathbf{X}_1\rangle, \qquad (11.1.11)$$

where $\hat{\tau} = \tau_2 - \tau_1 \geq 0$. The light-cone Hamiltonian P^-, defined in §2.3.1, is conjugate to $X^+ = \hat{\tau}/2$ and is represented in position space by the expression

$$P^- = \pi \int\limits_{0}^{2\pi|p^+|} \left\{ \mathbf{P}^2(\sigma) + \frac{1}{\pi^2}\left(\frac{\partial \mathbf{X}(\sigma)}{\partial \sigma}\right)^2 \right\} d\sigma, \qquad (11.1.12)$$

where $\mathbf{P}(\sigma) = \dot{\mathbf{X}}(\sigma)/\pi = -i\delta/\delta\mathbf{X}(\sigma)$. This is just a superposition of an infinite number of harmonic oscillator Hamiltonians for the normal-mode coordinates.

The string wave function Ψ, satisfying

$$(P^- + \frac{\partial}{\partial X^+})\Psi = 0, \qquad (11.1.13)$$

is just the product of an infinite set of harmonic oscillator wave functions. Likewise, the propagator is just the product of the Green functions for an infinite number of harmonic oscillators, where the Green function for a single harmonic oscillator for a normal mode x with frequency ω has the form (after a 'Wick' rotation to imaginary time)

$$G_\omega(x_1, 0; x_2, t) = \left(\frac{\omega}{2\pi \sinh \omega t}\right)^{1/2} \exp\Big(-\tfrac{1}{2}\omega(x_1^2 + x_2^2) \coth \omega t$$

$$+ \omega x_1 x_2 \operatorname{csch} \omega t \Big).$$

(11.1.14)

In this expression x_1 and x_2 are the initial and final values of x, and $G(x, t; x', t')$ satisfies the equation

$$\left(H + \frac{\partial}{\partial t}\right) G(x, t; x', t') = \delta(t - t')\delta(x - x')$$

(11.1.15)

with

$$H = -\frac{1}{2}\left(\frac{\partial}{\partial x}\right)^2 + \frac{1}{2}\omega^2 x^2.$$

(11.1.16)

$G(x_1, 0; x_2, t)$ is normalized to equal $\delta(x_1 - x_2)$ at $t = 0$, which corresponds to the normalization of the states in (11.1.7). The x-dependent factor in (11.1.14) is just the exponential of the classical action while the quantum-mechanical information resides in the x-independent factor.

The string-theory Green function corresponds to an infinite number of oscillators with a linear frequency spectrum. The product of the x-independent terms diverges due to the factor $(\prod_m \omega_m)^{1/2}$ (where $\omega_m \sim m$). We now describe a method, due to Giles and Thorn, of regulating this divergence by means of a high-frequency cutoff.

11.1.3 A Lattice Cutoff

In order to analyze the divergences carefully, a cutoff of the high frequencies can be introduced by replacing the continuous parameter σ by a one-dimensional lattice with lattice spacing a. (An alternative procedure would be to introduce a lattice in both the σ and the τ directions and treat them symmetrically. It leads to the same physical results.) We expect to find that the physical string excitations are those with wavelengths much larger than a, so that the string theory emerges in the continuum limit. The coordinates and the momenta at the lattice sites are defined by

$$X_I^i = X^i(Ia), \qquad I = 0, 1, \ldots, M,$$

(11.1.17)

and

$$P_I^i \equiv -i\frac{\partial}{\partial X_I^i} = aP^i(Ia), \qquad I = 0, 1, \ldots, M,$$

(11.1.18)

where the length of the string in σ is given by

$$2\pi p^+ = Ma. \tag{11.1.19}$$

The closed-string periodic boundary conditions require $\mathbf{X}_M = \mathbf{X}_0$, while for open strings the condition that the derivative of $\mathbf{X}$ vanish at the endpoints becomes $\mathbf{X}_0 = \mathbf{X}_{-1}$ and $\mathbf{X}_M = \mathbf{X}_{M-1}$. Integrals over σ are replaced by discrete lattice sums,

$$\int_0^{2\pi p^+} d\sigma \rightarrow a \sum_{I=0}^M, \tag{11.1.20}$$

and $\partial X(\sigma)/\partial\sigma \rightarrow (X_{I+1} - X_I)/a$. The lattice expression for the Hamiltonian is therefore given by

$$P^- = \frac{\pi}{a} \sum_{I=0}^{M-1} \left\{ \mathbf{P}_I^2 + \frac{1}{\pi^2}(\mathbf{X}_{I+1} - \mathbf{X}_I)^2 \right\} \tag{11.1.21}$$

in place of (11.1.12). This system of M coupled harmonic oscillators can be solved by transforming to a basis of normal modes.

Starting with closed strings the coordinates $\mathbf{X}_I$ and $\mathbf{P}_I$ can be written in terms of their modes $\hat{X}_m^i$ and $\hat{P}_m^i \equiv -i\partial/\partial\hat{X}_{-m}^i$ as

$$\mathbf{X}_I = \mathbf{x} + \sqrt{\frac{\pi}{M}} \sum_{m=1}^{(M-1)/2} \left(\hat{\mathbf{X}}_m e^{2\pi i mI/M} + \hat{\mathbf{X}}_{-m} e^{-2\pi i mI/M} \right)$$

$$\equiv \mathbf{x} + i\frac{\pi}{M} \sum_{m=1}^{(M-1)/2} \frac{1}{\omega_m^c} \left((\underline{\alpha}_m - \underline{\tilde{\alpha}}_{-m}) e^{2\pi i mI/M} \right.$$

$$\left. + (\underline{\tilde{\alpha}}_m - \underline{\alpha}_{-m}) e^{-2\pi i mI/M} \right), \tag{11.1.22}$$

and

$$\mathbf{P}_I = \frac{\mathbf{p}}{M} + \frac{1}{\sqrt{\pi M}} \sum_{m=1}^{(M-1)/2} \left(\hat{\mathbf{P}}_m e^{2\pi i mI/M} + \hat{\mathbf{P}}_{-m} e^{-2\pi i mI/M} \right)$$

$$\equiv \frac{\mathbf{p}}{M} + \frac{1}{M} \sum_{m=1}^{(M-1)/2} \left((\underline{\alpha}_m + \underline{\tilde{\alpha}}_{-m}) e^{2\pi i mI/M} \right.$$

$$\left. + (\underline{\tilde{\alpha}}_m + \underline{\alpha}_{-m}) e^{-2\pi i mI/M} \right), \tag{11.1.23}$$

where M is assumed to be odd. The closed-string frequencies are defined by

$$\omega_m^c = 2 \sin \frac{\pi m}{M}. \tag{11.1.24}$$

Reality of the coordinates requires that $\hat{\mathbf{X}}_{-m} = \hat{\mathbf{X}}_m^*$ and $\hat{\mathbf{P}}_{-m} = \hat{\mathbf{P}}_m^*$. The operators $\underline{\alpha}_m$ and $\underline{\tilde{\alpha}}_m$ defined by these equations satisfy harmonic oscillator commutation relations (appropriately normalized to reduce to the continuum ones in the limit $m/M \to 0$),

$$[\alpha_m^i, \alpha_n^j] = [\tilde{\alpha}_m^i, \tilde{\alpha}_n^j] = \frac{\omega_m^c M}{2\pi} \delta_{m+n} \delta^{ij}. \tag{11.1.25}$$

In terms of the normal-mode variables the Hamiltonian can be written in the form (using $Ma = 2\pi p^+$)

$$P^- = \frac{\mathbf{p}^2}{2p^+} + \frac{2}{a} \sum_{m=1}^{(M-1)/2} \left(|\hat{\mathbf{P}}_m|^2 + (\omega_m^c)^2 |\hat{\mathbf{X}}_m|^2 \right)$$

$$\equiv \frac{\mathbf{p}^2}{2p^+} + \frac{2}{p^+} \sum_{m=1}^{(M-1)/2} \left(\underline{\alpha}_{-m} \cdot \underline{\alpha}_m + \underline{\tilde{\alpha}}_{-m} \cdot \underline{\tilde{\alpha}}_m \right) \tag{11.1.26}$$

$$+ \frac{2(D-2)}{a} \sum_{m=1}^{(M-1)/2} \omega_m^c,$$

where the last term arises from the fact that the α's have been normal ordered (*i.e.*, the raising operators are to the left of the lowering operators). This term can be approximated for large M by using

$$\frac{2}{a} \sum_{m=1}^{(M-1)/2} \omega_m^c = \frac{2}{a} \cot(\pi/2M) \sim \frac{4M}{\pi a} - \frac{1}{6p^+} + O(1/M). \tag{11.1.27}$$

The divergent term is proportional to the number of lattice sites. When this discussion is generalized to scattering amplitudes this term is the same for any diagram contributing to a given process, since the total p^+, and hence the total number of sites, is fixed. As a result the factor $\exp(-P^-\hat{\tau}/2)$ in the amplitude contains the non-Lorentz-invariant factor $\exp[-2(D-2)M\hat{\tau}/\pi a]$. This is a physically irrelevant overall phase factor (when continued back to real time), where the phase is proportional to the area of the world sheet. It can can be eliminated by subtracting a term $4(D-2)/a\pi$ from the expression for the Hamiltonian at each lattice site, which is a *local* subtraction (*i.e.*, it is independent of M and of p^+). The nonleading term, $-1/6p^+$, cannot be changed by a local subtraction and therefore its value has absolute meaning. As a result, (11.1.26) defines the physical mass spectrum for the closed string. For example, the ground state (mass)^2 is

$$2p^+ P^- - \mathbf{p}^2 = -\frac{(D-2)}{3}, \tag{11.1.28}$$

(where P^- is now the subtracted Hamiltonian). Evidently this procedure gives the same expression for the tachyon mass as the ζ-function method of regularization described in chapter 2.

The excited modes are obtained as usual by applying the raising operators $\underline{\alpha}_{-m}$ and $\underline{\tilde{\alpha}}_{-m}$ to the ground state. These states must satisfy the discrete analog of the constraint in (11.1.8). This means that states must be eigenstates of the operator U that shifts $\mathbf{X}_I$ to $\mathbf{X}_{I+J}$ and $\partial/\partial \mathbf{X}_I$ to $\partial/\partial \mathbf{X}_{I+J}$, *i.e.*,

$$U^{-1} \alpha_m^i U = \alpha_m^i \exp\{2\pi i m J/M\}$$

$$U^{-1} \tilde{\alpha}_m^i U = \tilde{\alpha}_m^i \exp\{-2\pi i m J/M\}, \tag{11.1.29}$$

where

$$U = \exp\left\{ \frac{2\pi i J}{M} \sum_{m=1}^{(M-1)/2} \frac{2\pi m}{\omega_m^c M} \left(\alpha_{-m}^i \alpha_m^i - \tilde{\alpha}_{-m}^i \tilde{\alpha}_m^i \right) \right\}. \tag{11.1.30}$$

This reduces to (11.1.9) in the continuum limit.

The analysis for open strings is slightly different because of the boundary conditions. Open-string coordinates are expanded in a series of functions, $\cos[m\pi(I + \frac{1}{2})/M]$, where $1 \le m \le M - 1$. The frequencies of the open-string modes are given by

$$\omega_m^o = 2 \sin \frac{m\pi}{2M}, \tag{11.1.31}$$

and the expression for the Hamiltonian, defined by (11.1.21), is similar to (11.1.26) but the normal-ordering term is now $(D-2) \sum \omega_m^o / a$, where

$$\frac{1}{a} \sum_1^{M-1} \omega_m^o = \frac{1}{a} \left(\cot \frac{\pi}{4M} - 1 \right) \sim \frac{4M}{\pi a} - \frac{1}{a} - \frac{1}{24p^+} + O(1/M). \tag{11.1.32}$$

The resulting expression for P^- is given by

$$P^- \sim \frac{\mathbf{p}^2}{2p^+} + \frac{1}{p^+} \sum_{m=1}^{M-1} \underline{\alpha}_{-m} \cdot \underline{\alpha}_m$$

$$+ (D-2)\left(\frac{4M}{\pi a} - \frac{1}{a} - \frac{1}{24p^+}\right). \tag{11.1.33}$$

In addition to the term proportional to M there is another divergent term $(-(D-2)/a)$ that gives a non-Lorentz-invariant contribution to scattering amplitudes. This is associated with the free endpoints of the string and its contribution can be eliminated by a local modification of the Hamiltonian. This is achieved by adding a term $(D-2)/2a$ to the Hamiltonian at each string endpoint. The consistency of this procedure, suggested by the propagator calculation, must be checked for multistring amplitudes. The open-string spectrum of states coincides with that obtained in previous chapters, since the ground state has a (mass)^2 given by $-(D-2)/12$.

The string propagator is well-defined in the presence of the lattice regulator. It is given by the product of Green functions for the individual oscillators (where each complex mode $\hat{\mathbf{X}}_m^i$ describes two independent oscillators – corresponding to its real and imaginary part). Comparing the normalizations in (11.1.16) with (11.1.26), we see that each complex oscillator of frequency ω_m gives a factor $G_{\omega_m}(\hat{\mathbf{X}}_{1m}, 0; \hat{\mathbf{X}}_{2m}, \hat{\tau})$ of the form in (11.1.14) but with t replaced by $\hat{\tau}/a$, x_1 and x_2 replaced by the real or imaginary parts of $\hat{\mathbf{X}}_{1m}$ and $\hat{\mathbf{X}}_{2m}$ ($m \neq 0$) and an overall factor of 2. The zero-frequency Green function becomes $(a/2\pi\hat{\tau})^{(D-2)/2} \exp[-(x_1 - x_2)^2 p^+/\hat{\tau}]$ so the closed-string propagator has the form

$$G^c(\mathbf{X}_1, 0; \mathbf{X}_2, \hat{\tau}) = G^c(0, 0; 0, \hat{\tau}) \exp\left(-\frac{(x_1 - x_2)^2 p^+}{\hat{\tau}}\right)$$

$$\times \prod_{m=1}^{(M-1)/2} \exp\left(-\left(|\hat{\mathbf{X}}_{1m}|^2 + |\hat{\mathbf{X}}_{2m}|^2\right)\omega_m \coth(\omega_m\hat{\tau}/a)\right.$$

$$\left. + 2\mathrm{Re}(\hat{\mathbf{X}}_{1m} \cdot \hat{\mathbf{X}}_{2m})\omega_m \mathrm{csch}(\omega_m\hat{\tau}/a)\right). \tag{11.1.34}$$

The $\mathbf{X}$-independent factor $G^c(0, 0; 0, \hat{\tau})$ arises in the functional integral (using the Dirichlet conditions $\mathbf{X}_1 = \mathbf{X}_2 = 0$ at the initial and final times) from $(\det\Delta)^{-1/2}$, where Δ is the Laplace operator on the world sheet. Using the expression for the oscillator Green function (11.1.14), we see explicitly that this is

$$G^c(0,0;0,\hat{\tau}) = 2\left\{\left(\frac{a}{2\pi\hat{\tau}}\right)^{1/2}\prod_{m=1}^{(M-1)/2}\left(\frac{\omega_m}{2\pi\sinh\omega_m\hat{\tau}/a}\right)\right\}^{D-2}.$$

$$(11.1.35)$$

11.1.4 The Continuum Limit

The closed-string frequencies ω_m^c given by (11.1.24) satisfy

$$\prod_{m=1}^{(M-1)/2}\omega_m^c = \prod_{m=1}^{(M-1)/2}2\sin\frac{\pi m}{M} = \sqrt{M}, \qquad (11.1.36)$$

which leads to the expression for closed strings (using $\omega_m^c = \omega_{M-m}^c$)

$$G^c(0,0;0,\hat{\tau}) = \left\{\pi^{-(M-1)}\left(\frac{p^+}{\hat{\tau}}\right)\prod_{m=1}^{M-1}(1-\exp[-2\omega_m^c\hat{\tau}/a])^{-1}\right.$$
$$\left.\times\exp\left[-\hat{\tau}\sum_{m=1}^{M-1}\omega_m^c/a\right]\right\}^{(D-2)/2}$$

$$(11.1.37)$$

$$\sim\left\{(\pi)^{-(M-1)}\left(\frac{p^+}{\hat{\tau}}\right)\prod_{m=1}^{M-1}[1-\exp(-2m\hat{\tau}/p^+)]\right.$$
$$\left.\times\exp(-4M\hat{\tau}/\pi a+\hat{\tau}/6p^+)\right\}^{(D-2)/2}.$$

The divergent term in the exponent is just the one discussed earlier that can be absorbed into a redefinition of the Hamiltonian. The only other M dependence comes from the factor $(\pi)^{(2-D)M/2}$, which can be absorbed into a local redefinition of the external wave functions. The resulting expression then has a smooth continuum limit.

The subsidiary constraint on closed-string states (which are eigenstates of the operator U in (11.1.30)) demands that the physical propagator G_{phys}^c is constructed from G^c by symmetrization, *i.e.*,

$$G_{\text{phys}}^c = \frac{1}{M}\sum_{J=0}^{M-1}G^c(\{\mathbf{X}_{1,I}\},0;\{\mathbf{X}_{2,I+J}\},\hat{\tau}). \qquad (11.1.38)$$

In calculating a scattering amplitude the sum is to be replaced by an integral over σ so that the normalization of each external state must include the factor $(aM)^{-1/2} = (2\pi p^+)^{-1/2}$ discussed earlier.

In order to understand how the spectrum of states is coded into the propagator we may consider the simple example in which the initial and final states are 'pointlike', *i.e.*, $\hat{X}_{1m} = \hat{X}_{2m} = 0$ for all $m \neq 0$. These states automatically satisfy the subsidary condition $L_0 = \tilde{L}_0$, since they have the property that they are annihilated by $\alpha^i_m - \tilde{\alpha}^i_{-m}$, as is evident from (11.1.22). Upon substituting in (11.1.34) and converting the propagator into a function of the transverse momenta by Fourier transforming with respect to $\mathbf{x}_1$ and $\mathbf{x}_2$, the propagator has the form

$$\tilde{G}^c \sim e^{-\mathbf{p}^2/4p^+} G^c(\mathbf{0}, 0; \mathbf{0}, 0). \tag{11.1.39}$$

Equation (11.1.37) shows that $\tilde{G}^c$ can be expanded in an infinite series of exponential terms, $\exp(-2N\hat{\tau}/p^+)$ (where N is a nonnegative integer) so that all the dependence on the transverse momentum occurs in the combination $\exp\left(-(\mathbf{p}^2 + 8N - 8)\hat{\tau}/4p^+\right)$. This may be converted from a dependence on τ into a dependence on p^- by a Laplace transform

$$\int_0^\infty d\hat{\tau} e^{p^-\hat{\tau}/2} \tilde{G}^c(\mathbf{P}_1, 0; \mathbf{P}_2, \hat{\tau}). \tag{11.1.40}$$

Changing the integration variable to $\hat{\tau}/2p^+$ leads to a function of p^2 (where p^μ is the covariant D-dimensional momentum) multiplied by $2p^+$. This function has poles corresponding to all the states of the closed string that couple to the point-like external states. The masses of these states are given by the pole positions at $(\text{mass})^2 = -p^2 = -\mathbf{p}^2 + 2p^+p^- = 8N - 8$. In evaluating scattering amplitudes the factor of $2p^+$ in the numerator of any internal propagator is canceled by compensating factors of $(2p^+)^{-1/2}$ for each particle coupling to a vertex. This leaves a factor of $(2p^+)^{1/2}$ for each external particle attached to a vertex, which is appropriate for the external states to be normalized in the usual relativistic manner.

The simple example considered in the previous paragraph is singular due to the point-like nature of the initial and final states. Nevertheless, it illustrates in a formal manner the basic features, which also apply to the propagator between general physical states.

The calculation of the propagator for the open string makes use of the identity

$$\prod_{m=1}^{M-1} \omega^o_m \equiv \prod_{m=1}^{M-1} 2\sin\frac{m\pi}{2M} = \sqrt{M}, \tag{11.1.41}$$

analogous to (11.1.37), to give a propagator of similar structure with an

X-independent factor

$$G^o(\mathbf{0}, 0; \mathbf{0}, \hat{\tau}) \sim \left\{ (\pi)^{-(M-1)} \left(\frac{p^+}{\hat{\tau}} \right) (M)^{-1/2} \right.$$

$$\times \exp\{-\hat{\tau}(\frac{4M}{\pi a} - \frac{1}{a} - \frac{1}{24p^+})\} \qquad (11.1.42)$$

$$\left. \times \prod_{m=1}^{M-1} [1 - \exp(-m\hat{\tau}/p^+)] \right\}^{(D-2)/2} .$$

The divergent terms in the exponent are those discussed earlier, which can be removed by a local redefinition of the Hamiltonian. In addition, there is a normalization factor of $M^{(2-D)/4}$, which cannot be absorbed into a local renormalization of the Hamiltonian. It has to be associated with a normalization factor of $M^{(2-D)/8}$ in the open-string wave functions. Since the propagator was constructed so that its $\tau = 0$ limit gave correctly normalized states, this M-dependent factor is clearly a necessary factor in obtaining correctly normalized states.

It is possible to show, at least in simple examples, that in the calculation of on-shell scattering amplitudes these factors combine with other factors in the vertices to give a finite and Lorentz-invariant continuum limit only in the critical dimension $D = 26$. Therefore, in the critical dimension there is no need for infinite wave function or coupling constant renormalizations. In fact, there appears to be no inconsistency in making such momentum-independent renormalizations, so we generally will not try to keep track of the way the infinities cancel in the amplitude calculations in this chapter. (The critical dimension certainly plays a crucial role, however, in ensuring the Lorentz invariance of the amplitudes.)

11.2 Amplitude Calculations

The study of the propagator makes it plausible that the infinities associated with the high-frequency modes of the string can be absorbed into redefinitions of the phase of the amplitude and the normalization of the wave functions and the coupling constant. In calculating scattering amplitudes we shall adopt a continuum approach to evaluating the functional integral, the success of which relies on the consistency of this treatment of the high-frequency modes.

11.2.1 Interaction Vertices

The interactions between strings are described by splitting or joining processes that occur at points on the string. This means that the interactions are local, which is an important physical ingredient in the theory. Associated with this is the condition that the string coordinates evolve continuously so that the vertex is nonzero only between states I_1 and I_2 in which the initial and final strings overlap along their entire lengths. These statements are not only physically reasonable but are an important element in obtaining results that are causal and Lorentz invariant. The vertex therefore contains a Δ functional, which is an infinite product of Dirac δ functions identifying each element of the incoming and outgoing strings. By a Fourier transform, it can also be expressed in terms of the infinite product of δ functions for the normal modes. If we denote the time of an interaction by τ_I then the string states at $\tau^1 = \tau_I - \epsilon$ and $\tau^2 = \tau_I + \epsilon$ (where ϵ is an infinitesimal time) are coupled by the interaction vertex V, which has matrix elements $\langle 2|V|1 \rangle$. In the bosonic theory there are no other factors in the vertex other than this Δ functional so that the general interaction can be written in the coordinate basis as

$$\langle X_{I_2}^i |V| X_{I_1}^i \rangle \propto \Delta[X_{I_2}^i(\sigma) - X_{I_1}^i(\sigma)], \qquad (11.2.1)$$

where $X_{I_1}^i(\sigma)$ denotes the transverse position coordinates of the strings before the interaction and $X_{I_2}^i(\sigma)$ the coordinates after the interaction. The choice for the vertex is certainly simple, but is it correct and unique? The answer to both questions is yes. One way to prove this is by a careful study of the requirements of Lorentz invariance. Rather than doing that, which is technically rather demanding, we present explicit calculations of scattering amplitudes and show that they give results with the desired behavior (including Lorentz invariance). In superstring theories the vertices also have additional factors involving operators such as a functional derivative with respect to $X^i(\sigma)$ acting on the Δ functional at the interaction point $\sigma = \sigma_I$. This distinction between superstrings and bosonic strings is analogous to the fact that in Yang–Mills theory the cubic interaction between gauge particles has a derivative in it, whereas the cubic interaction of scalar fields does not.

   The interactions between strings can be divided into two classes in a simple geometrical way according to whether the joining or splitting occurs at interior points of the strings or involves the touching of string endpoints.

   (i) The first class consists of those interactions in which two open-string endpoints touch and join or the time-reversed reaction in which an internal point on a string breaks to form two new free ends. The vertex that describes the joining of string 1 and string 2 to form string 3, for

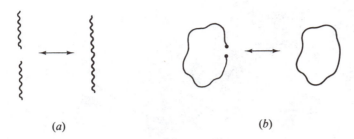

Figure 11.1. The breaking/joining string interaction: (a) The three open-string interaction. This is the string generalization of the cubic Yang–Mills interaction. (b)
The transition between an open and a closed string. This interaction couples the
gravitational sector and the Yang–Mills sector of the type I theories.

example, is given by

$$g\Delta \left[X_1^i(\sigma) + X_2^i(\sigma) - X_3^i(\sigma)\right], \qquad (11.2.2)$$

where $X_r^i(\sigma)$ denotes the coordinate of the rth string and the X_r's are
only nonzero over the appropriate range of σ (this parametrization will be
described in more detail in §11.4.5). This interaction, which is illustrated
in fig. 11.1 a, is the string generalization of the cubic Yang–Mills coupling,
which is why it is assigned a coupling constant g.

The same process of two endpoints joining is also responsible for the
interaction by which an open string turns into a closed one as illustrated
in fig. 11.1 b. Since this interaction only affects the string locally, the
consistency of the theory requires this coupling to occur with the same
strength as the cubic open-string interaction. For example, this process is
responsible for the occurrence of the closed-string poles in the nonplanar
loop calculation described in chapter 8. Without this contribution the
amplitudes would not be Lorentz invariant.

(ii) The second class consists of those interactions in which two internal
points touch so that the incoming strings instantaneously rearrange their
pieces to form the outgoing ones. An example of this is the interaction
illustrated in fig. 11.2 a in which two closed strings join to form a single
closed string or, conversely, a single closed string splits into two. This is
the string generalization of the cubic interaction between gravitons and
is therefore assigned the gravitational strength κ. This is the only interaction possible for any of the oriented closed-string theories (such as the
type II superstring theories or the heterotic theories).

Another interaction vertex generated by the touching of internal string
points is one in which two incoming open strings rearrange segments to

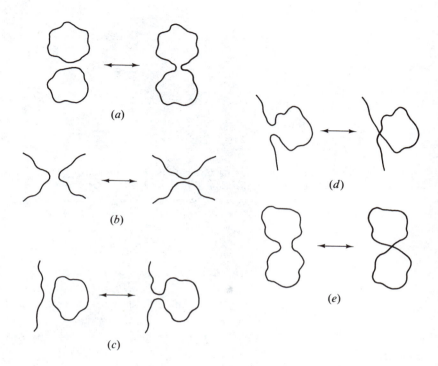

Figure 11.2. The exchange string interaction: (*a*) The interaction that joins two closed strings into one or splits one closed string into two. This is the string generalization of the gravitational interaction. It is the only interaction in the light-cone gauge formulation of oriented closed-string theories. (*b*) The interaction in which a pair of open strings form another pair of open strings. (*c*) The interaction that joins an open string to a closed string or the reverse reaction. (*d*) The interaction in which a single open string rearranges itself. (*e*) The interaction of a single closed string analogous to (*d*).

form two outgoing open strings (fig. 11.2*b*). Locally on the string this is the same interaction as in fig. 11.2*a*, and so it is also of gravitational strength. Although this is a contact interaction, it should not be confused with the contact interactions of point-particle field theory. For example, there is no contribution from this term to the process in which a single string breaks into three strings, whereas the Yang–Mills contact term contributes to any four-particle process. The contribution of this term is essential in obtaining Lorentz-invariant tree amplitudes. The requirement that the scattering amplitude be Lorentz invariant determines the strength of this interaction to be g^2, as will be apparent in the discussion in §11.3.1. This in turn is related to the gravitational coupling κ, which leads to the relationship between the gravitational and Yang–Mills coupling constants

$$\kappa \sim Tg^2. \tag{11.2.3}$$

There are three other possible interactions involving the touching of internal points on strings. One of these (fig. 11.2c) describes the joining of a closed and an open string to form a single open string. Figure 11.2d depicts the self-interaction of an open string in which two internal points touch and a segment of the string is reversed. The analogous process for a nonoriented closed string is depicted in fig. 11.2e. The interactions illustrated by both figs. 11.2c and 11.2e can only occur for theories with nonoriented closed strings. This can readily be deduced from the impossibility of attaching arrows to the strings to give them an orientation.

11.2.2 Parametrization of Scattering Processes

Let us consider amplitudes for the scattering of initial on-shell momentum eigenstates into final states that are also on-shell momentum eigenstates. The initial and final times, τ_i and τ_f, are taken to $-\infty$ and $+\infty$ at the end of the calculation.

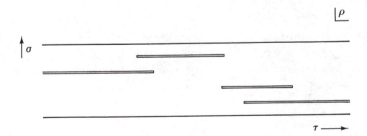

Figure 11.3. A representation of the world sheet for an open-string scattering amplitude in the ρ plane. The horizontal slits represent boundaries of the world sheet, *i.e.*, the world lines of string endpoints.

The world sheet for such a scattering process involving only open strings is conveniently represented by a strip as shown in fig. 11.3. In this parametrization the horizontal axis is τ, *i.e.*, it is proportional to the light-cone time, while the vertical axis defines the σ parametrization. The horizontal lines indicate string boundaries at which the string coordinates satisfy the condition that the normal derivative of X vanishes. Since the coordinate X on either side of a horizontal line lies on different strings, it is not continuous across these lines.

The diagram in fig. 11.3 illustrates strings joining at their endpoints and splitting to form new endpoints at the interaction 'times' τ_I. In evaluating the scattering amplitude it is necessary to integrate over these times. Horizontal lines that connect two interaction points represent loops, *i.e.*, they describe the splitting and subsequent recombination of the strings. The horizontal lines are therefore to be thought of as slits cut in the strip so that the external boundary of the strip is a continuous line that traces out the world lines of the endpoints of the external strings (where an incoming string at $\tau_i \to -\infty$ or an outgoing string at $\tau_f \to +\infty$ is considered to be a single point). The group-theory factor for an open-string diagram is incorporated by associating the charges with the boundaries of the strip (which is just the Chan–Paton method described in §6.1). Such factors are suppressed in this chapter.

By convention, the parameter σ_r on the rth string increases from the bottom to the top for incoming strings while it increases from the top to the bottom for outgoing strings. The range of the parameter σ_r is taken to be $0 \le \sigma_r \le \pi|\alpha_r|$, where α_r is defined by

$$\alpha_r = 2p_r^+, \tag{11.2.4}$$

which is positive for incoming strings and negative for outgoing strings. The advantage of this normalization of the parameters is that the total width of the strip is constant in time independent of the string interactions as a consequence of the conservation of p^+, which implies $\sum_i \alpha_i = -\sum_f \alpha_f$, where i and f refer to initial and final particles respectively. We usually want to use the single parameter σ, which has a range $0 \le \sigma \le \pi \sum_i \alpha_i$. On a given external incoming string r the parameter σ is related to σ_r by

$$\sigma = \sigma_r + \sigma_{0r}, \tag{11.2.5}$$

where σ_{0r} is the value of σ at the bottom of the string. For outgoing strings the relation is

$$\sigma = -\sigma_r + \sigma_{0r}, \tag{11.2.6}$$

where σ_{0r} now refers to the top of the outgoing string. An open-string loop correction to an open-string diagram, represented by the internal horizontal slit in the world sheet in fig. 11.3, is associated with the two τ_I variables at the ends of the slit and a σ variable (denoted by σ_L) labelling the distance of the slit from the bottom of the diagram. Thus for every loop there are three additional variables to integrate over. Loops of closed strings, which are handles attached to the world sheet, can also be represented in the $\rho = \tau + i\sigma$ plane in a more complicated manner. In this parametrization of the world sheet the metric on the surface has been chosen so that the two-dimensional curvature is everywhere flat apart from the interaction points, where the curvature (of the one-dimensional boundary for open-string world sheets or of the two-dimensional surface for closed strings) is infinite. This is the parametrization in which the integration variables have the physical interpretation described above. In evaluating the functional integral it is convenient to map the world sheet to the upper half complex plane or the whole complex plane, which makes contact with more standard parametrizations of Riemann surfaces.

11.2.3 Evaluation of the Functional Integral

A typical Green function has the structure of a succession of vertices between which strings propagate freely with propagators given by (11.1.2). The interaction times are to be integrated over their complete range. The momentum-space Green functions with off-shell external propagators attached are obtained by multiplying by $\exp[-\sum_r p_r^- \tau_r/2]$ and integrating over the times τ_r at which the external strings are created in prescribed configurations. Here p_r^- is a c number that is equal to the operator P_r^- at the residues of the poles in the propagators of the external strings. On the other hand, the connected part of an S-matrix element is obtained by multiplying the Green function by factors of $\exp[-P_r^- \tau_r/2]$ for each external particle r at the initial or final time $\tau_r (= \tau_i \to -\infty$ or $\tau_f \to \infty)$. This factor amputates the external propagators from the Green function. The connected part of the S matrix is not a relativistically covariant quantity, since it contains the flux factors needed to obtain the relativistic phase-space measure that enters into the calculation of the cross-section. In the light-cone coordinate frame the relativistic phase-space measure for an on-shell particle of momentum p^μ is $d^{D-2}\mathbf{p}\,dp^+/\alpha^{1/2}$ (with $p^- = (\mathbf{p}^2 + m^2)/\alpha$) so that the covariant amplitude A for the scattering of M scalar particles is given in terms of the S matrix by

$$\langle f|(S-1)|i\rangle = -i(2\pi)^D \delta^D(\sum_r p_r^\mu) \prod_{r=1}^{M} (\alpha_r)^{-1/2} A(1,2,\ldots,M). \quad (11.2.7)$$

Since the only effect of the vertices is to insert Δ functionals that identify coordinates just before and just after the interaction time, the result can be expressed as a single functional integral over all surfaces joining the initial strings at $\tau = \tau_i \to -\infty$ and the final strings at $\tau = \tau_f \to \infty$. Notice that the complete amplitude includes diagrams with both τ orderings of the interactions as depicted in fig. 11.3. For a given group-theory factor (*i.e.*, for a given ordering of the external particles along the string boundary) the result is a single expression, which has to be integrated over all possible values of the τ_I and σ_L variables. The fact that the integrands of diagrams with different τ_I orderings take the same form at the point where two τ_I's become equal is highly nontrivial and is only true in the critical dimension. The general expression for the L-loop scattering amplitude of M open-string particles (which has $M - 2 + 2L$ vertices) has the form

$$A(1, 2, \ldots, M) = G \int \underbrace{\prod \, d\tau_I}_{\text{vertices}} \underbrace{\prod \, d\sigma_L}_{\text{loops}}$$

$$\int \prod_{r,n,i} dP_{r,n}^i \, \Psi_r \left(P_{r,n}^i \right) W \left(P_{r,n}^i, p_r^i, \alpha_r, \tau_r \right),$$

$$(11.2.8)$$

where G is the group-theory factor appropriate for the process being described (if it involves open strings). In this expression the function W is the amplitude for scattering incoming and outgoing string states described in terms of the modes $P_{r,n}^i$ of the momenta $P_r^i(\sigma)$. They are defined by

$$P_r^i(\sigma_r) = \frac{1}{\pi \alpha_r} \left\{ p_r^i + \sum_n P_{r,n}^i \cos \frac{n \sigma_r}{\alpha_r} \right\}, \qquad (11.2.9)$$

where r labels the string and n labels the mode number. These external states are eigenstates of the energy operators P_r^-, which means that they are momentum eigenstates and states of definite occupation number. The wave functions for the external strings Ψ_r are therefore products of the oscillator eigenfunctions for each mode, *i.e.*, the wave function for a state with occupation numbers $k_{r,n}^i$, where n and i label the modes, is given by

$$\Psi_r \left(P_{r,n}^i \right) = \prod_{n=1}^{\infty} \prod_{i=1}^{D-2} H_{k_{r,n}^i} (P_{r,n}^i) \exp \left[-(P_{r,n}^i)^2 / 4n \right], \qquad (11.2.10)$$

where $H_{k_{r,n}^i}$ is a Hermite polynomial of degree $k_{r,n}^i$ in $P_{r,n}^i$. The wave function for a ground-state string (*i.e.*, the tachyon) is simply the product of gaussian factors.

The function W is determined by the functional integral

$$W(P_{r,n}^i, p_r^i, \alpha_r, \tau_r) = g^{M-2+2L} \int \prod \mathcal{D} X^i(\sigma, \tau) \prod_{r=1}^{M} (\alpha_r)^{1/2}$$

$$\times \exp \left[-\sum_{r=1}^{M} P_r^- \tau_r / 2 \right] \exp \left(i \sum_{r=1}^{M} \int P_r^i(\sigma) X^i(\sigma, \tau_r) d\sigma - \int \mathcal{L} d\tau d\sigma \right),$$

$$(11.2.11)$$

where the factor of $\exp(i \sum_r \int P_r^i X_r^i d\sigma)$ converts the external wave functions from momentum space to position space. The factors $\exp[-P_r^- \tau_r / 2]$ are required to convert the Green function to the on-shell amplitude and the factors of $\sqrt{\alpha_r}$ are the inverse flux factors that compensate for the $1/\sqrt{\alpha_r}$'s in (11.2.7) in the expression for the covariant amplitude. The light-cone gauge Lagrangian density for the bosonic theory is defined by

$$\mathcal{L} = \frac{1}{2\pi} \left[\left(\frac{\partial X^i}{\partial \tau} \right)^2 + \left(\frac{\partial X^i}{\partial \sigma} \right)^2 \right]. \tag{11.2.12}$$

The functions $X^i(\sigma, \tau)$, which are integrated in (11.2.11), are subject to the boundary conditions $\partial X^i / \partial \sigma = 0$ along the solid lines, *i.e.*, at the ends of the open strings. The whole of this discussion applies equally well to closed strings, subject only to the appropriate modification of the boundary conditions, namely that X^i is periodic on each closed string.

The functional integral can be defined by discretizing both the σ and τ variables to form a lattice with a finite number of points, which reduces the functional integral to the product of a finite number of integrals. Alternatively, the integral can be formulated in terms of the normal modes of the coordinates with a cutoff introduced on the high-frequency modes by a σ lattice, as was considered in the propagator calculation. In either case, the functional integral to be evaluated has the form of coupled gaussian integrals with external sources at $\tau = \pm\infty$ corresponding to the terms linear in $X^i(\sigma_r, \tau_r)$ in (11.2.11). As a result the functional integral can be evaluated by the usual procedure of completing a square. This is done by writing the terms involving $X^i(\sigma, \tau)$ in the exponent of (11.2.11) as

$$\sum_{r,s} \Big\{ \int d\sigma' d\sigma'' \sum_{r,s} \frac{1}{4} P_r^i(\sigma') N(\sigma', \tau_r; \sigma'', \tau_s) P_s^i(\sigma'')$$

$$- \pi \int d\sigma d\tau \Big(X^i(\sigma, \tau) + \frac{i}{2} \sum_r \int d\sigma' P_r^i(\sigma') N(\sigma', \tau_r; \sigma, \tau) \Big) \Delta$$

$$\Big(X^i(\sigma, \tau) + \frac{i}{2} \sum_s \int d\sigma'' P_s^i(\sigma'') N(\sigma'', \tau_s; \sigma, \tau) \Big) \Big\}, \tag{11.2.13}$$

where Δ is $-1/2\pi^2$ times the two-dimensional Laplacian, *i.e.*,

$$\Delta = -\frac{1}{2\pi^2} \Big\{ \Big(\frac{\partial}{\partial \sigma} \Big)^2 + \Big(\frac{\partial}{\partial \tau} \Big)^2 \Big\}. \tag{11.2.14}$$

In (11.2.13) surface terms at τ_r have been ignored. They are implicitly taken into account in the definition of $\det \Delta$ below.

The Green function $N(\sigma, \tau; \sigma', \tau')$, which is the inverse of Δ, is required to obey

$$\Delta N(\sigma, \tau; \sigma', \tau') = -\frac{1}{\pi}\delta(\sigma - \sigma')\delta(\tau - \tau') + f(\sigma, \tau), \qquad (11.2.15)$$

where f is an arbitrary function. In comparing (11.2.13) with (11.2.11) it is necessary to do several integrations by parts and use the Neumann boundary conditions on $X(\sigma, \tau)$ and the Green function

$$\frac{\partial}{\partial n}N(\sigma, \tau; \sigma', \tau') = g(\sigma, \tau), \qquad (11.2.16)$$

where n denotes the normal to the boundary of the world sheet (which is the σ direction at the string endpoints and is the τ direction for all values of σ at the initial and final times τ_i and τ_f). The function $g(\sigma, \tau)$ is constrained by Gauss's law. Integration of (11.2.15) and use of Green's theorem gives $2\pi + \int d\sigma d\tau f(\sigma, \tau) = \oint dl g(\sigma, \tau)$. Neither g nor f has any effect in comparing (11.2.13) with (11.2.11) due to overall momentum conservation implied by the integration over the zero modes of X^i. Note that N is only defined up to an arbitrary solution of Poisson's equation with source $f(\sigma, \tau)$. In the case of closed strings the Green function is periodic in the σ and σ' coordinates.

The functional integral in (11.2.13) is gaussian in the variables

$$X^{i'}(\sigma, \tau) \equiv X^i(\sigma, \tau) + \frac{i}{2}\sum_r \int d\sigma' P_r^i(\sigma')N(\sigma', \tau_r; \sigma, \tau) \qquad (11.2.17)$$

and therefore, using

$$\int \mathcal{D}X^{i'}(\sigma, \tau) \exp\left(-\pi X^{i'}(\sigma, \tau)\Delta X^{i'}(\sigma, \tau)\right) = [\det \Delta]^{-(D-2)/2},$$
$$(11.2.18)$$

we can write the amplitude $W(P_{r,n}^i, p_r^i, \alpha_r)$ in the form

$$W = g^{M-2+2L}[\det \Delta]^{-(D-2)/2} \prod_r (\alpha_r)^{1/2} \exp[-\sum_r P_r^- \tau_r/2]$$

$$\times \exp\left(\frac{1}{4}\sum_{r,s} \int d\sigma' d\sigma'' P_r^i(\sigma')N(\sigma', \tau_r; \sigma'', \tau_s)P_s^i(\sigma'')\right). \qquad (11.2.19)$$

The exponential factors in this expression can be combined into a manifestly covariant expression by re-expressing the time variables τ_r associated with the initial and final strings in terms of the Neumann function by using the relation

$$\tau = \frac{1}{2\pi} \sum_r \int d\sigma' N(\sigma, \tau; \sigma', \tau_r), \qquad (11.2.20)$$

where the integral is taken over the incoming and outgoing strings at the initial or final times. This equation can be verified by noting firstly that both sides satisfy Laplace's equation for general values of τ. The fact that the right-hand side of (11.2.20) has vanishing normal derivative at the string endpoints for general τ follows from the boundary condition on N (11.2.16) and conservation of p^+. The normal derivative of the right-hand side at $\tau \to \tau_r (\equiv \tau_i$ or τ_f and with σ on string r) requires careful consideration in the region of the σ' integration close to σ, where N is singular. By integrating (11.2.15) in a small region around the point $\sigma' = \sigma$ when $\tau \to \tau_r$ this contribution is seen to be $+1$ for $\tau_r = \tau_f$ and -1 for $\tau_r = \tau_i$ in accord with the left-hand side. This result allows us to make the replacement

$$\sum_r P_r^- \tau_r = \frac{1}{2\pi} \sum_{r,s} P_r^- \int d\sigma' N(\sigma, \tau_r; \sigma', \tau_s)$$

$$= -\frac{1}{\pi^2} \sum_{r,s} \frac{P_r^- p_s^+}{\alpha_r \alpha_s} \int d\sigma \int d\sigma' N(\sigma, \tau_r; \sigma', \tau_s) \qquad (11.2.21)$$

in (11.2.19).

11.2.4 Amplitudes with External Ground States

For many purposes the whole of an incoming or outgoing string can be considered to be a single point in the ρ plane. For example, when evaluated between points on different incoming or outgoing strings the function $N(\sigma, \tau_r; \sigma', \tau_s)$ $(r \neq s)$ turns out to be independent of σ and σ' up to terms that decrease exponentially with $|\tau_r|$ (As before, τ_r and τ_s are equal to τ_i for an incoming string or τ_f for an outgoing string and σ lies on string r and σ' on string s.) Such terms are negligible when the external states are ground states. They become important when the external states are excited because the factor $\exp(-\sum_r \tau_r P_r^-/2)$ provides a compensating exponentially increase. This will be apparent from the explicit form of the Neumann function in §11.4. When the external states are ground states, the σ' and σ'' integrations for the terms with $r \neq s$ in (11.2.19) and (11.2.21) can be evaluated trivially and just give factors of $\pi \alpha_r$ and $\pi \alpha_s$. These multiply the zero modes of $P_r^i(\sigma')$ and of $P_s^i(\sigma'')$ in the exponent of (11.2.19). Recalling that these zero modes have the form $p^i/\pi\alpha$

the net effect is to cancel the powers of $\pi\alpha$. The result combines with the exponent of (11.2.21) to give a term involving the Lorentz-invariant vector product $p_r \cdot p_s$.

The terms with $r = s$ involve the Green function evaluated between points on the same string at τ_i or τ_f. The Green function $N(\sigma', \tau_r; \sigma''; \tau_r)$ does depend on σ' and σ'', but in a way that does not depend on the shape of the diagram, *i.e.*, it is independent of the α_r's, since the points (σ', τ_r) and (σ'', τ_r) are far away from the interaction points. As a result, when inserted in (11.2.19) and (11.2.21) the σ' and σ'' integrations give terms involving $(P^i_{r,n})^2$ (where $n > 0$) that do not depend on the α_r's. The other source of $P^i_{r,n}$-dependence comes from the external wave functions of the form of (11.2.10). For external ground states the $P^i_{r,n}$ integrations (with $n > 0$) are therefore gaussian and give a change in the overall normalization without affecting the α_r dependence. The zero-mode pieces of the Green functions with $r = s$ in (11.2.19) and (11.2.21) combine to give a factor

$$\mathcal{M} = \prod_r \exp\left(\frac{D-2}{48(\pi\alpha_r)^2} \int_0^{\pi\alpha_r} d\sigma' \int_0^{\pi\alpha_r} d\sigma'' N(\sigma', \tau_r; \sigma'', \tau_r)\right), \quad (11.2.22)$$

where we have used the result for the square of the ground-state mass, $p_r^2 = (D-2)/12$ implied by the sum of zero-point energies in (11.1.33).

When the external states are excited the leading term in the Green function evaluated between points on different incoming or outgoing strings (*i.e.*, the term that is not exponentially decreasing) does not contribute. This is due to

$$\int dP^i_{r,n} H_{k^i_{r,n}}(P^i_{r,n}) \exp\left[-(P^i_{r,n})^2/2n\right] = 0 \qquad (11.2.23)$$

for $k^i_{r,n} \neq 0$, where one factor of $\exp\left[-(P^i_{r,n})^2/4n\right]$ comes from the wave function and another from the $r = s$ terms mentioned in the previous paragraph.[*] In this case the appropriate nonzero modes of the Green function must be evaluated, even though they are exponentially small. This is because the term $\exp(-P_r^-\tau_r/2)$ contains a compensating large factor $\exp\left(-N_{(r)}\tau_r/\alpha_r\right)$ (for open strings), where $N_{(r)}$ is the occupation number of the rth string, which compensates exactly the exponential decrease of the Green function.

---

[*] Mathematically, this is the orthogonality of ground-state and excited-state harmonic oscillator wave functions.

Combining all these terms gives the general expression for the L-loop amplitude describing the scattering of M open-string on-shell tachyons

$$A(1, 2, \ldots, M) = g^{M-2+2L} G \int \prod_{\text{vertices}} d\tau_I \prod_{\text{loops}} d\sigma_L \mathcal{M} \det \Delta^{-(D-2)/2}$$

$$\times \prod_r (\alpha_r)^{1/2} \exp\left(\frac{1}{2} \sum_{r<s} p_r \cdot p_s N(\sigma, \tau_r; \sigma', \tau_s)\right).$$

$$(11.2.24)$$

This formula does not yet cover all possible loop diagrams, however, since so far we have only described how to deal with holes in the world sheet. Other diagrams, obtained by adding handles as well, involve similar techniques. Thus similar formulas apply to the closed-string bosonic theory, as we shall see later.

The problem of calculating an arbitrary diagram has now been recast into a problem of evaluating the inverse of the Laplace operator Δ, specifically the Green function $\det \Delta$ on an appropriate two-dimensional manifold subject to a prescribed set of Neumann boundary conditions. This is a subject with a vast mathematical literature. For tree diagrams and one-loop diagrams the solution can be given explicitly and, needless to say, the results agree with those calculated in earlier chapters of this book. The real power of the functional approach becomes apparent in the study of the multiloop amplitudes. We begin, however, with the tree diagrams.

11.3 Open-String Tree Amplitudes

Figure 11.4 is an example of a string tree diagram for the interaction of M open strings. There are $M - 2$ interaction times τ_I of which one is trivial, since the amplitude only depends on τ differences. Overall τ translation invariance implies conservation of energy and is incorporated by the factor $\delta(\sum_r p_r^-)$ in the S matrix. If the number of external strings is not too large it is possible to make a judicious Lorentz transformation to set the p^+ components of all the momenta except two to zero, which makes all the strings have zero width in σ apart from one incoming and one outgoing one. In this case the diagram reduces to a strip that has no horizontal boundaries apart from its edges at $\sigma = 0$ and $\pi\alpha$. In this particular Lorentz frame the Green function and the functional determinant are particularly simple to evaluate. It is this observation that lies behind the simplicity of the string light-cone gauge operator formalism discussed in earlier chapters. However, this procedure requires faith that the light-cone gauge formalism is Lorentz invariant and, in any case, it cannot be

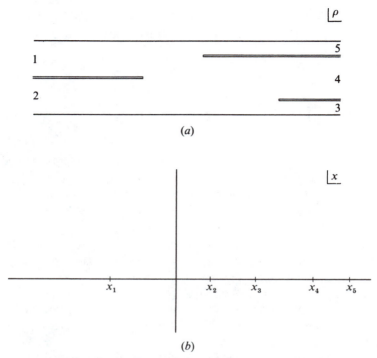

Figure 11.4. (a) The string diagram for a general tree diagram and its mapping (b) into the upper half z plane.

generalized to multiloop diagrams. For these reasons it is important to calculate the Green function for the general tree diagram as represented in fig. 11.4.

11.3.1 The Conformal Mapping

The calculation of the Green function is simplified by making use of its invariance under conformal transformations to map it to the upper half plane in some complex variable z at which point the Green function can be written down trivially. The boundary of the string diagram, including the slits along the horizontal lines is mapped continuously to the real z axis. The z-plane Green function (which has vanishing derivative normal to the real axis) is easy to evaluate. The conformal invariance of the Laplace equation guarantees that it transforms back to the ρ variables in a trivial manner. The fact that an arbitrary string tree diagram can be conformally mapped into the upper half z plane means that all string diagrams can be conformally mapped into each other by an appropriate choice of transformation. This is an example of the theorem of Riemann that we met in §1.4 in the context of our discussion of the covariant method of calculating string diagrams. This conformal equivalence of tree diagrams is a property not shared by the loop diagrams. Loop diagrams depend on 'moduli' that characterize conformally inequivalent surfaces of the same topology. The parameter τ of the torus that was described in chapters 8 and 9 is a simple example.

The appropriate mapping is given by a Schwarz–Christoffel transformation

$$\rho \equiv \tau + i\sigma = \sum_{r=1}^{M} \alpha_r \ln(x_r - z). \tag{11.3.1}$$

The boundary of the string diagram in the ρ plane maps into the real z axis as illustrated in fig. 11.4. The small semicircles in the upper half plane around the singular points of the transformation $z = x_r$ are mappings of the strings at the initial or final (very large) times, τ_i or τ_f, depending on the sign of α_r. At the end of the calculation these times are taken to $\pm\infty$ and the radii ϵ_r of these semicircles approach zero. As the variable z decreases and passes around the semicircle near a singular point x_r, the imaginary part of ρ changes by an amount $\pi i \alpha_r$, that corresponds to the increase or decrease in σ at the infinite incoming or outgoing time. The interaction points ($\rho_I = \tau_I + i\sigma_I$) are the points at which the boundary of fig. 11.4 turns back on itself. These points are therefore mapped into points y_I, which are determined by the turning points $\partial\rho/\partial z = 0$. Hence the y_I are solutions of

$$\sum_{r=1}^{M} \frac{\alpha_r}{z - x_r} = 0. \tag{11.3.2}$$

If one multiplies this equation by all the denominators, one obtains a polynomial of order $M - 1$ in y. However, the coefficient of the leading term is $\sum \alpha_r = 0$, so that the polynomial is actually only of order $M - 2$ and has $M - 2$ roots. Of these, only the $M - 3$ that correspond to τ differences are meaningful.

The transformation of a given string diagram is not uniquely specified, since the upper half plane can be mapped into itself by a real projective transformation, *i.e.*, an $SL(2, R)$ transformation with three arbitrary independent parameters. This takes z into z', where

$$z' \equiv R(z) = \frac{az + b}{cz + d}, \tag{11.3.3}$$

$ad - bc = 1$ and a, b, c and d are real. Such a transformation can be written in an alternative convenient form

$$\frac{R(z) - y_1}{R(z) - y_2} = \eta \frac{z - y_1}{z - y_2}. \tag{11.3.4}$$

The parameters y_1 and y_2 are the *invariant points* of the transformation while η is the *multiplier*. In the case of present interest the transformation is *hyperbolic*, which means that η, y_1 and y_2 are real with $\eta \neq 1$.[*] This ambiguity in the mapping means that three of the x_r's can be fixed arbitrarily. A conventional choice is to take these fixed x_r's to be at 0, 1 and ∞, as we did in chapter 7. This leaves $M - 3$ integration variables corresponding to the number of interaction times τ_I in the string diagram (after subtracting one, which is irrelevant because of overall time-translation invariance). In principle, the $M - 3$ x_r variables can be expressed as functions of the τ_I's (holding one of them fixed) by inverting the mapping, but this is very complicated for a general process.

Let us consider the mapping in some detail for the case of four external particles ($M = 4$). Setting $x_1 = 1$, $x_2 = \infty$, $x_3 = 0$ and $x_4 = x$ results in the mapping

$$\rho = \alpha_1 \ln(1 - z) + \alpha_3 \ln z + \alpha_4 \ln(z - x) + T_0, \tag{11.3.5}$$

where T_0 is an irrelevant infinite constant. Substituting this into (11.3.2), the images z_+ and z_- of the interaction points ρ_1 and ρ_2 can be determined as the solutions of a quadratic equation,

$$z_\pm = \frac{1}{2(1 - \gamma_1)} \left\{ 1 + (\gamma_2 - \gamma_1)x \pm \Delta^{1/2} \right\}, \tag{11.3.6}$$

where

$$\gamma_1 = -\frac{\alpha_1}{\alpha_1 + \alpha_2}, \qquad \gamma_2 = \frac{\alpha_3}{\alpha_1 + \alpha_2} \tag{11.3.7}$$

and

$$\Delta = x^2(\gamma_2 - \gamma_1)^2 + 2x(2\gamma_1\gamma_2 - \gamma_1 - \gamma_2) + 1. \tag{11.3.8}$$

Figures 11.5*a* and 11.5*b* show two of the light-cone string diagrams that contribute to the amplitude. These are the only contributions associated with the group-theory factor $\text{tr}(\lambda_1\lambda_2\lambda_3\lambda_4)$. These diagrams are mapped into the z plane with the variable x in the range $0 \leq x \leq 1$. The expression

---

[*] A projective transformation is called *elliptic* if $y_1 = y_2^*$ and $|\eta| = 1$ or *parabolic* if $y_2 - y_1 = \epsilon$, $\eta = 1 - \alpha\epsilon$ and $\epsilon \to 0$.

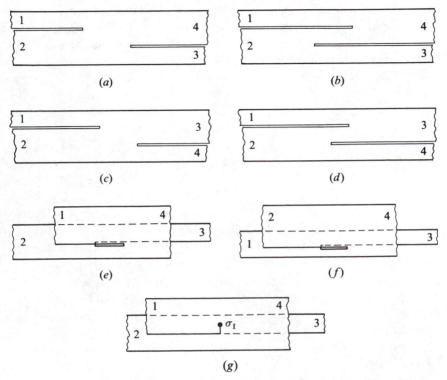

Figure 11.5. The tree diagrams contributing to four-particle scattering: (a) and (b) have the group-theory factor $\text{tr}(\lambda_1 \lambda_2 \lambda_3 \lambda_4)$ and poles in the s and t channels, respectively. (c) and (d) have the group-theory factor $\text{tr}(\lambda_1 \lambda_2 \lambda_3 \lambda_4)$ and poles in the s and u channels, respectively. The last three diagrams have a group-theory factor of $\text{tr}(\lambda_1 \lambda_3 \lambda_2 \lambda_4)$. (e) and (f) have poles in the t and u channels, respectively. The four-string contact term is shown in (g). The point σ_I separates the portion of string 1 that is joined to string 3 from the portion joined to string 4.

for the amplitude involves an integral over all positive values of the time difference between the two interaction points ($\hat{\tau} \equiv \tau_2 - \tau_1$) in fig. 11.5a and negative values in fig. 11.5b. This time difference is given by

$$\hat{\tau} = \text{Re}[\rho(z_+) - \rho(z_-)]. \tag{11.3.9}$$

The contribution from fig. 11.5a contains the poles in the invariant $s = -(p_1 + p_2)^2$, which arise from the limit of integration in which the intermediate string propagates for an infinite time. The endpoint of the $\hat{\tau}$ integration of fig. 11.5 at $\hat{\tau} = \infty$ corresponds to the point $x = 0$ (where $z_- = 0$). The other endpoint of the integration at $\hat{\tau} = 0$ is mapped into a value of x lying between 0 and 1, which we denote by x_0. The value of x_0 is not Lorentz invariant, since it depends on the values of the α_r's. Since the values of z_+ and z_- do not coincide at the point $x = x_0$, the values of $\sigma \equiv -i \operatorname{Im}\rho$ at the two interaction points are unequal. Similarly, the process in fig. 11.5b gives poles in the invariant $t = -(p_1 + p_4)^2$ from the integration limit $\hat{\tau} = -\infty$, which corresponds to the limit $x = 1$ (where $z_+ = 1$). These two diagrams have the same integrands when expressed in terms of the x variable and so the complete Lorentz-invariant integration region, $0 \leq x \leq 1$, is obtained after summing the two contributions.

The diagram of fig. 11.5c is obtained from fig. 11.5a by interchanging particles 3 and 4, and therefore contains poles in the s channel. The mapping of its boundary can be obtained by substituting $x/(x-1)$ for x in fig. 11.5a, which gives negative values of x in the interval $x_0/(x_0-1) \leq x \leq 0$. The contribution from fig. 11.5d fills in the rest of the negative x-axis, $-\infty \leq x \leq x_0/(x_0-1)$ and gives the poles in the u channel (where $u = -(p_1 + p_3)^2$). These diagrams piece together to form the Lorentz-invariant piece of the amplitude with group-theory factor $\operatorname{tr}(\lambda_1\lambda_2\lambda_3\lambda_4)$.

The last contributions are the diagrams in figs. 11.5e, 11.5f, and 11.5g with group-theory factor $\operatorname{tr}(\lambda_1\lambda_4\lambda_2\lambda_3)$, which are mapped into the region $1 \leq x \leq \infty$. In this case there is a contribution with poles in the s channel depicted in fig. 11.5e. As the $\hat{\tau}$ variable moves from ∞ to zero the variable x moves from 1 to a value x_-, which is a value at which the two interaction points coincide, in both σ and τ in this case. This occurs when the points z_- and z_+ coincide, *i.e.*, when $\Delta = 0$. The condition for this has the two solutions

$$x = x_\pm = \frac{1}{(\gamma_2 - \gamma_1)^2} \left(\gamma_1 + \gamma_2 - 2\gamma_1\gamma_2 \pm 2\sqrt{\gamma_1(\gamma_1 - 1)\gamma_2(\gamma_2 - 1)} \right).$$

$$(11.3.10)$$

The diagram with poles in the u channel, fig. 11.5f, is obtained from the region $x_+ \leq x \leq \infty$. The amplitudes described by figs. 11.5e and 11.5f do not merge continuously into each other at $\tau = 0$. In the region in between x_+ and x_- the values of z_+ and z_- become complex-conjugate pairs and only one of them (z_-) is in the upper half plane. The string configuration corresponding to this region is depicted in fig. 11.5g. It represents a four-string contact interaction in which two incoming strings touch at an internal point $(\sigma = \sigma_I)$ and swap string bits to form the final two strings. The value of σ_I is to be integrated between the bottom of string 1 and the top of string 3. The figure has two sheets and the vertical solid line indicates a cut so that below the point σ_I string 1 is joined to string 3 while the rest of string 1 is joined to string 4.

When expressed in terms of the x variable, all three of the contributions have Lorentz-invariant integrands of the same form. In order for the complete amplitude to be Lorentz invariant, it is necessary to include the diagram that fills in the region between $x = x_-$ and $x = x_+$ with a definite coefficient. This means that the normalization of the four-string instantaneous interaction is determined in terms of the interaction that joins two open strings into one. The integrands of figs. 11.5e and 11.5g have the same form at the point $x = x_-$, and so their contributions to the amplitude merge into each other in a continuous manner when they are normalized correctly. However, the contribution from fig. 11.5g involves an integration over σ_I (*i.e.*, $\text{Im}\,\rho_I$), whereas that of fig. 11.5e involves a $\hat{\tau}$ (*i.e.*, $\text{Re}\,\rho_I$) integral. As a result the coefficient of the four-string term must be ig^2 to compensate the factor of i in the measure. Locally in σ this four-string contact interaction is the same as the one associated with the joining of two closed strings to form a single closed string. That interaction is the string generalization of the gravitational interaction, and so we see that the gravitational coupling κ is determined in terms of the open-string (or Yang–Mills) coupling g.

11.3.2 Evaluation of Amplitudes

The Green function in the z plane is given by the solution of Laplace's equation with vanishing normal derivative along the real axis

$$N(z; z') = \ln|z - z'| + \ln|z - z'^{\star}|. \qquad (11.3.11)$$

In the expression for the amplitude (11.2.24), the Green function is to be evaluated between points at $\tau = \pm\infty$ on the various strings, *i.e.*, at values of z and z' equal to the x_r's so that the required Green functions are just $N(x_r; x_s)$. In order to make use of this result, the x_r's could be expressed in terms of the integration variables of the original string diagram, *i.e.*, the τ_I's, and substituted into (11.2.24), but this is a forbidding task. It is easier to change the integration variables in (11.2.24) from the τ_I's to the x_r's. This requires the evaluation of the Jacobian $|\partial\tau_I/\partial z_r|$ for the transformation.

After this change of integration variables the amplitude is given as an integral over $M - 3$ independent x_r variables. These are just the Koba–Nielsen variables described in §1.5.4 and in §7.1.4. The factors in the integrand still to be evaluated are the functional determinant $\det \Delta$, the exponential involving the Green function evaluated between points on the same string $\mathcal{M}$ and the Jacobian $|\partial\tau_I/\partial z_r|$. These expressions are all independent of the external transverse momenta. Since the exponential involving the Green function in (11.2.24) is Lorentz invariant when expressed in terms of the x_r variables, the product of these factors must be Lorentz invariant also and hence independent of the α_r's. This means that the resulting covariant scattering amplitude $A(1, 2, \ldots, M)$ must have the form

$$\int dx_2 dx_3 \ldots dx_{M-2} V(x_1, \ldots, x_M) \prod_{r<s} |x_r - x_s|^{p_r \cdot p_s}, \qquad (11.3.12)$$

where V is a measure factor that is independent of the α_r's. The notation implies that the variables x_1, x_{M-1} and x_M are held fixed at arbitrary values.

The explicit calculation of $\det \Delta$ is outlined in appendix 11.A, and the Jacobian for the transformation from the string diagram to the upper half plane in appendix 11.B. This leads to an expression for

$$|\partial \tau_I / \partial z_r| [\det \Delta]^{(2-D)/2} \mathcal{M}, \qquad (11.3.13)$$

and hence the integration measure V. The result is that in the critical dimension $(D = 26)$ the measure contributes precisely the factor $\prod_r (\alpha_r)^{-1/2}$ that is needed to compensate for the flux factors in obtaining a covariant expression. The fact that the correct flux factors only emerge from this calculation of the measure in 26 dimensions is yet another example of how the critical dimension enters the theory.

For tree diagrams, as well as for one-loop amplitudes, it is possible to take a short cut to the evaluation of V, which involves working in a special Lorentz frame. This assumes that the theory is Lorentz invariant (which is true in the critical dimension $D = 26$). This method does not, however, generalize to multiloop amplitudes (for which the calculation in appendices 11.A and 11.B must be generalized). Consider first the special limit in which all of the strings have very small p^+ apart from two that are arbitrarily chosen to be numbers 1 and M (*i.e.*, $\alpha_r \ll \alpha_1, \alpha_M$) and the external transverse momenta all vanish. The mass-shell condition then requires that $p_r^- \sim -2/\alpha_r$. If string 1 is in its ground state, then it is not possible for any of the intermediate states to be excited, since the coupling of any excited state to two ground states involves powers of the transverse momenta.

The covariant amplitude is then easy to write down by using ordinary quantum-mechanical rules

$$A(1, 2, \ldots, M) = \mathcal{V}(\alpha_1, \ldots, \alpha_M) \int_0^\infty d\tau_2 \ldots d\tau_{M-2} \exp(-\sum_{r=2}^{M-1} p_r^- \tau_r / 2),$$

$$(11.3.14)$$

where $\mathcal{V}$ is a normalization factor that is independent of the τ_r's. The equation for the interaction points (11.3.2) can be solved explicitly for this special case of infinitesimal α_r giving the $M-2$ solutions y_r

$$y_r \sim x_r - \frac{\alpha_r}{\alpha_1}\tilde{x}_r + O(\alpha_r^2) \quad r = 2,\ldots,M-1, \tag{11.3.15}$$

where

$$\tilde{x}_r = \frac{(x_r - x_1)(x_r - x_M)}{(x_1 - x_M)}. \tag{11.3.16}$$

Note that the $M-2$ solutions y_r are naturally associated with the strings whose α is infinitesimal. Upon substituting $z = y_r$ in (11.3.1) we obtain (after some straightforward algebra) the values of τ_r in terms of x_r

$$\tau_r \equiv \mathrm{Re}\rho_r \sim \alpha_1 \ln(x_r - x_1) - \alpha_1 \ln(x_r - x_M)$$
$$+ \alpha_r(\ln(\alpha_r \tilde{x}_r/\alpha_1) - 1) + \sum_{2 \le s \le M-1}' \alpha_s \ln|x_r - x_s|, \tag{11.3.17}$$

where the prime means that the $s = r$ term is not included. Thus

$$\exp\{-\sum_{r=2}^{M-1} p_r^- \tau_r/2\} \sim (x_M - x_1)^2 \prod_{r=2}^{M-1} \left(\frac{\alpha_r \tilde{x}_r}{e\alpha_1}\right)$$
$$\times \exp\left\{\sum_{1 \le r < s \le M} (\frac{\alpha_s}{\alpha_r} + \frac{\alpha_r}{\alpha_s}) \ln|x_r - x_s|\right\}, \tag{11.3.18}$$

where the factor of $(x_M - x_1)^2$ compensates for the term with $r = 1$ and $s = M$ in the exponential factor.

The Jacobian for the change of integration variables from τ_r to x_r (holding x_1, x_{M-1} and x_M fixed) can be deduced from (11.3.17), giving

$$\frac{\partial(\tau_2 \ldots \tau_{M-2})}{\partial(x_2 \ldots x_{M-2})} \sim \prod_{r=2}^{M-2} \left|\frac{\alpha_1}{\tilde{x}_r}\right|. \tag{11.3.19}$$

Substituting into (11.3.14) and using the fact that

$$p_r \cdot p_s \sim \alpha_r/\alpha_s + \alpha_s/\alpha_r, \tag{11.3.20}$$

gives the expression

$$
A(1, 2, \ldots, M) = \mathcal{V}(\alpha_1, \ldots, \alpha_M) \frac{1}{|\alpha_1|} \prod_{r=2}^{M-1} \left| \frac{\alpha_r}{e} \right|
$$
$$
\times |(x_{M-1} - x_1)(x_M - x_1)(x_M - x_{M-1})|
$$
$$
\times \int dx_2 dx_3 \ldots dx_{M-2} \prod_{1 \le r < s \le M} |x_s - x_r|^{p_r \cdot p_s}.
$$

$$(11.3.21)$$

This can be compared directly with the general result for the covariant string scattering amplitude in (11.3.12). We define

$$
\mathcal{V}(\alpha_1, \ldots, \alpha_M) = g^{M-2} G |\alpha_1| \prod_{r=2}^{M-2} |e/\alpha_r|, \qquad (11.3.22)
$$

where G is the appropriate group-theory factor and a coupling constant g is inserted for each interaction. Then

$$
V(x_1, \ldots, x_M) = g^{M-2} G \prod_{1}^{M-1} \theta(x_{r+1} - x_r)
$$
$$
\times |(x_{M-1} - x_1)(x_M - x_1)(x_M - x_{M-1})|.
$$

$$(11.3.23)$$

In this way the amplitude of (11.3.12) reduces to the Koba–Nielsen form derived in chapters 1 and 7.

11.4 Open-String Trees with Excited External States

Up to now our explicit calculations have been for processes in which the external states have been on-shell ground states, in which case the calculations simplify because only the leading behavior of the Green function enters the calculations. In this section we consider the structure of the Green functions in more detail and obtain expressions for open-string tree diagrams with arbitrary excited external on-shell string states. This includes the explicit expression for the cubic couplings between arbitrary on-shell states in terms of oscillators that act in the individual Fock spaces of the three strings. This vertex generalizes the emission vertices that have been used in earlier chapters (to describe the emission of a string in one particular on-shell state). The more general vertex provides the basis for an interacting field theory of strings in the light-cone gauge (a subject not covered in this book).

In this analysis it is useful to introduce separate parametrizations for the individual strings

$$\zeta_r = \xi_r + i\eta_r \equiv \frac{\tau_r}{\alpha_r} + i\frac{\sigma_r}{|\alpha_r|}, \tag{11.4.1}$$

so that

$$0 \le \eta_r \le \pi \tag{11.4.2}$$

and the value of ξ_r is negative on all of the strings, irrespective of whether they are incoming or outgoing. Each string has mode expansions

$$\mathbf{X}_r(\eta_r) = \mathbf{x}_r + 2\sum_{n=1}^{\infty} \mathbf{X}_{rn} \cos n\eta_r, \tag{11.4.3}$$

and

$$\mathbf{P}_r(\eta_r) = \frac{1}{\pi\alpha_r}\left(\mathbf{p}_r + \sum_{n=1}^{\infty} \mathbf{P}_{rn} \cos n\eta_r\right). \tag{11.4.4}$$

11.4.1 The Green Function on an Infinite Strip

Consider first the Green function for a single freely propagating open string, which is represented by a long strip of width $\pi|\alpha|$. In this case we define $\alpha = \alpha_1 = -\alpha_2$. The Green function is easy to evaluate explicitly by mapping the strip to the half plane using the transformation

$$\zeta = \rho/\alpha = \ln z, \tag{11.4.5}$$

which maps the incoming string at $\xi = \xi_i$ into a small semicircle in the upper half plane around $z = 0$ and the outgoing string at $\xi = \xi_f$ into a 'small' semicircle around the point $z^{-1} = 0$ (*i.e.*, a large circle). When evaluated between arbitrary points away from $z = 0$ and $z = \infty$ the initial and final values of ξ can be set equal to infinity and the Green function in the z plane is given by (11.3.11), which can be written in terms of the ζ variable by inverting the transformation that sets $z = \exp \zeta$, giving

$$N_{\text{strip}}(\sigma, \tau; \sigma', \tau') = \ln |e^\zeta - e^{\zeta'}| + \ln |e^\zeta - e^{\zeta'^*}|$$

$$= 2 \max(\xi, \xi') - \sum_{n=1}^{\infty} \frac{2}{n} e^{-n|\xi - \xi'|} \cos n\eta \cos n\eta'. \tag{11.4.6}$$

If the points ζ and ζ' are situated at the initial or final times, the Green function must be modified due to the fact that the Neumann boundary conditions should be imposed on the boundary of the half plane with the small semicircles cut out. For each exponentially decreasing term, $\exp\{n\xi\}$ or $\exp\{n\xi'\}$, there is an additional term corresponding to a wave reflected from the boundary at the initial or final times in order for the Green function to satisfy the boundary condition $\partial N / \partial \xi = 0$ at ξ_i or ξ_f (with a similar condition on the ξ' derivative). For ξ near ξ_i the reflected wave has $\exp\{n\xi\}$ replaced by $\exp\{2n\xi_i - n\xi\}$ with a similar replacement for ξ near ξ_f (and for the exponentials involving ξ'). The net effect as ξ and $\xi' \to \xi_i$ or ξ_f is to double each term involving $\exp\{n\xi\}$ or $\exp\{n\xi'\}$, which gives

$$N_{\text{strip}}(\sigma, \tau; \sigma', \tau') = 2 \max(\xi, \xi') - \sum_{n=1}^{\infty} \frac{4}{n} e^{-n|\xi - \xi'|} \cos n\eta \cos n\eta'. \quad (11.4.7)$$

11.4.2 Green Functions for Arbitrary Tree Amplitudes

We turn now to the calculation of the Green function for a general open-string tree diagram such as the one illustrated in fig. 11.4. The mapping that relates the z plane to the ρ plane is given in (11.3.1). Using (11.3.11), it is easy to read off the leading piece of the Green function between points ρ_r and ρ_s on the incoming or outgoing strings at τ_i or τ_f, which are large. Thus, if r and s are on different strings the leading term is the finite term,

$$N(\sigma_r, \tau_r; \sigma_s, \tau_s) = 2 \ln |x_r - x_s|, \quad (11.4.8)$$

which is independent of ρ_r and ρ_s. This term is familiar from the previous calculation of tree diagrams with external ground states.

When the two points are on the same string r the situation is a bit more complicated, since the Green function is singular at $\eta_r = \eta'_r$ in that case. The leading behavior is the same as if string r were noninteracting, since the points on the world sheet at $\xi = \xi_r$ are very far away from the interaction region. This fact was also used earlier in arguing that, with external tachyons, the terms with $r = s$ in the exponent of (11.2.19) only contributed to the measure factors. A slightly more careful analysis is needed in this case, which involves considering the points $z_r = x_r + \epsilon \exp\{i\phi_r\}$ and $z'_r = x_r + \epsilon \exp\{i\phi'_r\}$ on a small semicircle of radius ϵ around the point x_r. These points are mapped into the points $\xi_r = \xi'_r \equiv \tau_r / \alpha_r \sim \ln \epsilon + \sum_{s \neq r} \alpha_s / \alpha_r \ln |x_r - x_s|$ and $\eta_r = \phi_r$, $\eta'_r = \phi'_r$. The terms in the

Green function that do not decrease with ξ_r are given by

$$N(\sigma_r, \tau_r; \sigma'_r, \tau_r) = 2\xi_r - 2\ln|2(\cos\eta_r - \cos\eta'_r)|$$

$$- 2\sum_{s\neq r} \frac{\alpha_s}{\alpha_r} \ln|x_r - x_s| + O(\tau_r^{-1}). \qquad (11.4.9)$$

The first two terms on the right-hand side of this expression are the same as the expression for the Green function on the strip evaluated between two points at the same initial or final time (11.4.7).

By integrating (11.4.9) over both σ_r and σ'_r we obtain

$$\frac{\tau_r}{\alpha_r} = N_{00}^{rr} + \sum_{s\neq r} \frac{\alpha_s}{\alpha_r} \ln|x_r - x_s|. \qquad (11.4.10)$$

By combining (11.4.8) and (11.4.9), we find

$$\xi_r \equiv \frac{\tau_r}{\alpha_r} = \sum_s \frac{\alpha_s}{2\pi\alpha_r} \int d\eta_s N(\sigma_r, \tau_r; \sigma_s, \tau_s), \qquad (11.4.11)$$

which is the same equation for τ that was given earlier in (11.2.20).

We now turn to consider the complete Green function, including the exponentially decreasing terms. When evaluated between points away from the initial or final times on any of the strings, the Green function has the general expansion

$$N(\sigma_r, \tau_r; \sigma_s, \tau_s) = -\delta_{rs} \sum_{n=1}^{\infty} \frac{2}{n} \cos n\eta_r \cos n\eta_s$$

$$+ \sum_{n,m=0}^{\infty} 2N_{mn}^{rs} e^{m\xi_r + n\xi_s} \cos m\eta_r \cos n\eta_s. \qquad (11.4.12)$$

The first term and $2N_{00}^{rs}$ are the only terms that are not decreasing functions of ξ_r and ξ_s. They correspond to the terms in (11.4.8) and (11.4.9) so that

$$N_{00}^{rs} = \delta_{rs}\xi_r + (1 - \delta_{rs})\ln|x_r - x_s| - \delta_{rs} \sum_{s'\neq r} \frac{\alpha_{s'}}{\alpha_r} \ln|x_r - x_{s'}|. \qquad (11.4.13)$$

The infinite series with coefficients N_{mn}^{rs} that have m or $n \neq 0$ are the terms in the sum that are decreasing functions of ξ_r and ξ_s.

In order to deduce the Green function evaluated between points on the strings at the large initial or final times this expression again has to be modified to include reflected waves, which are implied by the Neumann boundary conditions on the small semicircles in the z plane. As we have seen this doubles the coefficient of every term that has a factor of $\exp\{n\xi_r\}$ or $\exp\{n\xi_s\}$. This means that when ξ_r and $\xi_s = \xi_i$ or $\xi_f \to -\infty$

$$
N(\sigma_r, \tau_r; \sigma_s, \tau_s) = -\,\delta_{rs} \sum_{n=1}^{\infty} \frac{4}{n} \cos n\eta_r \cos n\eta_s
$$
$$
+ 4 \sum_{m=1}^{\infty} \left(N_{m0}^{rs} e^{m\xi_r} \cos m\eta_r + N_{0m}^{rs} e^{m\xi_s} \cos m\eta_s \right)
$$
$$
+ 8 \sum_{m,n=1}^{\infty} N_{mn}^{rs} e^{m\xi_r + n\xi_s} \cos m\eta_r \cos n\eta_s + 2N_{00}^{rs}.
$$

$$(11.4.14)$$

11.4.3 The Amplitude in Terms of Oscillators

Upon substituting the expression for N in (11.4.14) into the formula for W in (11.2.19), the integrations over σ and σ', which project out the modes of $\mathbf{P}_r$, can be carried out. The amplitude for an arbitrary tree diagram (given by (11.2.8)) then becomes

$$
A(1, 2, \ldots, M) = g^{M-2} G \int \prod_{I=1}^{M-3} d\tau_I \mathcal{V}(\{\alpha_r\}, \{\tau_I\}) \int \prod_{r,n,i} dP_{r,n}^i
$$
$$
\times\ \Psi_r\!\left(P_{r,n}^i\right) \exp\left\{ -\sum_r \tfrac{1}{2} P_r^- \tau_r - \sum_r \sum_{n=1}^{\infty} \frac{(\mathbf{P}_{rn})^2}{4n} \right.
$$
$$
\left. + \sum_{r,s} \sum_{m,n=0}^{\infty} \tfrac{1}{2} N_{mn}^{rs} e^{m\xi_r + n\xi_s} \mathbf{P}_{rm} \cdot \mathbf{P}_{sn} \right\}.
$$

$$(11.4.15)$$

The volume element $\mathcal{V}$ is independent of the transverse momenta and the occupation numbers of the external strings, and therefore it can be determined by evaluating a simple special case, which is the method we shall use for the vertex calculation. In the open-string vertex calculation $M = 3$ and so there are no τ_I integration variables.

The oscillator representation of the integrand of (11.4.15) can be deduced by using the momentum-space wave functions

$$\Psi_r(P^i_{r,n}) = \prod_{n,i} \langle k^i_{r,n} | P^i_{r,n} \rangle, \tag{11.4.16}$$

where $k^i_{r,n}$ are the occupation numbers of string r. This also requires the fact that

$$\prod_{r,n} \exp\left[-(\mathbf{P}_{r,n})^2/4n\right] = \prod_{r,i,n} \Psi^0_r(P^i_{r,n}) = \prod_{r,i,n} \langle P^i_{r,n} | 0 \rangle \tag{11.4.17}$$

is the product of ground-state wave functions for the nonzero modes of the external strings. The states $|P^i_n\rangle$ are eigenstates of the momentum operator and can be expressed in terms of the annihilation and creation modes by

$$|P^i_n\rangle = \left(\frac{1}{\pi n}\right)^{1/4} \exp\left(\frac{P^i_n \alpha^i_{-n}}{n} - \frac{(\alpha_{-n})^2}{2n} - \frac{(P^i_n)^2}{4n}\right) |0\rangle, \tag{11.4.18}$$

which is an eigenstate of $\alpha^i_n + \alpha^i_{-n}$ with eigenvalue P^i_n. Also $\langle P'^i_n | P^i_n \rangle = \delta(P^i_n - P'^i_n)$.

Substituting these expressions for $\Psi_n(P^i_{r,n})$ and $\exp(-\sum(P_{rn})^2/4n)$ in (11.4.15) the momentum factors $\mathbf{P}_{rm}$ can be replaced by $\alpha^r_m + \alpha^r_{-m}$. The exponent in (11.4.15) then contains terms with factors like $(\alpha^r_m + \alpha^r_{-m})\exp\{m\xi_r\}$, where $e^{m\xi_r}$ is an exponentially small factor. These small factors can be compensated by the factors of $\exp(\sum P^-_r \tau_r/2)$, which are large when multiplying bra states of nonzero occupation number. As a result only the creation modes survive in the exponential and the result becomes (using (11.4.16))

$$A(1,2,\ldots,M) = g^{M-2} G \mathcal{V} \langle \{k^i_{r,n}\} | \exp\left(-\sum_r P^-_r \tau_r/2\right)$$

$$\times \exp\left\{\tfrac{1}{2} \sum_{r,s} \sum_{m,n=0}^{\infty} N^{rs}_{mn} e^{m\xi_r+n\xi_s} \underline{\alpha}^r_{-m} \cdot \underline{\alpha}^s_{-n}\right\} |0\rangle, \tag{11.4.19}$$

where the P^-_r's are written in terms of their oscillator expressions and $\underline{\alpha}^r_0 \equiv \mathbf{p}_r$.

The fact that the apparently exponentially small factors in the exponent contribute a finite amount is readily seen by commuting the factor of

$\exp\{-\sum P_r^- \tau_r/2\}$ to the right using

$$\exp\left(-P_r^- \tau_r/2\right) \underline{\alpha}_{-n}^r = \underline{\alpha}_{-n}^r e^{-n\tau_r/\alpha_r} \exp\left(-P_r^- \tau_r/2\right), \qquad (11.4.20)$$

which cancels the ξ_r and ξ_s dependence in the exponential. The excited modes inside P_r^- then annihilate on the ground state leaving factors of $\exp\left(\tau_r(\mathbf{p}_r^2 - 1)/2\alpha_r\right)$. Notice that if the annihilation modes had been kept their exponential factors would have become even smaller as a result of this operation so it was correct to drop them.

The terms in the exponent of (11.4.19) involving the coefficients N_{00}^{rs} can be treated explicitly by using (11.4.13). The contribution from the $\delta_{rs}\xi_r$ term gives a factor that cancels the factors of $\exp\{-\tau_r \mathbf{p}_r^2/2\alpha_r\}$. Using (11.4.10) to re-express the factor of $\exp\{\tau_r/\alpha_r\} \equiv \exp\{\xi_r\}$ results in a total contribution from the zero modes to the exponent in (11.4.19) of

$$\sum_{r \neq s}\left\{\tfrac{1}{2}\mathbf{p}_r \cdot \mathbf{p}_s \ln|x_r - x_s| - (1 - \tfrac{1}{2}\mathbf{p}_s^2)\frac{\alpha_s}{\alpha_r}\ln|x_r - x_s| + N_{00}^{rr}\right\}. \qquad (11.4.21)$$

Substituting into (11.4.19) and recalling that the factor of $\mathcal{M} = \exp\{N_{00}^{rr}\}$ combines with the volume element in a simple way gives

$$A(1, 2, \ldots, M) = g^{M-2} G \int dx_2 dx_3 \ldots dx_{M-2} \prod_{r<s} |x_r - x_s|^{\mathbf{p}_r \cdot \mathbf{p}_s}$$

$$\times \prod_{r<s} (x_r - x_s)^{(1-\mathbf{p}_r^2/2)\alpha_s/\alpha_r + (1-\mathbf{p}_s^2/2)\alpha_r/\alpha_s}$$

$$\times \langle\{k_{r,n}^i\}| \exp\left\{\tfrac{1}{2}\sum_{r,s}\sum_{m,n}{}' N_{mn}^{rs} \underline{\alpha}_{-m}^r \cdot \underline{\alpha}_{-n}^s\right\}|0\rangle,$$

$$(11.4.22)$$

where $-\mathbf{p}_r^2 = 2(N_r - 1)$ is the (mass)^2 of string r. The choices $x_1 = 1$, $x_{M-1} = 0$ and $x_M = \infty$ have been made by exploiting the residual elements of the $SL(2, R)$ invariance of the mapping. $\sum'$ denotes the sum without the term with $m = n = 0$. This expression manifestly reduces to the Koba–Nielsen form for the scattering of on-shell tachyons (for which $\mathbf{k}_{n_r} = 0$ and $\mathbf{p}_r^2 = 2$).

11.4.4 The General Form of the Neumann Coefficients

The Green function in the z plane is given by (11.3.11), which translates into the ρ plane by inverting the transformation in (11.3.1)

$$N(\rho; \rho') = \ln|z(\rho) - z'(\rho')| + \ln|z(\rho) - z'^{\star}(\rho')|. \qquad (11.4.23)$$

The Neumann coefficients N_{mn}^{rs} in (11.4.12) can be expressed in terms of a contour integral in the z plane by making use of the z-plane Green function of (11.3.11) in the following way. The nonzero Fourier coefficients $(m, n \neq 0)$ of (11.4.12) are given by

$$
\begin{aligned}
N_{mn}^{rs} =& \delta_{rs}\delta_{mn}\frac{1}{n}e^{-n|\xi_r-\xi_s|} \\
& + \frac{1}{\pi^2}e^{-m\xi_r-n\xi_s}\int_0^{2\pi} d\eta_r d\eta_s \cos m\eta_r \cos n\eta_s \ln(z_r - z_s),
\end{aligned} \tag{11.4.24}
$$

where the two terms of the Green function $(\ln|z_r - z_s|$ and $\ln|z_r - z_s^*|)$ have been combined by extending the range of the η_r and η_s integrals.

Expanding the cosine functions in terms of exponentials gives one term of the form

$$
\begin{aligned}
N_{mn}^{rs} &= \frac{1}{(2\pi i)^2}\int_0^{2\pi} d\eta_r d\eta_s e^{-m\zeta_r-n\zeta_s}\ln\left(z(\zeta_r) - z(\zeta_s)\right) \\
&= \frac{1}{mn(2\pi i)^2}\int_{x_r} dz_r \int_{x_s} dz_s \frac{e^{-m\zeta_r(z_r)-n\zeta_s(z_s)}}{(z_r - z_s)^2},
\end{aligned} \tag{11.4.25}
$$

where the second step involves integration by parts followed by a change of integration variables to integrals in the z plane surrounding the points x_r and x_s. Since the functions $\zeta_r(z_r)$ and $\zeta_s(z_s)$ are singular at $z_r = x_r$ and $z_s = x_s$, respectively, this integral gives a nonzero result. The other terms arising from the cosine involve the functions $\zeta_r(z_r^*)$ or $\zeta_s(z_s^*)$, which are not singular at these points and therefore do not contribute when $r \neq s$. When $r = s$ they do contribute in just the manner that cancels the δ_{rs} term in (11.4.24). Therefore (11.4.25) is the complete result for all r and s. The result is also true when either m or $n = 0$.

11.4.5 The Neumann Coefficients for the Cubic Open-String Vertex

The expression for the interaction vertex coupling three open strings in the oscillator basis can be extracted from (11.4.22) by considering the case in which $M = 3$ and fixing arbitrarily $x_1 = 1$, $x_2 = 0$ and $x_3 = \infty$. A vertex $|V\rangle$ can be defined in the tensor product of the oscillator spaces of the three strings by

$$
\begin{aligned}
A(1,2,3) &\equiv g\langle k_{1,n}^i, k_{2,n}^i, k_{3,n}^i|V\rangle \\
&= g\langle\{k_{r,n}^i\}|\exp\Big\{\tfrac{1}{2}\sum_{r,s=1}^{3}{\sum_{m,n}}' N_{mn}^{rs}\underline{\alpha}_{-m}^r \cdot \underline{\alpha}_{-n}^s\Big\}|0\rangle.
\end{aligned} \tag{11.4.26}
$$

In this equation the interaction time is fixed at some value $\tau = \tau_0$, determined by the mapping of (11.3.1) for the case $M = 3$ (using the special values of x_1, x_2 and x_3)

$$\rho = \alpha_1 \ln(z - 1) + \alpha_2 \ln z. \qquad (11.4.27)$$

The interaction point ρ_0 is the turning point of the mapping, which satisfies

$$\frac{\alpha_1}{z - 1} + \frac{\alpha_2}{z} = 0. \qquad (11.4.28)$$

This has the unique solution

$$z_0 = -\frac{\alpha_2}{\alpha_3}, \qquad (11.4.29)$$

which determines ρ_0 and hence the interaction time

$$\tau_0 \equiv \mathrm{Re}\rho_0 = \alpha_1 \ln\left(-\frac{\alpha_1}{\alpha_3}\right) + \alpha_2 \ln\left(-\frac{\alpha_2}{\alpha_3}\right) = \sum_{r=1}^{3} \alpha_r \ln |\alpha_r|. \quad (11.4.30)$$

The Neumann coefficients for the cubic vertex may be calculated explicitly from the expression in (11.4.25) by substituting the expressions for the ζ_r's that follow from the mapping (11.4.27)

$$e^{-\zeta_1} = \frac{z^{-\alpha_2/\alpha_1}}{z - 1}, \qquad (11.4.31)$$

$$e^{-\zeta_2} = \frac{(1 - z)^{-\alpha_1/\alpha_2}}{z}, \qquad (11.4.32)$$

$$e^{-\zeta_3} = -z\left(1 - \frac{1}{z}\right)^{-\alpha_1/\alpha_3}. \qquad (11.4.33)$$

The Neumann coefficients can be evaluated from (11.4.25) by expanding the integrand in a Taylor expansion to identify the residues of the poles at $z_r = x_r$ and $z_s = x_s$.

Next we define

$$N_m^r \mathcal{P} = \sum_s N_{m0}^{rs} \mathrm{P}_s \qquad (11.4.34)$$

where

$$\mathcal{P} = \alpha_1 \mathrm{p}_2 - \alpha_2 \mathrm{p}_1. \qquad (11.4.35)$$

Although it is not manifest, $\mathcal{P}$ is symmetric under cyclic permutations of the external states when the conservation conditions $\sum_r \alpha_r$ and $\sum_r \mathrm{p}_r$

are taken into account. Furthermore,

$$\sum_{r=1}^{3} \frac{\mathbf{p}_r^2}{\alpha_r} = -\frac{\mathcal{P}^2}{\hat{\alpha}}, \qquad (11.4.36)$$

where

$$\hat{\alpha} = \alpha_1 \alpha_2 \alpha_3. \qquad (11.4.37)$$

The coefficients, N_m^r and N_{mn}^{rs}, that result from (11.4.25) after substituting (11.4.31)-(11.4.33) are

$$N_m^r = \frac{1}{\alpha_r} f_m\left(-\frac{\alpha_{r+1}}{\alpha_r}\right), \qquad (11.4.38)$$

and

$$N_{mn}^{rs} = -\frac{mn\alpha_1\alpha_2\alpha_3}{n\alpha_r + m\alpha_s} N_m^r N_n^s, \qquad (11.4.39)$$

where $\alpha_4 = \alpha_1$. The functions f_m are defined by

$$f_m = \frac{1}{m!} \frac{\Gamma(m\gamma)}{\Gamma(m\gamma+1-m)} = \frac{(-1)^{m-1}}{m!} \frac{\Gamma(m-m\gamma)}{\Gamma(1-m\gamma)}$$
$$= \frac{1}{m!}(m\gamma-1)(m\gamma-2)\dots(m\gamma-m+1). \qquad (11.4.40)$$

These functions have an interpretation in terms of the matching of the incoming and outgoing strings at the interaction time described in appendix 11.C.

The expression for the vertex at $\tau = 0$ is obtained by shifting the vertex described by (11.4.22) by the τ translation operator $\exp\left[\sum P_r^- \tau_0/2\right]$. This has no effect when all three external states are on their mass shells, *i.e.*, when $\sum_r P_r^- = 0$, which is not surprising given that the on-shell amplitude is invariant under τ translations. The result is

$$|V\rangle = \exp\left(\sum P_r^- \tau_0/2\right) \exp\left\{ \frac{1}{2} \sum_{r,s} \sum_{m,n=1}^{\infty} N_{mn}^{rs} \underline{\alpha}_{-m}^r \cdot \underline{\alpha}_{-n}^s \right.$$
$$\left. + \sum_{r,s} \sum_{m} N_m^r \underline{\alpha}_{-m}^r \cdot \mathcal{P} \right\} |0\rangle \delta\left(\sum_r \mathbf{p}_r\right).$$
$$(11.4.41)$$

The factor of $\exp\left[\sum P_r^- \tau_0/2\right]$ can be moved to the right of the other exponential factors in $|V\rangle$ using (11.4.20), which converts N's into $\overline{N}$'s, where

$$\overline{N}_m^r = N_m^r e^{m\tau_0/\alpha_r}. \tag{11.4.42}$$

We have also used

$$\exp\left(\sum P_r^- \tau_0/2\right)|0\rangle = \exp\left(-(\mathcal{P}^2/2\hat{\alpha} + \sum_r 1/\alpha_r)\tau_0\right)|0\rangle. \tag{11.4.43}$$

The final form of the vertex at $\tau = 0$ is then given by

$$|V\rangle = \exp\left(-\tau_0 \sum_r 1/\alpha_r + \Delta_B\right)|0\rangle\delta(\sum_r \mathbf{p}_r), \tag{11.4.44}$$

where

$$\Delta_B = \tfrac{1}{2}\sum_{r,s}\sum_{m,n=1}^{\infty} \overline{N}_{mn}^{rs}\underline{\alpha}_{-m}^r \cdot \underline{\alpha}_{-n}^s + \sum_{r,s}\sum_m \overline{N}_m^r\underline{\alpha}_{-m}^r \cdot \mathcal{P} - \frac{\tau_0}{2\hat{\alpha}}\mathcal{P}^2. \tag{11.4.45}$$

The vertex $|V\rangle$ represents $\Delta[X_1(\sigma) + X_2(\sigma) - X_3(\sigma)]$ in the oscillator basis. This can be checked directly by noting that $|V\rangle$ satisfies the identities

$$\left(\hat{X}_1(\sigma) - \hat{X}_3(\sigma)\right)|V\rangle = 0, \qquad 0 \le \sigma \le \pi\alpha_1, \tag{11.4.46}$$

$$\left(\hat{X}_2(\sigma) - \hat{X}_3(\sigma)\right)|V\rangle = 0, \qquad \pi\alpha_1 \le \sigma \le \pi(\alpha_1 + \alpha_2). \tag{11.4.47}$$

Furthermore, since $P^i(\sigma) \equiv -i\delta/\delta X^i(\sigma)$ the interaction also satisfies the momentum conservation conditions

$$\left(\hat{P}_1(\sigma) + \hat{P}_3(\sigma)\right)|V\rangle = 0 \qquad 0 \le \sigma \le \pi\alpha_1, \tag{11.4.48}$$

$$\left(\hat{P}_2(\sigma) + \hat{P}_3(\sigma)\right)|V\rangle = 0 \qquad \pi\alpha_1 \le \sigma \le \pi(\alpha_1 + \alpha_2). \tag{11.4.49}$$

These equations are checked by expressing the coordinates and momenta in terms of the creation and annihilation modes (denoted by the hats) and eliminating the annihilation modes by moving them to the right past the factor $\exp \Delta_B$. In the form given in (11.4.41) it is easy to impose the on-shell condition on the external states, which sets $\sum_r P_r^-$ equal to zero.

11.5   One-Loop Open-String Amplitudes

Let us now consider the evaluation of loop amplitudes by functional methods. The simplest example is the planar one-loop diagram, which was discussed using the operator formalism in chapter 8. All other open-string one-loop diagrams can be treated in analogous fashion. The necessary changes are fairly straightforward and are not described in detail here, since the results were already given in chapter 8. Henceforth, we only consider processes with external ground states in order to keep the formulas from becoming too unwieldy.

11.5.1 The Conformal Mapping for the Planar Loop Diagram

Consider an amplitude with M external tachyons and a single open-string loop. The string diagram now has a slit joining two internal interaction points. Associated with this diagram is a group-theory factor $n\mathrm{tr}(\lambda_1 \ldots \lambda_M)$, where the factor of n arises from $\mathrm{tr}(1)$ on the inner boundary. The diagram can be mapped to the upper half complex plane in many ways. For example, in one of these the external boundary is again mapped to the real z axis while the slit is mapped into a circle in the upper half plane. This is a mapping that generalizes nicely to multiloop open-string diagrams. When there is only one loop, however, it is somewhat simpler to use a mapping that cuts an internal string propagator as shown in fig. 11.6. This mapping takes the external boundary into one segment of the positive real axis while the internal boundary of the loop is mapped into a segment of the negative real axis. The wiggly lines denoting the cut are mapped into two semicircles with radii r_1 and r_2, respectively (with $r_1 < r_2$), which intersect the positive real axis at the two endpoints of the mapping of the external boundary. The identification of these two lines in the string diagram means that the two semicircles are identified.

Defining the ratio of the radii by

$$\frac{r_1}{r_2} = w, \tag{11.5.1}$$

the required mapping is given by

$$\rho = \sum_{r=1}^{M} \alpha_r \ln \psi \left(x_r/z, w \right) + \text{constant}, \tag{11.5.2}$$

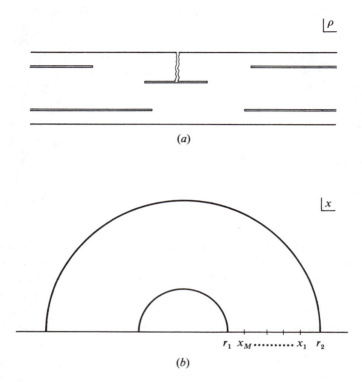

Figure 11.6. The mapping of the planar loop diagram from the ρ plane (a) to a semiannulus in the upper half z plane (b). The string boundary is mapped to a segment of the positive real axis between the two semicircles of radii r_1 and r_2, while the boundary of the ρ plane slit is mapped to the segment of the negative real z axis between the two semicircles. The mapping cuts one of the internal propagators along the wiggly lines, which are mapped into the inner and outer semicircles, and are therefore to be identified.

where the function ψ was defined in chapter 8. Its logarithm is given by

$$\ln \psi(x, w) = \ln \left(\frac{1 - x}{\sqrt{x}} \right) + \frac{\ln^2 x}{2 \ln w}$$
$$+ \sum_{n=1}^{\infty} [\ln(1 - w^n x) + \ln(1 - w^n/x) - 2 \ln(1 - w^n)],$$

$$(11.5.3)$$

which was seen in chapter 8 to be proportional to the correlation function $\langle X^i(z)X^j(z') \rangle$. The periodicity of the ψ function under $\ln x \to \ln x + \ln w$ implies that a point z is identified with any of the points $w^n z$ (where n is any integer). The mapping of the external boundary to the positive z axis is similar to the case of the tree diagrams, since $\psi \sim (z - x_r)$ when $x_r \sim z$.

The external particles are mapped into the points x_r lying between r_1 and r_2 (and repeated along the positive axis in the intervals $w^n r_1$ to $w^n r_2$). For negative z the term $\sum_{r=1}^{M} \alpha_r \ln^2 (x_r/z) / 2 \ln w$ develops an imaginary part so the negative z axis is mapped to the slit in the ρ plane, which has a value of σ given by

$$\sigma_a \equiv \mathrm{Im}\rho_a = \pi \sum_r \alpha_r \frac{\ln x_r}{\ln w}, \qquad (11.5.4)$$

so the value of σ_a (*i.e.*, the width of the internally propagating string that is not cut in fig. 11.6) is explicitly given in terms of w. One circuit of the boundary of the slit in the ρ variable is traced out between $z = -r_1$ and $z = -r_2$. This is repeated in intervals $-w^n r_1$ to $-w^n r_2$ along the negative axis. In this case the mapping from ρ to z has an arbitrariness associated with scale transformations of z (since it depends only on the ratio of the radii r_1 and r_2) rather than the full group of projective transformations. (In the notation of (11.3.4) the invariant points have been fixed to be $y_1 = 0$ and $y_2 = \infty$ leaving the single arbitrary parameter η.)

The integration variables in the ρ plane are the interaction times for the external particles (of which there are $M - 2$), the two interaction times associated with the loop (τ_a and τ_b) and the value of σ_a. One of the interaction times can be fixed due to overall τ translation invariance. This gives a total of M variables. After the conformal mapping these variables are taken to be w and the $M - 1$ variables x_r in the z plane (one x_r can be fixed using the scaling symmetry). The integration region is given by $0 \le w \le 1$, $wx_M \le x_1 \le \cdots \le x_{M-1} \le x_M$, where x_M is fixed arbitrarily and can be chosen equal to w for example (as was the case in chapter 8 where the x_r's were called ρ_r and we set $\rho_M = w$). Notice that string configurations corresponding to different values of w are conformally inequivalent.

11.5.2 The Green Function

The required Green function is a solution of Laplace's equation inside the annulus in fig. 11.6 that satisfies Neumann boundary conditions along the real axis as well as the condition that the two semicircles are identified. This can be viewed as an electrostatic problem (using the analog model §11.1.1) and solved by the method of images. In this case an infinite series of image charges are required. The result is the same as that of the calculation of the correlation function $\langle X^i(z) X^j(z') \rangle$ in chapter 8, namely

$$N(z, z') = \ln |\psi(z'/z)| + \ln |\psi(\bar{z}'/z)|. \qquad (11.5.5)$$

Near the singular point $z = z'$ this function reduces to the same Green function as the one for tree diagrams with no finite corrections. However, the terms in the amplitude that contribute to $\mathcal{M}$ in (11.2.22), namely the contributions containing the exponential of the Green function evaluated on a single string at $\tau = \infty$ or $-\infty$, are sensitive to the fact that the first term of (11.5.2) differs from the corresponding expression for tree diagrams. The Green function contains $\ln|(z' - z)(z' - \bar{z})/zz'|$ instead of $\ln|(z' - z)(z' - \bar{z})|$ so that, when $z \sim z'$, N has an extra term $-2\ln|z|$. The expression for $\mathcal{M}$ in (11.2.22) therefore gets an extra factor

$$\exp\left(-\sum_r \frac{1}{24}(D - 2)\ln|x_r|\right), \qquad (11.5.6)$$

which is equal to $\prod_r |x_r|^{-1}$ when $D = 26$.

11.5.3 The Planar One-Loop Amplitude

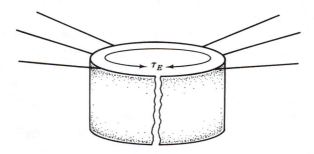

Figure 11.7. A representation of the planar loop diagram used in calculating the volume element. All the strings have $p^+ = 0$ apart from one incoming and one outgoing string. These two strings are sewn together and the time separation τ_E is integrated.

The calculation of one-loop amplitudes requires the evaluation of the product $|\partial\tau_I/\partial z_r|[\det \Delta]^{(2-D)/2}\mathcal{M}$. As in the case of tree diagrams, there is a quick way of calculating this volume factor (which is independent of the transverse momenta), provided we are willing to assume Lorentz invariance. The loop is first thought of as arising from sewing together an M-particle tree diagram, as described in chapter 8. This is depicted in fig. 11.7. The wide strings (with width $\pm\pi\alpha$) begin and end at finite times in arbitrary states, where the time separation between these states is τ_E.

The loop is constructed by taking the trace on these states (including an integration on the loop momentum) to form a cylinder of perimeter τ_E. (This is very similar to the construction used in calculating loop diagrams in chapters 8 and 9.) It is easy to see that this picture is conformally equivalent to the one we wish to calculate by mapping the string diagram from the ρ plane to the z plane using the transformation $z = e^\rho/\alpha$. The strip is then mapped to the semiannulus of fig. 11.6 with the parameter $w = r_1/r_2$ identified with $\exp\{-\tau_E/\alpha\}$. The volume element can again be calculated by considering the special case in which the external ground-state strings have very small α_r's and vanishing transverse momenta. As in the case of the open-string tree amplitudes, the states circulating around the loop cannot change between interactions with the external particles, since this would require powers of external transverse momenta.

Instead of pursuing that route, however, it is possible to give a direct of the volume element by an explicit calculation of $|\partial\tau_I/\partial z_r|[\det\Delta]^{(2-D)/2}\mathcal{M}$ as sketched in appendices 11.A and 11.B. Either method gives the same expression for the volume element. The resulting amplitude (for $D = 26$) is

$$
A_P(1, 2, \ldots, M) = g^M G \int\limits_0^1 \frac{dw}{w^2} \int\limits_{x_M}^{x_M/w} \frac{dx_1}{x_1} \int\limits_{x_M}^{x_1} \cdots \int\limits_{x_M}^{x_{M-2}} \frac{dx_{M-1}}{x_{M-1}}
$$
$$
\times \left(\frac{\pi}{\ln w}\right)^{13} f(w)^{-24} \prod_{1 \leq s < r \leq N} [\psi(x_s/x_r, w)]^{p_r \cdot p_s},
$$

$$(11.5.7)$$

which is the same expression as that obtained in §8.1.1, where x_M was set equal to w. The endpoint of the integration region at which the amplitude is divergent, $i.e.$, $w \to 1$, is the limit in which the length of the slit becomes small, $\tau_a - \tau_b \to 0$. This means that the divergence of the open-string loop appears as a short-distance effect in this formulation of the loop amplitude. As we saw in §8.1.1, there is an alternative interpretation of the divergence as an infrared effect due to the emission of a soft dilaton.

11.5.4 Other One-Loop Amplitudes

The treatment of the planar loop diagram can be generalized to the nonplanar and nonorientable open-string loop diagrams in a fairly obvious manner. The string diagrams now describe interactions in which the strings propagating within the loop twist before recombining. When both

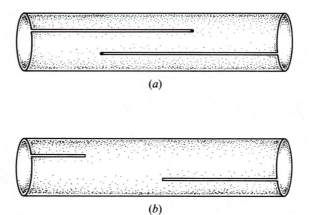

(a)

(b)

Figure 11.8. The two types of contributions to the one-loop nonplanar diagram with one initial and one final open string. (*a*) corresponds to the double-twisted string diagram. (*b*) is the contribution that has closed-string poles. Note that (*a*) has a two open-string intermediate state, whereas (*b*) has a one closed-string intermediate state. The two become the same when the time separation of the vertices vanishes.

the internal propagators have twists the result is an orientable nonplanar diagram that has particles attached to both boundaries. This diagram, fig. 11.8*a*, is a piece of the complete nonplanar amplitude considered in chapter 8. There we noted that the complete amplitude contains the closed-string poles in the channel formed by particles 1 and 2. In the light-cone gauge these are seen to emerge from the contribution shown in fig. 11.8*b*, which involves two successive interactions that convert an open string into a closed one by joining the ends of the open string.

Locally in σ the new interaction is identical to the one that joins two open strings into one. Just as one would expect, it is given by the overlap between an open and a closed string. The appropriate conformal transformation to the z plane for the processes in fig. 11.8 is given by

$$\rho = \sum_{r=1}^{K} \alpha_r \ln \psi \left(x_r / z, w \right) + \sum_{r=K+1}^{M} \alpha_r \ln \psi^T \left(x_r / z, w \right), \qquad (11.5.8)$$

where K particles are attached to the outer boundary and $M - K$ to the inner. In this case the x_r variables are mapped to the positive real z-axis for the external particles attached to one boundary of the string diagram and the negative real axis for those attached to the other boundary. The diagrams in fig. 11.8 give contributions to the amplitude that have the same integrands when expressed in terms of the x_r variables, but each is integrated over a non-Lorentz-invariant region. In order for the complete Lorentz-invariant integration region to be covered, it is necessary to add the contributions from both of the processes with the correct relative normalization. This determines the normalization of the open-string $\leftrightarrow$ closed-string vertex in terms of the three open-string vertex. By requiring the two diagrams in fig. 11.8 to match at their common integration endpoint, it is clear that the new vertex has strength g. This is similar to the way in which the normalization of the four-string interaction was determined by considering the four open-string amplitude. When only one of the internal propagators is twisted, the result is a nonorientable diagram that has a single boundary (and is a Möbius strip). Since the amplitudes associated with these diagrams were also calculated in chapter 8 in the operator formalism, we do not consider these cases any further here.

11.6   Closed-String Amplitudes

11.6.1 Tree Amplitudes

The world sheet for closed-string tree diagrams has periodic boundary conditions as shown in fig. 11.9, which illustrates the M closed-string scattering amplitude. In addition to integrating over the time of the interaction, it is necessary to integrate over the value of σ on the internal string to ensure that it only propagates states that are invariant under global shifts of the σ coordinate. Therefore the interaction point is now specified by two parameters τ_I and σ_I or one complex parameter $\rho_I = \tau_I + i\sigma_I$. (The external states are assumed to be in the physical subspace satisfying (11.1.10).) Due to invariance under rigid τ translations and σ translations one of the (two-dimensional) interaction points can be fixed while the other $2M - 6$ real parameters are to be integrated. The tree amplitudes for M external ground-state closed strings is given by

$$A_c(1, 2, \ldots, M) = 4\pi(\frac{\kappa}{4\pi})^{M-2} \int \prod_{I=1}^{M-3} d\tau_I d\sigma_I \mathcal{M} \det \Delta^{-(D-2)/2}$$

$$\times \exp\left(\frac{1}{4}\sum_{r \neq s} p_r \cdot p_s N(\sigma, \tau_r; \sigma', \tau_s)\right).$$

$$(11.6.1)$$

The Green function N can be evaluated by conformally mapping the string diagram to the whole complex plane. The appropriate mapping is

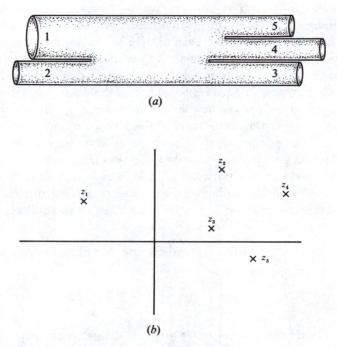

Figure 11.9. The mapping of an arbitrary closed-string tree diagram onto the complex plane. The initial and final strings are mapped to the points z_r, marked by crosses.

again given by (11.3.1) but with complex parameters z_r instead of the x_r

$$\rho \equiv \tau + i\sigma = \sum_{r=1}^{M} \alpha_r \ln(z_r - z). \qquad (11.6.2)$$

The ends of the string diagram at $\tau = \tau_i$ or τ_f are mapped into small circles surrounding the points z_r. The radii of the circles vanish as τ_i or $\tau_f \to \infty$. This picture is the same as the one considered in the §1.4.4. We saw there that the complex plane can be conformally mapped onto itself by $SL(2,C)$ transformations, and therefore there is enough arbitrariness to fix three of the complex z_r parameters. The remaining $M-3$ complex z_r variables are integrated over the complex plane. This is the number expected from the string picture as obtained above.

A particular string diagram, such as fig. 11.9, has a definite ordering of the τ_I variables. The differences between successive τ_I variables are to be integrated from 0 to ∞. When mapped into the z plane this corresponds to integrating the $M-3$ independent z_r variables over a non-Lorentz-invariant section of the complex plane. Different string diagrams are associated with different orderings of the τ_I variables. Adding these contributions together extends the integration region to the whole complex z plane. Just as in the case of the open-string diagrams it is nontrivial that these different diagrams piece together smoothly and, in fact, this is only true in the critical dimension.

Since the string world sheet has no boundaries, the Green function is given by

$$N(z; z') = \ln |z - z'|. \tag{11.6.3}$$

The volume element can again be found by considering the special case in which $M - 2$ strings have infinitesimal p^+ and zero transverse momenta (and assuming that the result is Lorentz invariant in 26 dimensions). The general calculation of $|\partial \tau_I / \partial z_r| [\det \Delta]^{(2-D)/2} \mathcal{M}$ is similar to that of the open-string as outlined in appendices 11.A and 11.B. In either case, fixing $x_1 = 0$, $x_{M-1} = 1$ and $x_M = \infty$, the volume element just contributes the flux factor $\prod_r \alpha_r^{-1/2}$. The resulting covariant scattering amplitude has the form

$$A(1, 2, \ldots, M) \sim 4\pi (\frac{\kappa}{4\pi})^{M-2} \int \left(\prod_{r=2}^{M-2} d^2 z_r \right) \prod_{1 \leq s < r \leq M-1} |z_r - z_s|^{p_r \cdot p_s / 2}$$

$$\tag{11.6.4}$$

for $z_1 = 0$, $z_{M-1} = 1$, and $z_M = \infty$, which is the same as the Shapiro–Virasoro model given in §7.2.1.

11.6.2 Closed-String One-Loop Amplitudes

Let us now consider one-loop closed-string amplitudes. The discussion in this case parallels that of the open-string loop amplitude of §11.5. The string diagram is shown in fig. 11.10a. Once again, the labelling indicates that the horizontal lines are to be identified in pairs. There are several different kinds of conformal transformations available for mapping this region into the complex plane. In one of these the diagram is mapped to the upper half complex plane with a hole removed. The boundary of the hole has four sectors, which are identified in opposite pairs to make a torus. Another method maps the world sheet into the complex plane with two nonconcentric circular holes removed, whose boundaries are identified with each other by a projective transformation. Both the methods generalize to multiloop diagrams but are more cumbersome than necessary for the one-loop case.

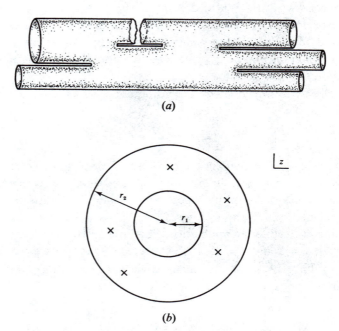

(a)

(b)

Figure 11.10. The mapping of the closed-string one-loop diagram (a) onto an annulus in the z plane (b). In (b) the ratio of the internal and external radii is $r_1/r_2 = |w|$, and any point z is identified with the points $w^n z$, where n is an arbitrary integer.

The most convenient method for treating a closed-string one-loop diagram entails cutting the string diagram along the dashed line and conformally transforming it to an annulus in the complex z plane shown in fig. 11.10b. This is similar to the transformation used for open-string loops. The inner and outer circles (with radii r_1 and r_2) are again identified, $i.e.$, any point z is identified with any of the points $w^n z$, where $|w| = r_1/r_2$. The phase of the complex parameter w corresponds to the fact that the two circles formed by cutting along the wiggly lines in fig. 11.10a only need to be matched up to a global shift in the σ parameter. The mapping is expressed in terms of the function $\psi(z_r/z, w)$ by

$$\rho = \sum_{r=1}^{M} \alpha_r \left\{ \ln \psi(z_r/z, w) - \frac{\ln^2(z_r/z)}{2 \ln w} - \frac{\ln z \ln |z_r|}{\ln |w|} \right\}, \qquad (11.6.5)$$

where the expression for ψ is given in (11.5.3). The width in σ of the internally propagating string (which is not cut in fig. 11.10) is determined by the last term in (11.6.5)

$$\sigma_a = -2\pi \sum_{r=1}^{M} \alpha_r \frac{\ln |z_r|}{\ln |w|}. \tag{11.6.6}$$

This corresponds to the imaginary part of ρ generated as z encircles the origin. (The sum of the first two terms in (11.6.5) is regular at $z = 0$.) In general the parametrization has a 'twist' in going around the handle, with a change in σ equal to

$$\phi_a = \text{Im}[\rho(wz) - \rho(z)] = \text{Im} \sum_{r=1}^{M} \alpha_r \left(\ln z_r - \ln |z_r| \frac{\ln w}{\ln |w|} \right), \tag{11.6.7}$$

where we have used $\psi(wx, w) = \psi(x, w)$. In fact, the difference $\rho(wz) - \rho(z)$ has no real part.

The amplitude is given by an integral over the $M - 1$ pairs of σ_r and τ_r values at the interaction points, where the invariance under overall τ translation and σ translation can be used to fix one of the interaction points. In addition, the width in σ of the string below the slit (σ_a) and the relative twist between the two internal propagators (ϕ_a) must be integrated over. This gives a total of $2M$ real variables, which are mapped into the M complex variables, z_r and w. As expected, this counting of integration variables agrees with the number of integration variables associated with a closed Riemann surface with M points attached and one handle.

The range of integration of the z_r variables shown in fig. 11.10 depends on the value of $w \equiv \exp\{2\pi i\tau\}$, since this determines the ratio of radii of the annulus. The range of integration of w (or τ) requires more careful consideration. It is important that the conformal mapping of (11.6.5) is invariant under the transformations

$$\tau \to \tau + 1, \tag{11.6.8}$$

and under

$$\tau \to -1/\tau, \quad \ln z_r \to 2\pi i \ln z_r / \ln w, \quad \ln z \to 2\pi i \ln z / \ln w, \tag{11.6.9}$$

which are modular transformations of the variable τ. The associated transformation of the $\ln z$ variables corresponds to $\nu \to \nu/\tau$ in the notation of chapter 8. We know from chapter 8 that in the correct expression for the closed string one loop integral, τ should be integrated only over a single fundamental region such as the one denoted F in fig. 8.21. It is very important test of the formalism developed in this chapter that the region of integration should turn out to be correctly described by the string diagram with no overcounting. Recall that in the earlier discussions of closed-string loop amplitudes it was necessary to truncate the integration to a single fundamental region in order to satisfy the unitarity restriction. Since the light-cone calculations of this chapter are an application of ordinary quantum mechanics, unitarity is guaranteed and it inevitable that the correct region of integration emerges. Any consistent field-theoretic formulation of string theory, such as that based on the light-cone gauge, is guaranteed to produce the correct integration region.

The Green function in the z plane (as we have already seen in chapter 8) is given by

$$N(z, z') = \ln \chi \left(z'/z, w \right), \qquad (11.6.10)$$

where the function χ, which is similar to $|\psi|$, was defined in §8.2.1). It is given by

$$\ln \chi(x, w) = \ln \left| \frac{1 - x}{\sqrt{x}} \right| + \frac{\ln^2 |x|}{2 \ln |w|}$$
$$+ \sum_{n=1}^{\infty} \left(\ln |1 - w^n x| + \ln |1 - w^n/x| - 2 \ln |1 - w^n| \right).$$
$$(11.6.11)$$

This function satisfies the required periodicity under $x \to wx$. The factors in the loop amplitude that are independent of the transverse momenta can again be evaluated either by treating the special case in which the external particles have vanishing α_r and p_r^i or by explicit evaluation using a generalization of the methods outlined in appendices 11.A and 11.B. The resulting amplitude is just the one given in §8.2.1.

Although we shall not be evaluating multiloop diagrams, it is easy to count the number of integration variables that would be associated with an arbitrary diagram. For example, consider a closed-string process with M external particles and L loops (*i.e.*, handles). The addition of a handle, corresponding to the insertion of a pair of identified horizontal lines in fig. 11.10, increases the number of internal propagators by three. A generalization of the argument given for the one-loop case shows that there are effectively two real parameters associated with each of these propagators so that there is a total of $6L + 2M - 6$ real parameters to be integrated over. These are precisely the modular parameters that we first encountered in a geometric language in §3.4. The conformal structure of a Riemann surface of genus L involves $3L - 3$ complex parameters or

$6L - 6$ real ones, and M complex parameters or $2M$ real ones specify the positions of the M external particles on the surface.

11.7 Superstrings

In this section we sketch the application of light-cone functional methods to the calculation of superstring amplitudes. Most of the discussion concerns open-string processes, but the extension to closed strings is straightforward. The major new feature that will emerge is that superstring interactions are not completely described by overlap delta functionals that ensure the continuity of the coordinates at the interaction time. In addition it is necessary to insert an operator at the interaction point in order for the interaction to be consistent with Lorentz invariance and supersymmetry.

We use the light-cone gauge formalism in which space-time supersymmetry is apparent. In that case the constraint that the interacting theory is invariant under space-time supersymmetry more or less determines the form of the interactions. The additional constraints imposed by Lorentz invariance merely determine a measure factor that is a function of the p_r^+'s (*i.e.*, of the α_r's).

11.7.1 The $SU(4) \times U(1)$ Formalism

The evaluation of the functional integral is easiest when expressed as an integral over a commuting (or anticommuting) set of coordinates. Since the fermionic $SO(8)$ spinor $S^a(\sigma)$ satisfies

$$\{S^a(\sigma), S^b(\sigma')\} = \pi \delta^{ab} \delta(\sigma - \sigma'), \tag{11.7.1}$$

it is simultaneously a coordinate and its conjugate momentum. The second spinor $\tilde{S}^a$ (or $\tilde{S}^{\dot{a}}$ in the case of the type IIA theory) satisfies a similar relation. In the case of the type IIB theory it is possible to define anticommuting Grassmann coordinates, $S_1^a + i\tilde{S}_2^a$, which are distinct from their conjugate momenta $S_1^a - i\tilde{S}_2^a$. However, for the other superstring theories there is no such decomposition that preserves manifest $SO(8)$ symmetry. (S and $\tilde{S}$ are inequivalent $SO(8)$ spinors in the type IIA theory, they are not independent spinors in type I open-string theories and there is no $\tilde{S}$ at all in the heterotic theory.) In these cases we can introduce distinct 'coordinates' and conjugate 'momenta' only if we break the manifest spin(8) symmetry. The most straightforward way of doing this is to decompose the variables with respect to a spin(6) × spin(2) $\sim SU(4) \times U(1)$ subgroup.

It is natural to think of the spin(6) group as representing rotations of six 'internal' space-time dimensions and the $U(1)$ as representing helicity in ordinary four-dimensional space-time. The embedding is chosen so that the three eight-dimensional spin(8) representations decompose as

$$8_v \to 6_0 + 1_1 + 1_{-1}, \tag{11.7.2}$$

$$8_s \to 4_{\frac{1}{2}} + \overline{4}_{-\frac{1}{2}}, \tag{11.7.3}$$

$$8_c \to 4_{-\frac{1}{2}} + \overline{4}_{\frac{1}{2}}, \tag{11.7.4}$$

where the subscripts indicate the $U(1)$ content, which can be thought of as the four-dimensional helicity. Using (11.7.2)

$$X^i \to \left(X^I, X^R, X^L \right), \tag{11.7.5}$$

where $I = 1, \ldots, 6$ and R and L refer to the right-handed and left-handed chirality, *i.e.*,

$$X^R = \frac{1}{\sqrt{2}}(X^7 + iX^8), \tag{11.7.6}$$

$$X^L = \frac{1}{\sqrt{2}}(X^7 - iX^8). \tag{11.7.7}$$

Also

$$S^A(\sigma) + iS^{A+4}(\sigma) \sim \theta^A(\sigma), \tag{11.7.8}$$

$$S^A(\sigma) - iS^{A+4}(\sigma) \sim \lambda_A(\sigma) \equiv \frac{\delta}{\delta\theta^A(\sigma)}, \tag{11.7.9}$$

where the superscript $A = 1, \ldots, 4$ labels a **4** of $SU(4)$. When written as a subscript it represents a $\overline{\mathbf{4}}$ of $SU(4)$. The symbol $\sim$ indicates that an arbitrary choice of normalization has been made so that θ^A and λ_A satisfy

$$\{\lambda_A(\sigma), \theta^B(\sigma')\} = \delta_A^B \delta(\sigma - \sigma'), \tag{11.7.10}$$

$$\{\theta^A(\sigma), \theta^B(\sigma')\} = \{\lambda_A(\sigma), \lambda_B(\sigma')\} = 0. \tag{11.7.11}$$

A second Grassmann $SU(4)$ spinor $\tilde{\theta}^A(\sigma)$ and its conjugate momentum $\tilde{\lambda}_A(\sigma)$ are similarly defined from $\tilde{S}^a(\sigma)$. This decomposition, which treats six of the transverse dimensions differently from the other two, is introduced as a mathematical device to define the two sets of Grassmann coordinates, θ^A and $\tilde{\theta}^A$. However, this may be a quite natural treatment of the space-time coordinates, since ultimately these theories can only make physical sense if six dimensions compactify and are therefore really distinguished from the other ones.

For open strings the boundary conditions satisfied by the Grassmann coordinates are

$$\theta^A(0) = \tilde{\theta}^A(0), \qquad \theta^A(\pi|\alpha|) = \tilde{\theta}^A(\pi|\alpha|), \tag{11.7.12}$$

so that the mode expansions are

$$\theta^A(\sigma) = \sum_{-\infty}^{\infty} \theta_m^A e^{im\sigma/|\alpha|}, \qquad \tilde{\theta}^A(\sigma) = \sum_{-\infty}^{\infty} \theta_m^A e^{-im\sigma/|\alpha|}. \tag{11.7.13}$$

The coordinates θ^A and $\tilde{\theta}^A$ are independent functions in the range of the string $0 \le \sigma \le \pi|\alpha|$, since the exponentials in (11.7.13) are only a complete basis over the doubled range. This is demonstrated by the fact that

$$\tilde{\theta}^A(\sigma) = \theta^A(-\sigma). \tag{11.7.14}$$

Similarly, the open-string Grassmann momenta have the expansions

$$\lambda_A(\sigma) = \sum_{-\infty}^{\infty} \lambda_{mA} e^{im\sigma/|\alpha|}, \qquad \tilde{\lambda}_A(\sigma) = \sum_{-\infty}^{\infty} \tilde{\lambda}_{mA} e^{-im\sigma/|\alpha|}, \tag{11.7.15}$$

where

$$\lambda_{mA} = \frac{1}{2\pi|\alpha|} \frac{\partial}{\partial \theta_{-m}^A}, \tag{11.7.16}$$

which ensures that (11.7.10) and (11.7.11) are satisfied.

It is convenient to normalize the modes of θ and λ by defining fermionic creation and annihilation operators R_m^A and R_{mA}, by

$$\theta_m^A = \frac{1}{\sqrt{2\alpha}} R_m^A, \qquad \lambda_{mA} = \frac{1}{\sqrt{2\pi|\alpha|}} R_{mA}, \tag{11.7.17}$$

where

$$\{R_m^A, R_{nB}\} = \alpha \delta_{m+n} \delta_B^A, \tag{11.7.18}$$

and

$$\{R_m^A, R_n^B\} = \{R_{mA}, R_{nB}\} = 0. \tag{11.7.19}$$

These modes contain a factor of $\sqrt{\alpha}$ relative to the dimensionless modes S_m^a. This factor is important when considering the action of the Lorentz generators J^{+-} and J^{i-}, since these generators contain factors of $x^- = -i\partial/\partial p^+$. Although S^a and $\tilde{S}^a$ transform as the components of a world-sheet spinor, θ^A and $\tilde{\theta}^A$ do not. The factors of $\sqrt{\alpha}$ have been inserted in order to avoid cumbersome notation in the subsequent development.

For closed strings the coordinates θ^A and $\tilde{\theta}^A$ have independent expansions

$$\theta^A(\sigma) = \sum_{-\infty}^{\infty} \theta_m^A e^{2im\sigma/|\alpha|}, \qquad \tilde{\theta}^A(\sigma) = \sum_{-\infty}^{\infty} \tilde{\theta}_m^A e^{-2im\sigma/|\alpha|}, \qquad (11.7.20)$$

as do the momenta λ_A and $\tilde{\lambda}_A$

$$\lambda_A(\sigma) = \sum_{-\infty}^{\infty} \lambda_{mA} e^{2im\sigma/|\alpha|}, \qquad \tilde{\lambda}_A(\sigma) = \sum_{-\infty}^{\infty} \tilde{\lambda}_{mA} e^{-2im\sigma/|\alpha|}. \qquad (11.7.21)$$

The relations $\lambda_A(\sigma) = \delta/\delta\theta^A(\sigma)$ and $\tilde{\lambda}_A(\sigma) = \delta/\delta\tilde{\theta}^A(\sigma)$ follow from the identifications

$$\theta_m^A = \frac{1}{\alpha} Q_m^A, \qquad \tilde{\theta}_m^A = \frac{1}{\alpha} \tilde{Q}_m^A \qquad (11.7.22)$$

$$\lambda_{mA} = \frac{1}{\pi|\alpha|} Q_{mA}, \qquad \tilde{\lambda}_{mA} = \frac{1}{\pi|\alpha|} \tilde{Q}_{mA}, \qquad (11.7.23)$$

where

$$\{Q_m^A, Q_{nB}\} = \{\tilde{Q}_m^A, \tilde{Q}_{nB}\} = \alpha\delta_{m+n}\delta_B^A \qquad (11.7.24)$$

with the other anticommutators vanishing.

The string wave functions are now functions of the θ_m^A and $\tilde{\theta}_m^A$ as well as the momentum modes P_m^i. They are therefore at most linear functions of each mode. Consider, for example, the ground state wave function $u(p^i, \theta_0^A)$. This wave function can be expanded in a power series

$$u(\mathbf{x}, \theta_0^A) = u^1 + u_A^{1/2}\theta_0^A + \alpha u_{AB}^0 \theta_0^A \theta_0^B + \frac{1}{3}\alpha\epsilon_{ABCD}u^{A,-1/2}\theta_0^B \theta_0^C \theta_0^D$$
$$+ \frac{1}{6}\alpha^2\epsilon_{ABCD}u^{-1}\theta_0^A \theta_0^B \theta_0^C \theta_0^D$$

$$(11.7.25)$$

The superscripts on the components denote their $U(1)$ helicities. These component wave functions satisfy TCP self-conjugacy conditions imposed by

$$u^{-1} = u^{1*}, \qquad u^{A,-1/2} = u_A^{1/2*}, \qquad (11.7.26)$$

$$u_{AB}^0 = u^{0AB*} = \frac{1}{2}\epsilon_{ABCD}u^{0CD}. \qquad (11.7.27)$$

After imposing these constraints we see that the ground-state wave function describes eight boson states (with integer helicities) and eight fermion states (with half-integer helicities) – precisely the states of the super Yang–Mills multiplet in ten dimensions.

11.7.2 The Super-Poincaré Generators

The super-Poincaré generators can be expressed in terms of the $SU(4) \times U(1)$ coordinates and momenta. This requires decomposing the expressions for the generators, which were written in the $SO(8)$ notation in §5.2.2, and normalizing σ to lie in the range $0 \leq \sigma \leq \pi|\alpha|$. The transverse momentum operator is just the integral of the momentum density

$$p^i = \int\limits_0^{\pi|\alpha|} P^i(\sigma)d\sigma, \tag{11.7.28}$$

where i takes the eight values R, L and $I(=1,\ldots,6)$. The two $SO(8)$ supercharges Q^a and $Q^{\dot a}$, which make up a single 16-component Majorana–Weyl ten-dimensional supercharge, each split into two $SU(4)$ spinor supercharges. Thus, in the type IIB theory the two $SO(8)$ supercharges decompose as

$$Q_M^a \rightarrow \left(Q_M^{+A}, Q_{MA}^+\right), \tag{11.7.29}$$

and

$$Q_M^{\dot a} \rightarrow \left(Q_M^{-A}, Q_{MA}^-\right), \tag{11.7.30}$$

where $M = 1, 2$. The significance of the superscripts $\pm$ will become clear shortly.

The components of the undotted supercharges are expressed as integrals over the Grassmann coordinates and momenta,

$$Q_1^{+A} = \int\limits_0^{\pi|\alpha|} Q_1^{+A}(\sigma)d\sigma, \qquad Q_2^{+A} = \int\limits_0^{\pi|\alpha|} Q_2^{+A}(\sigma)d\sigma, \tag{11.7.31}$$

where the charge densities are given by

$$Q_1^{+A}(\sigma) = \frac{1}{\pi}\epsilon(\alpha)\theta^A(\sigma), \qquad Q_2^{+A}(\sigma) = \frac{1}{\pi}\epsilon(\alpha)\tilde\theta^A(\sigma) \tag{11.7.32}$$

(with $\epsilon(\alpha) = 1$ if $\alpha > 0$ and $\epsilon(\alpha) = -1$ if $\alpha < 0$) and

$$Q_{1A}^+ = \int\limits_0^{\pi|\alpha|} Q_{1A}^+(\sigma)d\sigma, \qquad Q_{2A}^+ = \int\limits_0^{\pi|\alpha|} Q_{2A}^+(\sigma)d\sigma, \tag{11.7.33}$$

where

$$Q^+_{1A}(\sigma) = \lambda_A(\sigma), \qquad Q^+_{2A}(\sigma) = \tilde{\lambda}_A(\sigma). \tag{11.7.34}$$

The components of the dotted supercharges are more complicated, since they are bilinear in oscillators, as discussed in §5.2.2. In $SU(4)$ notation they are given by

$$Q^-_{1A} = \int_0^{\pi|\alpha|} \left\{ \sqrt{2} \rho^I_{AB} \left(P^I - \frac{1}{\pi} X'^I \right) \theta^B + 2\pi\epsilon(\alpha) \left(P^L - \frac{1}{\pi} X'^L \right) \lambda_A \right\} d\sigma, \tag{11.7.35}$$

$$Q^-_{2A} = \int_0^{\pi|\alpha|} \left\{ \sqrt{2} \rho^I_{AB} \left(P^I + \frac{1}{\pi} X'^I \right) \tilde{\theta}^B + 2\pi\epsilon(\alpha) \left(P^L + \frac{1}{\pi} X'^L \right) \tilde{\lambda}_A \right\} d\sigma, \tag{11.7.36}$$

$$Q^{-A}_1 = \int_0^{\pi|\alpha|} \left\{ 2 \left(P^R - \frac{1}{\pi} X'^R \right) \theta^A - \sqrt{2}\pi\epsilon(\alpha) \rho^{IAB} \left(P^I - \frac{1}{\pi} X'^I \right) \lambda_B \right\} d\sigma, \tag{11.7.37}$$

and

$$Q^{-A}_2 = \int_0^{\pi|\alpha|} \left\{ 2 \left(P^R + \frac{1}{\pi} X'^R \right) \tilde{\theta}^A - \sqrt{2}\pi\epsilon(\alpha) \rho^{IAB} \left(P^I + \frac{1}{\pi} X'^I \right) \tilde{\lambda}_B \right\} d\sigma. \tag{11.7.38}$$

In these expressions the matrices ρ^I_{AB} and ρ^{IAB} are the Clebsch–Gordan coefficients of $SU(4)$ that describe the coupling of the vector representation to a pair of spinors or antispinors. Some of their relevant properties are listed in appendix 11.D.

The supercharges satisfy the relevant (anti)commutation relations to form the supercharge algebra

$$\left\{ Q^{+A}_M, Q^+_{NB} \right\} = \alpha \delta^A_B \delta_{M,N} \tag{11.7.39}$$

$$\left\{ Q^{+A}_M, Q^{-B}_N \right\} = \sqrt{2} \rho^{IAB} p^I \delta_{M,N} \tag{11.7.40}$$

$$\left\{ Q^+_{MA}, Q^-_{NB} \right\} = \sqrt{2} \rho^I_{AB} p^I \delta_{M,N} \tag{11.7.41}$$

$$\left\{ Q^{+A}_M, Q^-_{NB} \right\} = 2 p^L \delta_{M,N} \delta^A_B \tag{11.7.42}$$

$$\left\{ Q^+_{MA}, Q^{-B}_N \right\} = 2p^R \delta^B_A \delta_{M,N} \tag{11.7.43}$$

$$\left\{ Q^{-A}_M, Q^-_{NB} \right\} = 2H \delta_{M,N} \delta^A_B, \tag{11.7.44}$$

with other anticommutators and commutators vanishing. This is the $SU(4) \times U(1)$ decomposition of the $N = 2$ superalgebra that was expressed in $SO(8)$ notation in §5.2.2. Covariantly it would read

$$\left\{ Q_M, Q_N \right\} \sim \Gamma_\mu P^\mu \delta_{M,N}. \tag{11.7.45}$$

The expression for the Hamiltonian on the right-hand side of (11.7.44) is

$$H \equiv P^- = \int_0^{\pi|\alpha|} \left\{ \epsilon(\alpha) \left(\pi \mathbf{P}^2 + \frac{1}{\pi} \mathbf{X}'^2 \right) - 2i \left(\theta^A \lambda'_A - \tilde{\theta}^A \tilde{\lambda}'_A \right) \right\} d\sigma. \tag{11.7.46}$$

To verify the closure of (11.7.44), it is important to impose the subsidiary condition $N = \tilde{N}$ for type II theories or $N = \tilde{N} - 1$ for the heterotic theory, where N and $\tilde{N}$ are the number operators made out of the appropriate kinds of oscillators.

The expressions for the generators in the heterotic string theory are obtained by deleting one of the Grassmann coordinates ($\tilde{\theta}$ say) in the above expressions and including extra terms, which describe the internal symmetry (as discussed in chapter 6). In the nonchiral type IIA theory the expressions for the generators are the same as in the type IIB theory but with the roles of $\tilde{\theta}^A$ and λ_A interchanged.

In type I theories not all of the supercharges are conserved (*i.e.*, they do not all commute with H). For open strings the boundary conditions truncate the supersymmetry so that only the average supercharges

$$Q^+ = \sqrt{\tfrac{1}{2}} (Q^+_1 + Q^+_2) = \sqrt{\tfrac{1}{2}} \int_{-\pi|\alpha|}^{\pi|\alpha|} Q^+_1(\sigma) d\sigma, \tag{11.7.47}$$

and

$$Q^- = \sqrt{\tfrac{1}{2}} (Q^-_1 + Q^-_2) = \sqrt{\tfrac{1}{2}} \int_{-\pi|\alpha|}^{\pi|\alpha|} Q^-_1(\sigma) d\sigma \tag{11.7.48}$$

are conserved, where the suppressed indices are either those of a **4** or $\overline{\mathbf{4}}$. In these expressions we have expressed the charges as integrals over the doubled region $-\pi|\alpha| \le \sigma \le \pi|\alpha|$. Likewise, for type I closed strings the condition that symmetrizes between the two Fock spaces leads to a corresponding truncation of the supersymmetry.

The expressions for the supercharges are easily written in terms of the modes by substituting the expansions for the coordinates and the momenta. For example, for open strings the expressions for the Q^- charges are

$$Q_A^- \equiv \sqrt{\tfrac{1}{2}}(Q_{1A}^- + Q_{2A}^-) = \frac{1}{\alpha} \sum_{m=-\infty}^{\infty} (\sqrt{2}\alpha_{-m}^I \rho_{AB}^I R_m^B + 2\alpha_{-m}^L R_{mA}),$$

(11.7.49)

$$Q^{-A} \equiv \sqrt{\tfrac{1}{2}}(Q_1^{-A} + Q_2^{-A}) = \frac{1}{\alpha} \sum_{m=-\infty}^{\infty} (-\sqrt{2}\alpha_{-m}^I \rho^{IAB} R_{mB} + 2\alpha_{-m}^R R_m^A)$$

(11.7.50)

and the Hamiltonian is

$$H_{op} \equiv P_{op}^- = \frac{2}{\alpha} \sum_{m=-\infty}^{\infty} (\tfrac{1}{2}\alpha_{-m}\alpha_m + \frac{m}{\alpha} R_{-m}^A R_{mA})$$
$$= \frac{1}{\alpha}(\mathbf{p}^2 + 2N_{op})$$

(11.7.51)

In these expressions the bosonic zero mode oscillator is defined by $\alpha_0^i = p^i$ and the operator N_{op} is defined by

$$N_{op} = \sum_{m=1}^{\infty} (\alpha_{-m}^i \alpha_m^i + \frac{m}{\alpha} R_{-m}^A R_{mA} + \frac{m}{\alpha} R_{-mA} R_m^A).$$

(11.7.52)

There is no normal-ordering constant, since this cancels mode by mode in the passage from (11.7.46) to (11.7.52).

In the type IIB theory the corresponding formulas are

$$Q_{1A}^- = \frac{2}{\alpha} \sum_{m=-\infty}^{\infty} (\sqrt{2}\alpha_{-m}^I \rho_{AB}^I Q_m^B + 2\alpha_{-m}^L Q_{mA})$$

(11.7.53)

$$Q_1^{-A} = \frac{2}{\alpha} \sum_{m=-\infty}^{\infty} (-\sqrt{2}\alpha_{-m}^I \rho^{IAB} Q_{mB} + 2\alpha_{-m}^R Q_m^A),$$

(11.7.54)

with equivalent formulas for the Q_2^-'s using the $\tilde{Q}$ operators. The bosonic zero-mode oscillator in the closed-string sector is defined by $\alpha_0^i = p^i/2$.

The closed-string Hamiltonian is given by

$$H_{cl} \equiv P_{cl}^- = \frac{1}{\alpha}\left(\mathbf{p}^2 + 4N_{cl} + 4\tilde{N}_{cl}\right), \qquad (11.7.55)$$

where

$$N_{cl} = \sum_{n=1}^{\infty}\left(\alpha_{-n}^i \alpha_n^i + \frac{n}{\alpha}Q_{-n}^A Q_{nA} + \frac{n}{\alpha}Q_{-nA}Q_n^A\right), \qquad (11.7.56)$$

with an identical expression for $\tilde{N}_{cl}$ made out of the $\tilde{\alpha}_m$'s and the $\tilde{Q}_m$'s.

The Lorentz generators $J^{\mu\nu}$, listed in §5.2.2, can also be transcribed into the $SU(4) \times U(1)$ formalism in a similar fashion. These generators have the form

$$J^{\mu\nu} = l^{\mu\nu} + E^{\mu\nu} + K^{\mu\nu}, \qquad (11.7.57)$$

where $l^{\mu\nu}$ and $E^{\mu\nu}$ are given by the same expressions as in chapter 5 with the transverse vector indices decomposed into the $SU(4)$ components **6**, L and R. The spin pieces of these generators are given by

$$K^{IJ} = -\frac{i}{2}\int_0^{\pi|\alpha|}\left(\lambda(\sigma)\rho^{IJ}\theta(\sigma) + \tilde{\lambda}(\sigma)\rho^{IJ}\tilde{\theta}(\sigma)\right)d\sigma$$

$$= -\frac{i}{2\alpha}\sum_{n=-\infty}^{\infty}R_{-n}^A(\rho^{IJ})_A{}^B R_{nB}, \qquad (11.7.58)$$

where the matrix $(\rho^{IJ})_A{}^B = \left(\rho_{AC}^I\rho^{JCB} - \rho_{AC}^J\rho^{ICB}\right)/2$,

$$K^{RI} = \frac{\pi i \epsilon(\alpha)}{2\sqrt{2}}\int_0^{\pi|\alpha|}\left(\lambda(\sigma)\rho^I\lambda(\sigma) + \tilde{\lambda}(\sigma)\rho^I\tilde{\lambda}(\sigma)\right)d\sigma$$

$$= \frac{i}{2\sqrt{2}\alpha}\sum_{n=-\infty}^{\infty}R_{-nA}\rho^{IAB}R_{nB}, \qquad (11.7.59)$$

$$K^{LI} = -\frac{i\epsilon(\alpha)}{2\pi\sqrt{2}}\int_0^{\pi|\alpha|}\left(\theta(\sigma)\rho^I\theta(\sigma) + \tilde{\theta}(\sigma)\rho^I\tilde{\theta}(\sigma)\right)d\sigma$$

$$= -\frac{i}{2\sqrt{2}\alpha}\sum_{n=-\infty}^{\infty}R_{-n}^A\rho_{AB}^I R_n^B \qquad (11.7.60)$$

and

$$K^{LR} = \frac{i}{2} \int\limits_{0}^{\pi|\alpha|} : \Big(\lambda(\sigma)\theta(\sigma) + \tilde{\lambda}(\sigma)\tilde{\theta}(\sigma) \Big) : d\sigma$$

$$= -\frac{i}{2\alpha} \sum_{n=-\infty}^{\infty} : R^A_{-n} R_{nA} : .$$

(11.7.61)

In this formula the symbol :: denotes normal ordering of the nonzero modes while the zero modes are antisymmetrized (*i.e.*, $: R_{0A} R_0^A := (R_{0A} R_0^A - R_0^A R_{0A})/2$). The analogous expressions for the closed-string theories are obtained by substituting the appropriate mode expansions.

The spin generators can easily be shown to satisfy the $SO(8)$ subalgebra of the full ten-dimensional Lorentz group,

$$\big[K^{ij}, K^{kl} \big] = -iK^{il}\delta^{jk} + \text{permutations,}$$

(11.7.62)

even though this is not a manifest symmetry of the theory in the $SU(4) \times U(1)$ formalism. In particular, the rotations generated by J^{LI} and J^{RI} are symmetries that need to be checked explicitly. These generate transformations that mix the two inequivalent $SU(4)$ spinors giving

$$\big[J^{LI}, Q^{-A} \big] = -i\sqrt{\tfrac{1}{2}} \rho^{IAB} Q_B^-$$

(11.7.63)

$$\big[J^{LI}, Q_A^- \big] = 0,$$

(11.7.64)

with similar relations involving J^{RI}. The Hamiltonian is invariant under these transformations so that

$$\big[J^{LI}, H \big] = \big[J^{RI}, H \big] = 0.$$

(11.7.65)

11.7.3 Supersymmetry Algebra in the Interacting Theory

In the bosonic theory the interactions were based on the simple *ansatz* that the string coordinates are continuous at the interaction time. This led to the same Lorentz-invariant amplitudes as were derived in the earlier chapters by operator methods. For superstrings continuity of the (super)space coordinates is not the whole story. In addition to the continuity delta functional, the vertex must contain an operator that acts at the joining or splitting point. For example, the interaction that couples three open superstrings, which was given by entirely by a delta functional for the bosonic theory, will be shown to be given by

$$g\hat{H}(\sigma_I)\Delta[Z_1(\sigma) + Z_2(\sigma) - Z_3(\sigma)] \qquad (11.7.66)$$

in type I superstring theory. The collection of superspace coordinates is denoted by $Z_r(\sigma) = (X_r^i(\sigma), \theta_r^A(\sigma), \tilde{\theta}_r^A(\sigma))$ and $\hat{H}(\sigma_I)$ is an operator made out of $P_r^i(\sigma)$, $\theta_r^A(\sigma)$ and $\tilde{\theta}_r^A(\sigma)$ acting at the interaction point $\sigma_I = \pi\alpha_1$ (with the choice of parametrization given in §11.2.2). The operator $\hat{H}$ will turn out to be linear in P^i, which makes it analogous to the cubic interaction between gauge particles in ordinary Yang–Mills theory.

In order to derive the form of the interaction, it proves useful to work in the oscillator basis in which the vertex can be written as a tensor product of ket vectors in the oscillator spaces of the three strings. This generalizes the expression for $|V\rangle$ of the bosonic theory, (11.4.45). The corresponding quantity in the superstring theory is denoted $|H\rangle$, and (11.7.66) is transcribed into the oscillator basis as

$$|H\rangle = \hat{H}|V\rangle_S, \qquad (11.7.67)$$

where $|V\rangle_S$ is the oscillator-basis representation of the superspace delta functional $\Delta[Z_1(\sigma) + Z_2(\sigma) - Z_3(\sigma)]$.

The interaction vertex $|H\rangle$ generates nonlinear transformations on the space of string states, since it maps a single string into a pair of strings. It is, in fact, equivalent to a cubic interaction in the *field theory* of strings. Although string field theory is not developed in this book, our approach to determining the form of $|H\rangle$ is equivalent to the method for determining the cubic vertex in light-cone gauge string field theory. Just as the Hamiltonian (or P^-) gets interaction corrections, the other generators in the superalgebra must also get interaction terms in order for the complete interacting theory to satisfy the supercharge algebra, which is isomorphic to (11.7.39) – (11.7.44). In particular, since the right-hand side of (11.7.44) now contains the interaction Hamiltonian, which acts nonlinearly on string states, the left-hand side must also act nonlinearly. This, in turn, shows that Q^{-A} and Q_A^- must have interaction terms that can be represented (to order g) by ket vectors

$$|Q^{-A}\rangle = \hat{Q}^{-A}|V\rangle_S \qquad (11.7.68)$$

and

$$|Q_A^-\rangle = \hat{Q}_A^-|V\rangle_S, \qquad (11.7.69)$$

respectively.

The complete set of interacting generators must satisfy the supersymmetry algebra. The various generators consist of the free-theory piece given earlier plus an order g interaction term now under consideration. (There could be higher-order terms, as well, but we only consider effects of order g here.) We can isolate the terms that map a single string into two strings (which are terms of order g). The equations that determine these terms are the supersymmetry algebra equations expanded to order g. The only terms of this order are the cross terms involving an anticommutator of a leading-order supercharge with an order g interaction term. Expanding the equations in this manner gives the conditions

$$\sum_r Q_r^{-A}|Q_B^-\rangle + \sum_r Q_{rB}^-|Q^{-A}\rangle = 2\delta_B^A|H\rangle, \tag{11.7.70}$$

$$\sum_r Q_r^{-A}|H\rangle + \sum_r H_r|Q^{-A}\rangle = 0, \tag{11.7.71}$$

$$\sum_r Q_{rA}^-|H\rangle + \sum_r H_r|Q_A^-\rangle = 0, \tag{11.7.72}$$

$$\sum_r Q_r^{-A}|Q^{-B}\rangle + \sum_r Q_r^{-B}|Q^{-A}\rangle = \sum_r Q_{rA}^-|Q_B^-\rangle + \sum_r Q_{rB}^-|Q_A^-\rangle = 0. \tag{11.7.73}$$

In these expressions the generators labelled with a subscript r refer to those defined in the previous subsection made out of the coordinates and momenta of string r. Since we are writing the interaction terms in a three Fock space notation it is necessary to represent each of the zeroth order generators as a sums of three terms, one for each of the three string Fock spaces. The Lorentz generators J^{+-} and J^{i-} also develop interaction corrections that act nonlinearly on the space of strings but we shall not pursue them here. Instead we shall outline how (11.7.70) – (11.7.73) determine the form of the interaction terms, including the interaction Hamiltonian $|H\rangle$.

11.7.4 The Continuity Delta Functional

The expression for $|V\rangle_S$ can be determined from the delta functional by similar methods to those that determined $|V\rangle$ in the bosonic theory (§11.4.5). The result is given by

$$|V\rangle_S = \exp\{\Delta_B + \Delta_F\}|0\rangle\delta(\sum_r \mathbf{p}_r)\delta(\sum_r \alpha_r \theta_r^A)\delta(\sum_r \alpha_r), \tag{11.7.74}$$

where Δ_B is the same expression (made out of the bosonic modes) as before (11.4.45), while the fermionic modes occur in the form

$$\Delta_F = \sum_{r,s=1}^{3} \sum_{m,n=1}^{\infty} U_{mn}^{rs} R_{-mA}^r R_{-n}^{sA} + \sum_{r=1}^{3} \sum_{m=1}^{\infty} V_m^r R_{-mA} \Theta^A. \qquad (11.7.75)$$

In this expression

$$\Theta^A = \frac{1}{\alpha_3} \left(\theta_1^A - \theta_2^A \right), \qquad (11.7.76)$$

which has cyclic symmetry by virtue of the conservation δ functions in (11.7.74). The matrices U_{mn}^{rs} are related to the N_{mn}^{rs} by

$$U_{mn}^{rs} = \frac{m}{\alpha_r} \overline{N}_{mn}^{rs}, \qquad (11.7.77)$$

while

$$V_m^r = -\hat{a}\sqrt{2} \frac{m}{\alpha_r} \overline{N}_m^r. \qquad (11.7.78)$$

Although (11.7.74) can be derived from the functional integral by the same kinds of arguments that led to $|V\rangle$ for the bosonic theory it is easier to check that it does represent the Δ functional by verifying the continuity of the coordinates

$$(Z_1(\sigma) - Z_3(\sigma)) |V\rangle_S = 0, \qquad 0 \le \sigma \le \pi\alpha_1, \qquad (11.7.79)$$

$$(Z_2(\sigma) - Z_3(\sigma)) |V\rangle_S = 0, \qquad \pi\alpha_1 \le \sigma \le \pi(\alpha_1 + \alpha_2) \quad (11.7.80)$$

and conservation of momenta

$$\left(\frac{\delta}{\delta Z_1(\sigma)} + \frac{\delta}{\delta Z_3(\sigma)} \right) |V\rangle_S = 0, \quad 0 \le \sigma \le \pi\alpha_1, \qquad (11.7.81)$$

$$\left(\frac{\delta}{\delta Z_2(\sigma)} + \frac{\delta}{\delta Z_3(\sigma)} \right) |V\rangle_S = 0, \quad \pi\alpha_1 \le \sigma \le \pi(\alpha_1 + \alpha_2) \quad (11.7.82)$$

where the conditions summarize conditions on both the bosonic and the Grassmann variables. The coordinates and the momenta in (11.7.79) - (11.7.82) are expressed in terms of the operator modes $(\alpha_m^i, R_m^A$ and $R_{mA})$.

The supercharges Q^{A+} and Q_A^+ are proportional to the integrals over the densities $\epsilon(\alpha)\theta^A(\sigma)$ and $\lambda_A(\sigma)$, respectively. This means that the Grassmann components of (11.7.79) - (11.7.82) are equivalent to the conservation of the $+$ components of the supercharge densities on the strings at the interaction time at all values of σ.

The situation with the $-$ components of the supercharges is less simple. Recall that Q^{-A} and Q_A^- are integrals of quadratic forms in the bosonic and fermionic coordinates and momenta. The Δ functional at the interaction vertex ensures that the densities of these charges are conserved for all values of σ away from the singular interaction point, $\sigma_I = \pi\alpha_1$. However, the region near $\sigma = \sigma_I$ must be considered more carefully, since various operators are singular there. The way in which $\sum_r Q_r^-$ acts on $|H\rangle$ is determined by the solution of (11.7.70) - (11.7.73). First, however, we must study the behavior of operators that are singular near $\sigma = \sigma_I$.

11.7.5 Singular Operators Near the Interaction Point

The fact that the evolution of the string coordinates is smooth everywhere except at splitting or joining points, where certain operators are singular, is not surprising from the point of view of the functional approach in the light-cone gauge. The curvature of the world sheet is infinite at the interaction points, while it vanishes everywhere else. The existence of these singularities, crucial for the interaction, is straightforward from the functional point of view. The conformal mapping to the upper half z plane gives a parametrization in which the curvature of the world sheet is nowhere singular, and all the operators are well-behaved. Thus the existence of singular operators can be traced to the mapping.

Consider the conformal mapping that takes the upper half z plane to the string diagram (the ρ plane, where $\rho = \tau + i\sigma$). In the vicinity of the interaction point (z_I or ρ_I) the mapping has a square-root branch point, since the string boundary doubles back on itself, so that

$$(z - z_I) \sim c(\rho - \rho_I)^{1/2}, \tag{11.7.83}$$

where c is a constant. This means that

$$\frac{\partial z}{\partial \rho} \sim c(\rho - \rho_i)^{-1/2}. \tag{11.7.84}$$

The operators in the theory pick up factors of $(\partial z/\partial \rho)^J$, where J is the conformal weight of the operator. Since P^i transforms with $J = 1$, this means that, for example,

$$P^i(\rho) = \frac{\partial z}{\partial \rho} P^i(z), \tag{11.7.85}$$

is divergent in the vicinity of the interaction point. There is a similar divergence for $\theta(\rho)$ and $\tilde{\theta}(\rho)$. On the other hand, $X^i(\rho)$ and the $\lambda(\rho)$'s have $J = 0$ and so they are nonsingular. (The rules for the θ's and λ's will be clarified below, where explicit calculations are cited.)

To be more explicit, consider first the behavior of the operator $(P_1^i(\sigma) - X_1'^i(\sigma)/\pi)$ defined on string #1 near the interaction point $\sigma = \pi\alpha_1$,

$$\left(P_1^i(\sigma) - \frac{1}{\pi}X_1'^i(\sigma)\right)|V\rangle_S = \frac{1}{\pi\alpha_1}\sum_{m=-\infty}^{\infty}\alpha_{-m}^{1i}e^{im\sigma/\alpha_1}|V\rangle_S. \tag{11.7.86}$$

Substituting the expression for $|V\rangle_S$ and commuting the annihilation operators through the factor of $\exp\Delta_B$ results in the expression

$$\frac{1}{\pi\alpha_1}\left\{p_1^i + \sum_{m=1}^{\infty}\alpha_{-m}^{1i}e^{-im\sigma/\alpha_1} + \mathcal{P}^i\sum_{n=1}^{\infty}ne^{in\sigma/\alpha_1}\overline{N}_n^1 \right.$$
$$\left. + \sum_{s=1}^{3}\sum_{m,n=1}^{\infty}n\overline{N}_{nm}^{1s}\alpha_{-m}^{si}e^{in\sigma/\alpha_1}\right\}|V\rangle_S, \tag{11.7.87}$$

where $\mathcal{P}^i$ was defined in (11.4.35).

The last two terms in the preceding equation are found to be divergent in the limit $\epsilon \to 0$, where ϵ is a positive parameter defined by

$$\epsilon = \pi\alpha_1 - \sigma. \tag{11.7.88}$$

Substituting the expression for $\overline{N}_m$ from (11.4.42) and (11.4.38), the leading behavior of the second series in (11.7.87) is given (using Stirling's approximation) for $\epsilon \to 0$ by

$$\frac{\mathcal{P}^i}{\pi\alpha_1}\sum_{n=1}^{\infty}ne^{in\sigma/\alpha_1}\overline{N}_n^1 = \frac{\mathcal{P}^i}{\pi\alpha_1}\sum_{n=1}^{\infty}\frac{(-1)^n}{n!}e^{in\sigma/\alpha_1}\frac{\Gamma(-n\alpha_2/\alpha_1)}{\Gamma(1-n-n\alpha_2/\alpha_1)}e^{n\tau_0}$$
$$\sim -\frac{2\mathcal{P}^i}{(2\pi)^{3/2}\alpha_1\sqrt{-\alpha_2\alpha_3}}\sum_{n=1}^{\infty}\frac{1}{\sqrt{n}}e^{in\epsilon/\alpha_1}$$
$$\sim \frac{\eta^\star\mathcal{P}^i}{\pi(-2\hat{\alpha})^{1/2}}\epsilon^{-1/2}, \tag{11.7.89}$$

where

$$\eta = e^{\pi i/4} \tag{11.7.90}$$

and $\hat{\alpha} = \alpha_1\alpha_2\alpha_3$ as before. In order to define the infinite sum properly, it is important that ϵ have a small positive imaginary part, which corresponds to considering the operator to be at a slightly negative value of τ.

Similarly, from (11.4.39), as $n \to \infty$

$$\overline{N}_{nm}^{1s} \sim -\frac{m\hat{\alpha}}{\alpha_s}\overline{N}_n^1\overline{N}_m^s,$$ (11.7.91)

so the last term in (11.7.87) is also divergent when $\epsilon \to 0$ and has the form

$$\frac{\alpha_{-m}^{si}}{\pi\alpha_1}\sum_{n=1}^{\infty} n\overline{N}_{nm}^{1s}e^{in\sigma/\alpha_1} \sim -\frac{\eta^*\alpha_{-m}^{si}(-\hat{\alpha})^{1/2}}{\sqrt{2\pi}}\epsilon^{-1/2}\frac{m}{\alpha_s}\overline{N}_m^s.$$ (11.7.92)

These results imply that

$$\epsilon^{1/2}\left(P_1^i(\sigma) - \frac{1}{\pi}X_1^{\prime i}(\sigma)\right)|V\rangle_S \sim \frac{1}{\pi}\eta^*Z^i|V\rangle_S,$$ (11.7.93)

where the linear combination of bosonic oscillators Z^i is defined by

$$Z^i = |2\hat{\alpha}|^{-1/2}\left(\mathcal{P}^i - \hat{\alpha}\sum_{r,m}\frac{m}{\alpha_r}\overline{N}_m^r\alpha_{-m}^{ri}\right).$$ (11.7.94)

In a similar manner it follows that

$$\epsilon^{1/2}\left(P_1^i(\sigma) + \frac{1}{\pi}X_1^{\prime i}(\sigma)\right)|V\rangle_S \sim \frac{1}{\pi}\eta Z^i|V\rangle_S.$$ (11.7.95)

The expression Z^i is symmetric under the cyclic interchange of the three strings, which is a reflection of the fact that, although we started by considering operators defined on string #1, the same result is obtained by considering the action of the operators in the vicinity of the interaction point on any of the strings.

Similar considerations apply to the action of the Grassmann coordinates at the interaction point. Essentially identical analysis gives

$$\epsilon^{1/2}\theta_1^A(\sigma)|V\rangle_S \sim \eta^*Y^A|V\rangle_S,$$ (11.7.96)

and

$$\epsilon^{1/2}\tilde{\theta}_1^A(\sigma)|V\rangle_S \sim \eta Y^A|V\rangle_S,$$ (11.7.97)

where

$$Y^A = |\tfrac{1}{2}\hat{\alpha}|^{1/2}\left(\Theta^A + \sqrt{\tfrac{1}{2}}\sum_{r,m}\frac{m}{\alpha_r}\overline{N}_m^r R_{-m}^{rA}\right).$$ (11.7.98)

It is straightforward to show by the same sort of analysis that the operators $X_r^i(\sigma)$, $\lambda_A(\sigma)$ and $\tilde{\lambda}_A(\sigma)$ are nonsingular at the interaction point.

11.7.6 The Interaction Terms

We now return to consider (11.7.70)–(11.7.73), which are the equations involving the Q^-'s. Substituting $|H\rangle = \hat{H}|V\rangle$ and $|Q_A^-\rangle = \hat{Q}^{-A}|V\rangle$, the equations to be solved that involve $\hat{H}$ can be rewritten as

$$\sum_{r=1}^{3} Q_r^{-A}\hat{Q}_B^-|V\rangle_S + \sum_{r=1}^{3} Q_{rB}^-\hat{Q}^{-A}|V\rangle_S = 2\delta_B^A\hat{H}|V\rangle_S, \quad (11.7.99)$$

$$\sum_{r=1}^{3} Q_r^{-A}\hat{H}|V\rangle_S + \sum_{r=1}^{3} H_r\hat{Q}^{-A}|V\rangle_S = 0, \quad (11.7.100)$$

$$\sum_{r=1}^{3} Q_{rA}^-\hat{H}|V\rangle_S + \sum_{r=1}^{3} H_r\hat{Q}_A^-|V\rangle_S = 0, \quad (11.7.101)$$

while (11.7.73) gives more constraints on the $|Q^-\rangle$'s.

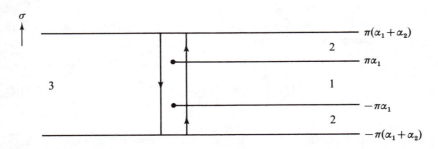

Figure 11.11. The contours of integration in obtaining $\sum_{r=1}^{3} Q_r^-|V\rangle_S$. The string parametrization is extended to the double range $-\pi(\alpha_1 + \alpha_2) \leq \sigma \leq \pi(\alpha_1 + \alpha_2)$ so that the interaction point occurs twice – at $\sigma = \pi\alpha_1$ and $\sigma = -\pi\alpha_1$.

The second terms in (11.7.100) and (11.7.101) vanish when matrix elements are considered with arbitrary on-shell states (since $\sum_r H_r = 0$ for on-shell states). It follows that for such states $\sum_r Q_r^- \hat{H}|V\rangle_S = 0$. This in turn implies that $\hat{H}$ cannot just be a constant, since $\sum_r Q_r^-|V\rangle_S \neq 0$. The easiest way to see this is by expressing the Q_r^-'s as integrals of densities. For example,

$$\sum_{r=1}^{3} Q_r^{-A}|V\rangle_S = \int_{1+2+3} \left\{ \left(P^R - \frac{1}{\pi}X'^R\right)\theta^A \right.$$
$$\left. - \frac{\pi\epsilon(\alpha)}{\sqrt{2}}\rho^{IAB}\left(P^I - \frac{1}{\pi}X'^I\right)\lambda_B \right\} d\sigma |V\rangle_S.$$
$$(11.7.102)$$

In this expression the contours on strings #1, 2 and 3 are taken over double the width of the strings (see fig. 11.11). The conservation of the momenta $P_r^i(\sigma)$, given in (11.7.81) and (11.7.82), and the continuity of the coordinates $X_r^i(\sigma)$ and $\theta_r^A(\sigma)$, given in (11.7.79) and (11.7.80), at the interaction time imply that the integration along strings #1 and #2 in (11.7.102) cancels that along string #3 everywhere, with the possible exception of the singular regions near $\sigma = \pm\pi\alpha_1$. Near these points the integrand is dominated by the leading $(\epsilon)^{-1/2}$ behavior of $(P - X'/\pi)$ and θ, given by (11.7.93), (11.7.95), (11.7.96) and (11.7.97). Substituting these expressions into (11.7.102) gives

$$\sum_{r=1}^{3} Q_r^{-A}|V\rangle_S = \oint \frac{d\epsilon}{2\pi i\epsilon} 4\sqrt{2} Z^R Y^A |V\rangle_S = 4\sqrt{2} Z^R Y^A |V\rangle_S. \quad (11.7.103)$$

The fact that the result is given by the residue of a pole in ϵ depends on the contour on string #3 being displaced to slightly positive τ and the contours on strings #1 and #2 to slightly negative τ, as shown in fig. 11.11. Similarly, we find that

$$\sum_{r=1}^{3} Q_{rA}^{-}|V\rangle_S = 4\rho_{AB}^I Z^I Y^B |V\rangle_S. \quad (11.7.104)$$

The $SO(8)$ components of the Lorentz generators J^{ij} are also not conserved when acting on $|V\rangle_S$. For example,

$$\sum_{r=1}^{3} J_r^{LI}|V\rangle_S = -i\sqrt{2} Y \rho^I Y |V\rangle_S. \quad (11.7.105)$$

We are now in a position to solve for the prefactors $\hat{Q}^{-A}$, $\hat{Q}_A^-$ and $\hat{H}$ that enter into the interaction terms in the superalgebra, (11.7.70) – (11.7.73). It is obviously important that the presence of these prefactors does not interfere with the conservation of the Q_A^+ and Q^{+A} supercharges. This means, for example, that we must require $\left[\sum_r Q_r^{+A}, \hat{H}\right] = 0$. Similar relations are also needed to ensure that the other symmetries of $|V\rangle_S$ are not destroyed by the prefactors. This can be achieved by building all the prefactors out of Z^i and Y^A, defined in (11.7.94) and (11.7.98).

The algebraic equations, (11.7.99) – (11.7.101), determine the form of the prefactors that define the interacting pieces of the superalgebra. The result is

$$\hat{Q}^{-A} = Y^A, \qquad (11.7.106)$$

$$\hat{Q}_A^- = \frac{2}{3}\epsilon_{ABCD}Y^BY^CY^D, \qquad (11.7.107)$$

and

$$\hat{H} = \frac{1}{2}Z^L - \sqrt{\frac{1}{2}}Z^I\rho^I_{AB}Y^AY^B + \frac{1}{3}Z^R\epsilon_{ABCD}Y^AY^BY^CY^D. \quad (11.7.108)$$

In principle, we should allow for a common overall factor that is an undetermined function of the α_r's in these expressions. Only the action of the generators J^{+-} and J^{i-} can distinguish the p^+ dependence (*i.e.*, the α dependence) in the vertex, since only these generators contain factors of $\partial/\partial p^+$. Thus this function can be deduced by considering the closure of the Lorentz algebra. However, since this overall function does not depend on the states being considered it is also determined by considering the special matrix element of $|H\rangle$, for example between ground-state vector particles.

Having obtained the form of the interactions in terms of the oscillator representation, it is possible to reconstruct the prefactors that multiply the Δ functional at the interaction point by using the relationship between Z^i and $(P^i(\sigma_I) \pm X'^I(\sigma_I)/\pi)$ in (11.7.94) and between $Y^A(\sigma_I)$ and $\theta^A(\sigma_I)$ in (11.7.98). As expected, the interaction vertex is linear in the momentum operator $P^i(\sigma_I)$, rather like the cubic Yang–Mills interaction.

The other kinds of interactions between open and closed superstrings that we described for the bosonic theory in §11.2.1 can be derived by the same techniques. The cubic coupling between oriented closed superstrings (type II or heterotic) is particularly interesting, since it generalizes the cubic coupling between gravitons in general relativity. In the type II theory this interaction (denoted $|H\rangle_{cl}$) is written in the oscillator basis in terms of a tensor product of ket vectors in the left-moving and right-moving spaces

$$|H\rangle_{cl} = |H\rangle \otimes |\tilde{H}\rangle, \qquad (11.7.109)$$

where the two factors on the right-hand side are just the expressions for the interactions in the untilded and tilded spaces. In writing this, it is assumed that the vertex acts on states that satisfy the subsidiary condition $N = \hat{N}$. In this way it is possible to deduce all the interaction vertices that enter into the functional calculation of scattering amplitudes in the light-cone gauge.

11.7.7 Tree Amplitudes for Open Superstrings

The evaluation of amplitudes follows by methods that extend those used for the bosonic string in the earlier part of the chapter. It is now necessary to include the integration over the Grassmann coordinates and to incorporate the operators at the interaction vertices. These operators act like additional sources so that the result now involves contractions between the external particles and the interaction points. We do not describe the details of these calculations but merely quote the results for tree diagrams with external open-string ground states obtained by using this method.

The four-particle amplitude has also been obtained by joining up a pair of $|H\rangle$'s by a propagator in the oscillator basis. This calculation illustrates the fact that operator methods are much more more complicated than functional methods for the most general calculations. Only when all but two of the external particles have vanishing $+$ components of their momenta are the operator methods simple.

The general tree amplitude is expressed as a function of the zero modes,

$$\theta_r \equiv \theta_{r,0}. \tag{11.7.110}$$

These enter the expression for the amplitude in the combinations Θ_{rs}, defined by

$$\Theta_{rs} = \frac{\theta_s - \theta_r}{\alpha_s + \alpha_r}, \tag{11.7.111}$$

where the lines r and s are joined to the same vertex in cyclic order (*i.e.*, s follows r in a clockwise direction around the vertex). We also define the following quantities associated with a given vertex involving the lines r and s:

$$\mathcal{P}_{rs}^i = \alpha_r p_s^i - \alpha_s p_r^i, \tag{11.7.112}$$

$$\hat{\alpha}_{rs} = -\alpha_r \alpha_s (\alpha_r + \alpha_s), \tag{11.7.113}$$

$$J_{rs} = \mathcal{P}_{rs}^L + \frac{\hat{\alpha}_{rs}}{\sqrt{2}} \rho_{AB}^I \mathcal{P}_{rs}^I \Theta_{rs}^A \Theta_{rs}^B + \frac{\hat{\alpha}_{rs}^2}{6} \mathcal{P}_{rs}^R \epsilon_{ABCD} \Theta_{rs}^A \Theta_{rs}^B \Theta_{rs}^C \Theta_{rs}^D. \tag{11.7.114}$$

In addition it is necessary to define a combination of factors associated with a pair of vertices containing particles (r, s) and (t, u)

$$K_{rs;tu} = -\frac{1}{24} \hat{\alpha}_{rs} \hat{\alpha}_{tu} \epsilon_{ABCD} \left(\frac{\mathcal{P}_{rs}^2}{\hat{\alpha}_{rs}} - \frac{\mathcal{P}_{tu}^2}{\hat{\alpha}_{tu}} \right)$$
$$\times \left\{ \hat{\alpha}_{rs} \Theta_{rs}^A \Theta_{rs}^B \Theta_{rs}^C \Theta_{tu}^D - \hat{\alpha}_{tu} \Theta_{rs}^A \Theta_{tu}^B \Theta_{tu}^C \Theta_{tu}^D \right\}. \tag{11.7.115}$$

A tree amplitude with M external ground-state particles in a given cyclic order has the form

$$A(1,\ldots,M) = g^{M-2}G \int \prod_{r=2}^{M-1} dx_r \Theta(x_{r+1} - x_r)(\text{K.N.}) \prod_{r=1}^{M-1} \frac{1}{(x_{r+1} - x_r)}$$

$$\times \sum_{perms} T(\{x_r, \theta_r\}) \delta^4(\sum_r \alpha_r \theta_r) \delta^{10}(\sum_r p_r),$$

$$(11.7.116)$$

where G is the Chan–Paton group-theory factor and (K.N.) denotes the usual Koba–Nielsen integrand (fixing $x_1 = 0$ and $x_M = \infty$)

$$\text{K.N.} = \prod_{r<s} (x_s - x_r)^{p_r \cdot p_s}. \qquad (11.7.117)$$

The sum $\sum_{perms} T$ is defined as follows. Each term in the sum is associated with a different tree diagram in which the end particles are always #1 and #M but the ordering of the other particles is permuted. The sum is over all possible permutations of the ordering (even though the complete amplitude describes the process in which the particles have the cyclic ordering asociated with the K.N. factor). Each term in the sum is itself a sum over all possible ways of pairing some of the vertices in the tree diagram with the rest left unpaired. For the vertices that are paired, called (r, s) and (t, u), there is a factor $K_{rs;tu}$, whereas there is a factor of J_{rs} for any unpaired vertex (r, s).

This expression for the amplitude is manifestly invariant under the supersymmetry transformations generated by $Q^{+A} = \sum_r \alpha_r \theta_r^A$ and $Q_A^+ = \sum_r \partial/\partial \theta_r^A$ due to the conservation δ functions in (11.7.116). It requires more work to see that the expression is also invariant under the Q^- supersymmetry transformations.

It is a virtue of this formalism that the result summarizes all the possible ground-state processes. In order to extract the amplitude for a particular set of massless external states one merely has to attach external wave functions, $u_r(k_r^i, \theta_r^A)$, and integrate the amplitude with respect to the θ_r's using the rules for Berezin integration, $\int d\theta_r^A = 0$ and $\int d\theta_r^A \theta_r^A = 1$. The resulting amplitudes can then be expressed, if one wishes, in terms of the wave functions of the component states that occur in the expansion of the wave function given in (11.7.25).

11.8 Summary

In this chapter we have developed the functional approach to calculations of string perturbation-theory diagrams in the light-cone gauge. In this gauge the calculation of scattering amplitudes amounts to an application of familiar quantum-mechanical principles. The extension of the path integral method from point particles to strings was applied to calculate the free string propagator and hence to deduce the spectrum of free string states. The method was then used to calculate on-shell tree diagrams and one-loop amplitudes for both open and closed bosonic strings. Most of these calculations were for processes with external ground-state strings but we also derived the general expression for open-string tree diagrams with arbitrary external states. This led to an explicit calculation of the oscillator expression for the vertex coupling three general open strings. This vertex, together with similar expressions for the other fundamental vertices in the theory, can be used to define the interactions between string fields in a second-quantized field theory of strings in the light-cone gauge, which is a subject that we do not develop.

The calculation of superstring scattering amplitudes was sketched in the supersymmetric light-cone gauge formalism. The major new features are the fact that the functional integral integral involves fermionic coordinates and the fact that the interaction vertices are no longer given as overlap Δ functionals. There is, in addition, an operator at the interaction point. This operator is linear in X^i derivatives for the cubic open-string interaction (just like the Yang–Mills cubic interaction) while it is bilinear in derivatives for the cubic interactions of type II strings (just like the cubic interaction of gravitons).

Appendix 11.A   The Determinant of the Laplacian

Quite generally, the determinant of the two-dimensional Laplace operator is singular. Therefore a regularization procedure is required to define a meaningful quantity. Singularities arise from several sources. There is a multiplicative infinity proportional to $\exp{(area)}$ that was encountered in discussing the propagator in §11.1.4. It can be absorbed into an overall phase of the scattering amplitude (when τ is continued back to real values). Since the area of the parameter space is independent of the number of external strings the phase is the same for all processes and the infinite phase has no effect on physics. This infinity is just a reflection of the normal-ordering infinity discussed in §11.1.3. In the case of a world sheet with boundary there is an additional infinity proportional to $\exp{(perimeter)}$, which we also saw in §11.1.4 can be absorbed into a renormalization of the string Hamiltonian at the string endpoints.

The Laplacian is also singular at points of infinite curvature. In the case of open-string diagrams the one-dimensional geodesic curvature of the boundary is infinite at the interaction points and must be carefully regularized. For closed-string diagrams the two-dimensional curvature is infinite at the interaction points. These divergences can be absorbed into a redefinition of the string coupling constant. In addition, the string wave functions require renormalization, as we saw in §11.1.4. These infinite coupling-constant and wave-function renormalizations actually cancel each other in calculations of physical processes in the critical dimension of space-time. This is yet another nontrivial aspect of the critical dimension. However, we choose to absorb them into renormalized parameters and rely on Lorentz invariance to determine the critical dimension.

All of the above infinities are aspects of two-dimensional field theory on the world sheet and have no connection with infinities of the quantum field theory of strings (which arise from limits of integrations over the positions of the splitting and joining points, *i.e.*, the shape of the two-dimensional surface). The infinities of the determinant occur for tree diagrams as well as for diagrams with an arbitrary number of loops. We are interested in the finite result that emerges after the various infinities are renormalized.

The conformal transformation of the world sheet to the upper half complex plane is also singular, and we are really interested in the product of the determinant and the Jacobian for this conformal transformation. The Jacobian will be described in the next appendix.

(a) Open-String Tree Diagrams

For tree diagrams the method for calculating the determinant is to exploit its anomalous transformation properties under conformal transformations. Since the determinant is easy to evaluate in the upper half plane, its form in the string diagram can be deduced if its behavior under conformal transformations is understood (since an arbitrary diagram can be conformally mapped to the upper half plane). A conformal transformation is expressed as a change in scale of the two-dimensional metric

$$\hat{h}^{\alpha\beta}(\sigma,\tau) \to h^{\alpha\beta}(\sigma,\tau) = e^{-2\phi(\sigma,\tau)}\hat{h}^{\alpha\beta}(\sigma,\tau), \qquad (11.A.1)$$

where $\phi(\sigma,\tau)$ is the arbitrary function that specifies the conformal transformation. Under this transformation the two-dimensional curvature $\hat{R}$ changes to

$$R = e^{-2\phi}(\hat{R} + 4\pi^2\hat{\Delta}\phi). \qquad (11.A.2)$$

The boundary of the string diagram is a curve $z^{\alpha}(s) = (\tau(s),\sigma(s))$ parametrized by s. The unit vector tangential to the curve is $t^{\alpha} = \dot{z}^{\alpha}/(h_{\beta\gamma}\dot{z}^{\beta}\dot{z}^{\gamma})^{1/2}$, while the normal is $n^{\alpha} = -\epsilon^{\alpha\beta}t_{\beta}$. The geodesic curvature k is a measure of how the curve deviates from being a geodesic and is defined by

$$k = n_\beta t^\alpha \nabla_\alpha t^\beta, \tag{11.A.3}$$

where ∇_α is the covariant derivative containing the Christoffel connection. The geodesic curvature of the boundary $\hat{k}$ changes under (11.A.1) to

$$k = e^{-\phi}(\hat{k} - \hat{n}^\alpha \partial_\alpha \phi), \tag{11.A.4}$$

where the hatted variables are those calculated from $\hat{h}$ (and n^α denotes the normal to the boundary).

The form of the Laplace operator is given, in general, by

$$\Delta = -\frac{1}{2\pi^2 \sqrt{h}} \partial_\alpha \sqrt{h} h^{\alpha\beta} \partial_\beta. \tag{11.A.5}$$

Its determinant may be calculated from the formula

$$\ln \det' \Delta = \operatorname{tr}' \ln \Delta, \tag{11.A.6}$$

where the prime indicates that the zero modes have been left out of the definitions of det and tr. The zero modes just give the $\delta^{D-2}(\sum_r \mathbf{p}_r)$ factor associated with the overall conservation of transverse momentum. It follows from (11.A.6) that

$$\ln \det' \Delta = -\int_\epsilon^\infty \frac{dt}{t} \operatorname{tr}' \left(e^{-t\Delta} \right), \tag{11.A.7}$$

where ϵ is an ultraviolet cutoff that defines the regularized determinant and spoils the conformal invariance of the classical theory. We denote the eigenvalues of Δ by λ_n, *i.e.*,

$$\Delta \Phi_n = \lambda_n \Phi_n, \qquad \lambda_n \geq 0, \tag{11.A.8}$$

where the wave functions Φ_n are normalized by

$$\langle \Phi_m, \Phi_n \rangle = \delta_{mn}. \tag{11.A.9}$$

Substituting in (11.A.7) gives

$$\ln \det' \Delta = -\int_\epsilon^\infty \frac{dt}{t} \sideset{}{'}\sum e^{-t\lambda_n}. \tag{11.A.10}$$

The standard way of solving this equation is to consider its variation under an arbitrary infinitesimal change of the conformal factor $\delta\phi(\sigma, \tau)$. Under such a variation the Laplacian changes by $-2\Delta\delta\phi$ so that the changes in the eigenvalues are given by

$$\delta\lambda_n = -2\lambda_n\langle\Phi_n\delta\phi\Phi_n\rangle. \tag{11.A.11}$$

Substituting into (11.A.10) gives

$$\delta\det'\Delta = 2\int_\epsilon^\infty dt \sum{}' \langle\Phi_n\delta\phi\Phi_n\rangle\frac{d}{dt}e^{-t\lambda_n} \tag{11.A.12}$$

$$= -2\mathrm{tr}'\left(\delta\phi e^{-\epsilon\Delta}\right),$$

so the result is determined by properties of the operator $\exp\{-t\Delta\}$ (the 'heat kernel') at a very short 'time', $t = \epsilon$. The expansion of this operator has the form for small t

$$\mathrm{tr}\left(fe^{-t\Delta}\right) = \int d^2z\sqrt{h(z)}f(z)\langle z|e^{-t\Delta}|z\rangle \tag{11.A.13}$$

$$\sim t^{-1}A(f) + t^{-1/2}B(f) + C(f) + O(t^{1/2}),$$

where the coefficients $A(f)$, $B(f)$ and $C(f)$ are known as the De Witt–Seeley coefficients and the formula applies for arbitrary functions f.

This expansion can be obtained by studying the diffusion operator

$$G(z, z'; t) = \langle z'|e^{-t\Delta}|z\rangle, \tag{11.A.14}$$

which satisfies the diffusion equation

$$\left(\frac{\partial}{\partial t} - \Delta\right)G = 2\pi\delta(t - t')\delta^2(z - z'). \tag{11.A.15}$$

This can be solved in perturbation theory around the initial fiducial metric $\hat{h}$. Since the result is only sensitive to the short-distance properties of the manifold, we can always consider a conformal gauge in which $h^{\alpha\beta} = \exp\{2\phi\}\delta^{\alpha\beta}$. The Laplacian can then be written as

$$2\pi^2\Delta = -\partial^2 - V, \tag{11.A.16}$$

where

$$V = \left(e^{-2\phi} - 1\right)\partial^2. \tag{11.A.17}$$

G is determined by the integral equation

$$G = G_0 + G_0 VG, \tag{11.A.18}$$

where G_0 is the diffusion operator in flat space, given by

$$G_0(z, z'; t) = (4\pi t)^{-1/2} \exp(-|z - z'|^2/4t), \tag{11.A.19}$$

in the case that there is no boundary (otherwise G_0 must satisfy the appropriate boundary conditions). To obtain the perturbative solution of (11.A.18) it is convenient to choose coordinates in which

$$\phi(0,0) = 0, \qquad \partial_\alpha \phi(0,0) = 0, \tag{11.A.20}$$

(when there is no boundary) in which case the expression for G truncates at terms quadratic in G_0 in the limit $t \to 0$, i.e.,

$$G = G_0 + G_0 VG_0 + O(t^{1/2}). \tag{11.A.21}$$

We do not present further details of the calculation of the coefficients (which is now quite straightforward). However, choosing $f = \delta\phi$ and substituting in (11.A.12) (and choosing Neumann boundary conditions), gives the expression

$$
\begin{aligned}
-\frac{1}{2}\delta \left(\ln \det' \Delta\right) \sim &\operatorname{tr}' \left(\delta\phi e^{-\epsilon\Delta}\right) \\
\sim &\frac{1}{4\pi\epsilon} \int d^2z \sqrt{\hat{h}}\delta\phi - \frac{1}{8\sqrt{\pi\epsilon}} \int d\hat{l}\delta\phi \\
&+ \frac{1}{12\pi} \left[\int d\hat{l}\hat{k}\delta\phi + \frac{1}{2}\int d^2z \sqrt{\hat{h}}\hat{R}\delta\phi\right] + O(\sqrt{\epsilon}),
\end{aligned}
\tag{11.A.22}
$$

where $d\hat{l}$ indicates a differential along the boundary.

Equations (11.A.1)–(11.A.4) determine the ϕ dependence of the various functions in (11.A.22) so that it can be integrated with respect to ϕ, giving

$$
\begin{aligned}
\ln \det'\Delta|_\phi - \ln \det'\Delta|_{\phi=0} = &-\frac{1}{4\pi\epsilon} \int d^2z \sqrt{\hat{h}}e^{2\phi} + \frac{1}{4\sqrt{\pi\epsilon}} \int d\hat{l}e^\phi \\
&- \frac{1}{6\pi}\Big(\frac{1}{2}\int d^2z \sqrt{\hat{h}}\hat{h}^{\alpha\beta} \frac{\partial\phi}{\partial z^\alpha}\frac{\partial\phi}{\partial z^\beta} \\
&+ \int d\hat{l}\phi\hat{k} + \frac{1}{2}\int d^2z \sqrt{\hat{h}}\phi\hat{R}\Big) \\
&+ \text{(terms independent of } \phi).
\end{aligned}
\tag{11.A.23}
$$

The first two terms on the right-hand side of this expression are the divergent terms proportional to the area and the perimeter of the world sheet that were referred to earlier. They are dropped from now on. The value of $\ln \det' \Delta$ at $\phi = 0$ is the same as the expression for $\ln \det' \Delta$ in the z plane with the small circles excised.

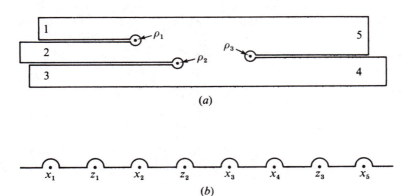

(a)

(b)

'igure 11.12. (*a*) Circular holes are cut out of the world sheet around the interaction points ρ_I. The incoming and outgoing strings are terminated at τ_r, which are large but finite times. (*b*) The images of the holes around ρ_I are semicircles with radii ϵ_I centered on the points z_I on the real z axis. The images of the strings at τ_r are the semicircles with radii ϵ_r centered on the points x_r.

Since (11.A.23) will be applied to transformations from the z to the ρ plane, we consider the the flat metric $\hat{h}^{\alpha\beta} = \delta^{\alpha\beta}$ appropriate to the z plane. In order to avoid the singular interaction points, we cut out small circles of radius r_I around each ρ_I, as shown in fig. 11.12a. These circles are mapped into semicircles (shown in fig. 11.12b) in the z plane centered on the points z_I. We also excise the points corresponding to the incoming and outgoing strings at $\tau = \pm\infty$ by terminating each string at $\tau = \tau_r$, where $|\tau_r|$ is large, which is also illustrated in fig. 11.12a. This corresponds to cutting semicircles out of the z plane centered on $z = x_r$ and with radii ϵ_r, which are small (fig. 11.12b). It also convenient to choose $x_M = \infty$, as usual, which means that the region to be excised around x_M is the region $|z| > 1/\epsilon_M$, where ϵ_M is small.

The surface is flat and so $\Delta\phi = 0$ everywhere. As a result, the whole effect of the conformal transformation is concentrated on the circles around the excluded regions. For a flat surface the scale factor is expressed in terms of the coordinates before and after the transformation (z and ρ respectively) by the formula

$$e^\phi = \left| \frac{\partial \rho}{\partial z} \right|. \tag{11.A.24}$$

Before considering the conformal transformation from the string diagram to the upper half z plane let us consider a $SL(2,R)$ transformation that takes the upper half plane into itself. As we saw in chapter 1, this is a nonsingular conformal transformation. A simple application of (11.A.23) leads to the expression for the determinant in the upper half plane with semicircles of radius ϵ_I cut out from around the image of the interaction points (z_I) and semicircles of radius ϵ_r cut out from around the points x_r (allowing for the special treatment of x_M). The calculation uses the fact that $\partial^2\phi = 0$, so that the first term of order ϵ^0 in (11.A.23) can be integrated by parts to give boundary terms

$$-\frac{1}{12\pi}\int d^2z\,\partial^\alpha\phi\partial_\alpha\phi = -\frac{1}{12\pi}\int d\hat{l}\hat{n}^\alpha\phi\partial_\alpha\phi, \qquad (11.A.25)$$

where the contour goes around the semicircles centered on $z = z_I$ or z_r. The resulting form of the determinant in the z plane, satisfying (11.A.23), is

$$\ln\det'\Delta\big|_{\phi=0} = \frac{1}{6}\sum_p \ln\epsilon_p + \text{terms independent of } \epsilon_p, \qquad (11.A.26)$$

where $\sum_p$ denotes the sum over all the values of I and r.

Now consider the conformal transformation from the z plane to the string diagram (*i.e.*, the ρ plane) defined by (11.3.1),

$$\rho = \sum_{r=1}^{M-1} \alpha_r \ln(x_r - z), \qquad (11.A.27)$$

where we have used τ-translation invariance to set $x_M = \infty$ and dropped an infinite constant from (11.3.1). The interaction points are solutions of $\partial\rho/\partial z = 0$, which gives

$$\sum_{r=1}^{M-1} \frac{\alpha_r}{(z_I - x_r)} = 0. \qquad (11.A.28)$$

Near any interaction point the transformation (11.A.27) has a square root branch point, so that near $\rho = \rho_I$

$$\rho - \rho_I = \frac{1}{2}c_I(z - z_I)^2, \qquad (11.A.29)$$

where

$$c_I = \frac{\partial^2\rho}{\partial z^2}\bigg|_{z=z_I}. \qquad (11.A.30)$$

The circles of radius r_I around the interaction points in the ρ plane are therefore mapped into circles of radius ϵ_I in the z plane, where

$$\ln \epsilon_I = \frac{1}{2}(\ln 2r_I - \ln c_I). \qquad (11.A.31)$$

The behavior of the function ϕ near the interaction points is given by

$$\phi \equiv \ln \left| \frac{\partial \rho}{\partial z} \right| \sim \ln |z - z_I| + \ln c_I = \frac{1}{2}\left(\ln(2|\rho - \rho_I|) + \ln c_I \right). \qquad (11.A.32)$$

Substituting this into (11.A.25) gives a contribution to $\ln \det' \Delta$ of the form

$$\frac{1}{12\pi} \sum_I \int d\hat{l} \frac{1}{\epsilon_I} (\ln \epsilon_I + \ln c_I) = \frac{1}{24} \sum_I (\ln 2r_I + \ln c_I). \qquad (11.A.33)$$

Only the second term depends on the shape of the string diagram *via* the c_I. The other contributions to (11.A.23) come from (11.A.26) and the term involving $\hat{k}$. The c_I dependence in these terms cancels out so that the complete c_I dependence of $\ln \det' \Delta$ is given by (11.A.33). The dependence on the radii, r_I, can be absorbed into a renormalization of the coupling constant, as mentioned earlier.

The other contributions to $\ln \det' \Delta$ come from the region in which the external particles are attached. Near any point $z = x_r$ ($r \neq M$) the mapping is well-approximated by

$$\rho \sim \alpha_r \ln(z - x_r), \qquad (11.A.34)$$

so that

$$\epsilon_r = e^{\tau_r/\alpha_r}, \qquad (11.A.35)$$

where τ_r is the large time at which the incoming or outgoing strings originate. Near any of these points

$$\phi = \ln \left| \frac{\partial \rho}{\partial z} \right| \sim \ln \frac{|\alpha_r|}{|z - x_r|}, \qquad (11.A.36)$$

In this case the contribution of the term in (11.A.25) to (11.A.23) is given by

$$-\frac{1}{12\pi} \int d\hat{l} \hat{n}^\alpha \phi \partial_\alpha \phi = -\frac{1}{12} \sum_r \ln \frac{|\alpha_r|}{\epsilon_r}, \qquad (11.A.37)$$

while there is also a nonzero contribution from the term in (11.A.23) of the form

$$-\frac{1}{6\pi}\int d\hat{i}\hat{k}\phi = \frac{1}{6}\sum_r \ln\frac{|\alpha_r|}{\epsilon_r}. \qquad (11.A.38)$$

The term with $r = M$ has similar form but the sign of $\ln|\alpha_M|$ is reversed due to the reversal of orientation of the contour integral around the point at infinity.

The final source of dependence on ϵ_r comes from $\mathcal{M}$, the contribution of the self-interaction on the rth string at $\tau = \tau_r$, defined in (11.2.22)

$$\ln\mathcal{M} = \sum_r \frac{D-2}{48(\pi\alpha_r)^2}\int_0^{\pi\alpha_r} d\sigma d\sigma' N(\sigma,\tau_r;\sigma',\tau_r)$$

$$\sim \frac{D-2}{24}\sum_r \ln\epsilon_r + \text{ terms independent of } \epsilon_r. \qquad (11.A.39)$$

In evaluating the integral we have made use of the formula for the Green function on a strip, (11.4.9).

The terms in (11.A.37), (11.A.38) and (11.A.26) combine with $\ln\mathcal{M}$ to cancel the dependence on ϵ_r for $r \neq M$. Due to the reversal in sign associated with the contour at infinity there is a net ϵ_M dependence, giving a factor of $x_M^{(D-2)/12}$. This is needed to cancel a singular factor of $\prod_{r<M}|x_M - x_r|^{p_r \cdot p_M} \sim x_M^{(2-D)/12}$ in the integrand of the amplitude (11.3.12) that arises when $x_M = \infty$. Together with (11.A.33) these equations determine $[\det'\Delta]^{(2-D)/2}\mathcal{M}$ so that

$$[\det'\Delta]^{(2-D)/2}\mathcal{M} = \left(x_M^{-2}\alpha_M^{-1}\prod_I \left|\frac{\partial^2\rho}{\partial z^2}\right|^{1/2}\prod_{r=1}^{M-1}\alpha_r\right)^{(2-D)/24}. \qquad (11.A.40)$$

This is the result that must be combined with the expression for the Jacobian to give the measure for open-string tree amplitudes.

(b) Closed-String Trees

For closed strings the calculations are very similar except that it is necessary to cut out complete circles around the interaction points z_I and the points $z = z_r$ where the external particles are attached. The result is that all contributions to $\ln\det'\Delta$ are doubled, and so the result for $[\det'\Delta]^{(2-D)/2}\mathcal{M}$ is squared.

(c) Loop Calculations

In the case of tree diagrams we were able to determine the measure by considering the effect of conformal transformations. As we have seen in earlier chapters the world sheet for a loop diagram depends on 'Teich-muller' parameters that are unchanged by infinitesimal conformal trans-formations of the metric. This means that it is not possible to determine the dependence of the measure factors on these parameters by just con-sidering conformal transformations as we did for the tree diagrams. The simplest example of a loop diagram is the planar open-string loop consid-ered in §8.1.1 and §11.5. In this case the dependence on the parameter $w = \exp\{2\pi i\tau\}$ cannot be determined by conformal transformations. We merely quote the result of the determinant calculation, which gives (choos-ing $x_M = 1$)

$$[\det'\Delta]^{(2-D)/2}\mathcal{M} = \prod_r x_r^{-1}\Big\{\frac{\pi}{\ln w}\prod_{n=1}^{\infty}(1 - w^n)^{-2}$$

$$\times \Big(\prod_I \frac{\partial^2\rho_I}{\partial z^2}\Big)^{-1/24}\prod_r(|\alpha_r|w)^{-1/12}\Big\}^{(D-2)/2}.$$

$$(11.\text{A}.41)$$

Appendix 11.B   The Jacobian for
the Conformal Transformation

(a) Tree Diagrams

The calculation of the Jacobian J is particularly simple if the three arbitrary x_r coordinates are chosen to be $x_1 = 0$, $x_{M-1} = 1$ and $x_M = \infty$. In that case

$$J = \Big|\det\frac{\partial\tau_I}{\partial x_r}\Big| \qquad (11.\text{B}.1)$$

has an x_r dependence that is exactly the same as that of

$$\prod_I |c_I|^{1/2} = \prod_I |\partial^2\rho/\partial z^2|^{1/2}. \qquad (11.\text{B}.2)$$

This can be seen from an analyticity argument based on the fact that the two expressions have the same singularities at the points where two x_r's touch or where two interaction points, *i.e.*, two ρ_I's, touch. These facts can be seen from the expression for the conformal mapping in (11.A.27). As a result, the combination $J\prod_I |c_I|^{-1/2}$ is independent of x_r. This is useful because when $D = 26$ this is just the combination that enters into the product of (11.A.40) and the Jacobian J. Since $J\prod_I |c_I|^{-1/2}$ does not depend on x_r it can be calculated by choosing a particularly convenient set of values $x_r \ll x_{r+1}$. In this case the relation between the $M-3$ interaction points and the x_r is particularly simple, namely,

$$\rho_I = \rho(z_I) \sim \ln x_I \sum_{s \leq I} \alpha_s + \cdots , \qquad (11.\text{B}.3)$$

where the dots represent terms that are independent of x_r for $r \leq I$. Therefore the Jacobian is given by the diagonal terms only. and by differentiating (11.A.27) we find that

$$J = \prod_{I=2}^{M-2} x_I^{-1} |\gamma_I|, \qquad (11.\text{B}.4)$$

where

$$\gamma_I = \sum_{s \leq I} \alpha_s. \qquad (11.\text{B}.5)$$

By differentiating (11.A.27) once again we find

$$|c_I| = \left| \frac{\partial^2 \rho_I}{\partial z^2} \right|_{z=x_I} = \left| \frac{(\gamma_I^3)}{x_I^2 \alpha_I \gamma_{I-1} \alpha_s} \right| . \qquad (11.\text{B}.6)$$

Noting that $\gamma_1 = \alpha_1$ and $\gamma_{M-1} = -\alpha_M$ it follows that

$$J \prod_{I=2}^{M-1} \left| \frac{\partial^2 \rho}{\partial z^2} \right|_{z=x_I}^{-1/2} = \left| \alpha_M^{-3/2} \prod_{r=1}^{M-1} \alpha_r^{1/2} \right| . \qquad (11.\text{B}.7)$$

This result evidently combines with (11.A.40) to give a result for the measure that is only simple in the critical dimension. Setting $D = 26$ the result is

$$J[\det'\Delta]^{(2-D)/2} \mathcal{M} = x_M^2 \prod_{r=1}^{M} |\alpha_r|^{-1/2} , \qquad (11.\text{B}.8)$$

which is the expected product of flux factors for the external particles multiplied by the factor of $(x_M - x_1)(x_M - x_{M-1})(x_{M-1} - x_1)$ evaluated with $x_M \to \infty$, $x_{M-1} = 1$ and $x_1 = 0$. The infinite factor of x_M^2 is needed to cancel a compensating zero factor in the integrand of the amplitude as explained above.

In the case of closed strings the Jacobian calculation is very similar but the result is squared, since the integration is two dimensional. Since the result in (11.A.40) was also squared for closed strings it appears that the measure involves the square of the required flux factors. However, we have already pointed out that the integration over the σ's of the external particles gives a factor of $2\pi|\alpha_r|$ for each particle. Taken together with the factor of $(2\pi|\alpha_r|)^{-1/2}$ in the normalization of each closed-string state the result is still proportional to the expected flux factor (when $D = 26$).

(b) The Planar Loop Diagram

As with the determinant calculation, we simply quote the result of the evaluation of the Jacobian for the planar loop diagram. The argument that leads to the result is a complicated generalization of the preceding argument for the tree diagrams. The result is

$$\left|\frac{\partial\tau}{\partial z}\right| \prod_I \left|\frac{\partial^2\rho}{\partial z^2}\right|^{-1/2} = \frac{\pi}{w|\ln w|} \prod_r |\alpha_r|^{1/2}. \tag{11.B.9}$$

Combining this expression with that of (11.A.41) gives a simple covariant result only in the critical dimension $D = 26$

$$J[\det'\Delta]^{(2-D)/2}\mathcal{M} = w^{-2} \left(\frac{\pi}{|\ln w|}\right)^{13} [f(w)]^{-24} \prod_r \alpha_r^{-1/2} x_r^{-1}. \tag{11.B.10}$$

When substituted into (11.2.24) (specialized to the case under consideration) this gives the same expression for the diagram as in chapter 8.

Appendix 11.C   Properties of the Functions f_m

The functions f_n defined in (11.4.40) can be rewritten in the form

$$f_n(\gamma) = \frac{1}{n!}(n\gamma - 1)(n\gamma - 2)\cdots(n\gamma - n + 1) \tag{11.C.1}.$$

They are related simply to the geometrical properties of the string diagram, as can be seen by inverting the mapping from the string diagram to the z plane as in (11.4.27). This allows z to be expressed in terms of the variables $\zeta_r \equiv \xi_r + i\eta_r$, in the appropriate regions of the ρ plane. In the region of string #3

$$\zeta_3 = \frac{\rho}{\alpha_3} + \pi i = \pi i - \gamma \ln\left(1 - \frac{1}{z}\right) - \ln z, \tag{11.C.2}$$

where $\gamma = -\alpha_1/\alpha_3$. This equation can be written in the form

$$y = \gamma \ln\left(1 + x e^y\right), \tag{11.C.3}$$

where

$$y = \pi i - \ln z - \zeta_3, \quad \text{and} \quad x = e^{\zeta_3}. \tag{11.C.4}$$

The function $y(x)$ can be obtained as a power series in x by using a formula due to Lagrange. Given any function ξ satisfying

$$\xi = a + x\phi(\xi), \tag{11.C.5}$$

an arbitrary function $f(\xi)$ can be written as

$$f(\xi) = f(a) + \sum_{n=1}^{\infty} \frac{x^n}{n!} \frac{d^{n-1}}{da^{n-1}} \left(f'(a)\phi^n(a)\right). \tag{11.C.6}$$

In order to solve (11.C.3) we make the identifications $\xi = e^{1/\gamma}$, $a = 1$, $\phi(\xi) = \xi^\gamma$, $f(\xi) = \log \xi = y/\gamma$. Substituting in (11.C.6) gives

$$\frac{y}{\gamma} = x + \frac{(2\gamma - 1)}{2}x^2 + \cdots + \frac{(n\gamma - 1)!}{n!(n\gamma - n)!}x^n + \cdots$$

$$= \sum_{n=1}^{\infty} f_n(\gamma)x^n, \tag{11.C.7}$$

where $f_n(\gamma)$ is the function defined in (11.4.40). This series converges for

$$|x| < |\gamma - 1|^{\gamma-1}|\gamma|^{-\gamma}. \tag{11.C.8}$$

The condition (11.C.4) gives

$$\ln z = -\zeta_3 + \pi i - \gamma \sum_{n=1}^{\infty} f_n(\gamma)e^{n\zeta_3}. \tag{11.C.9}$$

This series is convergent in the region in which string #3 is defined. In the same manner, the variable z can be related to the variable ζ_1 in the region of string #1, giving

$$\ln z = \sum_{n=1}^{\infty} f_n\left(1 - \frac{1}{\gamma}\right) e^{n\zeta_1}, \tag{11.C.10}$$

and to the variable ζ_2 by

$$\ln z = \zeta_2 + \sum_{n=1}^{\infty} f_n\left(\frac{1}{1 - \gamma}\right) e^{n\zeta_2}. \tag{11.C.11}$$

Appendix 11.D   Properties of the $SU(4)$ Clebsch–Gordan Coefficients

The ρ matrices are the Clebsch–Gordan coefficients that describe the coupling of two **4**'s or of two $\overline{\textbf{4}}$'s to a $\overline{\textbf{6}}$ of $SU(4)$. They are normalized, by analogy with Dirac matrices, to satisfy

$$\rho^{IAC}\rho^J_{CB} + \rho^{JAC}\rho^I_{CB} = 2\delta^A_B\delta^{IJ}. \tag{11.D.1}$$

The $SO(8)$ matrices γ^{ij}_{ab}, defined in appendix 5.B, form the spinor representation of $SO(8)$ (described in detail in appendix 5.A) They decompose into $SU(4)$ matrices according to

$$\gamma^{IJ} = \begin{pmatrix} (\rho^{IJ})^A{}_B & 0 \\ 0 & (\rho^{IJ})_A{}^B \end{pmatrix}, \tag{11.D.2}$$

$$\gamma^{IL} = \begin{pmatrix} 0 & 0 \\ \sqrt{2}\rho^I_{AB} & 0 \end{pmatrix}, \tag{11.D.3}$$

$$\gamma^{IR} = \begin{pmatrix} 0 & -\sqrt{2}\rho^{IAB} \\ 0 & 0 \end{pmatrix}, \tag{11.D.4}$$

$$\gamma^{RL} = \begin{pmatrix} \delta^A_B & 0 \\ 0 & -\delta^B_A \end{pmatrix}. \tag{11.D.5}$$

In the text we have also defined

$$\rho^{BIJ}_A = \tfrac{1}{2}\left(\rho^I_{AC}\rho^{CBJ} - \rho^J_{AC}\rho^{CBI}\right). \tag{11.D.6}$$

In order to prove the relations in §11.7 it is useful to note the following identities

$$\rho^I_{AB} = \tfrac{1}{2}\epsilon_{ABCD}\rho^{ICD}, \tag{11.D.7}$$

$$\rho^I_{A[B}\rho^J_{CD]} = \frac{1}{3}\epsilon_{BCDE}(\rho^I\rho^J)_A{}^E, \tag{11.D.8}$$

$$\rho^I_{[AB}\rho^J_{CD]} = -\frac{1}{3}\epsilon_{ABCD}\delta^{IJ}, \tag{11.D.9}$$

$$\rho^I_{ACP}\rho^{BDI} = 2\left(\delta^D_A\delta^B_C - \delta^B_A\delta^D_C\right). \tag{11.D.10}$$

12. Some Differential Geometry

In the first eleven chapters of this book we have attempted to introduce the reader to string theory as it is presently understood. Our focus now shifts to making contact with more familiar physics. In this chapter we develop some concepts in differential geometry that are useful in understanding general relativity and Yang–Mills theory even in four dimensions, but which are of particular utility in ten-dimensional physics. Our treatment in this chapter is comparatively elementary and aims mostly to develop the minimum material we require in chapters 13 and 14. In chapter 13, we will discuss supergravity theory in ten dimensions, which at least in perturbation theory is the low-energy limit of ten-dimensional superstring theory. In chapter 14, we will discuss some of the important ideas that arise in compactification from ten to four dimensions. The concluding chapters of this book, chapters 15 and 16, are devoted to more specialized mathematical background and more speculative ideas about compactification.

12.1 Spinors In General Relativity

A good place to start our discussion is to think about the coupling of spinors to a gravitational field. This problem is of great importance in string theories in which both fermions and gravity are present, and this alone would justify its consideration here. In addition, thinking about the coupling of spinors to a gravitational field forces us to examine issues whose analogs for Yang–Mills theory we will wish to consider later. The question of coupling spinors to general relativity was considered briefly in chapter 4, in connection with a discussion of two-dimensional supergravity, but here we will be more extensive.

In most of this chapter we consider a manifold M of dimension n; it may be endowed with a metric of Euclidean or Lorentzian signature. When we have to make a specific choice we usually consider a metric of Euclidean signature (positive definite). When we wish to consider global topological properties, we usually take M to be compact.

We would like to consider spinor fields on M. The reason that it is somewhat subtle to do so is the following. Under a change of coordinates from one set of coordinates x^μ to another set x'^μ, a vector field V^μ is transformed by

$$V^\mu \rightarrow V'^\mu = \frac{\partial x'^\mu}{\partial x^\nu} \cdot V^\nu. \qquad (12.1.1)$$

Here the matrix

$$Z = \partial x'^\mu / \partial x^\nu \qquad (12.1.2)$$

is, in general, an element of $GL(n, R)$, the group of invertible, real $n \times n$ matrices. The vector V^μ transforms in the fundamental, vector representation of this group. The most elementary formulas of general relativity for coupling of gravity to matter fields require that the matter fields form representations of $GL(n, R)$. Under a coordinate transformation, a physical field in a given $GL(n, R)$ representation is transformed by the matrix Z of (12.1.2) acting in that representation. The representations of $GL(n, R)$ are tensors. A $GL(n, R)$ representation always gives an $SO(n)$ representation by restriction, since $SO(n)$ is a subgroup of $GL(n, R)$, but the converse is not true. Spinors form a representation of $SO(n)$ that does not arise from a representation of $GL(n, R)$. To couple spinors to gravity requires finding a modified framework in which the Z matrix is replaced by an $SO(n)$ matrix (or an $SO(n-1, 1)$ matrix, depending on the signature).

A necessary step is to first introduce at each point x on M a basis of orthonormal tangent vectors $e^a_\mu(x)$, $a = 1, \ldots, n$. Here μ is an index labelling the components of a vector tangent to M at x, and a is merely the name of the vector $e^a_\mu(x)$. Orthonormality means that $e^a_\mu(x)e^{b\mu}(x) = \eta^{ab}$, with η^{ab} being the flat space metric. Equivalently,

$$e^a_\mu(x)e_{a\nu}(x) = g_{\mu\nu}(x) \qquad (12.1.3)$$

with $g_{\mu\nu}(x)$ being the metric tensor of M, and $e_{a\mu}(x) = \eta_{ab}e^b_\mu(x)$. The vectors e^a_μ form a basis for the tangent space T_x, which consists of all vectors tangent to the manifold M at the point x. This basis is usually called a vielbein or (in four dimensions) a vierbein or tetrad. The 'Lorentz' index a of the tetrad e^a_μ is raised and lowered with the Lorentz metric η_{ab}, while the space-time index μ is raised and lowered with the flat space metric tensor $g_{\mu\nu}$.

There is a large arbitrariness in the choice of a vielbein. Just as good as $e^a_\mu(x)$ would be $\tilde{e}^a_\mu(x) = \Lambda^a{}_b(x) \cdot e^b_\mu(x)$, with $\Lambda^a{}_b(x)$ being an arbitrary x dependent Lorentz transformation. Such a change of vielbein is called a local Lorentz transformation. It is the Lorentz transformation $\Lambda^a{}_b(x)$ that eventually replaces the $GL(n, R)$ matrix of the preceding discussion.

It is not self-evident at first sight that introducing a vielbein is a good idea; the utility of this will emerge in the course of the discussion. But it should be obvious that if we are going to introduce a vielbein, then we must ensure that the formalism is invariant under local Lorentz transformations, so that physical observables are independent of the arbitrary choice of a tetrad. As in any other discussion of local gauge invariance, to achieve local Lorentz invariance requires introducing a gauge field $\omega_\mu{}^a{}_b(x)$ of the Lorentz group $SO(1, n-1)$ (or $SO(n)$ in the positive signature case). Here μ is a vector index tangent to the manifold, while a and b are $SO(1, n-1)$ indices. The gauge field ω_μ transforms under local Lorentz transformations in the standard way $\omega_\mu \to \Lambda\omega_\mu\Lambda^{-1} - \partial_\mu\Lambda\cdot\Lambda^{-1}$. The gauge field of the local Lorentz group is usually called the spin connection.

We would like to find a minimal choice of the spin connection with the property that introducing the vielbein and spin connection does not change the content of general relativity. (Of course, in string theory general relativity is modified, and a nonminimal choice of spin connection may be preferred. But it is important to first understand the minimal choice.) As a preliminary, let us explain how the spin connection is used. The covariant derivative of a vector field V^μ is usually defined in general relativity by saying that $D_\lambda V^\mu = \partial_\lambda V^\mu + \Gamma^\mu_{\lambda\nu}V^\nu$, with $\Gamma^\mu_{\lambda\nu}$ being the Christoffel symbols. On the other hand, once a vielbein is introduced, one could work not with $V^\mu(x)$ but with $V^a(x) = e^a_\mu(x)V^\mu(x)$. The V^a contain the same information as V^μ since one can always reconstruct $V^\mu(x) = e^\mu_a V^a(x)$. In terms of V^a the natural covariant derivative would be $D_\mu V^a = \partial_\mu V^a + \omega_\mu{}^a{}_b V^b$. If we are to avoid modifying the standard content of general relativity, the two notions of the covariant derivative of a vector V must be equivalent. This will be so, in the sense that $D_\mu V^a = e^{a\nu}D_\mu V_\nu$, if we define the spin connection so that the covariant derivative of the vielbein is zero,

$$D_\mu e^a_\nu = 0. \qquad (12.1.4)$$

Here just as in conventional general relativity the covariant derivative of a field such as e^a_ν with several indices is defined by adding a connection term for each index: $D_\mu e^a_\nu = \partial_\mu e^a_\nu - \Gamma^\lambda_{\mu\nu}e^a_\lambda + \omega_\mu{}^a{}_b e^b_\nu$. Equation (12.1.4) has precisely enough information to uniquely determine both the Christoffel symbols (giving the standard formulas) and the spin connection. The easiest way to convince oneself of this is to note that (letting $\mu, \nu,$ and a vary) (12.1.4) contains n^3 independent equations, which is the combined number of independent components of the Christoffel symbols and spin connection. The explicit formula for the spin connection that follows from (12.1.4) is not needed in this book, but we present it anyway for completeness:

$$\omega_\mu^{ab} = \tfrac{1}{2}e^{\nu a}(\partial_\mu e_\nu^b - \partial_\nu e_\mu^b) - \tfrac{1}{2}e^{\nu b}(\partial_\mu e_\nu^a - \partial_\nu e_\mu^a)$$
$$- \tfrac{1}{2}e^{\rho a}e^{\sigma b}(\partial_\rho e_{\sigma c} - \partial_\sigma e_{\rho c})e_\mu^c. \tag{12.1.5}$$

A consequence of (12.1.3) and (12.1.4) (and the Leibniz rule $D_\mu(AB) = (D_\mu A)B + A(D_\mu B)$, which is obeyed by covariant derivatives) is the fact that $D_\lambda g_{\mu\nu} = 0$, as is indeed the case in general relativity.

Having defined the spin connection, we can form the gauge-covariant field strength $R_{\mu\nu}{}^a{}_b = \partial_\mu\omega_\nu{}^a{}_b - \partial_\nu\omega_\mu{}^a{}_b + [\omega_\mu,\omega_\nu]^a{}_b$. It has the same content as the Riemann tensor $R_{\mu\nu}{}^\sigma{}_\tau$ conventionally defined in terms of the Christoffel symbols and their derivatives; in fact, it follows from (12.1.4) that $R_{\mu\nu}{}^a{}_b = e^a_\sigma e^\tau_b R_{\mu\nu}{}^\sigma{}_\tau$.

With the aid of the spin connection, it is not difficult to couple spinors to general relativity. As with any gauge field, the spin connection can be coupled to a field $\psi(x)$ in any required representation of the gauge group. In this case, the gauge group is the Lorentz group, and we take $\psi(x)$ to be a field in the spinor representation of the Lorentz group. Letting $\Sigma^a{}_b$ be the generators of the Lorentz group in the spinor representation, the covariant derivative of ψ is defined in the standard way as

$$D_\mu\psi = \partial_\mu\psi + \frac{1}{2}\omega_\mu^{ab}\Sigma_{ab}\psi. \tag{12.1.6}$$

Under change of the vielbein by a local Lorentz transformation $\Lambda(x)$, we require $\psi(x) \to \Lambda(x)\psi(x)$; then the covariant derivative of $\psi(x)$ also transforms homogeneously, $D_\mu\psi(x) \to \Lambda(x)D_\mu\psi(x)$.

To define the Dirac equation, we also need gamma matrices. One first introduces the standard flat space gamma matrices Γ_a that obey $\{\Gamma_a,\Gamma_b\} = 2\eta_{ab}$.[*] Curved space gamma matrices are then defined as $\Gamma_\mu(x) = e^a_\mu\Gamma_a$. They obey $\{\Gamma_\mu(x),\Gamma_\nu(x)\} = 2g_{\mu\nu}(x)$. By virtue of (12.1.4) the gamma matrices are also covariantly constant in the sense that $D_\mu\Gamma_\lambda = \Gamma_\lambda D_\mu$. There is now no mystery about how to define the Dirac operator. It is simply $\slashed{D} = \Gamma^\mu D_\mu$, just as in gauge theories.

Several further remarks are in order. First, we have discussed the minimal spin connection, which obeys (12.1.4). In general it is useful to consider other, nonminimal spin connections, which normally depend on

---

[*] In our discussions of differential geometry in the remainder of this book we adopt (as is common in the physics literature) a convention for the Dirac matrices in which Γ^μ for space-like μ is hermitian.

physical fields other than the metric. The nonminimality of a nonminimal spin connection is conveniently measured by the so-called 'torsion' $T^a_{\mu\nu}$, defined as

$$T^a_{\mu\nu} = D_\mu e^a_\nu - D_\nu e^a_\mu. \qquad (12.1.7)$$

The torsion depends only on the spin connection, not on the Christoffel symbols, since (by virtue of symmetry of $\Gamma^\lambda_{\mu\nu}$ in the lower indices and the definition of covariant derivatives) the Christoffel symbols cancel out in the right-hand side of (12.1.7). For the same reason, the most efficient way to find the formula (12.1.5) for the spin connection in the minimal, torsion-free case is to start with (12.1.7) (with $T = 0$) rather than (12.1.4).

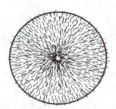

Figure 12.1. It is impossible to 'comb the hair on a sphere' or in other words to find an everywhere nonvanishing tangent vector field on the surface of an ordinary two-dimensional sphere.

We have already explained at the beginning of this section one reason that it is necessary to introduce a vielbein in order to couple spinors to general relativity. There is another and deeper reason. This arises if we consider a topologically nontrivial situation. We have assumed in the above that it is possible to find an orthonormal basis of tangent vectors, and this is certainly possible locally, but in general it is not possible globally on a topologically nontrivial manifold. For example, on the ordinary two sphere (fig. 12.1), a classic theorem asserts that it is not possible to find a smoothly varying, everywhere nonzero tangent vector field. This theorem is often described by saying that 'you can't comb the hair on a bowling ball'. If it is not possible to find a single everywhere nonzero vector field on the two sphere, it is certainly not possible to find a basis of orthonormal tangent vector fields – or vielbein. In fact, topologically nontrivial manifolds on which it *is* possible to define a single vielbein everywhere are comparatively scarce. They are called parallelizable manifolds and do not seem to be suitable for superstring compactification.

If topological problems prevent us from defining a vielbein everywhere, what do we do? We can certainly define a vielbein locally, so we cover M with open sets $O_{(\alpha)}$, on each of which we introduce a vielbein $e^a_{\mu(\alpha)}$. On overlap regions $O_{(\alpha\beta)} = O_{(\alpha)} \cap O_{(\beta)}$ the two vielbeins $e_{(\alpha)}$ and $e_{(\beta)}$ are necessarily related by a local Lorentz transformation

$$e_{(\alpha)}(x) = \Lambda_{(\alpha\beta)}(x)e_{(\beta)}(x). \tag{12.1.8}$$

The local Lorentz transformation $\Lambda_{(\alpha\beta)}(x)$ that appears in this formula is called a transition function. Of course, if we used different vielbeins $e_{(\alpha)}$ and $e_{(\beta)}$ in $O_{(\alpha)}$ and $O_{(\beta)}$, we would have different transition functions. But no matter what vielbeins and transition functions we use, it follows from (12.1.8) that in the triple intersection regions $O_{(\alpha\beta\gamma)} = O_{(\alpha)} \cap O_{(\beta)} \cap O_{(\gamma)}$, the transition functions obey

$$\Lambda_{(\alpha\beta)}\Lambda_{(\beta\gamma)}\Lambda_{(\gamma\alpha)} = 1. \tag{12.1.9}$$

Now, how do we introduce spinors in this more difficult situation? In each open set $O_{(\alpha)}$, we introduce a spinor field $\psi_{(\alpha)}$. On overlap regions $O_{(\alpha\beta)}$ we require that $\psi_{(\alpha)}$ and $\psi_{(\beta)}$ provide descriptions of the same physical situation in the sense that

$$\psi_{(\alpha)}(x) = \tilde{\Lambda}_{(\alpha\beta)}\psi_{(\beta)}(x). \tag{12.1.10}$$

Here $\tilde{\Lambda}$ is the $SO(n)$ or $SO(n-1,1)$ matrix Λ written in the spinor representation. Equation (12.1.10) only makes sense if on the triple overlap regions

$$\tilde{\Lambda}_{(\alpha\beta)}\tilde{\Lambda}_{(\beta\gamma)}\tilde{\Lambda}_{(\gamma\alpha)} = 1. \tag{12.1.11}$$

At first sight it may seem that all is well, since (12.1.11) seems to be a consequence of (12.1.9). However, we must realize that the spinor representation is double-valued and that this implies a sign ambiguity in going from Λ to $\tilde{\Lambda}$. In fact, $-\tilde{\Lambda}$ is just as good as $+\tilde{\Lambda}$. For this reason (12.1.11) does not quite follow from (12.1.9); the right-hand side of (12.1.11) may be -1 instead of $+1$. To define spinors on a manifold M that is not parallelizable, one must try to choose the signs of the $\tilde{\Lambda}_{(\alpha\beta)}$ so that (12.1.11) is obeyed. If this can be done, M is said to admit a spin structure or to be a spin manifold. (If M is not simply connected, there may be several inequivalent choices of signs compatible with (12.1.11), and in that case M is said to admit several spin structures. This is a phenomenon that we will encounter in the next section.) Superstring theories certainly have fermions, so they can only be defined on spin manifolds. In two or three dimensions, every orientable manifold is a spin manifold, and this is the reason that the concept of spin manifolds is not more well-known to physicists. In four or more dimensions, 'most' manifolds, by any reasonable measure, are not spin manifolds, so the requirement of a spin structure is an important restriction on superstring compactification.

12.2   Spin Structures On The String World Sheet

In the preceding section we have considered the propagation of spinors in curved space-time. A quite analogous mathematical problem arises in string theory in studying the propagation of spinors on a curved world sheet, as we have already discussed in §4.3.4. In that discussion we considered only local problems, but here we return to the subject and discuss some simple global properties.

An important simplifying feature is that the rotation group in two dimensions is $SO(2)$, which is abelian; in fact, it is isomorphic to $U(1)$, the group of complex numbers of modulus one. As in §4.4.1, let W be the one generator of $SO(2)$. A representation of $SO(2)$ is given by specifying the eigenvalue of W. Unlike the situation in more than two dimensions, a vector field V^μ does not give an irreducible representation of $SO(2)$ but decomposes into components V^+ and V^- of $W = 1$ and $W = -1$, respectively. If V^μ is real then V^+ and V^- are complex conjugates of each other. A two-component spinor field ψ^A likewise has components ψ^+ and ψ^- of positive and negative chirality, with $W = 1/2$ and $W = -1/2$, respectively. They are again complex conjugates of one another if ψ^A is real (Majorana).

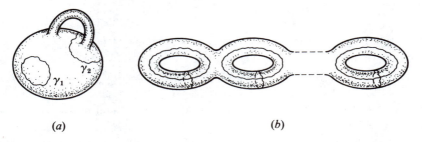

(a) (b)

Figure 12.2. A compact string world sheet is sketched in (a) with two closed curves γ_1 and γ_2 of which the first is topologically trivial and the second is not. In (b) we indicate the fact that there are $2k$ topologically independent closed curves on a surface of genus k.

For simplicity we restrict ourselves to orientable world sheets; in the nonorientable case a similar discussion can be carried out but requires much more care and much more precise definitions. Consider, as in fig. 12.2, a string world sheet that is a compact Riemann surface of genus k. We would like to discuss the parallel transport of vectors and spinors around closed curves such as the curves γ_1 and γ_2 indicated in fig. 12.2a. It suffices to consider V^+ and ψ^+ since V^- and ψ^- transform like the complex conjugate of these.

Under parallel transport around a closed curve γ, V^+ is transformed by some phase $e^{i\alpha}$. What happens to ψ^+ under parallel transport around γ? Except for a sign ambiguity, this question is easily answered by noting that the product $\psi^+ \cdot \psi^+$ transforms like V^+ and picks up a phase $e^{i\alpha}$ under parallel transport around γ. ψ^+, therefore, must transform like

$$\sqrt{e^{i\alpha}}. \qquad (12.2.1)$$

An ambiguity arises here because the square root has two signs. If the curve γ is contractible (like γ_1 in fig. 12.2a) then the ambiguity can be resolved in the following way. In the limit that γ is a very tiny closed curve, vectors and spinors must both be unchanged in transport around γ, assuming that the spin connection is nonsingular (which is certainly the case for the conventional spin connection defined in (12.1.5) to make the metric covariantly constant). Thus for a very tiny closed curve, $\alpha \approx 0$ and parallel transport of spinors gives the positive sign of the square root in (12.2.1). For any contractible closed curve γ, not necessarily infinitesimal, the sign of the square root in (12.2.1) is determined by the requirement that (12.2.1) be a continuously varying functional of γ that is positive if γ shrinks to zero. If on the other hand γ is topologically nontrivial, like γ_2 in fig. 12.2a, then there is no natural way to choose the sign in (12.2.1) and we must consider both possibilities.

More generally, as in fig. 12.2b, there are $2k$ independent noncontractible loops on a surface of genus k. Therefore, there are $2k$ choices of sign to be made in deciding how to carry out parallel transport of spinors on a surface of genus k. In all there are 2^{2k} possible choices for how to define parallel transport of spinors on a genus k surface. (If the discussion is recast in the language of the preceding section, it can be shown that all 2^{2k} choices are compatible with (12.1.11).) On a surface of genus k there are 2^{2k} spin structures.

We are now in a position to gain a better understanding of the GSO projection, which – as we saw in chapter 4 – gives supersymmetry in the RNS model. Let us first think about a genus one surface in detail. As in

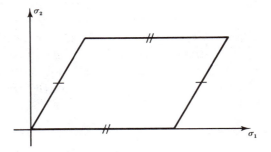

Figure 12.3. A genus one surface, regarded as a parallelogram in the (σ_1, σ_2) plane, with opposite sides identified.

chapters 8 and 9, it can be viewed as a parallelogram in the (σ_1, σ_2) plane, with opposite sides identified. This is depicted in fig. 12.3. We can choose a flat metric on the parallelogram, and if we do so a vector undergoes no change in parallel transport around closed loops, contractible or not. Spinors therefore change in sign by ± 1 in going around a closed loop in the σ_1 or σ_2 direction. There are four possible sets of boundary conditions on spinors, which we may call $(++)$, $(-+)$, $(+-)$ and $(--)$, with the first and second $+$ or $-$ sign referring respectively to the behavior under parallel transport in the σ_1 or σ_2 direction. These four sets of boundary conditions correspond to the four spin structures on the genus one surface.[*]

When we calculate a one-loop path integral in a string theory that contains fermions, which spin structure do we use? For superstrings in the RNS formalism, this was already discussed in §9.4 in the course of explaining why unitarity and modular invariance require the GSO projection (or some other related projection that eliminates the massless spin 3/2 particle). Fermions and bosons, respectively, appear in the RNS model upon choosing $+$ or $-$ boundary conditions in the σ_1 direction, so including both fermions and bosons amounts to summing over the possible boundary conditions in the σ_1 direction. The GSO projection amounts to an insertion of $(1 + (-1)^F)/2$ in the path integral. The operator $(-1)^F$ reverses the boundary conditions in the σ_2 direction, so the GSO insertion of $(1 + (-1)^F)/2$ is an instruction to sum over possible boundary conditions in the σ_2 direction. Combining these facts, the supersymmetric theory

---

[*] We have already discussed these boundary conditions on a genus one surface in chapter 9, though not in the language of spin structures. As was discussed there, if there are several different kinds of fermion in a theory they may each be governed by a different spin structure, perhaps with some correlations among the various choices.

with the GSO projection is obtained at the one-loop level by summing over all four possible boundary conditions or in other words over all four possible spin structures. Summing independently over boundary conditions in both the σ_1 and σ_2 directions would appear, formally speaking, to be a prescription invariant under modular transformations, which exchange σ_1 and σ_2. It is necessary to verify this and make sure that there is no anomaly under modular transformations; in the critical dimension this is true, as we have seen in chapter 9.

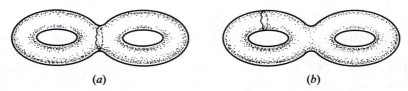

(a) (b)

Figure 12.4. A genus two surface, with (a) a cut that separates the two handles; (b) a cut revealing a representative state circling in one of the loops.

What is the generalization of this to higher genus surfaces? Since at the one-loop level we summed over all four spin structures, one might expect that at the k-loop level we should sum over all 2^{2k} spin structures. Indeed, once we decide to sum over all spin structures at the one-loop level, unitarity forces us to do the same in k-loop order. If one 'cuts' a two-loop diagram, as in fig. 12.4a, to separate it into two one-loop diagrams, then unitarity certainly forces one to sum separately over the four spin structures of each of the two subdiagrams if this was done at the one-loop level. Combining the two subdiagrams together, one is then summing over all $4 \times 4 = 16$ spin structures on the genus two surface.

We now can answer a question that may have puzzled the reader when the GSO projection was introduced in chapter 4. Granted that the theory has a discrete conserved quantum number $(-1)^F$, it is certainly possible in tree diagrams to carry out the GSO projection and only consider processes in which the external states are even under $(-1)^F$. But what ensures that particles of wrong G parity (odd under $(-1)^F$) are not pair produced in loop diagrams? To see that this does not occur, examine a typical string state circling around a loop λ in a diagram such as that of fig. 12.4b. The instruction to 'sum over all spin structures' means that we sum over fermion boundary conditions that are either periodic or antiperiodic in circumnavigating the loop λ. Thus, the string state cut along λ may be either an R or NS state – either a fermion or a boson. Summing over spin structures also means that the boundary conditions in the direction 'orthogonal' to λ may be either periodic or antiperiodic; this is the GSO projection. Thus, the intermediate states in loop diagrams are precisely those that we are allowing as external states; unwanted states do not appear in pairs.

12.3 Topologically Nontrivial Gauge Fields

Returning from the string world sheet to space-time concepts, our next goal is to think about topologically nontrivial gauge fields in space-time. The requisite concepts are quite similar to some that we encountered in our discussion of spin structures in n dimensions, and we therefore begin in §12.3.1 below by rethinking the notions of vectors and spinors in Riemannian geometry. The discussion of gauge fields in §12.3.2 will be cast in a less abstract form, and is logically independent, but the more abstract view is valuable, especially in applications to more difficult problems.

12.3.1 The Tangent Bundle

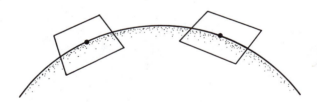

Figure 12.5. The tangent bundle at two points x and x' on a Riemannian manifold.

Given a smooth manifold M, at each point x we have a vector space T_x consisting of tangent vectors to M at x (fig. 12.5). A spin one or vector field V is at each point x an element $V^\mu(x)$ of T_x. In any natural sense, the tangent vector spaces T_x and $T_{x'}$ for different points x and x' in M are *different spaces*. One is the space of tangent vectors at x, and the other is the space of tangent vectors at x'.

Though it is obvious intuitively in fig. 12.5 that T_x and $T_{x'}$ are not really the same vector space, this point is of great importance and deserves careful exploration. One could compare tangent vectors at x to tangent vectors at x' by parallel transport along some path γ that leads from x to

x'. However, such parallel transport would depend on an arbitrary choice of path from x to x', and so is not a natural operation.

More generally, in what sense can two vector spaces such as T_x and $T_{x'}$ be said to be 'different'? In fact, any two vector spaces T and $\tilde{T}$ of the same dimension n are isomorphic. An isomorphism between them is simply an invertible linear mapping $\rho : T \to \tilde{T}$. Such mappings certainly exist, if T and $\tilde{T}$ have the same dimension, but there are many such mappings and no natural choice. A natural choice of isomorphism between T and $\tilde{T}$ would be possible if those spaces were endowed with an agreed upon choice of basis elements e^a, $a = 1, \dots, n$ for T and $\tilde{e}^a$, $a = 1, \dots, n$ for $\tilde{T}$. We would then define an isomorphism from T to $\tilde{T}$ by $\rho(e^a) = \tilde{e}^a$. In general, in Riemannian geometry there is no natural way to choose bases $e^a(x)$ for the vector spaces T_x. However, we have seen in §12.1 above that while there is no natural way to do so, it is nonetheless extremely useful to make an arbitrary choice of a basis $e^a(x)$,*i.e.*, a vielbein. In any region O in which a vielbein has been chosen, one can regard the vector spaces T_x as being the 'same' space in some sense. One simply picks an arbitrary point $x_0 \in O$ and maps T_x onto T_{x_0} by the invertible mapping $e^a(x) \to e^a(x_0)$. Though this is not really a natural operation, since it depends upon the arbitrary choice of vielbein, it gives some sense in which the various T_x might be considered the 'same'.

In general, however, as we mentioned at the end of §12.1, it is not possible to pick a smoothly varying vielbein throughout a topologically nontrivial manifold M. When this is not possible, we cannot view the vector spaces T_x as being the 'same' vector space even by making arbitrary choices; we must accept a view in which the vector field $V^\mu(x)$ takes values at each point x in a *different* (but smoothly varying) vector space T_x. This is then the prototype for an extremely important concept in mathematics. A 'vector bundle' V over a manifold M is a family of vector spaces V_x, one for each point $x \in M$, that varies smoothly with x. The vector space V_x is called the 'fiber' of the vector space V at the point x. The classic example of a vector bundle over M is the example we have been describing: the tangent bundle T, whose fiber at each point is the tangent space T_x. A vector bundle is said to be topologically trivial if by choosing a smoothly varying basis for each of the V_x one can identify them with V_{x_0} for any given x_0. According to this definition, the tangent bundle T is topologically trivial if it is possible to define a single vielbein throughout the manifold M; thus, what we earlier called parallelizable manifolds are manifolds whose tangent bundles are topologically trivial. Every vector bundle is trivial locally since locally the analog of a vielbein can always be introduced.

If the tangent bundle is *not* topologically trivial, then, as in §12.1, the best we can do is to cover M with open sets $O_{(\alpha)}$, and pick a vielbein $e_{(\alpha)}$ in each $O_{(\alpha)}$, related by suitable transition functions on the overlap regions. This formalism is quite similar to the analogous formalism for gauge fields, to which we turn next.

12.3.2 Gauge Fields and Vector Bundles

In the most elementary formulation of the concept of a gauge theory, one begins with a gauge group G and a matter field $\psi^i(x)$ transforming in some representation R of G. Here x labels a point in space-time M, and $i = 1, \ldots, n$ is an index in the R representation. One then considers local gauge transformations by an arbitrary space-time dependent gauge transformation $g(x)$. The transformation law is simply $\psi^i(x) \to g^i{}_j(x)\psi^j(x)$, where $g^i{}_j$ is written in the R representation.

This is a perfectly complete and satisfactory formulation for gauge theories in flat space or more generally in any situation that is topologically trivial. By a topologically trivial space we mean a space, such as Euclidean space or an open ball in Euclidean space, which can be shrunk to a point without tearing it. On a topologically nontrivial manifold M, however, such as a sphere or a Riemann surface with n handles, the conventional formulation of gauge theory admits a very important generalization.

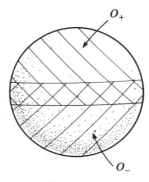

Figure 12.6. Covering a sphere with two open sets, O_+ and O_-.

The reasoning required is reminiscent of the principle of equivalence in general relativity. It is also reminiscent of our discussion in §12.1 of this chapter of how to introduce fermions in topologically nontrivial situations, but we now enter into somewhat more detail. We already know how to formulate gauge theories on a topologically trivial space. So we cover M

with open sets $O_{(\alpha)}$ each of which is topologically trivial. For instance, as in fig. 12.6, if M is a sphere then two open sets are adequate; we may pick them to be O_+ and O_-, where O_+ and O_- are, respectively, the complement of the south and north poles. On each open set $O_{(\alpha)}$ we formulate gauge theories in the usual way. Thus, on each $O_{(\alpha)}$ we introduce a field $\psi^i_{(\alpha)}(x)$, in the R representation of the gauge group. It admits a gauge transformation law

$$\psi^i_{(\alpha)} \to g^i{}_{j(\alpha)}\psi^j_{(\alpha)}, \tag{12.3.1}$$

where now we consider ourselves free to use a different gauge function $g^i{}_{j(\alpha)}$ in each set $\cdot O_{(\alpha)}$. Of course, we do not want the fields $\psi_{(\alpha)}$ on the various $O_{(\alpha)}$ to be completely independent; we must impose some conditions that correspond to the fact that the various $O_{(\alpha)}$ are glued together into a space M. To this end, let $O_{(\alpha\beta)} = O_{(\alpha)} \cap O_{(\beta)}$ be the region of overlap of $O_{(\alpha)}$ and $O_{(\beta)}$. We require that the field $\psi_{(\alpha)}(x)$ on $O_{(\alpha)}$ and the field $\psi_{(\beta)}(x)$ on $O_{(\beta)}$ should describe the same physics on the region where they are both defined, namely $O_{(\alpha\beta)}$. Of course, they describe the same physics if and only if they are related by some gauge transformation $g_{(\alpha\beta)}$. So we require that

$$\psi_{(\alpha)} = g_{(\alpha\beta)}\psi_{(\beta)} \qquad \text{on} \qquad O_{(\alpha\beta)} \tag{12.3.2}$$

for some $g_{(\alpha\beta)}$, where now we are suppressing the indices labeling the representation R that appeared in previous formulas. The $g_{(\alpha\beta)}$ are called transition functions, a logical name since they bring about a transition between two equivalent physical descriptions on the overlap region $O_{(\alpha\beta)}$. Analogous quantities appeared in our discussion of spinors in curved space-time.

There is a consistency condition that must be obeyed by the transition functions in order for (12.3.2) to make sense. Let $O_{(\alpha\beta\gamma)} = O_{(\alpha)} \cap O_{(\beta)} \cap O_{(\gamma)}$ be the region of triple overlap of $O_{(\alpha)}$, $O_{(\beta)}$ and $O_{(\gamma)}$. Then consistency of (12.3.2) requires that

$$g_{(\alpha\beta)}g_{(\beta\gamma)}g_{(\gamma\alpha)} = 1 \tag{12.3.3}$$

on $O_{(\alpha\beta\gamma)}$. How do the $g_{(\alpha\beta)}$ transform under a gauge transformation? Study of (12.3.1), (12.3.2) and (12.3.3) shows that the transformation law we must postulate is

$$g_{(\alpha\beta)} \to g_{(\alpha)}g_{(\alpha\beta)}g_{(\beta)}{}^{-1}. \tag{12.3.4}$$

If it is possible to choose the gauge functions $g_{(\alpha)}$ in (12.3.4) in order to set the transition functions $g_{(\alpha\beta)}$ to one, the various $\psi^i_{(\alpha)}$ are then equal on the overlap region, and the description reduces to the more usual one in which the matter field is just an ordinary field $\psi^i(x)$ defined throughout M. The novelty in this description arises if it is *not* possible to make a gauge transformation to set the transition functions to unity.

To illustrate what all of this means, consider in some detail the example of the sphere S^n, as in fig. 12.6. There is a single overlap region O_{+-}, consisting of everything except the north and south poles. Consequently, there is a single transition function $g_{+-} : O_{+-} \to G$. Of course, it is only defined up to a gauge transformation; g_{+-} is gauge equivalent to $g_{+}g_{+-}g_{-}^{-1}$ where g_{+} and g_{-} are any gauge transformations in O_{+} and O_{-}. At first sight one might think that with the two arbitrary G valued functions g_{+} and g_{-} one could set the single function g_{+-} to unity. If this were so, the generalization of conventional gauge theories that we are discussing would be vacuous, at least in the case of S^n. This fails only because of a topological subtlety.

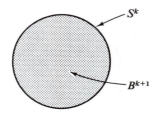

Figure 12.7. The sphere S^k is the boundary of a $(k+1)$-dimensional ball B. For any space G, a map $a : S^k \to G$ is said to be homotopically trivial if it can be extended to a map $a : B \to G$.

The overlap region O_{+-} contains, in particular, the equator $\Sigma \approx S^{n-1}$. Any of g_{+}, g_{-}, or g_{+-} can be restricted to the equator and gives a map $S^{n-1} \to G$. It is therefore important to briefly discuss the topological classification of such mappings. For any k, a map $a : S^k \to G$ is said to be topologically or homotopically trivial if it can be extended to a map $a : B \to G$ where B is a ball whose boundary is S^k (fig. 12.7). This leads to an equivalence relation, with two maps $a : S^k \to G$ and $b : S^k \to G$ considered equivalent if the product ab^{-1} is trivial in the sense just stated. The equivalence classes are called homotopy classes, and they form a group called the homotopy group $\pi_k(G)$. Now, in the above, g_{+} and g_{-} are trivial as elements of $\pi_{n-1}(G)$ since they have extensions over O_{+} and O_{-}, which include the northern and southern hemispheres, respectively. Nothing guarantees the same for g_{+-}. The transition function g_{+-} in general cannot be extended over either the northern or southern hemisphere and defines a nontrivial element of $\pi_{n-1}(G)$. The homotopy class of g_{+-} is invariant under (12.3.4) since g_+ and g_- are homotopically trivial. However, this topological information is the only property of g_{+-} that is invariant under (12.3.4). (In fact, if $f : O_{+-} \to G$ is in the same homotopy class as g_{+-}, then $g_+ = f \cdot g_{+-}^{-1}$ can be extended over O_+ and can be used in (12.3.4) to transform g_{+-} into f.) Therefore, gauge fields on S^n are classified topologically by specifying an element of $\pi_{n-1}(G)$. For $n = 2$ and $G = U(1)$, we have $\pi_1(U(1)) \approx Z$; the nontrivial gauge fields are called magnetic monopoles. For $n = 4$ and G any simple non-abelian group, we have $\pi_3(G) \approx Z$; the nontrivial gauge fields are called instantons. The analogs for higher n are less well-known to physicists. For spaces other than S^n the topological classification of gauge fields is quite complicated, in general. For a ten-dimensional manifold with gauge group E_8 a comparatively simple answer can be given but even so involves concepts beyond the scope of this book.

In the case of a topologically nontrivial gauge field, with transition functions $g_{(\alpha\beta)}$ that cannot all be set to one, one must work with a different gauge field $A_{(\alpha)}$ in each open set $O_{(\alpha)}$, just as one does with the charged matter fields $\psi_{(\alpha)}$. As for the charged fields, the various $A_{(\alpha)}$ are related to each other by gauge transformations,

$$A_{\mu(\alpha)} = g_{(\alpha\beta)} A_{\mu(\beta)} g_{(\alpha\beta)}^{-1} - \partial_\mu g_{(\alpha\beta)} g_{(\alpha\beta)}^{-1}. \tag{12.3.5}$$

In a topologically nontrivial situation, no one $A_{(\alpha)}$ can be extended throughout all space without meeting a Dirac string singularity. However, if the generalization of the Dirac quantization condition is properly obeyed, the Dirac string can be moved at will by means of a gauge transformation, and by choosing a different gauge in each $O_{(\alpha)}$, one can describe physics in each $O_{(\alpha)}$ in terms of a nonsingular gauge field $A_{(\alpha)}$. The analog of the Dirac quantization condition is actually the consistency condition obeyed by the transition functions on triple overlap regions, so it has been incorporated already in our discussion.

Let us now summarize this discussion in a way that aims to underscore the connection between the treatment of topologically nontrivial gauge fields and our discussion in §12.3.1 above of the tangent bundle. Consider a system with a topologically nontrivial gauge field A_μ, interacting with a charged field ψ. Cover space-time with simple open sets $O_{(\alpha)}$ as above. What the observers in $O_{(\alpha)}$ and $O_{(\beta)}$ can agree on is that at each point $x \in M$ the possible values of the charged field ψ make up a vector space V_x. The observers in $O_{(\alpha)}$ and $O_{(\beta)}$ are not able to agree on a choice of basis for the V_x, because a topologically nontrivial situation is precisely a situation in which no one smoothly varying basis for the V_x can exist. If one chooses a basis for V_x, then the field $\psi(x)$, which is an element of V_x, can be described by a wave function $\psi^i(x)$ consisting of the components relative to that basis. Since the observers in $O_{(\alpha)}$ and $O_{(\beta)}$ are not able to agree on a choice of basis, they describe the same field ψ by different wave functions $\psi^i_{(\alpha)}(x)$ and $\psi^i_{(\beta)}(x)$; these are related by transition functions that carry out the change of basis from $O_{(\alpha)}$ to $O_{(\beta)}$. But once again, what the observers in $O_{(\alpha)}$ and $O_{(\beta)}$ can all agree upon is the existence at each x of a vector space V_x of possible values of the charged field ψ. The intrinsic notion is thus the notion of a vector bundle, a smoothly varying family of vector spaces V_x, one for each point x in M.

Given a manifold M and a gauge group G, the classification of vector bundles on M with gauge group G is a very complicated problem in general, as we have noted already. For many purposes it is essential to understand the simplest topological invariants that can be associated with vector bundles. These are the characteristic classes. We will discuss characteristic classes in §12.5 below, but first it is necessary to develop the elementary aspects of the theory of differential forms.

12.4 Differential Forms

Consider an antisymmetric tensor field $B_{i_1 i_2 \ldots i_k}$ with k indices propagating on a compact manifold M of dimension n. Such a field is known by mathematicians as a k form or alternatively as a differential form of degree k. For the field B we postulate the gauge invariance

$$\delta B_{i_1 i_2 \ldots i_k} = \frac{1}{k} \Big\{ \partial_{i_1} \Lambda_{i_2 \ldots i_k} - (-1)^k \partial_{i_2} \Lambda_{i_3 \ldots i_k i_1} \\ \pm \text{ cyclic permutations} \Big\}, \tag{12.4.1}$$

where $\Lambda_{i_1 i_2 \ldots i_{k-1}}$ is antisymmetric in all $k - 1$ indices, and the signs on the right-hand side are chosen to ensure that δB is completely antisymmetric in all indices. It is convenient to avoid constantly repeating the cumbersome expression on the right-hand side of (12.4.1) by defining the exterior derivative operator d. Its definition is simply that acting on any differential form such as a p form ϕ it produces the $p + 1$ form whose components are

$$(d\phi)_{i_1 \ldots i_{p+1}} = \frac{1}{p+1} \Big\{ \partial_{i_1} \phi_{i_2 \ldots i_{p+1}} - (-1)^{p+1} \partial_{i_2} \phi_{i_3 \ldots i_{p+1} i_1} \\ \pm \text{ cyclic permutations} \Big\}. \tag{12.4.2}$$

The signs on the right-hand side are chosen to ensure the complete antisymmetry of $d\phi$. The definition is such that (12.4.1) can be succinctly rewritten $\delta B = d\Lambda$.

At first sight, it might appear that (12.4.1) and (12.4.2) are perfectly good formulas in flat space, but that in curved space all of the ordinary partial derivatives should be replaced by covariant derivatives. One might expect to replace (12.4.2) with

$$
(D\phi)_{i_1 \ldots i_{p+1}} = \frac{1}{p+1} \Big\{ D_{i_1} \phi_{i_2 \ldots i_{p+1}} - (-1)^{p+1} D_{i_2} \phi_{i_3 \ldots i_p i_1} \\
\pm \text{ cyclic permutations} \Big\},
$$

$$(12.4.3)$$

where D is the covariant derivative defined with respect to some Riemannian metric on M. However, (12.4.2) and (12.4.3) are equal. This fact is proved in textbooks on general relativity; it can be proved straightforwardly from the definition of the covariant derivative and the properties of the affine connection. For instance, if ϕ is a one form, then (12.4.3) would give

$$
2(D\phi)_{ij} = D_i\phi_j - D_j\phi_i = \partial_i\phi_j - \Gamma^k_{ij}\phi_k - \partial_j\phi_i + \Gamma^k_{ji}\phi_k
$$

$$
= \partial_i\phi_j - \partial_j\phi_i = 2(d\phi)_{ij}.
$$

$$(12.4.4)$$

We have used the definition of the covariant derivative $D_i\phi_j$ and the fact that in Riemannian geometry the affine connection Γ^k_{ij} is symmetric in its lower indices i and j. The equivalence of (12.4.3) and (12.4.2) for p forms with $p > 1$ is proved similarly, though it is necessary to keep track of more terms. The equivalence of (12.4.3) and (12.4.2), independent of the choice of metric used in defining (12.4.3), means that the operator d of (12.4.2) is a generally covariant operator even though it is defined without any choice of metric. This means that it depends only on the manifold M as a topological (or differentiable) manifold. Differential forms are the only fields on which it is possible to define a reasonable differential operator (namely d) that is independent of the choice of metric; this is the reason for their importance in mathematics. *Any* property of M that can be formulated purely in terms of the properties of the operator d is automatically a topological invariant, because of the facts just stated. Other differential operators studied in mathematics and physics (such as the Dirac operator, for example) depend on the choice of metric, and topological invariants (such as the Dirac index, which we will study in chapter 14) can be extracted from them only via rather special arguments.

Now, a differential form ϕ is said to be 'closed' if $d\phi = 0$. It is said to be 'exact' if $\phi = d\alpha$ for some differential form α. Of course, if ϕ is of

degree k (that is, a k form), then α must be of degree $k - 1$. We have
defined d so that it can operate on differential forms of any degree, so
given a p form ϕ, $d\phi$ is a $p + 1$ form to which d can be applied again; but
it is a fundamental fact that for any form ϕ, $d(d\phi) = 0$. This fundamental
fact may be expressed by saying that $d^2 = 0$ or by saying that every
exact form is closed. To verify that $d^2 = 0$ is straightforward given the
definition (12.4.2), requiring only some care to keep track of the various
terms; it may be wise to first practice with forms of low degree.

Using the fact that $d^2 = 0$, one can readily see that the $p + 1$ form

$$C = dB \tag{12.4.5}$$

is invariant under (12.4.1); moreover, it obeys the 'Bianchi identity' $dC = 0$. A special case of all of this is familiar from Maxwell theory. If B
happens to be a one form, it can be viewed as an Abelian gauge field,
and as such (12.4.1) is the standard gauge transformation law, Λ being in
this case a zero form. Moreover, (12.4.5) is in this case (up to a factor of
two) the standard definition $C_{ij} = (\partial_i B_j - \partial_j B_i)/2$ of the gauge-invariant
field strength, and $dC = 0$ is equivalent to the standard Bianchi identity
$\partial_i C_{jk} + \partial_j C_{ki} + \partial_k C_{ij} = 0$.

The Bianchi identity just stated has a kind of converse. Any differential
form C that obeys $dC = 0$ can locally be written in the form $C = dB$. For
general p this is known as the Poincaré lemma; for small p it is familiar
to physicists from electromagnetic theory. In electromagnetic theory it
is well-known that the Maxwell equation $dF = 0$ (F being the two form
that represents the electromagnetic field strength) implies the existence
of a one form A (the vector potential) such that $F = dA$. More precisely,
the existence of such a one form is only guaranteed locally; globally one
may run into troubles. The prototype of a situation in which $dF = 0$
but the vector potential cannot be defined everywhere without running
into singularities is the Dirac magnetic monopole. The general problem of
inverting the Bianchi identity – whether a p form C with $dC = 0$ can be
written globally as $C = dB$ – will occupy much of our attention in what
follows, for reasons that will soon become apparent.

Having found the gauge-invariant field strength C, it is straightforward
to formulate the generalization of the Maxwell action. It is

$$S = \frac{p+1}{2p!} \int_M g^{i_1 j_1} \ldots g^{i_{p+1} j_{p+1}} C_{i_1 \ldots i_{p+1}} C_{j_1 \ldots j_{p+1}} = \frac{\langle C, C \rangle}{2(p!)^2}, \tag{12.4.6}$$

where g is the metric tensor of the manifold M and the inner product is
defined by $\langle \phi, \psi \rangle = \int \phi \wedge *\psi$. Actually, this is the minimal generalization

of the Maxwell action; we will meet a further generalization in the next chapter.

Now, it is very useful, both for our present purposes and again in chapter 14, to understand the zero modes of the B field. By zero modes we mean modes that have zero action according to definition (12.4.6), but which are physical modes that cannot be gauged away by a gauge transformation (12.4.1). In chapter 14 we will discover that upon compactification of a higher-dimensional theory to four dimensions, the zero modes are manifested in four dimensions as massless particles. The mathematical tools we develop to understand the zero modes will also enable us to define the simplest topological invariants associated with vector bundles, namely the characteristic classes.

Since we require a zero mode B to have zero action, it is evident from (12.4.6) that it must obey $dB = 0$. At the same time, since we require that a physical zero mode is something that cannot be gauged away, it must be that $B \neq d\Lambda$ for any $p - 1$ form Λ. Thus, the number of linearly independent p form zero modes is the number of linearly independent p forms that are closed but not exact. This number is called the pth Betti number b_p. We have defined the b_p purely in terms of the properties of the operator d, which we know to be independent of the metric of M, so the integers b_p are topological invariants. They are among the most basic and most easily computed topological invariants of a manifold M.

An equivalent way to express the definition of the b_p is as follows. Let C_p be the vector space consisting of all closed p forms, that is all p forms α with $d\alpha = 0$. Let D_p be the vector space consisting of all exact p forms, that is all p forms α such that $\alpha = d\beta$ for some $p - 1$ form β. Then D_p is a subspace of C_p, so it is possible to define the quotient vector space

$$H^p(M; R) = C_p/D_p. \tag{12.4.7}$$

(Given a vector space C and a subspace D, the *quotient* space $H = C/D$ is defined by saying that it consists of equivalence classes of elements of C, two elements c and c' of C being considered equivalent if $c - c' \in D$.) The space $H^p(Z; R)$ is known as the pth cohomology group of M with real coefficients. [*] (The group structure of $H^p(M; R)$ is just its additive group structure as a vector space.) The connection between the spaces H^p and the Betti numbers b_p is very simple; b_p is just the dimension of the vector space $H^p(M; R)$.

---

[*] The reference to real coefficients reflects the fact that the H^p actually have a significant generalization – cohomology classes with integer coefficients. This concept is regrettably beyond the scope of our exposition here, though it plays a significant role in ten-dimensional physics.

It is not possible to give a complete account of the Betti numbers and the cohomology groups here, but we pause at this point to try to give a flavor of the subject. A p form α can be integrated over any p-dimensional submanifold T of M. For low p this is well-known to physicists. For instance, if $p = 1$ we are discussing a one form, which is an Abelian gauge field A. It can be integrated along any curve γ to give $I_\gamma = \int_\gamma A_i dx^i$. If $p = 2$ we are discussing a two form, such as the electromagnetic field strength, which can be integrated over any two surface S to give 'the magnetic flux through S', $I_S = \int_S F_{ij} d\Sigma^{ij}$, where $d\Sigma^{ij}$ is the area element of S. This generalizes to the integral of a p form α over any p-dimensional submanifold T, with or without boundary:

$$I_T = \int_T \alpha_{i_1 \ldots i_p} d\Sigma^{i_1 \ldots i_p}. \qquad (12.4.8)$$

Here $d\Sigma^{i_1 \ldots i_p}$ is the volume element of T. We henceforth abbreviate (12.4.8) as

$$I_T = \int_T \alpha. \qquad (12.4.9)$$

Integration of differential forms is governed by an important theorem known as Stokes' theorem. Given any compact submanifold W of M, let ∂W be its boundary. Suppose that W has dimension $p + 1$, and let β be a p form. Then Stokes' theorem asserts that

$$\int_W d\beta = \int_{\partial W} \beta. \qquad (12.4.10)$$

For low p this is familiar from electromagnetic theory.

Now we want to discuss the relation between the cohomology groups and the integrals of differential forms over closed submanifolds. By a closed submanifold, we mean simply a compact submanifold without boundary. Let α be a closed p form, and let T be a closed submanifold of M of dimension p. A closed p form α is an element of what we called C_p, so its equivalence class in C_p/D_p defines an element of $H^p(M; R)$ called the cohomology class of α. Two closed forms α and α' define the same cohomology class if $\alpha - \alpha' = d\beta$ for some $p - 1$ form β. If so

$$\int_T \alpha - \int_T \alpha' = \int_T d\beta = \int_{\partial T} \beta = 0, \qquad (12.4.11)$$

since $\partial T = 0$. Thus the integral of a closed form α over a closed manifold T depends only on the cohomology class of α. In particular, if the cohomology class of α is zero, in other words if $\alpha = d\beta$ for some β, then $\int_T \alpha = 0$ for all closed manifolds T. A celebrated theorem by de Rham establishes a converse to this. If the cohomology class of α is *not* zero, there is always some closed manifold T with $\int_T \alpha \neq 0$.

If α is such that $\int_T \alpha$ is an integer for all T, then α is said to be an integral cohomology class. Such classes form a link between the de Rham cohomology that we are considering here and the more subtle cohomology with integer coefficients, which we will unfortunately not be able to explore.

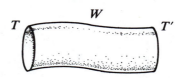

Figure 12.8. The manifold W interpolates between two manifolds T and T' that can be deformed into one another.

Now, fix a closed p form α, whose cohomology class may or may not be zero, and study $I(T) = \int_T \alpha$ as a functional of the closed manifold T. If T can be continuously deformed into T', then $I(T) = I(T')$ since by virtue of Stokes' theorem

$$\int_T \alpha - \int_{T'} \alpha = \int_{\partial W} \alpha = \int_W d\alpha = 0. \qquad (12.4.12)$$

Here W, indicated in fig. 12.8, is a manifold that interpolates between T and T' and obeys $\partial W = T - T'$ (the minus sign means that T and T' appear in ∂W with opposite orientation). Evidently, such a W exists whenever T can be deformed into T'. This is a special case of a much stronger restriction on $I(T)$. Let T be any closed p-dimensional manifold that is the boundary of some $(p+1)$-dimensional manifold W, $\partial W = T$. Then

$$I(T) = \int_T \alpha = \int_{\partial W} \alpha = \int_W d\alpha = 0. \qquad (12.4.13)$$

Thus, if T is a boundary, the integral over T of any closed form is zero. Again, de Rham's theorem establishes a kind of converse to this. If the closed submanifold T of M (and every multiple of it) is *not* the boundary of any submanifold of M of one dimension more, then according to de Rham's theorem there is always some closed differential form α on M such that $\int_T \alpha \neq 0$. We speak loosely of a closed submanifold T of space-time as being topologically nontrivial if T is not the boundary of any submanifold W of M.

Combining these statements, we find that the pth Betti number b_p of a manifold M is the number of independent closed p-dimensional surfaces in M that are topologically nontrivial.

There is another way of expressing some of these ideas that we sometimes find useful. Let α be a p form that is closed but not exact; this means that $d\alpha = 0$ but $\alpha \neq d\beta$ for any $p-1$ form β. In view of the latter statement, it is impossible to choose β to make the quantity

$$W = \langle \alpha - d\beta, \alpha - d\beta \rangle \qquad (12.4.14)$$

vanish. While we cannot make W vanish, we can choose β to make the positive definite quantity W as small as possible. The variational equation for minimizing W with respect to β is $d^*(\alpha - d\beta) = 0$, where d^* is the adjoint of d with respect to the inner product $\langle , \rangle$. Explicitly, d^* is easily seen to be the operator that, acting on any p form ψ, gives a $p-1$ form $d^*\psi$ defined by

$$(d^*\psi)_{i_1...i_{p-1}} = -pD^{i_0}\psi_{i_0...i_{p-1}}. \qquad (12.4.15)$$

Since $d^2 = 0$, the square d^{*2} of its adjoint must also be zero, and this can easily be checked explicitly from (12.4.15).

If we let $\gamma = \alpha - d\beta$, then what we have learned can be summarized by saying that any p-dimensional cohomology class (such as the class of an arbitrary closed p form α) can be represented by a differential form γ that obeys

$$d\gamma = d^*\gamma = 0. \qquad (12.4.16)$$

Now, while d maps p forms into $p+1$ forms, its adjoint d^* maps p forms into $p-1$ forms, so the operator

$$\Delta = dd^* + d^*d \qquad (12.4.17)$$

maps p forms into p forms. It is called the Hodge–de Rham Laplacian acting on differential forms. (The reader may wish to verify that acting on zero forms, which are ordinary functions, Δ is indeed the ordinary Laplacian.) A solution of $\Delta\alpha = 0$ is called a harmonic differential form. Evidently, any form γ that obeys (12.4.16) is harmonic, so since every cohomology class can be represented by a solution of (12.4.16), every cohomology class can be represented by a harmonic differential form.

The converse is also true; every harmonic differential form represents a nonzero cohomology class. To prove this, suppose first that γ obeys $\Delta\gamma = 0$. Then $0 = \langle\gamma, (dd^* + d^*d)\gamma\rangle = \langle d\gamma, d\gamma\rangle + \langle d^*\gamma, d^*\gamma\rangle$. Since $\langle\,,\,\rangle$ is a positive inner product on differential forms, this implies that γ obeys (12.4.16). In particular γ is closed and therefore represents a cohomology class. To see that the cohomology class represented by γ is not zero, suppose on the contrary that $\gamma = d\beta$ for some β. Then (since γ obeys (12.4.16)) $0 = \langle d^*\gamma, \beta\rangle = \langle\gamma, d\beta\rangle = \langle\gamma, \gamma\rangle$. This implies that $\gamma = 0$. Therefore, the cohomology class represented by a harmonic differential form γ can be zero only if γ is itself zero. This completes the demonstration that harmonic p forms are in one-to-one correspondence with p-dimensional cohomology classes. In particular, the pth Betti number b_p equals the number of linearly independent harmonic p forms.

Another way to express this result is frequently quite convenient. Since $d^2 = (d^*)^2 = 0$, $\Delta = S^2$ where $S = d + d^*$. Since S is hermitian, it has the same zero eigenvalues as its square Δ. In particular, zero modes of S can be chosen to be differential forms of definite degree (since this property is obvious for Δ, which maps p forms into p forms), and the pth Betti number or the number of harmonic p forms is the same as the number of independent p forms annihilated by S. As a first-order, hermitean operator, S is reminiscent of the Dirac operator. We will make the analogy much more precise in chapter 14.

We now describe an operation called the 'wedge product', which is essential in the study of differential forms. Given any two tensor fields A and B on a manifold M, it is possible to make new tensors by multiplying them and contracting indices in various ways. In the case of differential forms, there is a particularly important multiplication law that is suitable for studying topological properties of M because it does not depend on the use of a metric tensor. If A and B are a p form and a q form, respectively, we define the wedge product $A \wedge B$ to be the $p + q$ form

$$(A \wedge B)_{i_1 \ldots i_{p+q}} = \frac{p!q!}{(p+q)!}\left\{A_{i_1 \ldots i_p} B_{i_{p+1} \ldots i_{p+q}} \pm \text{permutations}\right\}, \quad (12.4.18)$$

where the right-hand side is summed over permutations so as to be completely antisymmetric in all $p+q$ indices. When confusion is not likely we often denote $A \wedge B$ simply as AB. The wedge product obeys

$$A \wedge B = (-1)^{pq} B \wedge A. \qquad (12.4.19)$$

This can be straightforwardly checked from the definition (12.4.18), the only slightly delicate point being the sign $(-1)^{pq}$, which arises from the complete antisymmetry in the definition of the wedge product. The wedge product and the exterior derivative obey

$$d(A \wedge B) = (dA) \wedge B + (-1)^p A \wedge (dB), \qquad (12.4.20)$$

as one can again see from the definitions. An important consequence of (12.4.20) is that if A and B are closed so is $A \wedge B$. If A and B are closed, they represent p- and q-dimensional cohomology classes, respectively, and their product $A \wedge B$, being closed, represents a $(p + q)$-dimensional cohomology class. Moreover, the cohomology class of $A \wedge B$ depends only on the cohomology classes of A and B. For if we replace, say, A by a differential form $\tilde{A} = A + d\alpha$ in the same cohomology class, then $\tilde{A} \wedge B = A \wedge B + (d\alpha) \wedge B = A \wedge B + d(\alpha \wedge B)$ and (since $d(\alpha \wedge B)$ is zero in cohomology) this means that $A \wedge B$ and $\tilde{A} \wedge B$ represent the same cohomology class. Thus, the wedge product gives a well-defined law of multiplying cohomology classes. Given closed p and q forms representing elements of $H^p(M; R)$ and $H^q(M; R)$, respectively, their wedge product defines an element of $H^{p+q}(M; R)$. This multiplication law for cohomology classes defines what is called the cohomology ring of the manifold M, and it contains much more information than is contained in the individual cohomology groups themselves. For instance, we will see in chapter 16 that under suitable conditions the cohomology groups, or rather their dimensions, which are the Betti numbers, determine the number of massless fermions, while the cohomology ring determines the Yukawa couplings.

Since the Betti numbers play an important role in many aspects of ten-dimensional physics, we pause here to discuss a few of their basic properties. The simplest property is that the zeroth Betti number b_0 is always one. This is easily shown by noting that a zero form is just an ordinary function ϕ, and a closed zero form ϕ obeying $d\phi = 0$ would have to be a constant function, say $\phi = 1$. While this is a closed zero form, it is not exact (for ϕ a zero form, $\phi = d\alpha$ would mean that α would have to be a (-1) form!), so there is up to normalization precisely one closed but not exact zero form, showing that $b_0 = 1$. Another basic property of the Betti numbers is that on a compact, orientable manifold M of dimension n, $b_p = b_{n-p}$ for any p. This is easily proved by thinking of the Betti number b_p as the number of zero eigenvalues of the Laplacian Δ acting on p forms. Let $\epsilon_{i_1 \dots i_n}$ be the Levi–Civita tensor of M; this is the completely antisymmetric tensor, which, apart from the metric tensor, is the only covariantly constant tensor field on a generic Riemannian manifold. There is a natural operator $*$ called the Poincaré duality operator, which maps

p forms into $n - p$ forms. It is defined by saying that for any p form α, $*\alpha$ is the $n - p$ form

$$(*\alpha)_{i_1 \ldots i_{n-p}} = \epsilon_{i_1 \ldots i_{n-p}}{}^{i_{n-p+1} \ldots i_n} \alpha_{i_{n-p+1} \ldots i_n}/p!. \qquad (12.4.21)$$

It obeys $(*)^2 = (-1)^{p(n-p)}$ and is related to the d and d^* operators defined earlier by $d^* = (-1)^{np+n+1} * d*$. These relations are nontrivial but easily verified. It follows from these properties that

$$\begin{aligned}
*\Delta &= *(dd^* + d^*d) = (-1)^{np+1} * ((-1)^n d * d * + * d * d) \\
&= (-1)^{np+1}((-1)^n * d * d + d * d*)* = (dd^* + d^*d)* = \Delta*,
\end{aligned}$$
$$(12.4.22)$$

so that $*$ commutes with the Laplacian Δ. Therefore, $*$ maps harmonic p forms into harmonic $n - p$ forms, and vice versa. This ensures that on a compact, orientable manifold of dimension n, $b_p = b_{n-p}$. One consequence of this is that on such a manifold $b_n = 1$.

Another fundamental property of the Betti numbers concerns their behavior when one considers a product of manifolds. Let Z_1 and Z_2 be two compact manifolds, and let $Z = Z_1 \times Z_2$. Let $b_p(Z_1), b_p(Z_2)$, and $b_p(Z)$ be respectively the pth Betti number of Z_1, Z_2 and Z. Then the Künneth formula asserts that

$$b_p(Z) = \sum_{k=0}^{p} b_k(Z_1) b_{p-k}(Z_2). \qquad (12.4.23)$$

The proof is quite simple. Let α be a harmonic k form on Z_1 and let β be a harmonic $p - k$ form on Z_2. Then $\alpha \wedge \beta$ is a harmonic p form on Z, which is always nonzero. This construction for various values of k and all possible choices of α and β gives enough harmonic p forms on Z to prove that $b_p(Z)$ is at least as large as the right-hand side of (12.4.23). To show that (12.4.23) holds exactly, not just as an inequality, we must show that wedge products $\alpha \wedge \beta$, where α and β are defined on Z_1 and Z_2, respectively, give a basis for the harmonic differential forms of Z. This follows easily from the fact that the Laplacians on Z, Z_1 and Z_2 obey $\Delta_Z = \Delta_{Z_1} + \Delta_{Z_2}$, a fact that is again nontrivial but not too difficult to verify.[*] Each of these operators is nonnegative (and Δ_{Z_1} and Δ_{Z_2} commute with each other) so a zero eigenvalue of Δ_Z must be a simultaneous zero eigenvalue of Δ_{Z_1} and Δ_{Z_2}. Hence a harmonic differential form on Z is

---

[*] The reader who wishes to verify this statement is urged to consider first the case in which Δ acts on zero forms.

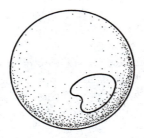

Figure 12.9. A circle on the two sphere is always topologically trivial; more generally on S^n, a p sphere S^p for $1 \leq p < n$ is always topologically trivial.

always a linear combination of expressions of the form $\alpha \wedge \beta$ with α and β harmonic forms on Z_1 and Z_2, respectively.

Let us consider briefly the Betti numbers of a few simple spaces. One general statement is that for a compact, orientable manifold of dimension n, $b_0 = b_n = 1$. Restricting ourselves to compact, orientable manifolds, the sphere S^n is the only space for which the other Betti numbers are zero. That the Betti numbers of S^n vanish except for b_0 and b_n can be proved by study of the Laplacian; alternatively, it is related to a different characterization of the Betti numbers that was sketched above. The Betti number b_p is the number of independent p-dimensional submanifolds Y that are topologically nontrivial in the sense that Y is not the boundary of R for any $(p+1)$-dimensional submanifold R. On the sphere S^n every p-dimensional submanifold for $1 \leq p \leq n-1$ is topologically trivial (this is sometimes described, as in fig. 12.9 by saying that 'you can't lasso a sphere'), so the b_p are zero for $1 \leq p \leq n-1$. (However, a point in S^n is a zero-dimensional submanifold that is not a boundary, so $b_0 = 1$, and S^n itself is an n-dimensional submanifold that is not a boundary, so $b_n = 1$.) The Betti numbers of a product of spheres can then be worked out from the Künneth formula. For instance, for the ordinary two-dimensional torus $T = S^1 \times S^1$ the Betti numbers are $b_0 = 1, b_1 = 2, b_2 = 1$. A classic result generalizes this to a compact Riemann surface (or string world sheet!) with g handles: $b_0 = 1, b_1 = 2g, b_2 = 1$. The value of b_1 reflects the same fact that entered in our discussion of spin structures on a Riemann surface of genus g: there are $2g$ independent noncontractible closed curves on such a surface. Actually, there are quite practical methods to calculate the Betti numbers of any reasonable space, but we will not be able to explore them in this book.

We conclude this introduction to differential forms by pointing out a very useful but slightly more abstract notation, which we have chosen

to avoid in the above. Let M be an n-dimensional manifold with local coordinates x^i, $i = 1, \ldots, n$. Although this viewpoint is not usually
adopted in physics literature, the 'coordinates' x^i on a manifold can be
viewed as real-valued functions on the manifold, or in other words zero
forms. Since the x^i are zero forms, their exterior derivatives are one forms
dx^i. Since the wedge product of one forms is antisymmetric, the dx^i obey
$dx^i \wedge dx^j = -dx^j \wedge dx^i$ or simply (if the wedge product is understood)
$dx^i dx^j = -dx^j dx^i$. The dx^i are a complete basis of one forms on M (at
least locally), in the sense that any one form A has an expansion

$$A = \sum_{i=1}^{n} A_i(x^j) dx^i, \qquad (12.4.24)$$

where the scalar functions $A_i(x^j)$ are the 'components' of A in the coordinate system given by the x^j. If we regard A as an abelian gauge field,
then the field strength is

$$F = dA = \frac{1}{2}(\partial_i A_j - \partial_j A_i) dx^i dx^j \qquad (12.4.25)$$

or simply $F = \frac{1}{2} F_{ij} dx^i dx^j$. Note that we could also have written $F = dx^i dx^j \partial_i A_j$ or $F = dx^i \partial_i A$ if it is understood that the derivative ∂_i acts
only on the components A_j and not the fundamental one forms dx^j in the
expansion $A = A_j dx^j$. More generally, if ϕ is a differential form of any
order, then we have in the same sense $d\phi = dx^i \partial_i \phi$.

12.5   Characteristic Classes

We now have the tools to describe the simplest topological invariants
characterizing topologically nontrivial gauge fields. These are the characteristic classes. Consider first the case of a $U(1)$ gauge field. As in §12.3.2
we cover our manifold M with open sets $O_{(\alpha)}$. In each open set $O_{(\alpha)}$ the
gauge field is a one form $A_{(\alpha)}$. Equation (12.3.5) reduces in this case (with
$g_{(\alpha\beta)} = \exp(i\phi_{(\alpha\beta)})$ and absorbing a factor of i in the gauge field to make
it real) to

$$A_{(\beta)} = A_{(\alpha)} + d\phi_{(\alpha\beta)}. \qquad (12.5.1)$$

Here the $\phi_{(\alpha\beta)}$ are smooth functions in $O_{(\alpha)} \cap O_{(\beta)}$.

The gauge-invariant field strength is $F = dA_{(\alpha)}$. It is a two form
which by virtue of (12.5.1) (and $d^2 = 0$) is independent of α and so is
uniquely defined throughout M. Moreover, the Bianchi identity $dF = 0$

asserts that F is closed and so defines an element of $H^2(M; R)$, the second cohomology group of M with real coefficients. The cohomology class of $F/2\pi$ in $H^2(M; R)$ is called the first Chern class of the $U(1)$ vector bundle under discussion. (A $U(1)$ vector bundle, that is a vector bundle in which the gauge group is $U(1)$, is frequently called a complex line bundle.) In a topologically trivial situation, $F = dA$ with the *same A* defined everywhere. In this case (by the definition of the cohomology groups), F is zero in H^2 and the first Chern class vanishes.

In a topologically nontrivial situation, a gauge field is really a family of gauge fields $A_{(\alpha)}$ that obeys (12.5.1). Two gauge fields, or rather two families $A_{(\alpha)}$ and $\tilde{A}_{(\alpha)}$, are of the same topological class if they obey (12.5.1) with the *same* transition functions $\phi_{(\alpha\beta)}$:

$$
\begin{aligned}
A_{(\alpha)} &= A_{(\beta)} + \phi_{(\alpha\beta)} \\
\tilde{A}_{(\alpha)} &= \tilde{A}_{(\beta)} + \phi_{(\alpha\beta)}
\end{aligned}
\qquad (12.5.2)
$$

Subtracting these equations, we note that $A_{(\alpha)} - \tilde{A}_{(\alpha)}$ is independent of α and so is an everywhere defined one form, which we may call B. Also, the respective field strengths are $F = dA$ and $\tilde{F} = d\tilde{A}$. Their difference is $F - \tilde{F} = dB$. Now the important point is that since $A_{(\alpha)}$ and $\tilde{A}_{(\alpha)}$ are not everywhere uniquely defined (but vary from $O_{(\beta)}$ to $O_{(\alpha)}$ according to (12.5.1)), the equations $F = dA$ and $\tilde{F} = d\tilde{A}$ do not imply that the cohomology class of F or $\tilde{F}$ is zero. But since B is everywhere uniquely defined, the equation $F - \tilde{F} = dB$ means that F and $\tilde{F}$ are of the same cohomology class. In other words, the first Chern class is a topological invariant.

If the first Chern class is not zero, so that F is not zero in $H^2(M; R)$, then by virtue of the theorems of de Rham that were surveyed in §12.4, there is always some closed two manifold T in space-time with

$$
I(T) = \frac{1}{2\pi} \int_T F \neq 0. \qquad (12.5.3)
$$

In this case, the Dirac quantization law for magnetic charge, applied to T, guarantees that $I(T)$ is an integer, so that the cohomology class of $F/2\pi$ is always an integral element of $H^2(M; R)$. The first Chern class can be thought of as a rule that assigns to each closed two manifold T in space-time the integer $I(T)$. Being integers, the $I(T)$ cannot change under continuous deformation of the gauge field, and this is another way to understand the topological invariance of the first Chern class.

Now we wish to generalize the notion of Chern classes to the case of a simple nonabelian gauge group G. In this case the discussion is slightly more complicated. The gauge field $A^a_{(\alpha)}$ carries an extra index a labeling the Lie algebra of G; it may be viewed as a one form with values in this Lie algebra. The gauge field strength is now $F^a = dA^a + f^a_{bc} A^b \wedge A^c$, where f^a_{bc} are the structure constants of the Lie algebra G. Here F^a may be characterized as a two form with values in the Lie algebra. Because of the Lie algebra index of F^a, it is not gauge invariant, only gauge covariant. The Bianchi identity asserts not that F^a is closed, but rather that

$$0 = dF^a + f^a_{bc} A^b \wedge F^c. \qquad (12.5.4)$$

Since F^a is not closed, there is no analog of the first Chern class. However, (12.5.4) is easily shown to imply that the four form

$$\Omega = \sum_{a,b} \delta_{ab} F^a \wedge F^b \qquad (12.5.5)$$

is closed, $d\Omega = 0$. If we normalize the generators of G to obey $\mathrm{tr} T^a T^b = \delta^{ab}$, then (12.5.5) is equivalent to $\Omega = \sum_{a,b} F^a \wedge F^b \cdot \mathrm{tr} T^a T^b$. By introducing matrix-valued forms

$$A = \sum T^a A^a \qquad F = \sum T^a F^a, \qquad (12.5.6)$$

this can be abbreviated as

$$\Omega = \mathrm{tr} F \wedge F. \qquad (12.5.7)$$

The 'trace' in (12.5.7) is then over the indices of the T matrices, which have not been explicitly indicated. Since $d\Omega = 0$, Ω defines an element of $H^4(M; R)$ known as the second Chern class in the case of an $SU(N)$ gauge field or the first Pontryagin class in the case of $SO(N)$.

To show that the second Chern class is zero in a topologically trivial situation, one introduces the Chern–Simons three form ω_3, defined in each $O_{(\alpha)}$ by saying that

$$\omega_{3(\alpha)} = \mathrm{tr} A_{(\alpha)} \wedge dA_{(\alpha)} + \frac{2}{3} \mathrm{tr} A_{(\alpha)} \wedge A_{(\alpha)} \wedge A_{(\alpha)}. \qquad (12.5.8)$$

Here, analogously to (12.5.7), $\mathrm{tr} A \wedge A \wedge A$ is an abbreviation for $A^a \wedge A^b \wedge A^c \cdot \mathrm{tr} T^a T^b T^c$. It follows from the definitions that $\mathrm{tr} F \wedge F = d\omega_3$.

In a topologically trivial situation, with ω_3 uniquely defined everywhere, this implies that the second Chern class is zero. To show that the second Chern class is a topological invariant, suppose that we are given two gauge fields A and $\tilde{A}$ that are in the same topological class; in other words the families $A_{(\alpha)}$ and $\tilde{A}_{(\alpha)}$ obey (12.3.5) with the *same* transition functions $g_{(\alpha\beta)}$. Then in the difference $\delta A = A - \tilde{A}$ the inhomogeneous term in (12.3.5) cancels out; δA is a gauge covariant one form just like the field strengths F and $\tilde{F}$. The three form $Z = \mathrm{tr}\delta A \wedge (F + \tilde{F}) - \frac{1}{3}\mathrm{tr}\delta A \wedge \delta A \wedge \delta A$ is uniquely defined, independent of α, and can be shown with a small amount of algebra to obey $\mathrm{tr}F \wedge F - \mathrm{tr}\tilde{F} \wedge \tilde{F} = dZ$. This shows that the cohomology classes of $\mathrm{tr}F \wedge F$ and $\mathrm{tr}\tilde{F} \wedge \tilde{F}$ are equal, or in other words that the second Chern class is the same for any two gauge fields in the same topological class.

As in the abelian case, if $\Omega = \mathrm{tr}F \wedge F$ is not zero in cohomology, there is always some four manifold T such that

$$I(T) = \frac{1}{8\pi^2} \int_T \mathrm{tr}F \wedge F \qquad (12.5.9)$$

is not zero. The integral $I(T)$ measures what in discussions of QCD would be called the 'total instanton number' of the gauge field on T, and, as is known from discussions of QCD, it is always an integer. Consequently, $\mathrm{tr}F \wedge F/8\pi^2$ always represents an integral class in $H^4(M; R)$. We can think of the second Chern class of an $SU(N)$ gauge field (or the first Pontryagin class of an $SO(N)$ gauge field) as a rule that assigns an integer to each closed four manifold T, namely the total number of instantons on T. Being integers, the $I(T)$ cannot change under continuous variation of the gauge field. As in our previous discussion, this is another way to understand that the second Chern class is a topological invariant.

The generalization to higher characteristic classes is relatively straightforward. If $d_{a_1 \ldots a_k}$ is any kth-order symmetric invariant tensor in the Lie algebra of G, let

$$\Omega_{2k} = \sum_{a_1 \ldots a_k} F^{a_1} \wedge \ldots \wedge F^{a_k} \cdot d_{a_1 \ldots a_k}. \qquad (12.5.10)$$

It follows from (12.5.4) that $d\Omega_{2k} = 0$ as long as $d_{a_1 \ldots a_k}$ is an invariant tensor, so that Ω_{2k} defines an element of $H^{2k}(M; R)$. A very important case is the case $d_{a_1 \ldots a_k} = \mathrm{str}T_{a_1} \ldots T_{a_k}$, where str denotes a symmetrized trace, symmetrized under all permutations of the indices $a_1 \ldots a_k$. In the case of a $U(N)$ gauge field, with str being the symmetrized trace in the fundamental representation, the cohomology class of Ω_{2k} is called the kth Chern class (with real coefficients). When confusion is not likely we write for this cohomology class $\Omega_{2k} = \text{tr} F^k$, the wedge product on the right-hand side being understood. In the case of a $U(N)$ gauge group, all characteristic classes can be expressed as a linear combination of wedge products of the Ω_{2k}. What is the analogous statement in the case of $SO(N)$? For $SO(N)$ the symmetrized trace of an odd number of generators vanishes in the fundamental representation, so Ω_{2k} is zero (with the above choice of $d_{a_1...a_k}$) unless $k = 2r$ for some r; in that case the cohomology class of Ω_{2k} is called the rth Pontryagin class of the $SO(N)$ bundle.

For $SO(2N)$ there is one other important choice of $d_{a_1...a_k}$ in the special case $k = N$. It arises as follows. For orthogonal groups the adjoint representation is the same as the second-rank antisymmetric tensor representation, so the generators T^a can be conveniently labeled as $T^{ij}(= -T^{ji})$, $i, j = 1, \ldots, 2N$, and likewise the Yang-Mills field strength F^a becomes $F^{ij}(= -F^{ji})$. The antisymmetric tensor $\epsilon_{i_1...i_{2N}}$ can therefore be viewed as an Nth-order invariant tensor of $SO(2N)$. Use of this invariant tensor in (12.5.10) leads to the definition of a closed $2N$ form

$$\eta = \frac{1}{N!(4\pi)^N} \epsilon_{i_1...i_{2N}} \cdot F^{i_1 i_2} \wedge \ldots \wedge F^{i_{2N-1} i_{2N}}, \tag{12.5.11}$$

whose cohomology class in $H^{2N}(M; R)$ is known as the Euler class of the vector bundle. The topological invariance of the higher Chern and Pontryagin classes and the Euler class can be proved as in our previous discussions.

The characteristic class $\text{tr} F \wedge F$ can be defined for any vector bundle whose structure group is a compact Lie group. This characteristic class can be defined, for example, for a vector bundle with gauge group E_8. In the case of E_8, after $\text{tr} F \wedge F$ the next characteristic class (apart from wedge products of $\text{tr} F \wedge F$) is the sixteen form $\text{tr} F^8$, since after the quadratic invariant used in defining $\text{tr} F \wedge F$, the next symmetric invariant in the E_8 Lie algebra is of eighth order.

12.5.2 Characteristic Classes of Manifolds

Given a manifold M and a vector bundle V, the simplest topological invariants that can be defined are the characteristic classes that we have been discussing. They can be defined for any choice of V. A particularly interesting case is the case in which V is the tangent bundle T of M. In that case, as explained in §12.1 and §12.3, the natural gauge field is the spin connection $\omega_\mu{}^a{}_b$, and the gauge-invariant field strength $R_{\mu\nu}{}^a{}_b$ is essentially the Riemann tensor. For the tangent bundle just as for any other vector bundle, the differential forms $\text{tr} R \wedge R$ and more generally $\text{tr} R^{2k}$ (a wedge product being understood in the last expression) are closed; their cohomology classes are topological invariants, independent of the choice of metric of M. The cohomology class of $\text{tr} R^{2k}$ is, according to our general discussion, the kth Pontryagin class of the tangent bundle of M; it is usually called simply the kth Pontryagin class of M. The Euler class (12.5.11) of the tangent bundle is likewise usually called the Euler class of M. If M has dimension $2n$, the Euler class is a $2n$ form that can be integrated over M to give a number,

$$\chi(M) = \int_M \eta. \tag{12.5.12}$$

This number is called the Euler characteristic of M; it will enter our story in chapter 14. If the dimension of M is divisible by four, it is likewise possible to make numbers (called characteristic numbers) by integrating certain characteristic classes or wedge products of characteristic classes over M. These are called Pontryagin numbers; they will be less useful for us. Euler, of course, originally defined the Euler characteristic (in two dimensions) in a more elementary way; the notions of curvature tensors and differential forms were developed centuries later.

One application of the notions we have been discussing in string theory is the following. In $N = 1$ supergravity in ten dimensions, one of the fields in the supergravity multiplet is the two form B. Considerations related to anomaly cancellation discussed in the next chapter show that the gauge-invariant field strength of B is not $H_0 = dB$ but $H = dB + \omega_{3L} - \omega_{3Y}$. Here $\omega_{3Y} = \text{tr} A \wedge dA + \frac{2}{3} \text{tr} A \wedge A \wedge A$ is the Yang–Mills Chern–Simons three form, A being the gauge field, and $\omega_{3L} = \text{tr} \omega \wedge d\omega + \frac{2}{3} \text{tr} \omega \wedge \omega \wedge \omega$ is the Chern–Simons three form constructed from the spin connection. They obey respectively $d\omega_{3Y} = \text{tr} F \wedge F, d\omega_{3L} = \text{tr} R \wedge R$. The Bianchi identity asserts that $dH_0 = 0$, from which it follows that H obeys the more subtle identity

$$dH = \text{tr} R \wedge R - \text{tr} F \wedge F. \tag{12.5.13}$$

Now, in cohomology dH is zero, so (12.5.13) asserts that the cohomology classes of $\text{tr} F \wedge F$ and $\text{tr} R \wedge R$ are equal. The cohomology class of $\text{tr} F \wedge F$ is the first Pontryagin class of the $SO(32)$ or $E_8 \times E_8$ bundle V, and the cohomology class of $\text{tr} R \wedge R$ is the first Pontryagin class of the tangent bundle T. (12.5.13) asserts that these cohomology classes are equal; this is an important restriction on the topological choices that are made in the course of compactification.

12.5.3 The Euler Characteristic of a Riemann Surface

For another interesting application of these results, consider the case in which M is actually a manifold of dimension two – the world sheet Σ of a string. In this case, p forms exist only for $p \leq 2$, and the only nontrivial characteristic class of the tangent bundle is the Euler class. Its integral over Σ is the Euler characteristic

$$\chi(\Sigma) = \frac{1}{8\pi} \int_\Sigma \epsilon^{ij} R_{ijkl} \epsilon^{kl}. \tag{12.5.14}$$

With the two-dimensional identity $\epsilon^{ij} \epsilon^{kl} = g^{ik} g^{jl} - g^{il} g^{jk}$, (12.5.14) becomes

$$\chi(\Sigma) = \frac{1}{4\pi} \int_\Sigma \sqrt{g}\, R, \tag{12.5.15}$$

where R is the Ricci curvature scalar and we have restored in (12.5.15) a factor $\sqrt{g}$, the square root of the determinant of the metric tensor, which has been suppressed in most of our formulas. We learn from (12.5.15) that the right-hand side is a topological invariant. This assertion was made in chapter 3 in explaining why the possible presence of a term proportional to the right-hand side of (12.5.15) in the world-sheet action is not important in the quantization of the free string. While the identification of (12.5.15) with the integral of the Euler class is a satisfying way to understand the fact that the right-hand side of (12.5.15) is a topological invariant, the machinery we have developed is certainly not necessary for establishing this particular point. Indeed, it is quite elementary to see that, locally, $\sqrt{g}R$ is a total derivative. For instance, in a coordinate system with $g_{ij} = \delta_{ij} e^\phi$, we learned in §3.1.2 that $\sqrt{g}R = \Box\phi$. The fact that $\sqrt{g}R$ is locally a total derivative means that it cannot change in any local variation of the metric and therefore is a topological invariant.

We can easily work out the value of the Euler characteristic for a Riemann surface of any genus. Let $\chi(g)$ be the Euler characteristic of a surface of genus g. For genus zero, by explicitly evaluating the integral in (12.5.15) for the ordinary round metric on S^2 (if the radius is r, the area is $4\pi r^2$ and the curvature is $R = 2/r^2$), we find $\chi(0) = 2$. For genus one, we note that the torus admits a flat metric. Evidently, with this choice of metric, (12.5.15) vanishes, so $\chi(1) = 0$. For $g > 1$, $\chi(g)$ is easily worked out from the recursion relation $\chi(g_1) + \chi(g_2) = \chi(g_1 + g_2) + \chi(0)$, or equivalently

$$\chi(g_1 + g_2) = \chi(g_1) + \chi(g_2) - 2. \tag{12.5.16}$$

The proof of (12.5.16) is carried out by a simple operation of cutting

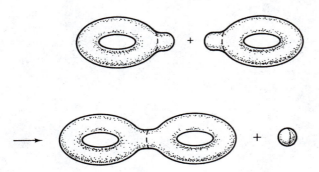

Figure 12.10. Given two surfaces Σ_1 and Σ_2, of genus g_1 and g_2, by removing a hemisphere from each and rearranging the pieces, we can make a surface of genus $g_1 + g_2$ and one of genus zero.

and pasting in which, without changing the total integrated curvature, a surface of genus g_1 and one of genus g_2 are recombined into a surface of genus $g_1 + g_2$ and one of genus zero. This is illustrated in fig. 12.10. Using the known values for genus zero and genus one, we can deduce from (12.5.16) that

$$\chi(g) = 2 - 2g \qquad (12.5.17)$$

for any genus. In chapter 14, this formula will be rederived by relating the Euler characteristic to the Betti numbers via an index theorem.

13. Low-Energy Effective Action

The particle spectrum of a string theory consists of a finite number of massless states and an infinite tower of massive excitations at a mass scale characterized by a fundamental parameter – the string tension or Regge slope. As has been explained in previous chapters, this parameter must be of order the Planck mass (10^{19} GeV) in order that the graviton interact with the usual Newtonian strength. If one wishes to give a phenomenological description of the consequences of string theory for low-energy physics, it should not be necessary to describe explicitly what the massive states are doing. It is natural, instead, to formulate an effective action based entirely on fields that correspond to massless, or at least very light, degrees of freedom only. Such a description turns out to be useful not only for a phenomenological analysis, but even as a framework for addressing certain theoretical issues, such as the occurrence of anomalies.

The infinite set of point-particle fields that arise in string theory consists of a finite number of massless fields, which we collectively represent for the moment by ϕ_0, and an infinite number of heavy fields collectively represented by ϕ_H. In principle, it must be possible to describe string theory by a classical action $S(\phi_0, \phi_H)$ (or, at the quantum level, a quantum effective action) governing these fields. At present, we do not have really satisfactory ways to formulate and understand the exact classical action $S(\phi_0, \phi_H)$. A proper understanding is likely to involve understanding the string-theoretic generalizations of general covariance and Yang–Mills gauge invariance, and so is clearly a very crucial problem, which in fact has attracted much renewed interest in the mid-1980s.

In principle, a low-energy effective action $S_{\text{eff}}(\phi_0)$ for the massless fields ϕ_0 could be obtained by integrating out the massive fields from $S(\phi_0, \phi_H)$:

$$e^{iS_{\text{eff}}(\phi_0)} \sim \int D\phi_H \; e^{iS(\phi_0, \phi_H)}.$$

In principle, this does not constitute an approximation, just a first step in evaluating the exact Feynman path integral. In the classical approximation, it would correspond to eliminating the heavy fields ϕ_H from the equations of motion leaving a smaller set of equations that only involve the massless fields; this amounts to carrying out the path integral in tree approximation. The complete path integral also includes diagrams involving loops of massive fields. It describes the proper elimination of these degrees of freedom from the quantum theory. Then the effective action $S_{\text{eff}}(\phi_0)$ can be regarded as series in powers of $\hbar$ corresponding to the number of ϕ_H loops that are involved. The leading $(\hbar)^0$ approximation corresponds to the classical elimination of ϕ_H described above.

The exact effective action for massless fields is bound to be horrendously complicated and nonlocal. Even in field theory, integrating out a massive field introduces nonlocality, and if anything an even richer structure of nonlocality can be expected in string theory, where the object being integrated out of the underlying classical action is really an extended body, the string. The whole idea of constructing an effective action for massless fields is that although the exact formula would be a complicated and artificial rewriting of the exact theory, useful formulas can be obtained by a systematic expansion in the number of derivatives. After all, each derivative corresponds to a suppression by a power of E/M, where E is a characteristic energy scale of a reaction and M is the characteristic mass scale of the string theory. Thus a truncation to the first few terms in such an expansion should represent an excellent approximation at ordinary energies, though one that sacrifices the good ultraviolet convergence properties of the exact string theory. In practice, since we do not know at present a usable form for $S(\phi_0, \phi_H)$, we cannot really construct a low-energy effective action for the massless fields. What we can do is to study string S matrix elements and simply construct a classical action for the massless fields that reproduces them. Moreover, in the extreme low-energy limit, the leading terms in the effective action can be constructed just from invariance principles – from gauge invariance and local supersymmetry.

13.1   Minimal Supergravity Plus Super Yang–Mills

Let us consider the long wavelength expansion of the effective action $S_{\text{eff}}^{(0)}$ more systematically. We would like to expand the effective action in powers of (length)$^{-1}$. In fact, every derivative represents a power of inverse length. On the other hand, it is necessary to take account of the length dimensions of the various fields. A free boson or free fermion kinetic energy $(\partial_M \phi)^2$ or $\overline{\psi} \Gamma^M \partial_M \psi$ are equally important at long wavelengths. We can accommodate this by assigning to fermions the dimension of (length)$^{-1/2}$. If N_∂ and N_f are the number of derivatives and the number of fermions in a given term in the Lagrangian, and

$$n = N_\partial + \tfrac{1}{2}N_f, \tag{13.1.1}$$

then counting n is a precise way to count powers of length in a long wavelength expansion of the theory. Terms of larger and larger n are less important at longer and longer wavelengths. The minimal supergravity Lagrangians in ten dimensions have only terms of $n = 2$, as we will see.

To see in another way why the parameter n is natural in supergravity, let us study the general form of the supersymmetry transformation laws. Let ϕ^m represents an arbitrary power of a Bose field and let ψ represent an arbitrary Fermi field. The supersymmetry transformations naturally take the form

$$\begin{aligned}
\delta_0 \phi &\sim \phi^m \psi \eta \\
\delta_0 \psi &\sim \partial \phi^m \eta + \phi^m \psi^2 \eta.
\end{aligned} \tag{13.1.2}$$

These preserve the n counting if one makes the assignment $n = -1/2$ for the supersymmetry parameter η. It is possible to consistently restrict the supersymmetry transformation laws and equations of motion to respect this n counting; in this way, one can consistently find supergravity Lagrangians in which all terms have $n = 2$. We call such a leading Lagrangian S_2; the $n = 2$ terms are important because they give the dominant behavior at long wavelength. The possible forms of S_2 can be determined just from supersymmetry, without recourse to string theory.

It is possible to add higher-order terms S_4, S_6, etc., to the supergravity Lagrangians and likewise to add higher-order terms δ_2, δ_4, etc. to the supersymmetry transformation laws. Indeed, in string theory such terms are certainly present. (It seems that only terms with even n arise.) Once one does so, one is no longer able to restrict to definite values of n; attempting to close the supersymmetry algebra forces the inclusion of terms with larger and larger n.

Before accepting the statement that the long wavelength behavior of supergravity is governed by terms of the type S_2, we should ask whether *lower*-order terms are possible. A term of type S_0 would have to be of the form $\int d^n x \sqrt{g} V(\phi)$, with ϕ being some scalar field. A term of type S_1 would have to be of the form $\int d^n x \sqrt{g} U(\phi)\overline{\psi}\psi$. In four-dimensional supergravity, such terms are possible and indeed play an important role in phenomenology. In chiral ten-dimensional supergravity such terms are incompatible with supersymmetry.[*] The attempt to modify ten-dimensional

---
[*] The situation is different for the non-chiral type IIA theory (see Romans, L. [432]).

supergravity by adding terms of type S_0 and S_1 to the Lagrangians below (together with supersymmetry variations of type δ_{-1}) is not successful. (The analysis is easy for $N = 1$ supersymmetry in ten dimensions. Terms of type S_1 and δ_{-1} are impossible because of chirality restrictions on the fields, and terms of type S_0 are then easily seen to violate supersymmetry. The four-dimensional constructions that do work involve an interplay between S_0, S_1, and δ_{-1}.) The impossibility of adding to ten-dimensional supergravity a term of type S_0 entered our analysis in an important way in chapter 10. A possible dilaton tadpole would be proportional to S_0 (more exactly, to $\partial V/\partial \phi$), and must be absent when anomalies cancel and supersymmetry is valid since S_0 violates supersymmetry. This was a crucial step in arguing that anomaly cancellation for type I superstrings would necessarily lead to cancellation of infinities.

13.1.1 $N = 1$ Supergravity in Ten and Eleven Dimensions

In this section we describe $N = 1$ supergravity in ten dimensions. The $N = 1$ superstring theories include type I superstring theory with gauge group $SO(32)$ and the heterotic string theories with gauge group $E_8 \times E_8$ or $SO(32)$. In each case the massless sector consists of the Yang–Mills supermultiplet (A_M^a, χ^a), each in the adjoint representation of the gauge group, and the $N = 1$ supergravity multiplet. The super Yang–Mills theory has already been described in appendix 4.A. We therefore turn straightaway to the supergravity theory after which the coupled super Yang–Mills plus supergravity system will be considered.

The supergravity multiplet consists of a graviton, described by a vielbein e_M^A, an antisymmetric tensor B_{MN}, which in the language introduced in chapter 12 is a two form, a scalar ϕ, a gravitino ψ_M, and a spinor λ. The spinors are all Majorana–Weyl with ψ_M left-handed and λ right-handed. Capital letters $M, N, \ldots$ denote ten-dimensional space-time coordinates, whereas indices $A, B, \ldots$ are local Lorentz indices or tangent space indices.[†] As usual, e_A^M denotes the inverse of e_M^A and e denotes the determinant of e_M^A. The Dirac matrices for ten-dimensional space-time are represented by

$$\Gamma^M = e_A^M \Gamma^A. \tag{13.1.3}$$

[†] The conventions here and in the following chapters differ somewhat from those of the first 12 chapters. Specifically, we previously wrote μ, ν, etc. for space-time coordinates, whereas we now write A, B, etc. The reason for this change is that in the following chapters we consider splitting the coordinates into four-dimensional ones x^μ and six-dimensional ones y^i.

The Γ^A are constant matrices obeying

$$\{\Gamma^A, \Gamma^B\} = 2\eta^{AB}, \qquad (13.1.4)$$

but the Γ^M are field dependent according to (13.1.3). The matrices Γ^A are 32-dimensional, since this is the minimum dimension required to realize the Dirac algebra for ten different matrices. However, when they multiply a chiral spinor, they are effectively multiplied by a projection operator $\frac{1}{2}(1 \pm \Gamma_{11})$, which projects them into a 16-dimensional subspace.

One way to derive the $N = 1$ supergravity action in ten dimensions is to first formulate supergravity in eleven dimensions and then make a truncation to $D = 10$. One advantage of this is that a different truncation gives the type IIA supergravity theory in ten dimensions. Also, the eleven-dimensional theory is interesting to formulate in its own right, even though it does not seem likely as a candidate theory of nature.

The eleven-dimensional supergravity theory only involves three different fields: the vielbein e_M^A, a Majorana gravitino ψ_M, and a three-form potential A_{MNP}. (The indices are 11-valued here.) Since there are fewer fields than in ten dimensions, the formulas are somewhat more concise. We settle here for simply quoting the formulas. The result (derived by Cremmer, Julia, and Scherk; the form of the multiplet was first indicated in an analysis by Nahm) is

$$L = -\frac{1}{2\kappa^2} e R - \frac{1}{2} e \overline{\psi}_M \Gamma^{MNP} D_N(\tfrac{1}{2}(\omega + \hat{\omega})) \psi_P - \frac{1}{48} e F^2_{MNPQ}$$

$$- \frac{\sqrt{2}\kappa}{384} e(\overline{\psi}_M \Gamma^{MNPQRS} \psi_S + 12 \overline{\psi}^N \Gamma^{PQ} \psi^R)(F + \hat{F})_{NPQR}$$

$$- \frac{\sqrt{2}\kappa}{3456} \epsilon^{M_1 \ldots M_{11}} F_{M_1 \ldots M_4} F_{M_5 \ldots M_8} A_{M_9 M_{10} M_{11}}. \qquad (13.1.5)$$

As usual, Γ's with multiple indices represent antisymmetrized products with unit weight. For example,

$$\Gamma^{MN} = \tfrac{1}{2}(\Gamma^M \Gamma^N - \Gamma^N \Gamma^M). \qquad (13.1.6)$$

The spin connection ω_{MAB} is understood to be given by the solution of the field equation that results from varying it as an independent field. (This includes a torsion part containing terms of the form $\overline{\psi}\Gamma\psi$.) The symbol $\hat{\omega}_{MAB}$ denotes the supercovariant connection, whose supersymmetry variation does not involve derivatives of the infinitesimal Grassmann parameter. It is given by

$$\hat{\omega}_{MAB} = \omega_{MAB} + \tfrac{1}{8}\overline{\psi}^P \Gamma_{PMABQ}\psi^Q. \tag{13.1.7}$$

F_{MNPQ} is the curl or invariant field strength of the field A_{MNP}, $F = 6dA$, as discussed in chapter 12. $\hat{F}_{MNPQ}$ is the supercovariantization of F_{MNPQ}. (This means that its supersymmetry variation does not contain derivatives of the parameter η.)

The action formed as the integral of the Lagrangian density (13.1.5) is invariant under local supersymmetry transformations. The transformations can be expressed in terms of an infinitesimal space-time dependent Grassmann parameter $\eta(x)$, which transforms as a Majorana spinor. They are given by

$$\delta e_M^A = \frac{\kappa}{2}\overline{\eta}\,\Gamma^A\psi_M,$$

$$\delta A_{MNP} = -\frac{\sqrt{2}}{8}\overline{\eta}\,\Gamma_{[MN}\psi_{P]},\tag{13.1.8}$$

$$\delta\psi_M = \frac{1}{\kappa}D_M(\hat{\omega})\eta + \frac{\sqrt{2}}{288}(\Gamma_M{}^{PQRS} - 8\delta_M^P\Gamma^{QRS})\eta\hat{F}_{PQRS}.$$

Given the result for $D = 11$, one can obtain $D = 10$ supergravity theories by dimensional reduction. All that is required is to drop the dependence of the fields on one of the spatial coordinates, and to decompose the eleven-dimensional fields into ten-dimensional pieces. In the reduction the eleven-dimensional vielbein decomposes to give the ten-dimensional vielbein, a vector and a scalar. Specifically,

$$e_M^A \rightarrow \begin{pmatrix} e_M^A & A_M \\ 0 & \phi \end{pmatrix}, \tag{13.1.9}$$

where we use the same labeling for the ten and eleven-dimensional vielbeins to keep things simple. The 10×1 block of zeros can be achieved by a gauge choice that uses up the gauge freedom associated with Lorentz transformations between the eleventh dimension and the first ten. The three form A_{MNP} decomposes in ten dimensions into a three form A_{MNP} and a two form B_{MN} (corresponding to A_{MN11}). The gravitino decomposes into a pair of Majorana–Weyl gravitinos ψ_M^i ($i = 1, 2$) and a pair of Majorana–Weyl spinors λ^i in $D = 10$. The two members of each pair have opposite chirality, reflecting the left–right symmetry of the original $D = 11$ theory.

The collection of fields obtained by this reduction describes type IIA supergravity in $D = 10$. The Lagrangian for this theory is obtained by carrying out the indicated decomposition of the $D = 11$ one. It is convenient to always express the resulting Lagrangian in a form where the Einstein term appears in the conventional form (proportional to the curvature scalar). However, the reduction described would give an extra factor of ϕ in this term due to the reduction of $e = \det e_M^A$. This can be removed by rescaling the ten-dimensional e_M^A by ϕ^γ, where $\gamma = -(D - 2)^{-1} = -1/8$. This causes peculiar powers of ϕ to occur in the other terms of the Lagrangian.

The type IIA supergravity theory described above can be truncated to give $N = 1$, $D = 10$ supergravity by setting $A_M = A_{MNP} = 0$ and also setting $\psi_M^R = \lambda^L = 0$. (R and L refer to right- and left-handed chiralities.) The Lagrangian that remains is then given by

$$
\begin{aligned}
e^{-1} L_{SG} = & -\frac{1}{2\kappa^2} R - \frac{1}{2}\overline{\psi}_M \Gamma^{MNP} D_N \psi_P - \frac{3}{4}\phi^{-3/2} H_{MNP}^2 \\
& -\frac{1}{2}\overline{\lambda}\Gamma^M D_M \lambda - \frac{9}{16\kappa^2}(\partial_M \phi/\phi)^2 - \frac{3\sqrt{2}}{8}\overline{\psi}_M \Gamma^N \Gamma^M \lambda(\partial_N \phi/\phi) \\
& +\frac{\sqrt{2}\kappa}{16}\phi^{-3/4} H_{NPQ}(\overline{\psi}_M \Gamma^{MNPQR}\psi_R + 6\overline{\psi}^N \Gamma^P \psi^Q \\
& -\sqrt{2}\,\overline{\psi}_M \Gamma^{NPQ}\Gamma^M \lambda) + (\text{Fermi})^4 \ .
\end{aligned}
$$

$$(13.1.10)$$

The terms with four Fermi fields are all known and can be found in the references. It is understood here that ψ_M is a left-handed Majorana–Weyl spinor and λ is a right-handed Majorana–Weyl spinor sometimes called a 'dilatino'. The three form H is the curl of B, *i.e.*, in differential form notation $H = dB$. The other fields that remain are the dilaton ϕ and the graviton, of course.

The supersymmetry transformations of this action can be deduced from those of the $D = 11$ theory. This requires some care, however. First of all, only the left-handed piece of η describes a surviving symmetry, since one of the $D = 10$ supersymmetries has been sacrificed in the truncation from the IIA to the $N = 1$ theory. The chirality of the supersymmetry parameters matches that of the gauge field ψ_M, of course. A more subtle point is that the transformation law one would naively deduce does not preserve the gauge choice of (13.1.9). In order to be consistent with this choice it is necessary to include in the definition of a $D = 10$ supersymmetry transformation a contribution from a $D = 11$ local Lorentz transformation that restores the gauge condition $e_{11}^A = 0$. The result of doing this is that

$$\delta e_M^A = \frac{\kappa}{2}\overline{\eta}\,\Gamma^A\psi_M$$

$$\delta\phi = -\frac{\sqrt{2}\kappa}{3}\phi\overline{\eta}\lambda$$

$$\delta B_{MN} = \frac{\sqrt{2}}{4}\phi^{3/4}(\overline{\eta}\,\Gamma_M\psi_N - \overline{\eta}\,\Gamma_N\psi_M - \frac{\sqrt{2}}{2}\overline{\eta}\,\Gamma_{MN}\lambda)$$

$$\delta\lambda = -\frac{3\sqrt{2}}{8}\frac{1}{\phi}(\Gamma\cdot\partial\phi)\eta + \frac{1}{8}\phi^{-3/4}\Gamma^{MNP}\eta H_{MNP} + (\text{Fermi})^2$$

$$\delta\psi_M = \frac{1}{\kappa}D_M\eta + \frac{\sqrt{2}}{32}\phi^{-3/4}(\Gamma_M{}^{NPQ} - 9\delta_M^N\Gamma^{PQ})\eta H_{NPQ} + (\text{Fermi})^2 \;.$$

$$(13.1.11)$$

The $(\text{Fermi})^4$ terms in (13.1.10) and the $(\text{Fermi})^2$ terms in (13.1.11) are known.

The ten-dimensional supergravity theories are, of course, the low-energy limits of certain string theories. Further dimensional reduction can give a variety of supergravity theories in various dimensions less than ten. If string theory proves to be correct, it could be regarded as 'explaining' the existence of supergravity theories for $D \leq 10$. Eleven-dimensional supergravity remains an enigma. It is hard to believe that its existence is just an accident, but it is difficult at the present time to state a compelling conjecture for what its role may be in the scheme of things.

13.1.2 Type IIB Supergravity

Dimensional reduction of $D = 11$ supergravity gives a theory in ten dimensions with two supersymmetries – type IIA supergravity. The two supersymmetries have opposite handedness, and the theory has an overall left–right symmetry, i.e., it is 'nonchiral'. There exists a second $N = 2$, $D = 10$ supergravity theory in which both supersymmetries have the same handedness. This theory, type IIB supergravity, is obviously chiral (left–right asymmetric). It is not obtainable by reduction or truncation of a theory in a larger number of dimensions. It describes the leading low-energy behavior of the classical effective action of type IIB superstring theory. This fact makes the explicit formulation of the supergravity theory of particular interest.

The massless spectrum of type IIB superstring theory was already given in §5.3.2, where we found that in terms of transverse $SO(8)$ representations the physical content is given by

$$(\mathbf{8_v} + \mathbf{8_c}) \otimes (\mathbf{8_v} + \mathbf{8_c}) = (1 + \mathbf{28} + \mathbf{35_v} + \mathbf{28} + \mathbf{35_c})_B$$
$$+ (\mathbf{8_s} + \mathbf{8_s} + \mathbf{56_s} + \mathbf{56_s})_F. \tag{13.1.12}$$

We recall that the $\mathbf{35_v}$ is the graviton and the $\mathbf{35_c}$ is a fourth-rank anti-symmetric self-dual tensor. The $\mathbf{56_s}$'s represent the two gravitinos, and the $\mathbf{8_s}$'s are a pair of Majorana–Weyl spinors. The $\mathbf{28}$'s are a pair of second-rank antisymmetric tensors.

Our purpose in this subsection is to describe the interacting type IIB supergravity theory in a manifestly covariant form. In doing this we represent the Bose fields by e_M^A, A^α, A_{MN}^α and A_{MNPQ}, where $\alpha = 1, 2$. The self-duality of the $SO(8)$ representation $\mathbf{35_c}$ is reflected in the *free* covariant theory by the $SO(9,1)$ self-duality of the field strength $F_{MNPQR} = 5\partial_{[M}A_{NPQR]}$. In the language of forms, $F = *F$. There does not appear to be a simple covariant action principle that gives rise to this equation of motion. The usual sort of kinetic term $\int d^{10}x (F_{MNPQR})^2$ describes both a self-dual and an anti-self-dual field strength. There is no simple way of modifying this action so that the self-dual piece corresponds to physical propagating degrees of freedom and the anti-self-dual piece does not. The theory does admit manifestly covariant equations of motion, however. Thus, for this particular theory it is much easier to attempt to derive covariant field equations than a covariant action principle. So this is the route that we follow.

The problem of formulating an action principle with manifest Lorentz covariance has an analog for supersymmetry. For many supersymmetric theories (including the one at hand) no 'off-shell' superspace formulation is known. This means that an action with manifest supersymmetry cannot be written down. However, it is always possible to introduce 'on-shell' superfields and write manifestly supersymmetric equations of motion. In fact, the type IIB supergravity theory is itself an example of such a theory. However, in order to keep things as simple as possible here, we will present the theory in terms of component fields rather than superfields. The superfield approach, while quite elegant, requires developing a rather elaborate formalism that is not required elsewhere in this book, and therefore is not presented here.

The derivation of the covariant field equations of type IIB supergravity hinges on two main ideas. The first one concerns the relationship between the field equations and the supersymmetry transformations of the fields. First of all, one has the obvious requirement that the supersymmetry variation of an equation of motion should give rise to an expression that also vanishes by the equations of motion. The conditions are stringent enough that if the supersymmetry variations are known and one of the field equations is known, all the others can be deduced. There is an even stronger requirement, however. When the commutator of two local supersymmetry transformations is applied to the fields, closure of the algebra requires that the result should correspond to a combination of the local symmetries of the theory – namely, a general coordinate transformation, a local Lorentz transformation, a local supersymmetry transformation, and additional local gauge transformations associated with the 'gauge fields' A^α_{MN} and A_{MNPQ}. However, since we are dealing with an on-shell formalism not possessing auxiliary fields required for off-shell closure of the algebra, this result only applies if the equations of motion are satisfied. This fact can be turned to our advantage. It means that in the process of constructing supersymmetry transformations of the fields with a consistent gauge algebra one can actually deduce some of the field equations at the same. The consistency aspects built into these conditions turn out (on calculation) to completely determine the theory!

While the consistency conditions described above do, as just asserted, completely determine the theory, the identification of another fundamental symmetry principle serves to organize concepts and formulas so as to make the calculations much simpler to carry out and the results much more elegant to describe. The symmetry in question is a global $SU(1,1)$. (This is a noncompact form of $SU(2)$, isomorphic to $SL(2,R)$, encountered earlier in the study of conformal mappings of the unit disk onto itself. Specifically, the mapping $z \to (az + b)/(cz + d)$ is a nonsingular and invertible mapping of the unit disk onto itself if $\begin{pmatrix} a & b \\ c & d \end{pmatrix}$ belongs to the group $SU(1,1)$.)

The $SU(1,1)$ global symmetry of type IIB supergravity is one example of a generic phenomenon in extended supergravity theories. In these theories there is a noncompact global symmetry group G with a maximal compact subgroup H. The scalar fields of the theory are associated with the coset G/H. This implies, in particular, that their number is $\dim G - \dim H$. In the case at hand, the maximal compact subgroup of $SU(1,1)$ is $U(1)$, and the theory possesses $3 - 1 = 2$ scalar fields. The most famous example of this phenomenon is $N = 8$ supergravity in four dimensions, which possesses the global symmetry $E_{7,7}$. The group $E_{7,7}$ is a noncompact form of E_7 whose maximal compact subgroup is $SU(8)$. Since E_7 has 133 generators and $SU(8)$ has 63 generators, it follows that $N = 8$ supergravity has $133 - 63 = 70$ scalar fields.

Since $\dim G - \dim H$ does not correspond to the dimension of a representation of G, one must be a little clever to exhibit the global G symmetry of such theories in a manifest form. However, it proves to be well worth the trouble of doing so. The appropriate method is closely (but not precisely) analogous to the vielbein formulation of relativity in which we describe the graviton by the field e_M^A. This field clearly transforms linearly under the global symmetry group $GL(D, R)$, which is a subgroup of the group of general coordinate transformations. (This group acts only on the base-space index M.) This is a noncompact group whose maximal compact subgroup is $SO(D)$. In the vielbein formalism this group is implemented as a separate local symmetry – local Lorentz invariance. There are no independent propagating gauge fields associated with this symmetry, since no kinetic term for the connection $w_M{}^{AB}$ (analogous to F^2 in Yang–Mills theory) is introduced. As a result, one can use the local symmetry to gauge away $D(D-1)/2$ components of e_M^A leaving only ones corresponding to the coset $GL(D, R)/SO(D)$. This still overcounts the number of physical polarization modes of a graviton because the local general coordinate transformations have not yet been taken into account. A careful analysis shows that they can be used to restrict the preceding analysis to $D-2$ transverse directions. This can be achieved by choosing light-cone gauge, for example. Thus the physical polarizations of a graviton actually correspond to a coset space $SL(D-2, R)/SO(D-2)$. This is the feature present for gravitation but not for the coset description of the scalar fields. Not having an analog of general coordinate invariance, the G/H description of the scalar fields is actually easier than general relativity.

For the reasons described above we wish to describe the two scalars of type IIB supergravity by the matrix V_a^α analogous to the vielbein e_M^A. The index $\alpha = 1, 2$ labels the **2** of $SU(1,1)$ whereas the index $a = \pm$ describes two $U(1)$ representations with 'charges' $U = \pm 1$. V_-^α is the complex conjugate of V_+^α. The 2×2 matrix of fields V_a^α belongs to the group $SU(1,1)$. Thus, in particular,

$$\epsilon_{\alpha\beta} V_-^\alpha V_+^\beta = \det V = 1. \tag{13.1.13}$$

Under a global $SU(1,1)$ transformation

$$\delta V_\pm^\alpha = m^\alpha{}_\beta V_\pm^\beta, \tag{13.1.14}$$

where $m^\alpha{}_\beta$ is a constant matrix belonging to the algebra of $SU(1,1)$. Under a local $U(1)$ transformation with infinitesimal parameter $\Sigma(x)$

$$\delta V_\pm^\alpha = \pm i \Sigma V_\pm^\alpha. \tag{13.1.15}$$

(In general, a field Φ_U of charge U satisfied $\delta\Phi_U = iU\Sigma\Phi_U$.)

The $SU(1,1)$-invariant combination

$$Q_M = -i\epsilon_{\alpha\beta}V_-^\alpha \partial_M V_+^\beta \tag{13.1.16}$$

acts as a $U(1)$ gauge field since it follows from (13.1.15) that

$$\delta Q_M = \partial_M \Sigma, \tag{13.1.17}$$

where we have used (13.1.13). Equation (13.1.16) is the analog of the formula for the spin connection in terms of the vielbein. There is yet another $SU(1,1)$-invariant quantity that can be formed from $V_\pm^\alpha$, namely

$$P_M = -\epsilon_{\alpha\beta}V_+^\alpha \partial_M V_+^\beta. \tag{13.1.18}$$

This expression clearly has $U = 2$, and its complex conjugate has $U = -2$. If our purpose were only to describe the nonlinear sigma model of $SU(1,1)/U(1)$ scalar fields, this would be easy to do. The Lagrangian would be proportional to $g^{MN}P_M P_N^*$. However, as already explained, this has no nice extension to the full type IIB supergravity theory.

How should we describe the other fields of the theory? We know from general relativity that the Fermi fields should transform under the local symmetry group rather than the global one. Thus associated with the $\mathbf{8_s}$ and $\mathbf{56_s}$ representations, we introduce fields λ and ψ_M whose $U(1)$ charges are $U = 3/2$ and $U = 1/2$, respectively. The fact that there are two of each is incorporated simply by virtue of the fact that they are complex. (The complex conjugates have $U = -3/2$ and $U = -1/2$.) The fact that λ and ψ_M are Weyl spinors is expressed by $\Gamma_{11}\lambda = \lambda$ and $\Gamma_{11}\psi_M = -\psi_M$.

For the Bose fields it is more convenient to use base-space indices in the case of general relativity, and hence $SU(1,1)$ indices as regards the sigma-model structure. Thus, the two $\mathbf{28}$ fields are represented by an $SU(1,1)$ doublet A_{MN}^α. The four form field A_{MNPQ} is a singlet of $SU(1,1)$ and neutral under $U(1)$, of course. From the fields A_{MN}^α and A_{MNPQ}, we can form the field strengths

$$F_{MNP}^\alpha = 3\partial_{[M}A_{NP]}^\alpha \tag{13.1.19}$$

$$F_{MNPQR} = 5\partial_{[M}A_{NPQR]} + \tfrac{5}{8}i\kappa\epsilon_{\alpha\beta}A_{[MN}^\alpha F_{PQR]}^\beta. \tag{13.1.20}$$

In the latter expression we have included an interaction term that will be convenient later. Equation (13.1.19) obviously defines a field strength that is invariant under $\delta A_{MN}^\alpha = 2\partial_{[M}\Lambda_{N]}^\alpha$, as usual. However, (13.1.20) at first sight is not invariant under this transformation. In fact, it is invariant provided that one simultaneously transforms the four-form potential according to the rule

$$\delta A_{MNPQ} = -\tfrac{1}{4} i \kappa \epsilon_{\alpha\beta} \Lambda^{\alpha}_{[M} F^{\beta}_{NPQ]}. \tag{13.1.21}$$

This is in addition to the usual gauge transformation

$$\delta A_{MNPQ} = 4\partial_{[M} \Lambda_{NPQ]}, \tag{13.1.22}$$

of course.

The $SU(1,1)$ doublet of $U(1)$ neutral fields strengths F^{α}_{MNP} can be replaced by an equivalent expression that is an $SU(1,1)$ singlet with charge $U = 1$. The appropriate formula is

$$G_{MNP} = -\epsilon_{\alpha\beta} V^{\alpha}_{+} F^{\beta}_{MNP}. \tag{13.1.23}$$

One more definition that we require is the $U(1)$-covariant derivative. A field Φ_U with charge U has a covariant derivative, also of charge U, given by

$$D_M \Phi_U = (\partial_M - iU Q_M)\Phi_U, \tag{13.1.24}$$

where Q_M is the connection given in (13.1.16). This is the usual construction of the covariant derivative, identical to the one of electromagnetism.

We have now described the ingredients required to present the supersymmetry transformation formulas and the field equations in a form with manifest $SU(1,1) \times U(1)$ symmetry, the $SU(1,1)$ being global and the $U(1)$ local. The actual derivation or verification of the formulas is straightforward (but tedious) and will not be presented here. For the local supersymmetry transformations one finds

$$\delta e^A_M = -2\kappa \mathrm{Im}(\bar{\eta}\,\Gamma^A \psi_M)$$

$$\delta V^{\alpha}_{+} = \kappa V^{\alpha}_{-}\bar{\eta}^* \lambda$$

$$\delta V^{\alpha}_{-} = \kappa V^{\alpha}_{+}\bar{\eta}\,\lambda^* \tag{13.1.25}$$

$$\delta A^{\alpha}_{MN} = V^{\alpha}_{+}(\bar{\eta}^* \Gamma_{MN}\lambda^* + 4i\bar{\eta}\,\Gamma_{[M}\psi^*_{N]}) + c.c.$$

$$\delta A_{MNPQ} = 2\mathrm{Re}(\bar{\eta}\,\Gamma_{[MNP}\psi_{Q]}) - \tfrac{3}{8} i\kappa \epsilon_{\alpha\beta} A^{\alpha}_{[MN}\delta A^{\beta}_{PQ]}$$

and

$$\delta\lambda = \frac{i}{\kappa}\Gamma^M \eta^* P_M - \frac{i}{24}\Gamma^{MNP}\eta\, G_{MNP} + (\text{Fermi})^2$$

$$\delta\psi_M = \frac{1}{\kappa}D_M\eta + \frac{i}{480}\Gamma^{M_1\ldots M_5}\Gamma_M\eta\, F_{M_1\ldots M_5} \qquad (13.1.26)$$

$$+ \frac{1}{96}(\Gamma_M{}^{NPQ}G_{NPQ} - 9\Gamma^{NP}G_{MNP})\eta^* + (\text{Fermi})^2.$$

Note that the supersymmetry parameter η has $U(1)$ charge $U = 1/2$, just like the gravitino field ψ_M.

The equations given above have forms that are easily checked to be correct, giving the usual supersymmetry algebra, in the linearized and global limit, *i.e.*, for the free theory. The embellishments occurring in the formulas above are the only possible structures compatible with the various symmetries of the theory – general coordinate invariance, local Lorentz invariance, Λ_M and Λ_{MNP} gauge invariances, local $U(1)$ symmetry and global $SU(1,1)$ symmetry. The nontrivial exercise is to verify that the numerical coefficients of the various terms are precisely the ones given. In fact, closure of the algebra completely determines these coefficients, as well as the equations of motion. Moreover, most coefficients are determined several times by the formulas – fortunately, always with the same result!

Closure of the supersymmetry algebra gives equations of motion for the chiral fields of the theory, each of which satisfies a first-order field equation. These fields are the Fermi fields λ and ψ_M, and the four-form potential A_{MNPQ}. The equation of motion for the four form A_{MNPQ} is basically the statement $F = *F$, where F differs from dA by interaction terms. The formula F is given in (13.1.20). (There is also a (Fermi)^2 term, which we omit.)

The other equation of motion can be obtained by applying supersymmetry transformations to the ones already obtained. One finds (neglecting Fermi field contributions) that

$$D^M P_M = \tfrac{1}{24}\kappa^2 G_{MNP}G^{MNP}$$

$$D^P G_{MNP} = P^P G^*_{MNP} - \tfrac{2}{3}i\kappa F_{MNPQR}G^{PQR}$$

$$-R_{MN} = P_M P_N^* + P_M^* P_N + \tfrac{1}{6}\kappa^2 F_{MP_1\ldots P_4}F^{P_1\ldots P_4}{}_N \qquad (13.1.27)$$

$$+ \tfrac{1}{8}\kappa^2(G_M{}^{PQ}G^*_{NPQ} + G_M^*{}^{PQ}G_{NPQ}$$

$$- \tfrac{1}{6}g_{MN}G^{PQR}G^*{}_{PQR}).$$

A potentially important use of these equations is the study of spontaneous compactification. In doing this it is generally assumed that the Fermi fields can be set equal to zero. (This is all one can do classically, although one could imagine a 'condensate' forming quantum mechanically.) For this purpose the equations in the form given (without Fermi field contributions) are sufficient.

Let us conclude this brief introduction to type IIB supergravity with one final remark. The $SU(1,1)/U(1)$ symmetry of the theory, which proved to be so useful in the construction described above, is not preserved by the type IIB superstring extension of the theory. In particular, the $U(1)$ group rotates the two supersymmetries into one another, but not even this symmetry is preserved by the superstring extension. The simplest way to understand this remark is to refer back to the supersymmetric superstring action $S = S_1 + S_2$ given in §5.1.2. There we introduced two superspace Grassmann coordinates $\theta^A(\sigma, \tau)$, $A = 1, 2$. The term S_1 manifestly has $SO(2)$ rotational symmetry, which is the $U(1)$ in question, but the term S_2 (which was crucial in the superstring analysis) does not share the symmetry.

13.1.3 The Coupled Supergravity Super Yang–Mills System

We have completed our survey of pure supergravity theories in ten dimensions. Our next problem is to couple the $N = 1$, $D = 10$ supergravity theory described in §13.1.1 to $N = 1$, $D = 10$ super Yang–Mills theory. Recall that in appendix 4.A we showed that the super Yang–Mills theory in isolation is given simply by

$$L_{YM} = -\frac{1}{4} F^a_{MN} F^{MNa} - \frac{1}{2} \overline{\chi}^a \Gamma^M (D_M \chi)^a. \tag{13.1.28}$$

This action is invariant under the global supersymmetry transformations

$$\delta A^a_M = \frac{1}{2} \overline{\eta} \, \Gamma_M \chi^a$$
$$\delta \chi^a = -\frac{1}{4} \Gamma^{MN} F^a_{MN} \eta. \tag{13.1.29}$$

The challenge is to couple this to the supergravity system so that the combined system is invariant under local supersymmetry transformations. As usual, the Yang–Mills field strength is defined by

$$F^a_{MN} = \partial_M A^a_N - \partial_N A^a_M + g f^a{}_{bc} A^b_M A^c_N \tag{13.1.30}$$

and the gauge-covariant derivative by

$$(D_M \chi)^a = \partial_M \chi^a + g f^a{}_{bc} A^b_M \chi^c, \tag{13.1.31}$$

where $f^a{}_{bc}$ are the structure constants of a semisimple Lie group. The Yang–Mills coupling constant g has dimensions (length)^3 for $D = 10$. At the quantum level, anomaly cancellation imposes restrictions on the allowed symmetry groups, but at the classical level, any semisimple group is acceptable.

As in §12.5.1, we introduce matrices T^a that represent the group generators, normalized so that $\text{tr}(T^a T^b) = \delta^{ab}$, and define a matrix of potentials

$$A = \sum_a T^a A^a_M \, dx^M, \tag{13.1.32}$$

and similarly for the 'gauginos' χ. The Yang–Mills field strength is described in this notation by a matrix-valued two form $F = \sum F^a T^a$, where

$$F^a = dA^a + g f^a{}_{bc} A^b \wedge A^c, \tag{13.1.33}$$

which we sometimes abbreviate as

$$F = dA + gA^2. \tag{13.1.34}$$

The coupling of the super Yang–Mills multiplet to the supergravity multiplet (with $n = 2$ terms only) is uniquely determined by the requirement of local supersymmetry. The result was worked out in the abelian $[U(1)]$ case by Chamseddine and by de Wit *et al.* and generalized to the non-abelian case by Chapline and Manton. The analysis involves much tedious algebra, but there is one real novelty that arises. The gauge-invariant field strength of the two form B_{MN} must be generalized from $H = dB$ to

$$H = dB - \frac{\kappa}{\sqrt{2}} \omega_3, \tag{13.1.35}$$

where ω_3 is the Chern–Simons three form

$$\omega_3 = A^a F^a - \tfrac{1}{3} g f_{abc} A^a A^b A^c = A^a dA^a + \tfrac{2}{3} g f_{abc} A^a A^b A^c, \tag{13.1.36}$$

or equivalently,

$$\omega_3 = \text{tr}(AF - \tfrac{1}{3} g A^3) = \text{tr}(A dA + \tfrac{2}{3} g A^3), \tag{13.1.37}$$

as described in the preceding chapter. While the modified H is obviously invariant under $\delta B = d\Lambda$, it is less obvious that it possesses Yang–Mills gauge invariance. In fact, to achieve gauge invariance, it is necessary to postulate a nontrivial gauge transformation law for the field B. This is unexpected, because B is neutral. Under a gauge transformation

$$\delta A = d\Lambda + [A, \Lambda],\qquad(13.1.38)$$

where Λ is a matrix of infinitesimal parameters, the Chern–Simons term has a variation

$$\delta\omega_3 = \text{tr}(d\Lambda dA) = d\omega_2^1,\qquad(13.1.39)$$

where

$$\omega_2^1 = \text{tr}(\Lambda dA).\qquad(13.1.40)$$

The subscript denotes the degree of the form and the superscript refers to the number of occurrences of the parameter Λ. It is now obvious how to achieve gauge invariance. If we assign the transformation law

$$\delta B = \frac{\kappa}{\sqrt{2}}\text{tr}(\Lambda dA) = \frac{\kappa}{\sqrt{2}}\omega_2^1\qquad(13.1.41)$$

to the two-form potential, then the three form $H = dB - \frac{\kappa}{\sqrt{2}}\omega_3$ is obviously gauge invariant. (The factor $\kappa/\sqrt{2}$ in (13.1.35) will be removed by a scaling of variables described below.)

The Lagrangian coupling supergravity and super Yang–Mills theory turns out to be

$$e^{-1}L = e^{-1}L_{SG}(\text{with modified } H_{MNP})$$

$$- \tfrac{1}{4}\phi^{-3/4}F^a_{MN}F^{MNa} - \tfrac{1}{2}\overline{\chi}^a\Gamma^M(D_M(\hat{\omega})\chi)^a$$

$$- \tfrac{1}{8}\kappa\phi^{-3/8}\overline{\chi}^a\Gamma^M\Gamma^{NP}(F^a_{NP} + \hat{F}^a_{NP})(\psi_M + \tfrac{1}{12}\sqrt{2}\Gamma_M\lambda)$$

$$+ \tfrac{1}{16}\sqrt{2}\kappa\phi^{-3/4}\overline{\chi}^a\Gamma^{MNP}\chi^a H_{MNP}$$

$$- \tfrac{1}{1536}\sqrt{2}\kappa^2\overline{\chi}^a\Gamma_{MNP}\chi^a\overline{\psi}_Q(4\Gamma^{MNP}\Gamma^Q + 3\Gamma^Q\Gamma^{MNP})\lambda$$

$$- \tfrac{1}{512}\kappa^2\overline{\chi}^a\Gamma_{MNP}\chi^a\overline{\lambda}\Gamma^{MNP}\lambda - \tfrac{1}{384}\kappa^2\overline{\chi}^a\Gamma_{MNP}\chi^a\overline{\chi}^b\Gamma^{MNP}\chi^b.$$
$$(13.1.42)$$

The last term in this expression vanishes when $a = b$ due to an identity valid for $D = 10$ Majorana–Weyl spinors. Therefore it is absent in the abelian case.

The infinitesimal supersymmetry transformations of the super Yang–Mills fields, which give the curved-space generalization of (13.1.29), are given by the formulas

$$\delta A_M^a = \frac{1}{2}\phi^{3/8}\overline{\eta}\,\Gamma_M\chi^a$$

$$\delta\chi^a = -\frac{1}{4}\phi^{-3/8}\Gamma^{MN}\hat{F}_{MN}^a\eta + \frac{\sqrt{2}}{64}\kappa[3(\overline{\lambda}\chi^a)\eta$$

$$-\frac{3}{2}(\overline{\lambda}\Gamma^{MN}\chi^a)\Gamma_{MN}\eta - \frac{1}{24}(\overline{\lambda}\Gamma^{MNPQ}\chi^a)\Gamma_{MNPQ}\eta].$$

$$(13.1.43)$$

The transformations of the supergravity fields are as given in (13.1.11) with the modified three form H substituted in the equations for $\delta\lambda$ and $\delta\psi_M$. Also, the variations acquire extra pieces

$$\delta'\lambda = \frac{\sqrt{2}}{432}\kappa\overline{\chi}^a\Gamma^{MNP}\chi^a\Gamma_{MNP}\eta$$

$$\delta'B_{MN} = \frac{\sqrt{2}}{2}\kappa\phi^{3/8}\overline{\eta}\Gamma_{[M}\chi^a A_{N]}^a \qquad\qquad (13.1.44)$$

$$\delta'\psi_M = -\frac{1}{256}\kappa\overline{\chi}^a\Gamma^{NPQ}\chi^a(\Gamma_{MNPQ} - 5g_{MN}\Gamma_{PQ})\eta.$$

By making these modifications, the local supersymmetry of the pure supergravity system is extended to the coupled supergravity plus super Yang–Mills system.

Before commenting further on the formulas, let us make some field redefinitions that make them a bit more elegant. Specifically, we set

$$\phi^{new} = (\phi^{old})^{3/4} \qquad\qquad (13.1.45)$$

$$A_M^{new} = gA_M^{old}, \qquad\qquad F_{MN}^{new} = gF_{MN}^{old} \qquad\qquad (13.1.46)$$

$$B_{MN}^{new} = \frac{\sqrt{2}\,g^2}{\kappa}B_{MN}^{old}, \qquad H_{MNP}^{new} = \frac{\sqrt{2}\,g^2}{\kappa}H_{MNP}^{old}. \qquad (13.1.47)$$

Equation (13.1.35) then becomes

$$H = dB - \omega_3. \qquad\qquad (13.1.48)$$

Making the notational changes listed in the preceding paragraph the Lagrangian for the coupled supergravity super Yang–Mills system takes the form

$$
e^{-1}L = -\frac{1}{2\kappa^2}R - \frac{1}{4g^2\phi}\mathrm{tr}(F^2_{MN}) - \frac{1}{\kappa^2}(\partial_M\phi/\phi)^2 - \frac{3\kappa^2}{8g^4\phi^2}H^2_{MNP}
$$
$$
-\frac{1}{2}\overline{\psi}_M\Gamma^{MNP}D_N\psi_P - \frac{1}{2}\overline{\lambda}\Gamma^M D_M\lambda - \frac{1}{2}\mathrm{tr}(\overline{\chi}\Gamma^M D_M\chi)
$$
$$
-\frac{1}{\sqrt{2}}\overline{\psi}_M\Gamma^N\Gamma^M\lambda(\partial_N\phi/\phi) + \frac{\kappa^2}{16g^2\phi}\mathrm{tr}(\overline{\chi}\Gamma^{MNP}\chi)H_{MNP}
$$
$$
-\frac{\kappa}{4g\sqrt{\phi}}\mathrm{tr}[\overline{\chi}\Gamma^M\Gamma^{NP}(\psi_M + \frac{\sqrt{2}}{12}\Gamma_M\lambda)F_{NP}]
$$
$$
+\frac{\kappa^2}{16g^2\phi}(\overline{\psi}_M\Gamma^{MNPQR}\psi_R + 6\overline{\psi}^N\Gamma^P\psi^Q
$$
$$
-\sqrt{2}\,\overline{\psi}_M\Gamma^{NPQ}\Gamma^M\lambda)H_{NPQ} + (\mathrm{Fermi})^4.
$$

$$(13.1.49)$$

In formulating this theory, we have introduced a gravitational coupling constant κ, which in ten dimensions has dimensions of $(\text{length})^{-4}$, and a Yang–Mills coupling constant g, which in ten dimensions has dimensions of $(\text{length})^{-3}$. At first sight, then, it appears that the theory is characterized by an arbitrary dimensionless parameter $\lambda = g^4/\kappa^3$. More careful reflection shows this is not so. Equation (13.1.49) is invariant under the rescaling $\phi \to c\phi$, if g transforms as $g \to g/\sqrt{c}$. If we wish, we can remove g from the above Lagrangian altogether by writing $\phi' = \phi(g^2/\kappa^{3/2})$. (The factor of $\kappa^{3/2}$ is inessential; it is included to make ϕ' dimensionless.) Thus, the supergravity theory has no free dimensionless parameter. The essence of the matter is that nothing in the classical Lagrangian (13.1.49) determines the expectation value of the field ϕ. Instead of a one-parameter family of theories labeled by the value of λ, what we really have is a single theory with a one-parameter family of vacuum states. At least, this degeneracy holds classically. It may or may not be lifted at the quantum level.[*] Knowing how the degeneracy is lifted or how the expectation value of ϕ is determined is of utmost importance in understanding the physical content of the theory, since (for example) the physical gauge coupling, which is really $g^2\phi$, depends on which of the one-parameter family of classical vacua is physically relevant.

If one considers superstring theory rather than supergravity theory, an additional quantity α' plays a crucial role, but there is still no arbitrary dimensionless parameter in the fundamental laws, since α' is determined

---

[*] It cannot be lifted unless supersymmetry is spontaneously broken, since a potential energy $V(\phi)$ – necessary to lift the degeneracy – violates supersymmetry, as we learned earlier.

in terms of g and κ by $\kappa \sim g^2/\alpha'$ (type I) or $\kappa^2 \sim g^2\alpha'$ (heterotic). Thus, there really is no fundamental adjustable dimensionless constant in the equations of string theory. In §3.4.6, we made essentially this observation in a different way, showing that the arbitrary adjustable parameter seemingly present in the string perturbation expansion is an illusion, and actually can be absorbed in shifting the value of the dilaton field ϕ.

In order to keep the formulas from becoming excessively messy, in (13.1.49) we have not written out all the (Fermi)^4 terms explicitly. One of them, the last term in (13.1.42) is of particular interest, however. It combines with the $\overline{\chi}\chi H$ and H^2 terms to give a perfect square

$$-\frac{3\kappa^2}{8g^4\phi^2}[H_{MNP} - \frac{g^2\phi}{12}\mathrm{tr}(\overline{\chi}\Gamma_{MNP}\chi)]^2. \tag{13.1.50}$$

It has been suggested in connection with supersymmetry breaking that dynamical effects could cause the bilinear $\overline{\chi}\Gamma_{MNP}\chi$ to acquire a vacuum expectation value. When this happens H_{MNP} could also be expected to obtain a compensating expectation value for which the square vanishes. This induces an inhomogeneous term in the supersymmetry transformation formula of the 'dilatino' λ. This provides an interesting scheme for achieving supersymmetry breaking while retaining a vanishing cosmological constant, at least at a certain level of approximation. It might even provide a rationale for why six dimensions (three complex dimensions) must curl up. There is at present no convincing basis for these speculations, however.

13.2   Scale Invariance of the Classical Theory

Following the formulation of (13.1.49), we have already made some remarks about the significance of the expectation value of the dilaton field. Here we pursue this subject further. Let us first think briefly about some simple but perhaps unfamiliar properties of ordinary general relativity, described in D dimensions by the action

$$S = -\frac{1}{2\kappa^2}\int d^Dx\,\sqrt{g}g^{MN}R_{MN}. \tag{13.2.1}$$

This theory is not scale invariant as a quantum theory.[†] It is, however,

---

[†] Except in two dimensions, where (13.2.1) is a topological invariant.

scale invariant as a classical theory. Under

$$g_{MN} \to t^{-2}g_{MN} \tag{13.2.2}$$

(this is a scale transformation, since all lengths are scaled by a factor of t^{-1}), standard formulas of general relativity show that R_{MN} is invariant, so S transforms as

$$S \to t^{-(D-2)}S. \tag{13.2.3}$$

At the classical level, the normalization of S is irrelevant – it scales out of the classical equations. So general relativity is scale invariant as a classical theory. At the quantum level a transformation such as (13.2.2), which rescales the action, is not a symmetry. (We can consider this question in a formal way even though standard general relativity probably does not make sense as a quantum theory.) This follows, for instance, from the description of the quantum theory as a path integral:

$$Z = \int e^{iS/\hbar}. \tag{13.2.4}$$

Evidently, a rescaling of S does not leave (13.2.4) invariant. Though not a symmetry, (13.2.2) has a consequence in quantum gravity that may seem provocative. To describe the *classical* theory requires the fundamental constants c (the speed of light) and κ (the gravitational constant). One might expect that a new fundamental constant $\hbar$ is needed in the quantum theory, but this is not really so; from (13.2.3) and (13.2.4) it is evident that a change in the value of $\hbar$ can be absorbed in a transformation of the type (13.2.2), so there really is no fundamental constant $\hbar$ in quantum general relativity.

To couple to fermions, we must introduce a vielbein e_{MA}, which obeys $g_{MN} = \eta^{AB}e_{MA}e_{NB}$ and so must transform under (13.2.2) as

$$e_{MA} \to t^{-1}e_{MA}. \tag{13.2.5}$$

The coupling to fermions

$$S = -\int d^D x \; e(\frac{1}{2\kappa^2}R + \frac{1}{2}\bar{\psi}e^{MA}\Gamma_A D_M\psi) \tag{13.2.6}$$

possesses the same classical scale invariance that we have just discussed if the transformation law of ψ is

$$\psi \to t^{1/2}\psi. \tag{13.2.7}$$

What happens in supergravity? Considering as an example the $N = 1$ supergravity theory formulated in (13.1.49), the reader will note that this theory possesses the same classical scale invariance, but now we must assign a nontrivial transformation law to the dilaton field ϕ. The transformation law of ϕ is in fact

$$\phi \to t^2 \phi. \qquad (13.2.8)$$

Thus, the supergravity theory possesses the same classical scale invariance as minimal general relativity. There is a basic difference, however. Although the expectation value of ϕ is arbitrary in the classical supergravity theory, we must attribute *some* expectation value to ϕ.[*] Therefore, in ten-dimensional supergravity, the scale invariance of the classical theory is spontaneously broken. At the classical level, the massless dilaton ϕ can be understood as a Goldstone boson of this spontaneously broken scale invariance.

Quantum mechanically, the scale invariance is not really a symmetry, so ϕ is better described as a pseudo-Goldstone boson. As such, one might expect it to gain a mass. This is prevented by supersymmetry as long as supersymmetry is unbroken, since ϕ is in the same supermultiplet with the graviton, which certainly cannot receive a mass. More generally, as was noted earlier, supersymmetry forbids not only a mass term for ϕ, but an arbitrary potential $V(\phi)$.

Let us ignore momentarily the constraints associated with supersymmetry and discuss instead constraints due to the classical scale invariance. At the classical level, scale invariance would permit a potential of the special form

$$V(\phi) \sim \phi, \qquad (13.2.9)$$

though because of supersymmetry such a term is actually absent in the Lagrangian that we are considering. Suppose that a potential is generated by a one-loop effect. A one-loop effect is of order $\hbar$. Under a scale transformation, $\hbar$ is rescaled; in ten dimensions the behavior is $\hbar \to t^8 \hbar$, according to (13.2.3) and (13.2.4). The scaling pseudosymmetry thus requires that a one-loop potential transform like $\hbar$ times (13.2.9), or like ϕ^5. More generally, an n-loop contribution $V^{(n)}$ to an effective potential would be proportional to $\hbar^n$ times ϕ, so, if not zero would behave as

$$V^{(n)}(\phi) \sim \phi^{4n+1}. \qquad (13.2.10)$$

Notice that contributions to the effective potential of higher and higher order vanish more and more rapidly for $\phi \to 0$. A nonperturbative effect

---

[*] The formulas are singular at $\phi = 0$; more precisely, the form of the kinetic energy $(\partial_M \phi / \phi)^2$ shows that $\phi = 0$ is 'infinitely far away' in field space.

would presumably vanish for $\phi \to 0$ faster than any power of ϕ; after all, nonperturbative effects should vanish more rapidly for weak coupling (small ϕ) than effects of any finite order of perturbation theory.

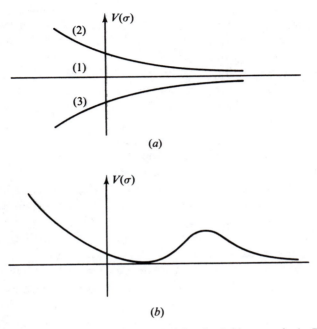

Figure 13.1. Possible forms of the potential for the field $\sigma = -\ln \phi$. This potential may be identically zero (curve (1) in (a)); it might approach zero as $\sigma \to \infty$ from above (curve (2) in (a)) or from below (curve (3) in (a)) while having no other zeros. It may also have an isolated zero in addition to the zero at infinity (b).

What behavior of $V(\phi)$ do we desire or expect? (This discussion should really be carried out in the context of a discussion of the reduction to four dimensions, but for brevity we work directly in ten dimensions.) It is possible that $V(\phi)$ is identically zero, as depicted by curve (1) in fig. 13.1a. In this case, ϕ is a massless particle. As it has coherent couplings to matter (for instance in the F^2/ϕ coupling), ϕ would then contribute corrections to large-scale gravitational phenomena that might be observable in astronomy. It would in fact be subject to fairly stringent experimental bounds, somewhat like the restrictions on the Brans–Dicke scalar.[†]

[†] Apart from direct astronomical tests, the most stringest bounds on coupling of a scalar such as ϕ arise from the fact that in cosmological models, because of its coupling to matter, the expectation value of ϕ would change in time. The mechanism for this is explained at the end of §14.5. The change in time of

If the potential $V(\phi)$ is not identically zero, we must face a host of new questions. The minimum of $V(\phi)$ is what is usually known as the cosmological constant. One of the real mysteries in superstring theory is undoubtedly the question of why the cosmological constant vanishes after supersymmetry breaking. This is perhaps the question on which our present ideas are most obviously deficient. The facts cited above concerning the classical scaling symmetry make the vanishing of the cosmological constant even more puzzling. Equation (13.2.10) seems to indicate that no matter what else happens, the potential $V(\phi)$ vanishes for $\phi \to 0$. Since $\phi = 0$ is really 'infinitely far away', our following remarks can be brought out more clearly by working with $\sigma = -\ln \phi$. A potential that vanishes for $\sigma \to \infty$ may approach zero either from above or from below. If V approaches zero from below, as depicted by curve (3) in fig. 13.1a, then the cosmological constant (the absolute minimum of V) is negative, an unhappy state of affairs. So we hope that V approaches zero from above. If so, one obvious possibility is that $V = 0$ is achieved only at $\sigma \to \infty$, as depicted by curve (2) in fig. 13.1a. If so, there is no stable vacuum in nature; σ will 'roll down the hill' indefinitely. Such an idea was essentially first proposed by Dirac in the 1930s to account for what is now called the gauge hierarchy problem; Dirac's idea was that large numbers (such as the ratio of the Planck mass to the proton mass) grow in time as σ rolls down the hill. Although this idea has many attractions, experimental bounds on the rate of change in time of coupling constants make it seem unlikely that this is the way nature works. Alternatively, it could be that in addition to the zero at $\sigma \to \infty$, V has an isolated minimum (at which $V = 0$) for some finite σ, as depicted in fig. 13.1b. There is nothing wrong with this option except that it is a mystery why V should have the stated form. Also, many physicists find it counterintuitive to postulate that our world (the zero at finite σ) is degenerate in energy with another world that is 'rolling down the hill'.

13.3 Anomaly Analysis

We now turn our attention to the analysis of anomalies, a subject that we met for the first time in chapter 10.

Symmetries of classical field theories can be broken by quantum effects known as anomalies. The origin of these effects can be traced to certain ill-behaved Feynman diagrams, with classically conserved currents attached,

---

the expectation value of ϕ would, because of the F^2/ϕ coupling, bring about a change in time of the fine-structure constant, a possibility that is severely limited by observation.

that do not admit a regulator compatible with simultaneous conservation of all the attached currents. Anomalies that spoil global conservation laws affect the physical content of a theory, but do not cause it to be inconsistent. Anomalies in local conservation laws, such as gauge invariance or general covariance, cause a theory to be inconsistent. Such anomalies can arise only in parity-violating amplitudes (gauge-invariant Pauli–Villars regularization is always possible for parity-conserving amplitudes).

Figure 13.2. Anomalous triangle diagram with $V - A$ currents at each vertex.

The classic example of restrictions on gauge couplings that are required to avoid anomalies is the $SU(2) \times U(1)$ model of electroweak interactions in four dimensions. The quarks and leptons are assigned to left-handed doublets and right-handed singlets of the weak $SU(2)$. In this case there is an anomalous fermion triangle diagram with $V-A$ currents at each vertex, as shown in fig. 13.2. Conservation cannot be imposed for the currents at all three vertices simultaneously, so the only consistent possibility is for a cancellation to occur when the contributions of the various fermions of the theory are summed. For example, if one chooses three currents carrying the $U(1)$ (weak hypercharge Y) quantum numbers the anomaly is proportional to Y^3. Hence we must get zero when we sum Y^3 for all the left-handed quarks and leptons. Since each generation has the same pattern of quantum numbers it suffices to consider the first one. The usual assignments are

$$
\begin{aligned}
Y(u_L) &= Y(d_L) = 1/3 \\
Y(\overline{u}_L) &= -4/3, \qquad Y(\overline{d}_L) = 2/3 \\
Y(e_L^-) &= Y(\nu_L) = -1, \qquad Y(e_L^+) = 2.
\end{aligned}
\qquad (13.3.1)
$$

Including a factor of three for color, one finds for the quarks $\mathrm{tr}(Y^3) = -6$ and for the leptons $\mathrm{tr}(Y^3) = +6$. Thus the standard generations do give the required cancellation, whereas a truncation of the theory to include only the quarks or the leptons would be inconsistent. Anomalies of the type YT^2 and $Y\lambda^2$, with T and λ being generators of $SU(2)_L$ and color, respectively,. cancel in a similar way. The standard model also has a potential anomaly in a triangle diagram with one external hypercharge generator and two external gravitons. This anomaly is proportional to tr$Y$ (the trace again being taken among left-handed fermions) and again vanishes for a standard generation of quarks and leptons. This too is a nontrivial restriction on the quantum numbers of the quarks and leptons.

Anomaly cancellation was originally understood as a requirement for renormalizability of the standard electroweak theory. If the requirement of anomaly cancellation is understood in this way, one may well wonder why the quark and lepton anomalies do cancel in nature, assuming string theory to be correct. String theory improves on the ultraviolet behavior of field theory by so much that if anomalies were basically a question of renormalizability, the field theory analysis of anomalies would not seem relevant! After all, it does not worry us to include at least one unrenormalizable theory – general relativity – as part of the low-energy approximation to string theory. So why is it necessary to cancel quark and lepton anomalies?

The real answer to this question is that although anomalies can be understood as an ultraviolet effect, related to the absence of a gauge-invariant regularization of certain diagrams, they can also be understood as an infrared effect. Even the low-energy part of the anomalous triangle diagram (and its higher-dimensional cousins) cannot be reconciled with gauge invariance and unitarity. This viewpoint came to be properly appreciated only under the stimulus of comparatively recent work on the role of anomalies in composite models of quarks and leptons. It is, however, no more than a refinement of what, since the early work on the subject, has been recognized as an essential part of the whole anomaly story: the anomaly is a failure of gauge invariance that cannot be removed by adding any local counterterm to the effective action and therefore cannot depend on unknown modifications of the short distance physics.

Our goal here is to obtain restrictions for supersymmetric theories in ten dimensions analogous to those described above for the standard model. In chapter 10 we calculated the hexagon loop amplitude in type I superstring theory shown in fig. 13.3. The conclusion there was that the only classical Yang–Mills group for which the nonabelian gauge anomaly could cancel is $SO(32)$. We now make this result more transparent by means of a low-energy analysis, which also will reveal that anomalies can cancel for a ten-dimensional $N = 1$ supergravity theory with gauge group $E_8 \times E_8$. The low-energy analysis is adequate for reasons that were explained in

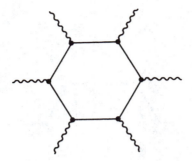

Figure 13.3. The hexagon diagram that gives anomalies in ten dimensions.

the previous paragraph.[*] The analysis will require considering potentially anomalous processes with external gauge bosons and external gravitons alike. The results can presumably be checked by full-fledged one-loop string calculations, but in practice only the calculation reported in chapter 10 has actually been completed so far. The analysis is certainly much easier for the low-energy effective action, and this approach has the advantage of not requiring any knowledge or assumptions about the nature or existence of possible superstring theories. In fact, this is how the group $E_8 \times E_8$ was first encountered, before it was known how to incorporate it in a string theory.

13.3.1. Structure of Field Theory Anomalies

Anomalies are a breakdown of gauge invariance and general covariance. To investigate the possible occurrence of anomalies, one investigates the effective action $\Gamma(A_M, g_{MN})$ for the gauge field A_M and the gravitational field g_{MN} obtained by integrating out all other fields, and one asks whether this effective action is gauge invariant and generally covariant. (The effective action Γ does not coincide with the object S_{eff} considered earlier, since it is obtained by integrating out massless as well as massive modes.)

The current induced by a given background field A is defined in terms of the effective action Γ as

$$ J_M = \frac{\delta \Gamma}{\delta A^M}. \tag{13.3.2} $$

---

[*] More precisely, the low-energy analysis is adequate for determining when anomalies *might* cancel with the addition of suitable counterterms, but not for determining when they *do* cancel, which depends as we will see on the coefficients of certain possible terms in the low-energy Lagrangian.

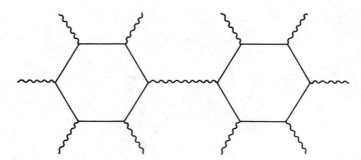

Figure 13.4. If longitudinal modes do not decouple from one loop diagrams, they will appear as poles in two loop diagrams.

Consider the variation of the effective action $\Gamma(A)$ under a gauge transformation $\delta A_M = D_M \Lambda$

$$G = \delta_\Lambda \Gamma = \text{tr} \int dx D_M \Lambda(x) \frac{\delta}{\delta A_M(x)} \Gamma(A). \qquad (13.3.3)$$

A partial integration shows that $G = 0$ if the gauge current (13.3.2) is covariantly constant. Thus, when there is an anomaly, the gauge currents are not really conserved. The loss of conservation of such gauge currents implies that unphysical polarization states of gauge fields show up as poles in S matrix elements, which implies a breakdown of unitarity. The calculation of chapter 10 described the anomaly as a coupling of a longitudinally polarized gauge field to five transversely polarized ones. This in turn implies that the longitudinal mode occurs as a pole in a two-loop diagram with ten transversely polarized gauge fields, as shown in fig. 13.4, which is inconsistent with unitarity.

Consider the hexagon diagram in ten dimensions with external gauge bosons only. There are many invariant ways to combine the gauge and Lorentz indices of the external lines, and consequently there are many invariant amplitudes characterizing the hexagon diagram. Anomalous behavior actually arises only in amplitudes that violate parity (others can be regularized in a gauge-invariant fashion) and only in amplitudes that are completely symmetric under permutations of the gauge indices of the external lines. We will not try to explain the latter feature systematically, except to note that it arises in every perturbative calculation of anomalies; it arose, for instance, in chapter 10. Because of the complete symmetry in gauge indices, we lose no essential information if we characterize all external gauge bosons by the same gauge generator T.[*] The anomalous loop diagrams with six external gauge bosons are proportional to sixth-order invariants made from T. The ten-dimensional CPT theorem forces us to deal with real fermion representations in which the trace of an odd power of T vanishes, so the relevant sixth-order invariants are

$$\text{tr}T^6, \quad \text{tr}T^4\text{tr}T^2, \quad (\text{tr}T^2)^3. \tag{13.3.4}$$

Let us describe more concretely (following, for instance, our experience in chapter 10) the form of anomalous hexagon amplitudes. It is convenient to let $F_0 = dA$ represent the linearized approximation to the Yang–Mills field strength; the linearized approximation is adequate since we are discussing the lowest number of external gauge bosons for which an anomaly arises. Under a linearized gauge transformation $A_M \to A_M + \partial_M \Lambda$ on one of the external lines, the anomalous variation of the hexagon amplitude Γ is[†]

$$\delta\Gamma = \int d^{10}x \left(c_1 \text{tr}\Lambda F_0^5 + c_2 \text{tr}\Lambda F_0 \text{tr}F_0^4 + c_3 \text{tr}\Lambda F_0 (\text{tr}F_0^2)^2 \right). \tag{13.3.5}$$

Wedge products of differential forms are understood in (13.3.5); the integrand, made from five two forms F_0, is a ten form, which can indeed be integrated over ten-dimensional space-time. The analogous formula in four dimensions is the familiar $\delta\Gamma = \text{tr}\Lambda F^2$ from the triangle diagram. The three coefficients c_1, c_2, c_3, which correspond to the three symmetric invariants in (13.3.4), depend on the gauge quantum numbers of the particles circulating in the hexagon loop. The generalization of the anomaly to terms with more than six external gauge bosons is *not* given simply by replacing F_0 by the gauge-invariant field strength F in (13.3.5). However, since the higher-order anomalies cancel when and only when the anomalies cancel in the six-point function, the form of the generalization of (13.3.5) is for some purposes not essential. The proof that the anomaly in any theory is of the general form (13.3.5), with only the three coefficients c_k depending on the theory, involves showing that by adding suitable local counterterms, if necessary, the anomaly in any Bose-symmetric action

---

[*] Anomalous amplitudes with more than six external gauge bosons in ten dimensions are not completely symmetric in gauge indices, but are uniquely determined in terms of anomalous six-point amplitudes by the Wess–Zumino consistency conditions described in §13.3.5.

[†] By adding to Γ a local counterterm of the form $\text{tr}AF_0\text{tr}AF_0^3$, we could if desired take c_2 to multiply $\text{tr}F_0^2\text{tr}\Lambda F_0^3$ instead of the more convenient form we have chosen.

functional can be put in the form of (13.3.5). The analysis is somewhat technical, and we content ourselves with noting that explicit calculations, such as those in chapter 10 or familiar four-dimensional anomaly calculations, always give answers of this form.

13.3.2 Gravitational Anomalies

It is also possible to have anomalies in Feynman diagrams with external gravitons. These are known as gravitational anomalies and represent a breakdown of general covariance. We have already met gravitational anomalies in two dimensions (the string world sheet) in §3.2.3. Now we consider the ten-dimensional case.

Parity-conserving diagrams can be regularized while preserving general covariance, so gravitational anomalies arise only for fields whose gravitational couplings violate parity and more specifically for fields whose Lorentz quantum numbers are such that their interactions *must* violate parity. In Euclidean space of D dimensions, this occurs for particles that are in a complex representation of the $SO(D)$ group, so that Lorentz invariance forbids mass terms and Pauli–Villars regularization is impossible. $SO(D)$ has complex representations only if D is of the form $4k + 2$ for some k, so these are the only dimensions in which gravitational anomalies can occur.

Like the gauge anomaly, the gravitational anomaly in ten dimensions arises first in a diagram with six external lines. The analysis of the Feynman diagrams is somewhat lengthier than we wish to delve into here, so we will try to state the result in a way that should sound plausible; we leave the interested reader to explore the references.

The analogy between Yang–Mills theory and general relativity is, of course, strongest if one introduces the spin connection ω_{MAB}, which in ten dimensions is an $SO(10)$ gauge field. In fact, the analogy is perhaps strongest if one considers a weak field gravitational wave of the special form

$$h_{MN} = h_M h_N e^{ik \cdot x}.\tag{13.3.6}$$

Every plane wave can be written as a linear combination of plane waves of this special kind, so we lose nothing essential by considering the scattering of gravitational waves of this particular kind. For a wave of this special kind, the linearized spin connection is

$$\omega_{MAB} = (h_M e^{ik \cdot x}) M_{AB},\tag{13.3.7}$$

where M_{AB} is the $SO(10)$ generator

$$M_{AB} = h_A k_B - h_B k_A. \tag{13.3.8}$$

We can think of the spin connection in (13.3.7) as an $SO(10)$ gauge field A_{MAB} with spatial dependence $h_M e^{ik\cdot x}$ and $SO(10)$ content given by the $SO(10)$ generator M_{AB}. The linearized Riemann tensor, for instance, is

$$R_{0AB} = d\omega_{AB}. \tag{13.3.9}$$

If one looks at things this way, then the general kinematics of gravitational anomalies turns out to be similar to that of gauge anomalies. Indeed, the anomaly only arises in a channel that is completely symmetric in the 'group indices' of the external gravitons, so there are three possible combinations, involving $\mathrm{tr}M^6$, $\mathrm{tr}M^4\mathrm{tr}M^2$, and $(\mathrm{tr}M^2)^3$, as in the case of gauge anomalies. The general form of the anomaly in the effective action is also similar to (13.3.5). Consider an infinitesimal general coordinate transformation of the special kind

$$x^M \rightarrow x^M + \epsilon_0 h^M e^{ik\cdot x} \tag{13.3.10}$$

with ϵ_0 a small parameter. Transforming the metric according to the standard rules of general relativity, one finds that the gravitational field remains in the form of (13.3.6), but h undergoes a gauge transformation

$$h_M \rightarrow h_M + k_M \epsilon_0. \tag{13.3.11}$$

In the analogy that we are trying to draw between gauge theories and gravity, h_M is the 'polarization vector' of the 'gauge field' in (13.3.7). Thus, (13.3.11) corresponds to the correct transformation law of this gauge field under gauge transformations. On the other hand, the charge matrix (13.3.8) is invariant under the gauge transformation (13.3.11), just as in the linearized limit of a gauge theory. Thus, it is plausible that for general coordinate transformations of the form (13.3.10), the general form of the anomalous variation of the effective action is very similar to what we encountered in the gauge theory case. In fact, if one defines the $SO(10)$ matrix $\Theta = M\epsilon_0$, then the general form of the variation of the effective action comes out to be

$$\delta\Gamma = \int d^{10}x(d_1\mathrm{tr}\Theta R_0^5 + d_2\mathrm{tr}\Theta R_0\mathrm{tr}R_0^4 + d_3\mathrm{tr}\Theta R_0(\mathrm{tr}R_0^2)^2). \tag{13.3.12}$$

Here R_0 is the linearized Riemann curvature two form given in (13.3.9); as before a wedge product of the five two forms is understood, and the trace is over the $SO(10)$ indices. While this formula is quite similar to the gauge theory formula (13.3.5), the detailed expression for the d_i in, say, the case of a hexagon anomaly due to a massless spin $1/2$ particle is quite different from the expression for the c_i, as we will discuss in more detail later.

Whether or not we have succeeded by our heuristic derivation in making this seem plausible, expression (13.3.12) is the general form of gravitational anomalies that emerges from actual perturbative calculations. The reader who would like to see (13.3.12) emerge from Feynman diagrams is referred to the references. We would like to stress, however, that (13.3.12) is the correct general expression for the change in the six-point amplitude under an infinitesimal diffeomorphism, not necessarily of the special form (13.3.10), which was used to facilitate the heuristic discussion. If one considers instead of (13.3.10) a general infinitesimal diffeomorphism

$$x^M \rightarrow x^M + \eta^M(x^N) \qquad (13.3.13)$$

then (13.3.12) is still valid, with

$$\Theta_{AB} = D_A \eta_B - D_B \eta_A. \qquad (13.3.14)$$

Gravitational anomalies represent a breakdown of energy–momentum conservation, just as gauge anomalies represent a breakdown of current conservation. Under an infinitesimal coordinate transformation $x^M \rightarrow x^M - \xi^M$, the variation of the metric is $\delta g_{MN} = D_M \xi_N + D_N \xi_M$. Therefore the variation of the effective action is

$$\delta \Gamma = \int dx \sqrt{g} (D_M \xi_N + D_N \xi_M) \delta \Gamma / \delta g_{MN}. \qquad (13.3.15)$$

But $\delta \Gamma / \delta g_{MN}$ is $\frac{1}{2} \langle T_{MN} \rangle$, where $\langle T_{MN} \rangle$ is the expectation value of the energy–momentum tensor of the gravitational fields. Thus, just as we argued in the case of (13.3.3), by a partial integration one obtains

$$\delta \Gamma = - \int dx \sqrt{g} \xi_N D_M \langle T^{MN} \rangle, \qquad (13.3.16)$$

so that a gravitational anomaly corresponds to a breakdown of conservation of the energy–momentum tensor.

13.3.3 Mixed Anomalies

Finally, it is possible to consider 'mixed anomalies' with both external gauge bosons and gravitons. As in the above, the gauge bosons are characterized by a charge matrix T and the gravitons by an $SO(10)$ matrix

M. Just as for purely gauge anomalies or purely gravitational anomalies, the mixed anomalies arise in channels that are completely symmetric in the charge matrices of external gauge bosons, and likewise in the $SO(10)$ matrices of external gravitons. The relevant group-theoretic invariants (bearing in mind that the trace of an odd power of T or M is zero in the relevant representations) are

$$\text{tr}T^4\text{tr}M^2, \quad \text{tr}T^2\text{tr}M^4, \quad \text{tr}T^2(\text{tr}M^2)^2, \quad (\text{tr}T^2)^2\text{tr}M^2. \qquad (13.3.17)$$

The general form of mixed terms in the variation in the effective action under a gauge and coordinate transformation is (with notation as above)

$$\begin{aligned}
\delta\Gamma = \int d^{10}x (e_1\text{tr}\Lambda F_0 \ \text{tr}R_0^4 + e_2\text{tr}\Theta R_0 \ \text{tr}F_0^4 \\
+ e_3\text{tr}\Lambda F_0(\text{tr}R_0^2)^2 + e_4\text{tr}\Theta R_0(\text{tr}F_0^2)^2),
\end{aligned} \qquad (13.3.18)$$

with the e_i being coefficients that depend on the details of a particular theory. Anomaly cancellation in ten dimensions means that the ten c_i, d_j, and e_k must all vanish.

The four terms in (13.3.18) correspond to the invariants in (13.3.17). All the same, the reader may wonder why we have omitted some other seemingly possible terms such as

$$\int d^{10}x \ \text{tr}\Lambda F_0 \ \text{tr}F_0^2 \ \text{tr}R_0^2 \qquad (13.3.19)$$

and a similar term with the role of gauge bosons and gravitons exchanged. The answer to this is that (13.3.19) can be eliminated in favor of the e_4 term in (13.3.18) by adding to the effective action Γ a local counterterm

$$\tilde{\Gamma} = \int d^{10}x \ \text{tr}AdA \ \text{tr}F_0^2 \ \text{tr}\omega d\omega, \qquad (13.3.20)$$

where ω is the spin connection and A is the gauge field. Using the gauge transformation laws

$$\delta(\text{tr}AdA) = d(\text{tr}\Lambda dA), \quad \delta(\text{tr}\omega d\omega) = d(\text{tr}\Theta d\omega) \qquad (13.3.21)$$

(and of course the gauge invariance of F_0 in the linearized approximation we are using) the reader should be able to see that by adding the local term $\tilde{\Gamma}$ of (13.3.20), (13.3.19) can be eliminated and the mixed anomalies can be put in the form given in (13.3.18).

Physically, we are permitted to add (13.3.20), or any other local expression, to the effective action, because being local this respects unitarity and all other physical principles. A theory in which anomalies can be eliminated by addition of a term such as (13.3.20) is not really anomalous, since gauge invariance and general covariance can be achieved without violating any physical principles. The freedom to adjust at will some of the seeming potential anomalies by adding terms such as (13.3.20) will be important later.

13.3.4 The Anomalous Feynman Diagrams

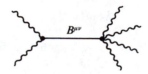

Figure 13.5. An anomalous diagram in which a massless two form B is exchanged between two gauge bosons (or gravitons) on one side and four gluons (or two gauge bosons and two gravitons or four gravitons) on the other side.

Postponing the detailed formulas, let us now discuss the general format of anomaly cancellation, as seen from the low-energy point of view. On general grounds, only Feynman diagrams with massless internal lines are relevant in discussing anomalies. Diagrams with massive internal lines give local amplitudes whose anomalous behavior, if any, is irrelevant, since it can be absorbed in adding a physically acceptable local counterterm to the action. Which diagrams with massless external lines are relevant? One choice is the hexagon diagram of fig. 13.3; it is this that corresponds in the most obvious way to the anomalous triangle diagram in four dimensions. In type IIB superstring theory, the hexagon diagrams are the only ones that are relevant. We will learn, however, that another anomalous diagram, not familiar from four-dimensional experience, plays a role in anomaly cancellation in the type I and heterotic theories. This is the diagram of fig. 13.5 in which a massless two form B is exchanged between gluons and gravitons. The coupling of B to two gluons appeared already in our discussion of $N = 1$ supergravity theory in ten dimensions. We noted that the naive form of the field strength of B, namely $H_0 = dB$, was actually replaced by $H = dB - \mathrm{tr}(AdA + \frac{2}{3}A^3)$, the addition being the Yang–Mills Chern–Simons form. The H^2 term includes a coupling of

B to two gluons, namely

$$S_1 = \int d^{10}x \sqrt{g}(H_0)^{MNP}\mathrm{tr}A_M \partial_N A_P. \tag{13.3.22}$$

To draw a diagram such as fig. 13.5 is only possible if we add a term to the Lagrangian – not present in the minimal supergravity theory – that couples B to four gauge bosons. The relevant coupling turns out to be

$$S_2 = \int d^{10}x B \wedge \mathrm{tr}F^4. \tag{13.3.23}$$

Notice that S_2 is invariant under the gauge transformation $\delta B = d\Lambda$ of the two form B; to prove this one must integrate by parts and use the fact that the form $\mathrm{tr}F^4$ is closed.

The idea of an anomaly in a tree diagram such as fig. 13.5 may seem surprising if we believe that anomalies only arise from ultraviolet divergences in loop diagrams. Actually, (13.3.23) is invariant under Yang–Mills gauge transformations provided B is invariant under such transformations. This is the natural, obvious state of affairs for the neutral field B. On the other hand, we learned in §13.1.3 that the H^2 term, from which (13.3.22) derives, is gauge invariant only if B is *not* invariant under gauge transformations. Taking the two together, we find that although either (13.3.22) or (13.3.23) is gauge invariant by itself, a Lagrangian containing both of them is not gauge invariant. Thus, the interference term between these two interactions, which appears in fig. 13.5, violates gauge invariance.

It is easy to determine the form of the violation. Under a gauge transformation $\delta A = d\Lambda$, (13.3.22) changes by

$$\delta S_1 = - \int d^{10}x \sqrt{g}\mathrm{tr}\left(\Lambda F^{MN}\right)D^P H_{MNP}. \tag{13.3.24}$$

If the only term in the Lagrangian in which B appears is the H^2 term, then the equation of motion for B is $D^P H_{MNP} = 0$. In this case, (13.3.24) vanishes, reflecting the fact that in and of itself (13.3.22) respects gauge invariance. In the presence, however, of other interactions, it is no longer true that $D^P H_{MNP} = 0$. The interaction (13.3.23) modifies this, in fact, to $D^P H_{MNP} = \epsilon_{MNA_1 A_2 \dots A_8}\mathrm{tr}(F^{A_1 A_2} \dots F^{A_7 A_8})$, so (13.3.24) becomes

$$\delta S_1 = - \int d^{10}x \sqrt{g}\mathrm{tr}\left(\Lambda F\right) \wedge \mathrm{tr}F^4, \tag{13.3.25}$$

which has precisely the form of some of the anomalies that we wish to cancel. An equivalent, diagrammatic way to express all of this is that when inserted in the diagram of fig. 13.5, the $D^P H_{MNP}$ in (13.3.24) cancels the B propagator that is present in that diagram, leaving us with (13.3.25).

We see, then, that an exotic phenomenon, which from a low-energy point of view would be described as an anomalous tree diagram, can cancel some of the anomalies we surveyed above. This requires the presence of the interaction (13.3.23), which is not present in the minimal supergravity model, but whose presence in string theory would not be very surprising.

Trying to cancel anomalies with diagrams such as fig. 13.5 has a dramatic consequence. The basic anomaly cancellation depends on the existence of a massless boson B that can couple to two or four gluons (or gravitons, in subsequent generalizations). Upon reduction to four dimensions, some modes of the B field are still massless particles, as we will see in detail in the next chapter. We will obtain in this way a massless spin zero particle a with a coupling to two four-dimensional massless gauge bosons. This coupling turns out, as we will see, to have the form $a\mathrm{tr}F\tilde{F}$ that enters in the axion solution of the strong CP problem, so it is quite conceivable that the solution of the strong CP problem is to be found in the mechanism described here for canceling anomalies.

Which anomalies can be canceled this way? The anomaly cancellation mechanism requires a B propagator that carries neither gauge nor $SO(10)$ quantum numbers, so we can only cancel anomalies in which the group-theory factors encountered above factorize as a product of traces. Thus, in the Yang–Mills case, for example, we can hope to cancel a $\mathrm{tr}T^2\mathrm{tr}T^4$ or $(\mathrm{tr}T^2)^3$ term in this way, but we cannot hope to cancel a $\mathrm{tr}T^6$ term. With a sufficiently large number of independent B fields in the above construction, we could cancel in this way all anomalies except the two irreducible ones $\mathrm{tr}T^6$ and $\mathrm{tr}M^6$. In fact, in the ten-dimensional $N = 1$ supergravity multiplet there is only a single two form B, and by suitably adjusting unknown couplings such as (13.3.23) it is possible to cancel only some of the factorizable anomalies. In fact, for the factorizable anomaly to be expressible as the residue of the pole of a single B field in fig. 13.5, it must have the general form

$$\delta\Gamma = \int d^{10}x(\mathrm{tr}\Lambda F_0 + u\mathrm{tr}\Theta R_0) \wedge (v\mathrm{tr}F_0^4 + w\mathrm{tr}R_0^4$$
$$+ x(\mathrm{tr}F_0^2)^2 + y(\mathrm{tr}R_0^2)^2 + z\mathrm{tr}F_0^2\mathrm{tr}R_0^2) \tag{13.3.26}$$

with unknown coefficients u, v, w, x, y, z. We are entitled, of course, to use the freedom in (13.3.20) in putting the anomalies in this form. Thus, anomaly cancellation requires that the irreducible anomalies $\mathrm{tr}T^6$ and $\mathrm{tr}M^6$ should cancel from hexagon diagrams, and that the eight factorizable anomalies parametrized earlier in terms of $c_2, c_3, d_3, d_4, e_1, e_2, e_3$, and e_4 should be expressible in terms of only six independent coefficients

u, v, w, x, y, z. Of course, it is necessary in addition that string theory should generate those latter parameters with the correct coefficients.

13.3.5 Mathematical Characterization of Anomalies

In this subsection (which is not strictly necessary for understanding the rest of the chapter), we would like to give at least a flavor of how anomalies with more than six external gauge bosons or gravitons are determined from the hexagon anomalies. Again, the interested reader may wish to look elsewhere for more detail.

Let $G(\Lambda)$ be the change in the effective action under the gauge transformation or general coordinate transformation Λ. Thus, with Γ being the effective action, G is

$$G(\Lambda) = \frac{\delta}{\delta\Lambda}\Gamma. \qquad (13.3.27)$$

In D dimensions, $G(\Lambda)$ is always given as the integral over all space-time of a D form I_D^1. We encountered some examples of such forms I_D^1 in the above. In the integral $G = \int I_D^1$, the D form I_D^1 is only well-defined modulo an exact form.

There is an important consistency condition for anomalies, called the Wess–Zumino condition. It states that

$$\delta_{\Lambda_1} G(\Lambda_2) - \delta_{\Lambda_2} G(\Lambda_1) = G(\Lambda), \qquad (13.3.28)$$

where

$$\Lambda = [\Lambda_1, \Lambda_2]. \qquad (13.3.29)$$

The necessity of this condition on G is evident from $G(\Lambda) = \delta_\Lambda\Gamma$ and (13.3.29).

There is an elegant way of solving the consistency condition (13.3.28) in terms of a $D + 2$ form I_{D+2} that is gauge invariant. Since we are supposing that space-time is only D-dimensional, there are no $D+2$ forms in the usual sense, so I_{D+2} must be regarded as a formal expression. (In fact, there is a sophisticated mathematical interpretation of the extra two dimensions that are required for the existence of a $D + 2$ form, but we do not need to know about that for our purposes here.)

Before explaining the relation between the $D + 2$ form I_{D+2} and the anomaly G, let us recall from chapter 12 some facts about the generic structure of gauge-invariant differential forms. Under a Yang–Mills gauge transformation, the two-form field strength transforms according to the rule $\delta F = [F, \Lambda]$. Therefore a $2m$ form of the type $\text{tr}(F^m)$ is gauge invariant as a consequence of the identity $\text{tr}(AB) = \text{tr}(BA)$, which is valid for forms of even degree. If the matrices F are expressed in a real representation of the gauge group, then the generators are represented by antisymmetric matrices. In this case $\text{tr}(F^m)$ vanishes unless m is even. In fact, this is the case of primary interest since in $D = 4k + 2$ dimensions CPT conjugates the group assignment of a field without changing its chirality. Thus the sum of an arbitrary representation and its conjugate (if different) is described by a real representation.

As we know, the gravitational field in the vielbein formalism can be treated very similarly to the Yang–Mills fields. Specifically, the spin connection $\omega_M{}^{AB}$ can be regarded as a Yang–Mills potential analogous to A_M. It can be expressed as a one form

$$\omega^{AB} = \omega_M{}^{AB} dx^M \tag{13.3.30}$$

and, by lowering an index with the Minkowski metric, regarded as a $D \times D$ matrix ω in the fundamental representation of the Lorentz algebra $SO(D-1, 1)$. Using this connection one forms a $D \times D$ matrix of curvature two forms

$$R = d\omega + \omega^2, \tag{13.3.31}$$

which is entirely analogous to (13.1.39). The expression ω^2 represents a matrix product as well as a wedge product of one forms. Under an infinitesimal Lorentz transformation with parameters given by an $SO(D-1, 1)$ matrix Θ, the transformations are

$$\delta\omega = d\Theta + [\omega, \Theta] \tag{13.3.32}$$

and

$$\delta R = [R, \Theta]. \tag{13.3.33}$$

This is completely analogous to the Yang–Mills gauge transformations.

We can construct Lorentz-invariant differential forms $\text{tr}(R^m)$ out of the curvature forms, just as in the Yang–Mills case. Since R is an antisymmetric matrix, these vanish unless m is even. We now can describe the most general possible expressions I_{D+2} that can arise. They are given by polynomials made from the gauge-invariant combinations $\text{tr}(F^m)$ and $\text{tr}(R^m)$. Containing both the Yang–Mills and gravitational curvatures, they suffice to describe the most general anomalies involving Yang–Mills and gravitational fields. Since F and R are two forms, the total number of F's and R's in each term of a $D+2$ form is $1+D/2$. This only makes sense if D is even, of course. But chiral fields, parity violation, and anomalies only occur for even D anyway.

Let us now explain how the $D + 2$ form I_{D+2} is related to the anomaly G. Any form of the type $\text{tr}(F^m)$ can be written locally in the form

$$\text{tr}(F^m) = d\omega_{2m-1}, \tag{13.3.34}$$

where the $2m - 1$ form ω_{2m-1}, which is a certain polynomial in the gauge field and its derivatives, is called a Chern–Simons form. Chern–Simons forms were introduced in §12.5. The Chern–Simons form is defined explicitly for the case $m = 2$ in (13.1.37). The formula for the general case is

$$\omega_{2m-1}(A) = m \int_0^1 dt \, t^{m-1} \text{tr}(A(dA + tA^2)^{m-1}). \tag{13.3.35}$$

Since we wish to introduce analogous expressions for the Lorentz case we introduce the symbols $\omega_{2m-1,Y}$ and $\omega_{2m-1,L}$, where the subscripts denote Yang–Mills and Lorentz. The fact that $\text{tr}(F^m)$ and $\text{tr}(R^m)$ are exact implies that any polynomial made from them is also exact. For example,

$$\text{tr}(F^m)\text{tr}(R^n) = d[\omega_{2m-1,Y}\text{tr}(R^n)] = d[\text{tr}(F^m)\omega_{2n-1,L}]. \tag{13.3.36}$$

We are using here the fact that

$$d\text{tr}\,F^m = d\text{tr}\,R^n = 0, \tag{13.3.37}$$

which follows from (13.3.34) and the fact that $d^2 = 0$.

Now suppose that I_{D+2} is a $D + 2$ form given as some polynomial in $\text{tr}F^m$ and $\text{tr}R^m$, so that it is gauge invariant. Since it is also exact, we may write

$$I_{D+2} = dI_{D+1}, \tag{13.3.38}$$

where the $D+1$ form I_{D+1} is well-defined modulo a closed form. Although I_{D+2} is gauge invariant, I_{D+1} in general is not. For example, the Chern–Simons three form was shown to have a gauge variation $d\text{tr}(\Lambda dA)$. While the gauge variation of I_{D+1} is not zero, it is always exact. Thus we may write

$$\delta I_{D+1} = dI_D^1. \tag{13.3.39}$$

This ensures the gauge invariance of I_{D+2}, in particular. The expression I_D^1 is a D form that is linear in the gauge parameters Λ and Θ (hence the superscript). This quantity is ambiguous up to a closed form, but it determines a unique integral

$$G = \int I_D^1, \tag{13.3.40}$$

which is the anomaly corresponding to I_{D+2}.

The construction ensures that G satisfies the Wess–Zumino condition (13.3.28). To prove this let us suppose that the D-dimensional space-time manifold M can be represented as the boundary of a $(D+1)$-dimensional region Σ. (In mathematical notation $M = \partial\Sigma$.) One way of achieving this, if we are working in Euclidean space, is to add a point at infinity so that M becomes topologically a D-sphere S^D. Then Σ can be taken as a $(D+1)$-dimensional ball having $M = S^D$ as its surface. In this case the anomaly can be re-expressed as follows

$$G(\Lambda) = \int_M I_D^1 = \int_\Sigma dI_D^1 = \delta_\Lambda \int_\Sigma I_{D+1}. \tag{13.3.41}$$

Then $[\delta_{\Lambda_1}, \delta_{\Lambda_2}] = \delta_\Lambda$ implies that (13.3.28) is satisfied. This has the following significance. In $D = 2n$ dimensions, the $D+2$ form involves $n+1$ field strengths F and/or R in each term. Since $F = dA + A^2$ and $R = d\omega + \omega^2$, the minimum number of fields that can enter in such an expression is also $n+1$. Thus the 'smallest' anomalous one-loop diagrams are ones with $n+1$ external gauge fields. However, because of the nonlinearity of F and R the formulas also determine anomalies for diagrams with more external gauge fields. In fact, it is evident that the term with $n+1$ gauge fields uniquely determines the complete formula by the substitutions $dA \rightarrow F$ and $d\omega \rightarrow R$. Thus if the anomalies with $n+1$ external gauge particles are derived by explicit Feynman diagram calculations, the general formulas can be deduced without further calculation.

13.3.6 Other Types of Anomalies

In addition to gauge and gravitational anomalies, one may wonder if it is possible to have anomalies that spoil local supersymmetry in some of the ten-dimensional supergravity theories that are otherwise anomaly free. Although most workers in the field consider this improbable, a convincing treatment has not yet appeared.

In our discussion above, we have considered only infinitesimal gauge and coordinate transformations. Absence of anomalies under such infinitesimal transformations implies absence of anomalies under any gauge and coordinate transformations that can be reached continuously from the identity. There remains the question of possible anomalies in gauge transformations or coordinate transformations that cannot be reached continuously from the identity. Such 'global' anomalies are the space-time analog of anomalies in world-sheet modular transformations. It has been shown that space-time global anomalies always vanish in ten-dimensional supergravity theories in which the perturbative anomalies cancel, but the analysis is rather lengthy and will not be pursued here.

13.4 Explicit Formulas for the Anomalies

In the above, we have discussed the general form of the anomalies in any ten-dimensional theory. Let us now discuss the form of the anomalies that actually arise from hexagon diagrams in the relevant cases.

Let us begin with pure gravitational anomalies. They arise in hexagon diagrams only from fields that transform in complex representations of $SO(10)$. For example, the Majorana–Weyl spinor λ in the $N = 1$, $D = 10$ supergravity multiplet belongs to the 16-dimensional spinor representation. This is a complex representation, its complex conjugate being the $\overline{16}$. Similarly chiral gravitinos can belong to the 144 or $\overline{144}$. For $D = 4k + 2$, $SO(D)$ also has complex bosonic representations. The simplest example is given by a tensor of rank $2k + 1$ that is self-dual or anti-self-dual

$$F_{M_1 \dots M_{2k+1}} = \pm \frac{i}{(2k+1)!} \epsilon_{M_1 \dots M_{2k+1} N_1 \dots N_{2k+1}} F^{N_1 \dots N_{2k+1}}. \qquad (13.4.1)$$

A significant example is the self-dual fifth-rank field strength that occurs in type IIB supergravity.

The explicit calculation of gravitational anomalies can be carried out in the 'minimal' case by considering a loop with $2k + 2$ external gravitons (for $D = 4k + 2$). Pauli–Villars regularization of the linear divergence is convenient, and then the anomaly can be calculated by the same general approach as in the case of Yang–Mills gauge anomalies. In practice a number of tricks are required to do the combinatorics effectively. Here we settle for reporting the results.

We have learned that the gravitational anomaly in ten dimensions involves a sixth-order invariant of an $SO(10)$ matrix M, the possibilities being $\text{tr}M^6$, $\text{tr}M^4\text{tr}M^2$, and $(\text{tr}M^2)^3$. More generally, in $4k + 2$ dimensions we would encounter invariants of order $2k + 2$ such as $\text{tr}M^{2k+2}$ and so forth. Henceforth, in order to bring our discussion more in line with usual mathematical terminology, we use the name R (for curvature two form) for the matrix M. We wish to emphasize, however, that although it arises in some mathematical discussions (as in §13.3.5) as a curvature two form whose two form indices are then suppressed, for our purposes R is just a matrix.[*]

Since the anomaly formulas are functions only of the invariant expressions $\text{tr}(R^{2m})$, they only depend on the eigenvalues of the antisymmetric $D \times D$ matrix R. By an orthogonal transformation such a matrix can be brought to the form

$$
R = \begin{pmatrix}
0 & x_1 & & & & & & 0 \\
-x_1 & 0 & & & & & & \\
& & 0 & x_2 & & & & \\
& & -x_2 & 0 & & & & \\
& & & & \ddots & & & \\
& & & & & 0 & x_{2k+1} & \\
0 & & & & & -x_{2k+1} & 0 & \\
& & & & & & & \ddots
\end{pmatrix}
\tag{13.4.2}
$$

In this basis it is easy to see that

$$
\text{tr}(R^{2m}) = 2(-1)^m \sum_{i=1}^{2k+1} x_i^{2m}.
\tag{13.4.3}
$$

Therefore if we are given a symmetric polynomial in $2k + 1$ variables x_i, even in each of them separately, we can associate an mth-order invariant of R by the rule

$$
\sum_i x_i^{2m} \to \frac{1}{2}(-1)^m \text{tr} R^{2m}.
\tag{13.4.4}
$$

Let us denote the anomalies of a chiral spin 1/2 field, a chiral spin 3/2 field, or a self-dual $2k + 1$ rank tensor as $I_{1/2}$, $I_{3/2}$, and I_A, respectively. Before writing down the expressions for these quantities, it is convenient to remove a common factor from each of the formulas, so we define

$$
I_{1/2} = -i(2\pi)^{-D/2} \hat{I}_{1/2}
\tag{13.4.5}
$$

with similar definitions of $\hat{I}_{3/2}$ and $\hat{I}_A$.

---

[*] We stress this point because we are about to discuss the eigenvalues of R – a concept that makes sense for matrices but must be interpreted more abstractly for matrix-valued two forms.

The formula for the gravitational anomaly of, say, a spin 1/2 particle in some given dimension of the form $4k + 2$ is rather complicated. Remarkably, it is easiest to write down a formula that simultaneously encompasses the $(4k + 2)$-dimensional anomalies for all k. The formula is

$$\hat{I}_{1/2} = \prod_{i=1}^{2k+1} \left(\frac{\frac{1}{2}x_i}{\sinh \frac{1}{2}x_i} \right). \tag{13.4.6}$$

The meaning of this formula is that for $D = 4k + 2$ dimensions, one must extract the terms that are homogeneous of degree $k + 1$ in order that the substitution (13.4.4) give a polynomial in R of order $2k + 2$, the correct value in $4k + 2$ dimensions. By actually carrying out the process, we can find the explicit form of the anomaly for small k. Letting $y_i = \frac{1}{2}x_i$, we have

$$\hat{I}_{1/2} = \prod_i (1 + \frac{y_i^2}{3!} + \frac{y_i^4}{5!} + \cdots)^{-1}$$

$$= \prod_i (1 - \frac{1}{6}y_i^2 + \frac{7}{360}y_i^4 - \frac{31}{15,120}y_i^6 + \cdots)$$

$$= 1 - \frac{1}{6}Y_2 + \frac{1}{180}Y_4 + \frac{1}{72}Y_2^2$$

$$- \frac{1}{2835}Y_6 - \frac{1}{1080}Y_2Y_4 - \frac{1}{1296}Y_2^3 + \cdots, \tag{13.4.7}$$

where we have defined

$$Y_{2m} = \sum_{i=1}^{2k+1} y_i^{2m} \sim \frac{1}{2} \left(-\frac{1}{4}\right)^m \text{tr} R^{2m}. \tag{13.4.8}$$

Thus, for example, the gravitational anomaly due to a complex Weyl spinor in $D = 6$ is given by $\hat{I}_{1/2} = \frac{1}{180}Y_4 + \frac{1}{72}Y_2^2$, which implies that

$$I_{1/2} = -i(2\pi)^{-3} \frac{1}{16} \left[\frac{1}{2 \cdot 180} \text{tr} R^4 + \frac{1}{4 \cdot 72}(\text{tr} R^2)^2 \right]. \tag{13.4.9}$$

The anomalies $I_{3/2}$ and I_A are described by analogous symmetric functions. Specifically

$$\hat{I}_{3/2} = \hat{I}_{1/2}(-1 + 2 \sum_{i=1}^{2k+1} \cosh x_i)$$
$$= \hat{I}_{1/2}(D - 1 + 4Y_2 + \frac{4}{3}Y_4 + \frac{8}{45}Y_6 + \cdots). \tag{13.4.10}$$

In the important case of $D = 10$, it is useful for our later analysis to note that

$$(\hat{I}_{3/2} - \hat{I}_{1/2})_6 = \frac{496}{2835}Y_6 + \left(\frac{496}{1080} - \frac{2}{3}\right)Y_2Y_4 + \left(\frac{496}{1296} - \frac{1}{3}\right)Y_2^3$$

$$= \tfrac{1}{2835}(496Y_6 - 588Y_2Y_4 + 140Y_2^3). \qquad (13.4.11)$$

Finally, the antisymmetric tensor anomaly is described by

$$\hat{I}_A = -\tfrac{1}{8} \prod_{i=1}^{2k+1} \frac{x_i}{\tanh x_i}$$

$$= -\tfrac{1}{8} - \tfrac{1}{6}Y_2 + \left(\tfrac{7}{45}Y_4 - \tfrac{1}{9}Y_2^2\right)$$

$$+ \tfrac{1}{2835}(-496Y_6 + 588Y_2Y_4 - 140Y_2^3) + \cdots. \qquad (13.4.12)$$

In addition to the pure gravitational anomalies described above, we also require formulas for anomalies in diagrams that contain both external gravitons and external gauge fields or external gauge fields only. These are characterized by invariants made from a charge matrix T – henceforth we call it F to achieve better agreement with mathematical terminology – as well as the $SO(10)$ matrix R. In $4k + 2$ dimensions, the mixed anomalies involve terms of total order $2k + 2$ in F and R. One-loop anomalies in diagrams containing external gauge fields arise only from loops made of chiral fields that carry the Yang–Mills charge. In our applications the only massless chiral field that carries gauge quantum numbers are Weyl spinors. Hence only the formula for $I_{1/2}$ needs to be generalized to include a dependence on the Yang–Mills field strength. The rule turns out to be surprisingly simple

$$\hat{I}_{1/2}(F, R) = \operatorname{tr}(e^{iF})\hat{I}_{1/2}(R), \qquad (13.4.13)$$

where $\hat{I}_{1/2}(R)$ refers to the expression formed from (13.4.6). Again, the meaning of this formula is that in $4k + 2$ dimensions, one is to expand in powers of F and R and keep precisely the terms of order $2k + 2$.

The trace in $\operatorname{tr} e^{iF}$ is to be taken in whatever representation the chiral spin $1/2$ fields may lie in. If $\operatorname{tr} F^{2m+1} = 0$, as happens for real representations, one can replace $\operatorname{tr} e^{iF}$ by $\operatorname{tr} \cos F$. Using the expansion

$$\operatorname{tr} \cos F = \sum_{m=0}^{\infty} \frac{(-1)^m}{(2m)!} \operatorname{tr}(F^{2m}) \qquad (13.4.14)$$

and the expansion in (13.4.7), the expansion of $\hat{I}_{1/2}(F, R)$ can be worked out. Note that (13.4.13) contains the pure gravitational anomaly as a special case. The first term in (13.4.14) is $\operatorname{tr}(1) = n$, where n is the dimension of the representation to which the chiral fermions belong. As we would expect, for the pure gravitational anomaly only the number and spin of these fields matters. Their other group-theoretic properties are irrelevant.

We learn from (13.4.7), (13.4.13) and (13.4.14) that the anomaly for $D = 2$ is given by

$$\hat{I}_{1/2}(F, R) = \frac{n}{48}\mathrm{tr}R^2 - \frac{1}{2}\mathrm{tr}F^2 \qquad (13.4.15)$$

and for $D = 6$ is given by

$$\hat{I}_{1/2}(F, R) = \frac{1}{24}\mathrm{tr}F^4 - \frac{1}{96}\mathrm{tr}F^2\mathrm{tr}R^2 \\ + \frac{n}{128}\Big(\frac{1}{45}\mathrm{tr}R^4 + \frac{1}{36}(\mathrm{tr}R^2)^2\Big). \qquad (13.4.16)$$

Similarly, in $D = 10$ one finds that

$$\hat{I}_{1/2}(F, R) = -\frac{1}{720}\mathrm{tr}F^6 + \frac{1}{24 \cdot 48}\mathrm{tr}F^4\mathrm{tr}R^2 \\ - \frac{1}{256}\mathrm{tr}F^2\Big[\frac{1}{45}\mathrm{tr}R^4 + \frac{1}{36}(\mathrm{tr}R^2)^2\Big] \\ + \frac{n}{64}\Big[\frac{1}{2 \cdot 2835}\mathrm{tr}R^6 + \frac{1}{4 \cdot 1080}\mathrm{tr}R^2\mathrm{tr}R^4 \\ + \frac{1}{8 \cdot 1296}(\mathrm{tr}R^2)^3\Big]. \qquad (13.4.17)$$

The formulas in this section correspond to the anomalies for complex Weyl fermions. If in $8k + 2$ dimensions one wishes to consider Majorana–Weyl fermions, it is necessary to divide by a factor of 2.

13.5 Anomaly Cancellations

We have seen that matter fields with chiral couplings can give rise to pure gravitational anomalies for space-time dimension $D = 4k + 2$. The chiral fields that occur in $D = 10$ supergravity and superstring theories are of the three types described in the previous section.

The relevant terms for anomalies in ten dimensions are sixth order in the charge matrices F and R (in the language of §13.3.5 the anomalies are described by 12 forms). We wish to investigate whether or not anomaly cancellations take place when the various contributions are combined. Therefore we may drop a common factor from the various contributions and focus our attention on the expressions $\hat{I}_{3/2}$, $\hat{I}_{1/2}$, and $\hat{I}_A$ defined in the previous section. We consider first the purely gravitational anomalies. We learn from the formulas given in (13.4.7), (13.4.11) and (13.4.12) that the sixth-order terms are

$$(\hat{I}_{1/2})_6 = -\tfrac{1}{2835}Y_6 - \tfrac{1}{1080}Y_2Y_4 - \tfrac{1}{1296}Y_2^3$$

$$(\hat{I}_{3/2})_6 = \tfrac{495}{2835}Y_6 - \tfrac{225}{1080}Y_2Y_4 + \tfrac{63}{1296}Y_2^3 \qquad (13.5.1)$$

$$(\hat{I}_A)_6 = -\tfrac{496}{2835}Y_6 + \tfrac{224}{1080}Y_2Y_4 - \tfrac{64}{1296}Y_2^3.$$

13.5.1 Type I Supergravity Without Matter

An easy case to consider is the type I supergravity theory without gauge interactions (or the type I superstring theory with closed strings only). In these theories the fields with anomalous couplings are a massless Majorana–Weyl gravitino field and a massless Majorana–Weyl spin 1/2 field of the opposite chirality. The sum of their contributions to the anomaly is characterized by $\tfrac{1}{2}(I_{3/2} - I_{1/2})$. (The factor of 1/2 was explained at the end of the preceding section.) Since $(\hat{I}_{3/2} - \hat{I}_{1/2})_6 \neq 0$, the anomaly does not cancel for the pure $N = 1$ supergravity theory. To the hexagon contributions listed above, it is necessary to add the contributions from exchange of the B field as described in §13.3.4 above and further explored below in §13.5.3, but this mechanism cannot cancel the Y_6 term. Consequently, there cannot be a string theory (or any other type of theory for that matter) that is approximated by pure $N = 1$, $D = 10$ supergravity at low energies and is anomaly free.

13.5.2 Type IIB Supergravity

In the case of type IIB supergravity there is a complex gravitino of one chirality, a complex spinor of the opposite chirality, and a self-dual antisymmetric tensor field strength. The total anomaly contribution of these fields is given by $I_{3/2} - I_{1/2} + I_A$. This combination exhibits a 'miraculous' cancellation, the first nontrivial anomaly cancellation to be discovered in ten-dimensional supergravity or superstring theory:

$$(\hat{I}_{3/2})_6 - (\hat{I}_{1/2})_6 + (\hat{I}_A)_6 = 0. \qquad (13.5.2)$$

The coefficients are just right to cancel separately each of the three different types of terms in the formula. The cancellation in (13.5.2), where nothing can be arbitrarily adjusted, is rather remarkable, though perhaps it is less surprising in light of the excellent convergence properties of the oriented one-loop superstring diagram.

13.5.3 Allowed Gauge Groups for $N = 1$ Superstring Theories

We saw in §13.5.1 that pure $N = 1$, $D = 10$ supergravity has anomalies that prevent it from being a low-energy approximation to an anomaly-free string theory. Let us now include the coupling of Yang–Mills supermultiplets. In this case there are n additional left-handed Majorana–Weyl spinors belonging to the adjoint representation of the group. (n is the dimension of this representation.) The total anomaly coming from one-loop hexagon diagrams is proportional to

$$\hat{I} = \hat{I}_{3/2}(R) - \hat{I}_{1/2}(R) + \hat{I}_{1/2}(F, R). \qquad (13.5.3)$$

Combining (13.4.11) and (13.4.17) (using the correspondence (13.4.8)) this is given by the sixth-order polynomial

$$\hat{I}_{12} = -\frac{1}{720}\mathrm{Tr}F^6 + \frac{1}{24\cdot 48}\mathrm{Tr}F^4\mathrm{tr}R^2$$

$$- \frac{1}{256}\mathrm{Tr}F^2\left[\frac{1}{45}\mathrm{tr}R^4 + \frac{1}{36}(\mathrm{tr}R^2)^2\right]$$

$$+ \frac{n-496}{64}\left[\frac{1}{2\cdot 2835}\mathrm{tr}R^6 + \frac{1}{4\cdot 1080}\mathrm{tr}R^2\mathrm{tr}R^4 + \frac{1}{8\cdot 1296}(\mathrm{tr}R^2)^3\right]$$

$$+ \frac{1}{384}\mathrm{tr}R^2\mathrm{tr}R^4 + \frac{1}{1536}(\mathrm{tr}R^2)^3. \qquad (13.5.4)$$

We have used the symbol Tr rather than tr for the gauge field traces to emphasize that the trace in question is a trace in the adjoint representation. The curvature matrices are ten-dimensional, of course, corresponding to the fundamental representation of $O(9,1)$ or $O(10)$. The actual anomalies are given by $G = \int I_{10}^1$, where I_{10}^1 is constructed from I_{12} by the steps described in §13.3.5.

The anomaly in (13.5.4) is certainly nonvanishing for every choice of n. However, it is possible for anomalous diagrams involving exchange of the B field to give a canceling anomaly contribution provided that I_{12} can be factorized in the form

$$I_{12} = (\mathrm{tr}R^2 + k\mathrm{Tr}F^2)X_8, \qquad (13.5.5)$$

where k is a constant and X_8 is a fourth-order polynomial in F and R. We have already discussed this matter in §13.3.4 above, and we will enter into more detail below. Let us first examine the conditions under which factorization occurs.

A necessary condition for (13.5.4) to factorize in the form (13.5.5) is for the coefficient of $\mathrm{tr}R^6$ to vanish. The point here is that $\mathrm{tr}R^6$ cannot be re-expressed as a combination of $\mathrm{tr}R^2\mathrm{tr}R^4$ and $(\mathrm{tr}R^2)^3$. The reason for this is that $SO(10)$ has an independent sixth-order Casimir invariant that contributes to $\mathrm{tr}R^6$ evaluated in the fundamental representation. So a necessary condition for the factorization of the anomaly (13.5.4) is that $n = 496$, *i.e.*, that the Yang–Mills group have 496 generators. This offers a possible answer, for the first time, to the question 'how many Yang–Mills symmetries are there in nature?' If $N = 1$ superstring theory is relevant, the analysis requires 484 additional ones beyond the 12 for which there is already experimental evidence. Putting $n = 496$ and multiplying by 48, (13.5.4) takes the form

$$I_{12} \propto -\tfrac{1}{15}\mathrm{Tr}F^6 + \tfrac{1}{24}\mathrm{Tr}F^4\mathrm{tr}R^2$$

$$-\tfrac{1}{960}\mathrm{Tr}F^2[4\mathrm{tr}R^4 + 5(\mathrm{tr}R^2)^2] \tag{13.5.6}$$

$$+\tfrac{1}{8}\mathrm{tr}R^2\mathrm{tr}R^4 + \tfrac{1}{32}(\mathrm{tr}R^2)^3.$$

This can factorize in the form (13.5.5) only if $\mathrm{Tr}F^6$ can be re-expressed as a linear combination of $\mathrm{Tr}F^2\mathrm{Tr}F^4$ and $(\mathrm{Tr}F^2)^3$. Assuming this is the case it is straightforward arithmetic to show that the unique possibility is for $k = -1/30$ and

$$\mathrm{Tr}F^6 = \tfrac{1}{48}\mathrm{Tr}F^2\mathrm{Tr}F^4 - \tfrac{1}{14,400}(\mathrm{Tr}F^2)^3. \tag{13.5.7}$$

Then one finds that

$$X_8 = \tfrac{1}{24}\mathrm{Tr}F^4 - \tfrac{1}{7200}(\mathrm{Tr}F^2)^2 - \tfrac{1}{240}\mathrm{Tr}F^2\mathrm{tr}R^2 + \tfrac{1}{8}\mathrm{tr}R^4 + \tfrac{1}{32}(\mathrm{tr}R^2)^2. \tag{13.5.8}$$

We must now ask for what 496-dimensional gauge groups is (13.5.7) satisfied. First we discuss the $SO(n)$ groups. We denote traces in the fundamental representation of $SO(n)$ by the symbol 'tr', while traces in the adjoint representation are denoted 'Tr'. It is possible to work out concrete formulas relating the two kinds of traces. Suppose we are given a generator F of $SO(n)$. In the fundamental representation, F is simply an antisymmetric $n \times n$ matrix F_{ac}. In the adjoint representation, F is represented by

$$F_{ab,cd} = \tfrac{1}{2}(F_{ac}\delta_{bd} - F_{bc}\delta_{ad} - F_{ad}\delta_{bc} + F_{bd}\delta_{ac}). \tag{13.5.9}$$

Using this formula one finds easily that

$$\mathrm{Tr}F^2 = (n - 2)\mathrm{tr}F^2 \tag{13.5.10}$$

$$\text{Tr}F^4 = (n-8)\text{tr}F^4 + 3(\text{tr}F^2)^2 \qquad (13.5.11)$$

$$\text{Tr}F^6 = (n-32)\text{tr}F^6 + 15\text{tr}F^2\text{tr}F^4 \qquad (13.5.12)$$

and so forth. It follows, for example, that in the case of $SO(32)$

$$\text{Tr}F^6 = 15\text{tr}F^2\text{tr}F^4. \qquad (13.5.13)$$

Here is a case where $\text{Tr}F^6$ does factorize even though the group has an independent sixth-order invariant. It just does not contribute to $\text{Tr}F^6$. It does contribute to $\text{tr}F^6$, which cannot be factorized, but that will not be relevant.

Among the $SO(n)$ groups we see from (13.5.12) that $SO(32)$ is the only one for which a sixth-order Casimir exists but $\text{Tr}F^6$ can still be decomposed.[*] In fact, using (13.5.10) and (13.5.11), as well, one finds that it satisfies (13.5.7) with exactly the right coefficients! Moreover, the dimension of $SO(32)$ is $31 \times 32/2 = 496$, as required. Thus this is one solution.

Remarkably, there is a second 496-dimensional group that satisfies (13.5.7). Since E_8 has 248 dimensions, a direct product of two of them, $E_8 \times E_8$, has the correct dimension. The group E_8 is known not to have independent fourth- or sixth-order Casimir invariants. This implies that $\text{Tr}F^4$ is proportional to $(\text{Tr}F^2)^2$ and $\text{Tr}F^6$ is proportional to $(\text{Tr}F^2)^3$. The problem is to work out the constants of proportionality. An elementary method of doing this is note that E_8 has an $SO(16)$ subgroup, with respect to which the adjoint can be decomposed according to $\mathbf{248} = \mathbf{120} + \mathbf{128}$, as explained in appendix 6.A. The $\mathbf{120}$ is the adjoint of $SO(16)$ and the $\mathbf{128}$ is a spinor. It is easy to calculate the various traces for specific $SO(16)$ matrices in both the $\mathbf{120}$ and the $\mathbf{128}$ representations. Using these results one can deduce the constants of proportionality for the sum, corresponding to the $\mathbf{248}$ of E_8. In this way one finds that

$$\text{Tr}F^4 = \tfrac{1}{100}(\text{Tr}F^2)^2$$
$$\text{Tr}F^6 = \tfrac{1}{7200}(\text{Tr}F^2)^3. \qquad (13.5.14)$$

Now what does this imply for the direct product $E_8 \times E_8$? In this case $\text{Tr}F^{2m}$ means $\text{Tr}F_1^{2m} + \text{Tr}F_2^{2m}$, where the subscripts 1 and 2 refer to the

---

[*] $\text{Tr}F^6$ can also be decomposed for $SO(n)$ for small n when there is no sixth-order Casimir so that $\text{tr}F^6$ can itself be expressed in terms of $\text{tr}F^2$ and $\text{tr}F^4$, but the number of generators is then much less than 496.

two E_8 factors. Thus to verify (13.5.7) we need that

$$\tfrac{1}{7200}[\mathrm{Tr}F_1^2)^3 + (\mathrm{Tr}F_2^2)^3]$$

$$=\tfrac{1}{48}(\mathrm{Tr}F_1^2 + \mathrm{Tr}F_2^2)\tfrac{1}{100}[(\mathrm{Tr}F_1^2)^2 + (\mathrm{Tr}F_2)^2] \qquad (13.5.15)$$

$$- \tfrac{1}{14,400}(\mathrm{Tr}F_1^2 + \mathrm{Tr}F_2^2)^3.$$

Remarkably, the cross terms cancel and the equation is satisfied.

The relation $\tfrac{1}{30}\mathrm{Tr}F^2 = \mathrm{tr}F^2$ is an identity for $SO(32)$ (see (13.5.10)). We find it useful to use this formula as the definition of the symbol tr for $E_8 \times E_8$. (In the latter case it coincides with the standard definition of 'tr' for generators in the $SO(16) \times SO(16)$ subgroup, which has a 32-dimensional fundamental representation.)

One may wonder whether there are other dimension 496 groups that satisfy (13.5.7). The answer is that there are two more. They are $[U(1)]^{496}$ and $E_8 \times [U(1)]^{248}$. These satisfy (13.5.7) rather trivially, since the traces vanish for the $U(1)$ factors. No string theories are known that correspond to either of these groups, and it appears extremely unlikely that any interesting theories can be based on either of them.

Let us now turn our attention to the origins of the factorization condition (13.5.5), which we now rewrite in the form $I_{12} = (\mathrm{tr}R^2 - \mathrm{tr}F^2)X_8$. We have already discussed the significance of the factorization requirement in §13.3.4 above, but in view of its fundamental importance we return here to consider this in a more formal way. In this more formal discussion we speak of the anomalies as formal twelve forms, as in §13.3.5.

The fact that I_{12} is an exact form, $I_{12} = dI_{11}$, implies that X_8 is exact, and hence closed. (Global considerations are not relevant here.) Now, we know that $\mathrm{tr}R^2 = d\omega_{3L}$ and $\mathrm{tr}F^2 = d\omega_{3Y}$, where ω_{3L} and ω_{3Y} are the Lorentz and Yang–Mills Chern–Simons forms given by (13.1.37) in the Yang–Mills case and an analogous expression obtained by replacing A by the spin connection ω in the Lorentz case. Therefore a possible choice for I_{11} is

$$I_{11}^{(a)} = (\omega_{3L} - \omega_{3Y})X_8. \qquad (13.5.16)$$

Under a local gauge transformation X_8 is invariant, but, according to (13.1.39), $\delta\omega_{3Y} = d\omega_{2Y}^1$. Similarly, under a local Lorentz transformation, $\delta\omega_{3L} = d\omega_{2L}^1$, where $\omega_{2L}^1 = \mathrm{tr}(\Theta d\omega)$. Another possible choice for I_{11} is

$$I_{11}^{(b)} = (\mathrm{tr}R^2 - \mathrm{tr}F^2)X_7 \qquad (13.5.17)$$

where $dX_7 = X_8$. Which choice is correct? Bose statistics requires treating each of the field strengths equally. (This gives the 'consistent' form of the anomaly that satisfies the Wess–Zumino consistency condition (13.3.28).) It means choosing the linear combination

$$I_{11} = \tfrac{1}{3}(\omega_{3L} - \omega_{3Y})X_8 + \tfrac{2}{3}(\mathrm{tr}R^2 - \mathrm{tr}F^2)X_7 + \alpha\, d[(\omega_{3L} - \omega_{3Y})X_7].\quad (13.5.18)$$

Here α is an arbitrary parameter; the last term is the ambiguity in the definition of I_{11}. We then define I_{10}^1 by $\delta I_{11} = dI_{10}^1$ and X_6^1 by $\delta X_7 = dX_6^1$. This implies that

$$I_{10}^1 = (\tfrac{2}{3} + \alpha)(\mathrm{tr}R^2 - \mathrm{tr}F^2)X_6^1 + (\tfrac{1}{3} - \alpha)(\omega_{2L}^1 - \omega_{2Y}^1)X_8,\quad (13.5.19)$$

modulo an exact form that does not matter for the anomaly $G = \int I_{10}^1$. A partial integration gives

$$G = (\tfrac{2}{3} + \alpha)\int(\omega_{3L} - \omega_{3Y})dX_6^1 + (\tfrac{1}{3} - \alpha)\int(\omega_{2L}^1 - \omega_{2Y}^1)X_8.\quad (13.5.20)$$

Along the lines of the discussion in §13.3.4, we now wish to construct local contributions to the effective action of massless fields that can contribute anomalies in tree diagrams.* As in our previous discussion, a crucial clue is provided by (13.1.41). The supersymmetry structure of the coupled supergravity plus super-Yang–Mills system was found to imply that the two form B has the Yang–Mills gauge transformation $\delta B = \omega_{2Y}^1$. What we need now is for this transformation law to be replaced by

$$\delta B = \omega_{2Y}^1 - \omega_{2L}^1,\quad (13.5.21)$$

because then the addition to the Lagrangian of the counterterm

$$\Delta\Gamma = \int BX_8 - (\tfrac{2}{3} + \alpha)\int(\omega_{3L} - \omega_{3Y})X_7\quad (13.5.22)$$

cancels the anomalies. This expression is unique up to gauge-invariant terms.

If the variation δB must contain ω_{2L}^1 as well as ω_{2Y}^1, why did we not find that term in §13.1.2? The reason is that ω_{2L}^1 is of higher-order in the low-energy expansion than ω_{2Y}^1 and therefore does not appear in the minimal $n = 2$ truncation of the theory. In passing from Yang–Mills

---

* The string theoretic interpretation of these new local contributions is that they arise from integrating massive modes out of one-loop diagrams. Thus, the anomalous tree diagrams of the low-energy theory are, as in chapter 10, derived at a more microscopic level as a loop contribution.

expressions to their Lorentz counterparts we replace the gauge field A by the spin connection ω. But the spin connection is linear in derivatives of the vielbein, so we obtain a term of higher order according to the rule in (13.1.1). This fact was manifested already in the minimal theory, where the kinetic terms R and F^2 were each counted as $n = 2$. The modified transformation formula (13.5.21) implies that the gauge-invariant three-form field strength in (13.1.48) must be modified to the form

$$H = dB + \omega_{3L} - \omega_{3Y}. \tag{13.5.23}$$

Let us point out an interesting consequence of (13.5.23). By taking the exterior derivative of (13.5.23) we learn that

$$dH = \operatorname{tr} R^2 - \operatorname{tr} F^2. \tag{13.5.24}$$

The three form H must be globally well defined, since $H_{MNP} H^{MNP}$ contributes to the energy density. It follows that for any closed four-dimensional submanifold M_4 of the ten-dimensional space-time M_{10}

$$\int_{M_4} dH = \int_{M_4} (\operatorname{tr} R^2 - \operatorname{tr} F^2) = 0. \tag{13.5.25}$$

In other words, the four form $\operatorname{tr} R^2 - \operatorname{tr} F^2$ must have trivial cohomology.

The choice of $\Delta\Gamma$ required for anomaly cancellation may seem *ad hoc*. The point, of course, is that exactly this form must necessarily arise in the effective action of the $SO(32)$ and $E_8 \times E_8$ superstring theories.

13.5.4 The $SO(16) \times SO(16)$ Theory

The $SO(16) \times SO(16)$ heterotic string theory was introduced in §9.5.3 and shown to satisfy one-loop modular invariance. Here we wish to examine the cancellation of gauge and gravitational anomalies in the corresponding low-energy effective action. This string theory, unlike the other tachyon-free ones, is not supersymmetric (or is perhaps to be interpreted as a theory in which supersymmetry is spontaneously broken).

In contrast to the other string theories, the $SO(16) \times SO(16)$ theory does not have finite one-loop amplitudes, since in the absence of super-symmetry dilaton tadpoles contribute infrared divergences. Therefore one cannot expect the anomaly cancellations to work as a consequence of the finiteness of the string-theory loop diagram, as one might in the other cases. This could be a cause for concern. However, as we demonstrated in

§9.3.4, it does satisfy one-loop modular invariance, the absence of global diffeomorphism anomalies on the world sheet. In the supersymmetric form of the heterotic string, modular invariance holds in precisely the cases ($SO(32)$ and $E_8 \times E_8$) in which hexagon anomalies cancel. This relationship suggests that the $SO(16) \times SO(16)$ theory should be anomaly-free. Let us now check it and find out.

The chiral fields of the $SO(16) \times SO(16)$ theory consist of Majorana–Weyl spinors only. In terms of physical polarizations described by the transverse group $SO(8)$ and the internal symmetry group $SO(16) \times SO(16)$, they belong to the representations

$$(\mathbf{8_s}; \mathbf{16}, \mathbf{16}), \quad (\mathbf{8_c}; \mathbf{128}, \mathbf{1}), \quad (\mathbf{8_c}; \mathbf{1}, \mathbf{128}). \tag{13.5.26}$$

Since these are the only chiral fields, the total anomaly is proportional to the 12-form part of

$$\hat{I}_{1/2}(F, R) = \mathrm{Tr}(\cos F)\hat{I}_{1/2}(R). \tag{13.5.27}$$

The expression $\mathrm{Tr}(\cos F)$ here means

$$\mathrm{tr}_{16 \times 16}(\cos F) - \mathrm{tr}_{128 \times 1}(\cos F) - \mathrm{tr}_{1 \times 128}(\cos F), \tag{13.5.28}$$

where the subscripts denote the representation of $SO(16) \times SO(16)$ in which the matrix of two forms F is to be expressed and the relative signs reflect the relative chirality of the space-time spinors.

The first thing to note about (13.5.28) is that when $\cos F$ is expanded as $1 - \frac{1}{2}F^2 + \ldots$, the term coming from the 1 vanishes because $16 \times 16 - 128 - 128 = 0$. This represents the fact the number of left-handed and right-handed spinors is equal and the pure gravitational anomaly cancels as a result. Unlike the $N = 1$ theories, which required 496 spinors to balance the gravitational anomaly due to the gravitino, cancellation of purely gravitational anomalies does not determine the total number of spinors, only that there should be an equal number of both chiralities. The theory is still chiral, of course, since the left-handed and right-handed spinors belong to different representations of the gauge group.

Expanding (13.5.27) using (13.4.7) and (13.4.8) gives the result

$$-\tfrac{1}{2}\mathrm{Tr}F^2[\frac{1}{180 \cdot 512}\mathrm{tr}R^4 + \frac{1}{72 \cdot 64}(\mathrm{tr}R^2)^2]$$
$$+ \frac{1}{24 \cdot 48}\mathrm{Tr}F^4\mathrm{tr}R^2 - \frac{1}{15 \cdot 48}\mathrm{Tr}F^6. \tag{13.5.29}$$

As in preceding section, let us attempt to factorize this into an expression of the form $(\mathrm{tr}R^2 + k\mathrm{tr}F^2)X_8$, where $\mathrm{tr}F^2$ is evaluated in some suitable representation. Since $\mathrm{tr}R^4$ cannot be related to $\mathrm{tr}R^2$ and there is no $\mathrm{tr}R^2\mathrm{tr}R^4$ term, it is necessary that $\mathrm{Tr}F^2 = 0$. Let us examine this first and then return to the rest of the formula.

Consider a single $SO(16)$ gauge group. In this case one can show by elementary calculations analogous to those discussed in the preceding section that

$$\mathrm{tr}_{128}F^6 = 16\mathrm{tr}F^6 + \frac{15}{4}(\mathrm{tr}F^2)^3 - 15\mathrm{tr}F^2\mathrm{tr}F^4$$

$$\mathrm{tr}_{128}F^4 = 6(\mathrm{tr}F^2)^2 - 8\mathrm{tr}F^4 \tag{13.5.30}$$

$$\mathrm{tr}_{128}F^2 = 16\mathrm{tr}F^2,$$

where the traces of the right-hand sides of these formulas are evaluated in the fundamental (16) representation. Using indices 1 and 2 to refer to the first and second $SO(16)$ factors,

$$\mathrm{tr}_{16\times16}F^2 = 16\mathrm{tr}F_1^2 + 16\mathrm{tr}F_2^2$$

$$\mathrm{tr}_{128\times1}F^2 = 16\mathrm{tr}F_1^2 \tag{13.5.31}$$

$$\mathrm{tr}_{1\times128}F^2 = 16\mathrm{tr}F_2^2.$$

It follows from this that

$$\mathrm{Tr}F^2 = \mathrm{tr}_{16\times16}F^2 - \mathrm{tr}_{128\times1}F^2 - \mathrm{tr}_{1\times128}F^2 = 0. \tag{13.5.32}$$

Analogous calculations give

$$\frac{1}{15}\mathrm{Tr}F^6 = \frac{1}{4}[(\mathrm{tr}F_1^2)^3 + (\mathrm{tr}F_2^2)^3] \\ - (\mathrm{tr}F_1^2 + \mathrm{tr}F_2^2)(\mathrm{tr}F_1^4 + \mathrm{tr}F_2^4) \tag{13.5.33}$$

and

$$\frac{1}{24}\mathrm{Tr}F^4 = \frac{1}{4}[(\mathrm{tr}F_1^2)^2 + (\mathrm{tr}F_2^2)^2 - \mathrm{tr}F_1^2\mathrm{tr}F_2^2] \\ - \mathrm{tr}F_1^4 - \mathrm{tr}F_2^4. \tag{13.5.34}$$

Inserting (13.5.32), (13.5.33), and (13.5.34) into (13.5.29), and multiplying by 48, gives the factorized result

$$(\mathrm{tr}R^2 - \mathrm{tr}F_1^2 - \mathrm{tr}F_2^2)X_8, \tag{13.5.35}$$

where $X_8 = \frac{1}{24}\mathrm{Tr}F^4$ is given in (13.5.34).

Since the $SO(16) \times SO(16)$ theory does contain the massless two form B_{MN} in its spectrum, we are now in a position to complete the anomaly cancellation argument by exactly the same reasoning as in the previous subsection. We take the two form to have Yang–Mills and Lorentz transformations given by

$$\delta B = \omega_{2Y}^1 - \omega_{2L}^1 \tag{13.5.36}$$

and construct the gauge-invariant field strength

$$H = dB + \omega_{3L} - \omega_{3Y}. \tag{13.5.37}$$

The traces that are used in the definitions of ω_{2Y}^1 and ω_{3Y} are now understood to be evaluated in the $(16,1) + (1,16)$ representation of $SO(16) \times SO(16)$. As before, X_7 is defined by $X_8 = dX_7$, and the counterterm in the effective action is given as in (13.5.22) by

$$\Delta\Gamma = \int BX_8 - (\tfrac{2}{3} + \alpha) \int (\omega_{3L} - \omega_{3Y})X_7. \tag{13.5.38}$$

In §9.3 we described techniques for constructing quite a few modular-invariant ten-dimensional string theories with tachyons. By computations similar to those that we have just described, those theories all turn out to be free of hexagon anomalies.

14. Compactification Of Higher Dimensions

Since superstring theories are necessarily ten-dimensional theories, any discussion of phenomenology must begin with a discussion of how apparent four-dimensional physics is related to underlying ten-dimensional physics. The present chapter is devoted to this question. We will carry out the discussion in the context of field theory, but with an emphasis on properties that depend only on qualitative assumptions, not numerical details, and so can remain valid in string theory. What we will try to accomplish in this chapter is not to develop detailed models of compactification but to set the stage and introduce some of the essential concepts.

14.1 Wave Operators in Ten Dimensions

Most of the preceding chapters have been devoted to string propagation in ten-dimensional flat Minkowski space M^{10}, but henceforth we will consider ten-dimensional space-time to be some more general ten manifold **M**. We take **M** to be of the form $M^4 \times K$, where M^4 is four-dimensional Minkowski space and K is a compact six manifold which is, unfortunately, as yet unknown. More precisely, we take the *vacuum* state to be a product $M^4 \times K$; it must have this form if we wish to maintain four-dimensional Poincaré invariance. Of course, physical fluctuations will not necessarily respect the product form of the vacuum configuration, but as in so many other areas of physics, understanding the ground state is the key to understanding the low-energy excitations. In fact, we will see that a vast assortment of physical questions amount to questions about the topology and geometry of K.

14.1.1 Massless Fields in Ten Dimensions

We will denote ten-dimensional indices as $M, N, P = 1, \ldots, 10$; the ten space-time coordinates will be called X^M, $M = 1, \ldots, 10$. Indices tangent to M^4 will be $\mu, \nu, \lambda = 1, \ldots, 4$; the coordinates of M^4 will be named x^μ, $\mu = 1, \ldots, 4$. Indices tangent to K will be labeled $i, j, k = 5, \ldots, 10$; we will denote the coordinates of K as y^i, $i = 5, \ldots, 10$.

In differential geometry, given an n-dimensional Riemannian manifold Q, the tangent vectors at any point $p \in Q$ make up an n-dimensional vector space T_p, the tangent space at p. The orthogonal transformations of T_p form a group, the tangent space group, isomorphic (if Q has positive signature) to $SO(n)$. The tangent space groups of $M^4 \times K, M^4$ and K are respectively $SO(1,9)$, $SO(1,3)$ and $SO(6)$. Of course $SO(1,3) \times SO(6)$ is a subgroup of $SO(1,9)$, with $SO(1,3)$ acting on the first four components and $SO(6)$ acting on the last six components of the ten vector of $SO(1,9)$. The tangent space group of a manifold Q is in no way the same thing as the symmetry group of Q, which may be larger or smaller. Q may have no symmetry at all, but its tangent space group depends only on the signature and dimension. The tangent-space group is a symmetry of measurements carried out in a small space-time region, and as such enters in Einstein's 'principle of equivalence'.

The 'spin' of a physical field Ψ in n-dimensional space-time is specified by choosing a representation X of the tangent space group. Given any point p, the fields at p transform in the X representation of the tangent space group. This means that in ten-dimensional physics, it is possible to introduce a field whose components (at any given point p) transform in any desired representation of $SO(1,9)$. The first step in relating ten-dimensional physics to four-dimensional physics is to realize that a field with a simple spin content in ten dimensions may have a rather complicated spin content as seen in four dimensions, since a simple representation of $SO(1,9)$ may be rather complicated and highly reducible as a representation of $SO(1,3) \times SO(6)$. In practice, there are five cases to consider:

(i) *Gauge Fields*

A gauge field A_M transforms in the vector representation of $SO(1,9)$, and we have already noted that this decomposes under $SO(1,3) \times SO(6)$ as $(\mathbf{4},\mathbf{1}) \oplus (\mathbf{1},\mathbf{6})$. This amounts to saying that the gauge field A_M, $M = 1, \ldots, 10$ splits in the four-dimensional sense into a vector field A_μ, $\mu = 1, \ldots, 4$, and scalars A_i, $i = 5, \ldots, 10$. Assuming that A_M was a gauge field for some gauge group G in ten dimensions, and so transformed in the adjoint representation of G, both A_μ and A_i transform in this representation. This means that the ten-dimensional theory has the potential to unify what in four-dimensional terms are seen as a gauge field A_μ and Higgs fields A_i in the adjoint representation. Actually, here (and in the other examples discussed momentarily) we are only taking account of the spin; when we allow for the y dependence of, for instance, $A_\mu(x^\nu, y^j)$, we will encounter an infinite number of four-dimensional fields.

(ii) *The Metric Tensor*

The decomposition of the metric tensor g_{MN}, $M, N = 1, \ldots, 10$ is similar. The components $g_{\mu\nu}$, $\mu, \nu = 1, \ldots, 4$ make up the four-dimensional metric tensor, while $g_{\mu k}$, $\mu = 1, \ldots, 4$, $k = 5, \ldots, 10$ are spin one fields as seen in four dimensions, and g_{mn}, $m, n = 5, \ldots, 10$ are observed in four dimensions as spin zero fields.

(iii) *Differential Forms*

The other massless Bose fields that arise from superstring theories are antisymmetric tensor fields $\Phi_{M_1 M_2 \ldots M_k}$ which are antisymmetric in all their indices. Such a field, as we have discussed in chapter 12, is called a k form in the mathematical literature, or a differential form of degree k. A gauge field can be viewed as a one form (with values in some Lie algebra), but we discussed this case separately because of its special importance. The decomposition of a k form under $SO(1,3) \times SO(6)$ is very simple; the components with q indices in the range $1, \ldots, 4$ and $k - q$ indices in the range $5, \ldots, 10$ transform as a q form under $SO(1,3)$ and a $k - q$ form under $SO(6)$.

(iv) *Dirac Fields*

The spinor representation of $SO(1,9)$ is constructed by introducing ten gamma matrices Γ_M, $M = 1, \ldots, 10$ which obey $\Gamma_M \Gamma_N + \Gamma_N \Gamma_M = 2\eta_{MN}$. The $SO(1,9)$ generators are then $\Sigma_{MN} = [\Gamma_M, \Gamma_N]/4$. To construct a spinor of $SO(1,3)$ would require only four gamma matrices γ_μ, $\mu = 1, \ldots, 4$. We can regard the first four Γ_M as gamma matrices of $SO(1,3)$. Then the $SO(1,9)$ definition of the generators Σ_{MN} for $M, N = 1, \ldots, 4$ coincides with the standard $SO(1,3)$ formula $\sigma_{\mu\nu} = [\gamma_\mu, \gamma_\nu]/4 = \gamma_{\mu\nu}/2$, so a spinor field of $SO(1,9)$ transforms as a spinor of $SO(1,3)$. The same argument (focusing on the last six Γ_M only) shows that a spinor of $SO(1,9)$ transforms as a spinor of $SO(6)$. The above remarks should really be supplemented by a discussion of chirality; we turn to that important question later.

(v) *Rarita–Schwinger Fields*

A Rarita–Schwinger field $\psi_{M\alpha}$ has an $SO(1,9)$ vector index M and spinor index α. We have already discussed the $SO(1,3) \times SO(6)$ decomposition of the vector and spinor indices separately, and it is easy to combine the two. The components with $M = 1, \ldots, 4$ transform as a vector-spinor (or Rarita–Schwinger field) of $SO(1,3)$ and a spinor of $SO(6)$, and the components with $M = 5, \ldots, 10$ transform as a spinor of $SO(1,3)$ and a vector-spinor of $SO(6)$.

14.1.2 Zero Modes of Wave Operators

After decomposing ten-dimensional fields according to their spin content in four dimensions, the next step is to determine the masses of the fields as seen in four dimensions. As a representative case, suppose we wish to consider a Dirac field Ψ which in ten dimensions obeys

$$0 = i\slashed{D}_{10}\Psi = i\sum_{M=1}^{10} \Gamma^M D_M \Psi, \tag{14.1.1}$$

where $\slashed{D}_{10}$ is the ten-dimensional Dirac operator. This equation can be rewritten

$$0 = i(\slashed{D}_4 + \slashed{D}_K)\Psi, \tag{14.1.2}$$

where we have introduced the four-dimensional and internal Dirac operators

$$\slashed{D}_4 = \sum_{\mu=1}^{4} \Gamma^\mu D_\mu, \quad \slashed{D}_K = \sum_{p=5}^{10} \Gamma^p D_p, \tag{14.1.3}$$

which obey the obvious identity

$$\slashed{D}_{10} = \slashed{D}_4 + \slashed{D}_K. \tag{14.1.4}$$

Equation (14.1.2) makes it immediately apparent that $\slashed{D}_K$ is a kind of 'mass' operator whose eigenvalues are fermion masses as seen in four dimensions.

In view of (14.1.2), one's first thought might be to solve the ten-dimensional Dirac equation by separation of variables in terms of simultaneous eigenstates of $\slashed{D}_4$ and $\slashed{D}_K$. This does not work because $\slashed{D}_4$ and $\slashed{D}_K$ do not commute but rather anticommute, so they cannot be simultaneously diagonalized (except in the sector consisting of zero modes of $\slashed{D}_K$). The proper procedure is to introduce the four-dimensional chirality operator $\Gamma^{(4)} = i\Gamma_1\Gamma_2\Gamma_3\Gamma_4$ and rewrite (14.1.2) in the obviously equivalent form

$$0 = i(\tilde{\slashed{D}}_4 + \tilde{\slashed{D}}_K)\Psi \tag{14.1.5}$$

with $\tilde{\slashed{D}}_4 = \Gamma^{(4)}D_4$, $\tilde{\slashed{D}}_K = \Gamma^{(4)}D_K$. Now $\tilde{\slashed{D}}_4$ and $\tilde{\slashed{D}}_K$ commute with each other and can be simultaneously diagonalized. Introducing a complete set of normalized solutions $\phi_i(y^k)$ of the eigenvalue problem

$$i\tilde{\slashed{D}}_K\phi_i(y^k) = \lambda_i\phi_i(y^k), \tag{14.1.6}$$

the Dirac equation (14.1.5) can be solved by writing

$$\Psi(x^\mu, y^k) = \sum_i \psi_i(x^\mu)\phi_i(y^k), \tag{14.1.7}$$

where now the ψ_i must obey

$$0 = (i\tilde{\not{D}}_4 + \lambda_i)\psi_i(x^\mu). \tag{14.1.8}$$

Thus, each ψ_i is observed in four dimensions as a fermion of mass λ_i. It is important to note that in these equations $\tilde{\not{D}}_4$ and $\tilde{\not{D}}_K$ are equivalent to the standard Dirac operators $\not{D}_4$ and $\not{D}_K$ since $\tilde{\Gamma}_\mu = i\Gamma^{(4)}\Gamma_\mu$, $\mu = 1,\ldots,4$ and $\tilde{\Gamma}_k = \Gamma^{(4)}\Gamma_k$, $k = 5,\ldots,10$ obey, respectively, the proper anticommutation relations of gamma matrices of M^4 and K.

With or without the refinement that leads to the separation of variables in (14.1.7), the main point here is that it is zero modes of $\not{D}_K$ that correspond to massless fermions in four dimensions. Nonzero eigenvalues of the Dirac equation on K, on the other hand, will presumably be of order $1/R$, R being the radius of the compact space K. Presumably $1/R$ is of order the Planck mass, 10^{19} GeV; if so, these particles are scarcely observable. The known fermions must all be particles which in the approximation under discussion here are massless; they must correspond to zero modes of $\not{D}_K$. Much of the following discussion is therefore devoted to understanding whether and why zero modes of $\not{D}_K$ exist.

Comments similar to those made above for Dirac fields can be made about Rarita–Schwinger fields and about the various Bose fields of relevance. In each case the masses, as seen in four dimensions, are determined by the eigenvalues of a suitable wave operator Ω_K on K; its zero eigenvalues are manifested in four dimensions in the existence of massless particles. A separation of variables similar to that in (14.1.7) can be carried out in each case without too much difficulty for the massless states and with somewhat more difficulty for the massive ones. The zero eigenvalues of $\not{D}_K$ and its analogs for particles of other spin play an absolutely central role in all phenomenological discussions, so they will be our main interest in the rest of this chapter. In what follows we will consider, in turn, the zero modes of fermion operators, the zero modes of differential forms, and the zero modes of disturbances in gauge and gravitational fields. We will also develop many mathematical tools that are important in understanding the phenomenology of compactification. Toward the end of this chapter we will apply what we have learned to formulate what may prove to be a realistic picture of the fermion quantum numbers in nature.

14.2 Massless Fermions

Now that we have defined the Dirac operator, we would like to discuss its zero modes – which, by virtue of the introductory remarks at the beginning of this chapter, are related to massless fermions in four dimensions.

Before entering into the mathematics of the fermion zero modes, let us first briefly review the experimental facts which we would like to describe. Several generations of spin one-half quarks and leptons are observed in nature. We know them mainly through their gauge interactions, which are specified by a choice of representations V_L and V_R for fermions of left- or right-handed helicity, respectively. One of the most important facts about the gauge interactions is that the representations V_L and V_R are different. For instance, in terms of an assumed $SU(5)$ unification group, V_L (for one generation) consists of $\bar{5} + 10$ while V_R consists of $5 + \overline{10}$. Since the CPT theorem implies that V_R is isomorphic to the complex-conjugate representation of V_L, the statement that V_L is different from V_R is equivalent to the statement that V_L furnishes a complex representation of the gauge group (a complex representation being simply one that is not isomorphic to its complex conjugate).

The observed fermions are extremely light, of course, compared to the mass scale of grand unification or gravity; otherwise they would not be observed. Why is this so? As long as the gauge symmetries are conserved, a left-handed fermion can gain a mass only by pairing up with a right-handed fermion of the same gauge quantum numbers (since a massive spin one-half particle has two helicity states, which must have the same gauge charges). The fact that the left- and right-handed fermions transform differently under the gauge group means that they must remain massless as long as the gauge group is unbroken. What seems to happen in nature is that the observed quarks and leptons remain massless down to a mass scale of a few hundred GeV at which the electroweak gauge group is broken down to a subgroup – electromagnetism. At that point, the fermions are in a *real* representation of the remaining gauge symmetries, and they can and do get masses, except possibly for the neutrinos. We do not understand why the scale of weak interaction symmetry breaking is so tiny compared to the Planck mass; this is the hierarchy problem. But we do at least understand the lightness of the quarks and leptons in terms of the lightness of the W and Z; it follows from the chiral asymmetry that we have been discussing between left- and right-handed fermions.

There may be additional fermions in nature that form a real representation of the gauge group. (Such particles certainly exist if superstring theory is correct.) If so, very large gauge-invariant masses are possible for these fermions, and such masses have presumably arisen, since we have not observed these particles. There is every indication that the chiral asymmetry is the only reason that we observe any fermions at all, the fermions we observe being precisely those that are light because of a chiral asymmetry. Our goal in studying the zero modes of the Dirac operator in ten dimensions will be to understand the origin of the chiral asymmetry in four dimensions.

14.2.1 The Index of the Dirac Operator

We turn now to the study of the ten-dimensional Dirac operator. As we have learned in appendix 5.A, the spinor representation of $SO(10)$ or $SO(1,9)$ is of dimension $2^5 = 32$. Moreover, with Lorentzian signature $\{-+++++++++\}$, the Dirac matrices Γ_M, which obey $\{\Gamma_M, \Gamma_N\} = 2g_{MN}$, can be chosen to be real, 32×32 matrices. Of fundamental importance are the chirality operators

$$
\begin{aligned}
\Gamma^{(10)} &= \Gamma_1 \Gamma_2 \ldots \Gamma_{10}, \\
\Gamma^{(4)} &= i\Gamma_1 \Gamma_2 \Gamma_3 \Gamma_4, \\
\Gamma^{(K)} &= -i\Gamma_5 \Gamma_6 \ldots \Gamma_K.
\end{aligned}
\tag{14.2.1}
$$

These three operators will be referred to as the ten-dimensional, four-dimensional, and internal chirality operators, respectively. The factors of $\pm i$ have been included to ensure that

$$(\Gamma^{(10)})^2 = (\Gamma^{(4)})^2 = (\Gamma^{(K)})^2 = 1 \tag{14.2.2}$$

and also that

$$\Gamma^{(10)} = \Gamma^{(4)} \Gamma^{(K)}. \tag{14.2.3}$$

Both of these equations will be of considerable importance.

Now, as we learned in chapter 4 (in studying the GSO projection) and in chapter 13 (in studying ten-dimensional supergravity), massless fermions in ten dimensions obey a chirality condition, which we will here take to be $\Gamma^{(10)} = +1$. That this condition can be imposed depends, among other things, on the reality of the gamma matrices, which ensures that $\Gamma^{(10)}$ is real and that its eigenstates can be CPT eigenstates. Imposing this ten-dimensional chirality condition means in terms of reduction on $M^4 \times K$ that

$$\Gamma^{(4)} = \Gamma^{(K)}. \tag{14.2.4}$$

Equation (14.2.4) establishes a correlation between four-dimensional and internal chirality.

Now we must turn to our basic question: Why would the Dirac operator on K have zero eigenvalues? In fact, the question of topological conditions for zero eigenvalues of Dirac operators is a vast and fascinating one.

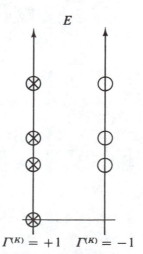

Figure 14.1. The H eigenvalues of nonzero energy are paired. Such pairing need not hold for the zero eigenvalues. In the figure, modes of positive or negative chirality are indicated by $\otimes$ or $\bigcirc$, respectively; the vertical scale is the H eigenvalue of E.

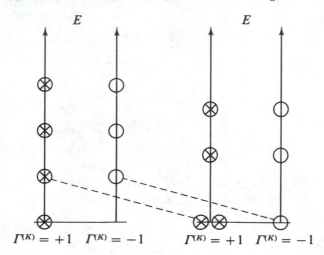

Figure 14.2. A continuous deformation of the spectrum of H that preserves the pairing at nonzero energy can change n_+ and n_-, but cannot change the index, which is the difference $\eta_+ - \eta_-$. Sketched is a process in which a pair of states originally at nonzero energy migrate down to zero energy, changing both η_+ and η_- by $+1$.

We will have use in this book for only the simplest topological invariant related to zero eigenvalues of Dirac operators; this is the notion of the Dirac index. It is this notion that turns out to be directly related to the chiral asymmetry in four dimensions. To describe this notion, define a 'Hamiltonian'

$$H = (i \slashed{D}_K)^2. \tag{14.2.5}$$

Since $[H, \Gamma^{(K)}] = 0$, H eigenstates can be chosen to be at the same time $\Gamma^{(K)}$ eigenstates. If $H\psi = E\psi$, then $H \cdot i\slashed{D}_K \psi = E \cdot i\slashed{D}_K \psi$, so ψ and $i\slashed{D}_K \psi$ are always degenerate in energy. Since $\slashed{D}_K \Gamma^{(K)} = -\Gamma^{(K)} \slashed{D}_K$, ψ and $i\slashed{D}_K \psi$ have opposite eigenvalues of $\Gamma^{(K)}$ and are linearly independent unless $i\slashed{D}_K \psi = 0$. Consequently (fig. 14.1), the H eigenvalues of nonzero energy are paired. For every state of $\Gamma^{(K)} = +1$ there is a state of $\Gamma^{(K)} = -1$, and vice versa. However, states with zero eigenvalues need not be paired in this way. Let us denote the number of zero eigenvalues of $i\slashed{D}_K$ with $\Gamma^{(K)} = \pm 1$ as $n_\pm$. Then it is customary to define the index of the operator $i\slashed{D}_K$ as the difference $n_+ - n_-$. We will denote it as index $\slashed{D}_K$.

The index is an important concept because it is invariant under arbitrary smooth deformations of $\slashed{D}_K$ that preserve its hermiticity and the basic relation $\Gamma^{(K)} \slashed{D}_K = -\slashed{D}_K \Gamma^{(K)}$. The reason for this is that no smooth deformation of fig. 14.1 that preserves the pairing at nonzero E can disturb whatever chirality imbalance may exist at $E = 0$. Under a smooth deformation of the spectrum which preserves the pairing at nonzero E, n_+ and n_- can change, but they necessarily change by the same amount, as in fig. 14.2, leaving the index invariant. The invariance of the index under smooth deformations of the spectrum means, in particular, that it is a topological invariant.

We have so far not mentioned gauge-field expectation values, and in fact little in the discussion until this point depends on whether they are present or not. If, however, we now specialize to the case in which gauge field expectation values are zero, it is easy to prove that the index is zero on a six-dimensional compact manifold K. (The same argument would work in $4k + 2$ dimensions for any k, and indeed we will also be interested in the case of two dimensions, $k = 0$, corresponding to the study of wave operators on the string world sheet.) The point is that since the gamma matrices are real, the equation for zero modes of $\slashed{D}_K$ is invariant under complex conjugation. However, $\Gamma^{(K)}$ is purely imaginary, and changes sign under complex conjugation. So if ψ is a zero mode of $\slashed{D}_K$ with one chirality, its complex conjugate ψ^* is a zero mode with the opposite chirality. Complex conjugation therefore gives a pairing between zero modes of opposite chirality, ensuring that $n_+ = n_-$ and that the index is zero. There is a simple 'physical' explanation for this result in four-dimensional terms, which we will come to later.

14.2.2 Incorporation of Gauge Fields

So far we have not considered gauge fields, but their incorporation is essential. Consider gauge fields of a gauge group J, with spinors in some representation Q. The index of the Dirac operator $\not{D}_K$ can be defined in the same way and is still a topological invariant, except now this means that the index depends on the topology of K and on whatever topological choices were made in defining the gauge-field expectation value. We can still consider the operation of complex conjugation of the Dirac wave functions. Complex conjugation still reverses the chirality, but now it exchanges the Q representation with its complex conjugate Q^*. Consequently, the number of positive- (or negative-) chirality zero modes in the Q representation is the same as the number of negative- (or positive-) chirality zero modes in the Q^* representation. This means that the Dirac index in the Q representation is the negative of the index in the Q^* representation:

$$\mathrm{index}_Q \not{D}_K = -\mathrm{index}_{Q^*} \not{D}_K. \tag{14.2.6}$$

In particular, if Q is a real or pseudoreal representation, so that Q and Q^* are equivalent, the index is zero, in $4k+2$ dimensions. But if Q is a complex representation, there is absolutely no reason for $\mathrm{index}_Q \not{D}_K$ to be zero. In fact, the Atiyah–Singer index theorem gives a formula for the index. In two dimensions (the case which is relevant for studying wave operators on a string world sheet Σ) the formula is

$$\mathrm{index}_Q \not{D}_K = \frac{1}{2\pi} \int_\Sigma \mathrm{tr}_Q F. \tag{14.2.7}$$

Here F is the Yang–Mills field strength regarded as a two form, and tr_Q refers to a trace in the Q representation of J. If the group J is semisimple, then $\mathrm{tr}_Q F = 0$, and the index vanishes. In general, (14.2.7) is a topological invariant by virtue of arguments in the last section of chapter 12. In six dimensions, which is the case of most interest in the study of compactification, the index formula is

$$\mathrm{index}_Q \not{D}_K = \frac{1}{48(2\pi)^3} \int_K [\mathrm{tr}_Q F \wedge F \wedge F - \frac{1}{8}\mathrm{tr}_Q F \wedge \mathrm{tr} R \wedge R]. \tag{14.2.8}$$

Here R is the Riemann tensor regarded (as in the concluding remarks in chapter 12) as a two form with an extra Lie algebra index. The integrand on the right-hand side of (14.2.8), being the wedge product of three two forms, is a six form, which can be integrated over the six manifold K to give a number. This number is, again, a topological invariant by virtue of the arguments presented in the last section of chapter 12. The second term in (14.2.8) drops out if the gauge group J is semisimple, and both terms vanish if the representation Q is real, since the symmetrized trace of an odd number of generators is zero in such a representation.

We will not give a proof of (14.2.8) in this book, but we will pause here to explain qualitatively how such a formula arises. To this end, an alternative way of formulating the concept of the index is useful. It can be written

$$\text{index}_Q \, D\!\!\!/_K = \text{tr} \left(\Gamma^{(K)} \exp -\beta H \right), \qquad (14.2.9)$$

where β is an arbitrary positive number. The idea in (14.2.9) is that states of nonzero energy occur in pairs with a state of $\Gamma^{(K)} = -1$ for every state of $\Gamma^{(K)} = +1$, and consequently they cancel out of the trace in (14.2.9). The trace receives contributions from the zero-energy states only, and the contribution of these states is precisely our previous definition of the index, $n_+ - n_-$. The trace is of course to be taken only for states in the Q representation. One of the reasons that (14.2.9) is useful is that if β is taken to be very small, (14.2.9) can be evaluated by means of the high-temperature expansion of quantum statistical mechanics. Indeed, let x and y be two points on the manifold K, and let

$$G(x, y; \beta) = \langle x | e^{-\beta H} | y \rangle. \qquad (14.2.10)$$

Then from (14.2.9),

$$\text{index}_Q \, D\!\!\!/_K = \int_K \text{tr} \left(G(x, x; \beta) \Gamma^{(K)} \right). \qquad (14.2.11)$$

The small β behavior of $G(x, x; \beta)$ can be computed by various techniques familiar to physicists. For example, one can write a Feynman path integral to express $G(x, x; \beta)$ in terms of paths that propagate from x to x in imaginary time β. For small β such a path integral is dominated by paths that do not stray far from x. This leads to an expansion of $G(x, x; \beta)$ in local functionals of the metric and the Yang–Mills field, multiplied by powers of β. Inserting this in (14.2.11), $\text{index}_Q \, D\!\!\!/_K$ will inevitably then be given as $\int_K O$, with O being a local functional of the metric and Yang–Mills fields. But the index is known to be a topological invariant. The only topological invariants which can be written as integrals of local functionals are the characteristic numbers, the integrals of polynomials in the Yang–Mills field strength and the Riemann curvature tensor. Consequently, the Dirac index must be given by such an expression. Working out the small β expansion of (14.2.11) to get the precise formula (it turns out that a supersymmetric path integral is the most efficient tool in the calculation), one arrives at (14.2.7) and (14.2.8) in two or six dimensions.

14.2.3 The Chiral Asymmetry

Returning to our physical problem, at first sight one might think that the gauge group J in the discussion of (14.2.8) should be taken as the underlying gauge group of the ten-dimensional theory. If so we would hardly get an interesting answer, since the CPT theorem requires that in ten dimensions fermions of given chirality should be in a real representation of the gauge group; supersymmetry in fact requires that this should be the adjoint representation. But (14.2.8) is nontrivial only in the case of a complex representation.

The proper interpretation is somewhat different. We begin in ten dimensions with a unified gauge group G. We turn on expectation values of gauge fields in some subgroup J of G. To preserve Lorentz invariance in the uncompactified dimensions, only the components of the gauge field A_M with $M = 5, \ldots, 10$ have expectation values. As we have discussed in the last section, these appear in four-dimensional terms as Higgs bosons. Their expectation value will break G to the subgroup that commutes with J; let us call this H. H is observed in four dimensions as the unbroken gauge group (or at least, as the group that is left unbroken at the compactification scale). Ten-dimensional fermions are in the adjoint representation A of G. Under $H \otimes J$ this has a decomposition of the form

$$A \approx \oplus_i L_i \otimes Q_i, \qquad (14.2.12)$$

where L_i and Q_i are representations of H and J respectively.

A four-dimensional physicist is not interested in how fermions transform under J but in how they transform under the unbroken group H. There may be a correlation between the two, however, since the J representation Q_i may depend on the H representation L_i. Four-dimensional massless fermions in the L_i representation of H arise from zero modes of the Dirac operator in the Q_i representation of J. A four-dimensional physicist is also not directly interested in the eigenvalue of $\Gamma^{(K)}$, but is quite interested in the eigenvalue of $\Gamma^{(4)}$, which is measured as ordinary fermion helicity. Happily, the two are equal because of equation (14.2.4).

For reasons that we have discussed, we are interested largely in the chiral asymmetry of the massless fermions. (By massless fermions we really mean fermions that only get their mass from low-energy gauge symmetry breaking.) Thus let $n^+_{L_i}$ and $n^-_{L_i}$ be the number of positive- and negative-chirality multiplets in the L_i representation of H. The chiral asymmetry is defined as $N_{L_i} = n_{L_i}^+ - n_{L_i}^-$. This asymmetry is important, because it is the existence of a nonvanishing chiral asymmetry that prevents unpaired fermions from having gauge-invariant bare masses. If $N_{L_i} = 0$, so that there are equal numbers of left- and right-handed fermions of given quantum numbers, then gauge invariance would not prevent the fermions from pairing up and gaining masses. In fact, experiment suggests that for each i, either $n_{L_i}^+$ or $n_{L_i}^-$ is zero, so that fermions in nature apparently *have* paired up and gained masses to the extent that this was permitted by the existence of a nonzero chiral asymmetry. In the case where L denotes a standard generation of quarks and leptons, the chiral asymmetry is often referred to as the number of generations N_{gen}.

We are now, finally, in a position to express the chiral asymmetry in four dimensions in terms of topological invariants of K. Fermions that transform as L_i under H transform as Q_i under J, and fermions of $\Gamma^4 = \pm 1$ have $\Gamma^K = \pm 1$, so the relation is simply

$$N_{L_i} = \text{index}_{Q_i} \not{D}_K. \qquad (14.2.13)$$

This relation is of far-reaching importance in phenomenological applications of superstrings. After developing some more mathematical machinery in the course of our discussion below of the zero modes of Bose fields, we will return toward the end of this chapter to evaluate (14.2.13) in a very interesting context.

14.2.4 The Rarita–Schwinger Operator

One can carry out a rather similar discussion for the Rarita–Schwinger operator. Its index can be defined in the same way as the Dirac index, and it is a topological invariant for the same reasons. As in the Dirac case, complex conjugation can be used to prove that the Rarita–Schwinger index is zero in $4k + 2$ dimensions for Rarita–Schwinger fields that do not have gauge couplings. Interesting ten-dimensional theories do not seem to have charged Rarita–Schwinger fields, so a nontrivial index does not emerge. Zero modes of Rarita–Schwinger fields can be expected only under more detailed assumptions such as low-energy supersymmetry, which is the subject of chapter 16.

14.2.5 Outlook

In the above discussion, we have found a general formula (14.2.13) only for the asymmetry between the quantum numbers of positive- and negative-chirality fermions. This, however, is enough to predict the massless fermions whose masslessness is determined by the unbroken gauge symmetries. As we have already noted, all of the known light fermions are of this kind. Massless fermions not predicted by (14.2.13) would be, for instance, family–antifamily pairs that could have had gauge-invariant bare masses but do not.

The reason that formula (14.2.13) for the chiral asymmetry is not just a field theoretic answer but is expected to be valid in the string theory context is that, as we have tried to stress, the chiral asymmetry which it governs is invariant under any smooth deformation of the theory. In the language of statistical physics, the chiral asymmetry depends only on the 'universality class' of a theory with gauge group H in four dimensions. It is for this reason that, along with general covariance and nonabelian gauge invariance, the chiral asymmetry observed in nature is probably one of the best clues that experiment has provided us about physics at the shortest distances.

It is conceivable, of course, that the string coupling constant is so strong that it pushes the vacuum state into a 'universality class' that cannot be seen from the field-theory viewpoint. If so, there may be very little phenomenology that can be extracted from our present knowledge of string theory. If, however, nature is willing to let us gain some understanding of phenomenology from our present knowledge of string theory, the chiral asymmetry is almost certainly one of the things that we can hope to understand, precisely because it only depends upon the universality class of the theory.

Finally, let us note here the 'physical' content of equation (14.2.6) which asserts that $\text{index}_Q = -\text{index}_{Q^*}$. In view of (14.2.13), this implies that the chiral asymmetry of four-dimensional fermions in a representation L of the unbroken group H is the negative of the chiral asymmetry in the complex conjugate representation L^*. But that was to be expected; it is a consequence of CPT invariance, since fermions in the Q representation have antiparticles of opposite chirality in the Q^* representation.

14.3   Zero Modes of Antisymmetric Tensor Fields

In this section we turn our attention to the zero modes of antisymmetric tensor fields, known to mathematicians as differential forms. We have already worked out most of the necessary mathematical preliminaries in chapter 12.

14.3.1 Antisymmetric Tensor Fields

Let $B_{i_1 i_2 \ldots i_p}$ be a p form, or pth rank antisymmetric tensor field. For such a field to be present in a physically sensible theory, there must be a gauge invariance to decouple the timelike modes. The requisite gauge invariance is simply

$$\delta B = d\Lambda, \tag{14.3.1}$$

with Λ being a $p-1$ form, and d being the exterior derivative that we discussed in chapter 12. Under certain conditions, as we learned in our study of anomaly cancellation in chapter 13, B must be taken to transform nontrivially under Yang–Mills gauge transformations and general coordinate transformations. For our present problem of identifying the massless states that arise upon compactification to four dimensions, consideration of that issue can be postponed.

As we learned in chapter 12, the gauge-invariant field strength associated with (14.3.1) is simply the $p+1$ form

$$C = dB, \tag{14.3.2}$$

and the analog of the Maxwell action is

$$S = \frac{p+1}{2p!} \int_M g^{i_1 j_1} \ldots g^{i_{p+1} j_{p+1}} C_{i_1 \ldots i_{p+1}} C_{j_1 \ldots j_{p+1}} = \frac{\langle C, C \rangle}{2(p!)^2}. \tag{14.3.3}$$

Here g is the metric tensor of the manifold M and the last part of the equation defines the inner product $\langle \, , \, \rangle$ for $p+1$ forms. In §12.4 we introduced the Hodge–de Rham Laplacian

$$\Delta = dd^* + d^*d \tag{14.3.4}$$

and studied its zero modes. As we discussed, the number of such zero modes (for p forms) is called the Betti number b_p. As we learned in chapter 12, a zero mode of Δ on a compact manifold K must have zero field strength,

$$dB = 0. \tag{14.3.5}$$

Of course, we are interested in modes that obey (14.3.5) but cannot be gauged away:

$$B \neq d\Lambda. \tag{14.3.6}$$

It was seen in chapter 12 that equivalence classes of p forms B that obey (14.3.5) (with B and B' being considered equivalent if $B - B' = d\Lambda$) are in one-to-one correspondence with the zero eigenvalues of Δ. In view of what we have already learned, it is natural to believe that in reduction of a ten-dimensional theory to four dimensions, massless four-dimensional fields will arise from zero eigenvalues of the Laplacian Δ. We will now see that this is the case.

The components of B transform in the pth rank antisymmetric tensor representation of the tangent space group $SO(1,9)$ of $\mathbf{M}$. Under $SO(1,3) \times SO(6)$ (the two factors being the tangent space groups of M^4 and K, respectively), components of B with n indices ranging from 1 to 4 (tangent to M^4) and $p-n$ ranging from 5 to 10 (tangent to K) transform as n forms of M^4 and $p-n$ forms of K. Let us refer to such components as $(n, p-n)$ forms.

Define the Hodge–de Rham Laplacians on the ten-dimensional manifold $\mathbf{M} = M^4 \times K$ and on the factors M^4 and K to be Δ_{10}, Δ_4, and Δ_K, respectively. They obey

$$\Delta_{10} = \Delta_4 + \Delta_K, \qquad (14.3.7)$$

an identity which is the analog of (14.1.4) for spinors. The field equation derived from (14.3.3) is

$$d^* dB = 0. \qquad (14.3.8)$$

Here d^* is the adjoint of d, discussed in chapter 12, which maps p forms to $p-1$ forms. If B is a one form, (14.3.8) is the Maxwell equation which in flat space is $\partial^\mu (\partial_\mu B_\nu - \partial_\nu B_\mu) = 0$. Just as in Maxwell theory, (14.3.8) can be simplified with a proper choice of gauge. The gauge choice

$$d^* B = 0 \qquad (14.3.9)$$

is the generalization of the Lorentz gauge condition $\partial^\mu B_\mu = 0$ in electrodynamics. The Lorentz gauge condition is well known to reduce the (flat-space) Maxwell equations to the simpler form $\Box B_\mu = 0$, where $\Box = \partial_\nu \partial^\nu$ is the d'Alembertian or wave operator. Likewise, the gauge condition (14.3.9) reduces (14.3.8) to the form $\Delta B = 0$, since $\Delta = dd^* + d^* d$. In view of (14.3.7), this becomes

$$(\Delta_4 + \Delta_K)B = 0. \qquad (14.3.10)$$

The wave equation (14.3.10) decouples the $(n, p-n)$ forms for various n, so we can discuss each case separately.[*]

---

[*] Both for one forms (abelian gauge fields) and for p forms of any p, $dd^* + d^* d$ coincides with $\Box$ in flat space.

Equation (14.3.10) implies that Δ_K behaves as the mass operator for four-dimensional fields. The number of zero eigenvalues of Δ_K for $(n, p - n)$ forms is the Betti number b_{p-n} of K, and this will be the number of massless n forms that arise in four dimensions upon reduction of a ten-dimensional theory on $M^4 \times K$.

In fact, suppose that we are given on K a harmonic differential form β obeying $d\beta = d^*\beta = 0$. We can suppose that β is normalized so that $\langle \beta, \beta \rangle_K = 1$, where $\langle , \rangle_K$ is the inner product defined in (14.3.3) for differential forms on K. Let us try an ansatz

$$B = \alpha \wedge \beta, \tag{14.3.11}$$

where α is to be a differential form on M^4 whose dynamics we will investigate. (Saying that α is a differential form on M^4 means that all indices are tangent to M^4, and that the components of α are independent of the coordinates of K.) α must obey

$$0 = \Delta_4 \alpha = (dd^* + d^*d)\alpha \tag{14.3.12}$$

if $B = \alpha \wedge \beta$ is to obey (14.3.10). Likewise, the gauge condition (14.3.9) reduces to the requirement that

$$d^*\alpha = 0. \tag{14.3.13}$$

Combining (14.3.12) and (14.3.13), we see that the conditions on α are

$$0 = d^*d\alpha = d^*\alpha. \tag{14.3.14}$$

The first is the equation that we would derive from the standard action

$$S^{(4)} = \frac{n+1}{2n!} \int_{M^4} g^{i_1 j_1} \dots g^{i_{n+1} j_{n+1}} c_{i_1 \dots i_{n+1}} c_{j_1 \dots j_{n+1}}$$
$$= \langle d\alpha, d\alpha \rangle_{M^4} / 2(n!)^2 \tag{14.3.15}$$

for an n form α propagating on M^4; here $c = d\alpha$. The second equation in (14.3.14) is (as we have already discussed in the ten-dimensional context) a condition that can be imposed to fix the gauge freedom of (14.3.15). Thus, in the ansatz $B = \alpha \wedge \beta$, with β being a harmonic $p - n$ form on K, α is governed by the standard action for an n form on M^4. Equation (14.3.15) could also have been obtained, in a perhaps more direct way, by simply inserting the ansatz $B = \alpha \wedge \beta$ into (14.3.3). These observations confirm the contention that the number of massless n forms that arise in four dimensions upon compactification on $M^4 \times K$ is the Betti number $b_{p-n}(K)$.

For $n = 0$, (14.3.15) is the standard formula for a massless scalar field propagating on M^4. For $n = 1$ it is the standard Maxwell action for a massless spin one meson. For larger values of n, (14.3.15) may be less familiar, but in fact is easily analyzed. For instance, if $n = 2$, (14.3.15) is an unusual way to describe a free massless scalar field in four dimensions. To see this, define a one form y_μ by $y_\mu = \epsilon_\mu{}^{\nu\sigma\tau} c_{\nu\sigma\tau}$. This relation between y and c is essentially the Poincaré duality operation that we discussed in §12.4. The equation of motion for the two form α – namely $d^*c = 0$ – becomes in terms of y the equation $dy = 0$. This implies that $y = d\phi$ for some scalar field ϕ. (One does not run into global problems in defining ϕ in the case of uncompactified four-dimensional Minkowski space.) The Bianchi identity $dc = 0$ is equivalent to $d^*y = 0$, and in terms of ϕ this becomes

$$\Delta\phi = 0, \qquad (14.3.16)$$

which is the desired equation of motion for the massless scalar ϕ. If $n > 2$ instead, it can be shown that (14.3.15) does not describe any propagating degrees of freedom in four dimensions.

14.3.2 Application to Axions in $N = 1$ Superstring Theory

We would now like to discuss some applications of the above results. Since the main applications involve axions and the strong CP problem, we first make a brief digression to review the latter.

Quantum chromodynamics, the $SU(3)$ gauge theory that is generally considered to be the correct theory of the strong nuclear force, allows a topological term of the form

$$L_{\text{top}} = \frac{\theta}{32\pi^2} \text{tr}(\tilde{F}_{\mu\nu} F^{\mu\nu}) \qquad (14.3.17)$$

to be added to the Lagrangian. Such a term does not contribute to the classical equations of motion, but does influence the quantum theory by its contributions to the path integral in sectors where the gauge fields have nonzero winding number ($\int L_{\text{top}} d^4x \neq 0$), the so-called instanton sectors. Such a term, if present, would be very important because it gives P and T (or CP) violation. The experimental limit, based on the absence of an electric-dipole moment for the neutron, is $\theta/32\pi^2 \leq 10^{-9}$. This is a strikingly small value for a dimensionless natural constant, and the strong CP problem is the question of why θ is so small.

A promising suggestion for solving the 'strong CP problem' is the idea of axions, especially the weakly coupled 'invisible axion'. An axion is simply a massless scalar a with a coupling

$$L_{axion} = \frac{1}{F} \int d^4x \; a \, \text{Tr}\tilde{F}_{\mu\nu} F^{\mu\nu} \qquad (14.3.18)$$

More precisely, an axion is a scalar that would have been massless except for a coupling (14.3.18); standard analysis shows that the coupling (14.3.18) will give the axion a mass of order $m_\pi f_\pi / F$. It is awkward to talk about a scalar that 'would have been massless', and a better formulation of things is to say that an axion is a spin zero field a with a global symmetry

$$a \to a + c \qquad (14.3.19)$$

(c being an arbitrary constant), which is violated only (or primarily) by the coupling (14.3.18). Experimental searches for axions depend very much on the value of F, since the axion coupling is proportional to F^{-1} while the mass also scales like F^{-1}. The potential axions we will find below have F of order the Planck mass, so they are very weakly coupled.

We now return from this digression to an analysis of the zero modes of ten-dimensional wave operators. We will concentrate on the particularly interesting case of ten-dimensional theories with $N = 1$ supersymmetry. In this case, there is a two form B in the ten-dimensional supergravity multiplet. Harmonic zero, one or two forms on the compact space K will lead, via the considerations described in the last subsection, to the appearance of massless two, one or zero forms in the effective four-dimensional theory. On K there is always a single harmonic zero form (the constant function 1), as was discussed in §12.4, so one always gets precisely a single two form in the effective four-dimensional theory. We will learn presently that it has axion-like couplings. The numbers of harmonic one forms and harmonic two forms on K are the Betti numbers b_1 and b_2, which are subject to no universal restrictions. The effective four-dimensional theory contains b_1 abelian gauge fields and b_2 massless scalars originating as components of the ten-dimensional two form B. We will learn presently that the scalars just mentioned may also have axion-like couplings under suitable conditions. Perhaps it is interesting to emphasize that since (with a suitable choice of the topology of K) the numbers b_1 and b_2 can be arbitrarily large, there is no general upper bound on how many massless bosons in four dimensions can originate from a single field B in the ten-dimensional theory.

Let us now explain why the massless scalars just encountered can have axion-like couplings. The basic reason was already noted in §13.5.3, where we pointed out that the mechanism for anomaly cancellation in $N = 1$

supergravity in ten dimensions requires the existence of massless bosons that couple to two or four gluons. The anomaly cancellation requires a generalization of the minimal action whose dimensional reduction was considered above. The gauge-invariant field strength of the two form B is not $C = dB$, as in (14.3.2), but

$$H = dB - \omega_Y + \omega_L, \qquad (14.3.20)$$

where ω_Y and ω_L are the Yang–Mills and Lorentz Chern–Simons three forms, which were introduced in §12.5.2 and included in the definition of the field strength in §13.1.3 and §13.5.3. The ten-dimensional action (14.3.3) is replaced by

$$\overline{S} \propto \langle H, H \rangle. \qquad (14.3.21)$$

Let us first consider the massless scalar which arises from $(2,0)$ forms, that is, from components of B_{MN} with $M, N = 1, \ldots, 4$. The massless mode is the one whose wave function is independent of the compactified coordinates, and it is governed in four dimensions precisely by the action (14.3.21), with all indices being now tangent to M^4. We now repeat precisely the steps which led to (14.3.16), but with a modification because of the form of (14.3.20). We write first $y_\mu = \epsilon_\mu{}^{\nu\sigma\tau} H_{\nu\sigma\tau}$. The equation of motion derived from (14.3.21) is still $d^* H = 0$, and this still gives $dy = 0$, so that $y = d\phi$, for some scalar field ϕ. Now, however, the Bianchi identity is not $dH = 0$, but

$$dH = \mathrm{tr}\, R \wedge R - \mathrm{tr}\, F \wedge F. \qquad (14.3.22)$$

In terms of ϕ, this gives not $\Box \phi = 0$, but

$$\Box \phi = \mathrm{tr}\, R \tilde{R} - \mathrm{tr}\, F \tilde{F}, \qquad (14.3.23)$$

which contains the standard coupling of an axion.

We call ϕ the model-independent axion, since it arises in a way that does not depend on the details of compactification. Now we move on to a consideration of model-dependent modes which arise when the second Betti number of K is positive, say $b_2(K) = n$. If so, there are harmonic two forms β_k, $k = 1, \ldots, n$ on K, and we consider as in (14.3.11) the ansatz

$$B = \sum_k a_k(x^\mu) \beta_k(y^m), \qquad (14.3.24)$$

where the a_k are spin zero fields in four dimensions. Our previous analysis gave the minimal equation $\Delta a_k = 0$ for the a_k. Under what conditions can the a_k obtain axion-like couplings?

For notational simplicity we take $k = 1$ and consider a single harmonic two form β and correspondingly a single four dimensional field a. The modification (14.3.20) of the field strength does not give axion-like couplings to a. Such couplings, however, can arise because of the other terms introduced in §13.5.3 to cancel anomalies. These included terms like

$$\Delta S = \int_M B \wedge \mathrm{tr} F^2 \wedge \mathrm{tr} F^2. \tag{14.3.25}$$

Equation (14.3.25) modifies the field equation of B from

$$d^* dB = 0 \tag{14.3.26}$$

to

$$d^* dB = *(\mathrm{tr} F^2 \wedge \mathrm{tr} F^2), \tag{14.3.27}$$

where $*$ is the Poincaré duality operator which makes a two form from the eight form $\mathrm{tr} F^2 \wedge \mathrm{tr} F^2$.

The question at this point is whether the four-dimensional field a gets an axion-like coupling from (14.3.27). An axion couples to *two* Yang–Mills field strengths, while on the right-hand side of (14.3.27) there are four. If, however, components of F tangent to K have expectation values (something desirable anyway to obtain a nontrivial answer for the chiral asymmetry of fermions), then by replacing one factor of $\mathrm{tr} F^2$ in (14.3.27) by its vacuum expectation value, we can obtain a coupling of the axionic form. Indeed, if we write

$$k = \frac{\int_K \beta \wedge \mathrm{tr} F^2}{\int_K |\beta|^2}, \tag{14.3.28}$$

then with insertion of the ansatz (14.3.24), (14.3.27) reduces in four dimensions to

$$\Delta a = 2k \, \mathrm{tr} F \tilde{F} \tag{14.3.29}$$

showing the promised axionic coupling.

To claim that a scalar is an axion, it is not enough to exhibit a coupling to $\mathrm{tr} F \tilde{F}$; it is necessary to show that this scalar is massless except for the effects of this coupling. A better way to say this is to note that the equation $\Delta a = 0$ is invariant under the symmetry

$$a \to a + c, \tag{14.3.30}$$

with c being a constant. Equation (14.3.30) is the analog in this context of the symmetry introduced by Peccei and Quinn in their original work on the strong CP problem. The Peccei–Quinn-like symmetry (14.3.30) is violated by the $\mathrm{tr}F^2$ coupling in (14.3.29). The four-dimensional scalar a will behave as an axion and can contribute to the solution of the strong CP problem only if the $\mathrm{tr}F^2$ coupling is the dominant effect that violates the axionic symmetry.

Very tiny effects, other than the $F\tilde{F}$ coupling, that violate the Peccei–Quinn symmetry could spoil the axion-like behavior of a would-be axion. To address this question, one must examine the form of the axion vertex operators to determine whether the axionic symmetry is really valid in string theory. In the case of the model independent mode ϕ, whose indices are all tangent to M^4, it seems likely that gauge theory instantons are the main effects that violate the axionic symmetry. For the model dependent axion a, the situation is more complicated and is currently under active investigation. We will discuss some aspects of this question here.

The mode a would arise in studying oriented bosonic strings, and we will consider this case since complications associated with superstrings are not essential in the present discussion.[*] To explore the validity of the symmetry (14.3.30), an exact string theoretic treatment is needed, since symmetry-violating effects of very high order in α' would be enough to prevent a from behaving as an axion. A suitable framework for an exact discussion is the formalism of string propagation in background fields which was introduced in §3.4. The string propagation on $M^4 \times K$ is governed by a suitable nonlinear sigma model. The vertex operator for a fluctuation in the B_{MN} field is

$$V = \int d^2\sigma \epsilon^{\alpha\beta} B_{MN}(X^K) \partial_\alpha X^M \partial_\beta X^N. \qquad (14.3.31)$$

This is an operator in the nonlinear sigma model of maps of the string world sheet Σ into $M^4 \times K$. Inserting in (14.3.31) the ansatz $B(X^K) = a(x^\mu)\beta(y^k)$, with a being a scalar field on M^4 and β a harmonic two form on K, (14.3.31) becomes

$$V = \int d^2\sigma\, a(x^\mu) \epsilon^{\alpha\beta} \partial_\alpha y^i \partial_\beta y^j \beta_{ij}. \qquad (14.3.32)$$

To investigate the validity of the symmetry $a \to a + c$, we must ask whether (14.3.32) is invariant under this operation. The change in

---

[*] In the heterotic theory, the extra terms in the vertex operator V considered below vanish at zero momentum and are irrelevant. In the type I theory, the vertex operator has a different form but again vanishes at zero momentum. Indeed, in that case, the axionic symmetry is not spoiled by world-sheet instantons, and seems to be exact.

(14.3.32) is

$$\delta V = c \int d^2\sigma \epsilon^{\alpha\beta} \partial_\alpha y^i \partial_\beta y^j \beta_{ij}. \qquad (14.3.33)$$

Does this vanish? The answer turns out to be that (14.3.33) vanishes to all finite orders in sigma-model perturbation theory and thus to all finite orders in α', but *not* exactly. To show this, note that *locally* the harmonic two form β can be written

$$\beta_{ij} = \partial_i \lambda_j - \partial_j \lambda_i, \qquad (14.3.34)$$

where λ is some one form on K. Inserting (14.3.34) in (14.3.33) gives

$$\delta V = 2c \int d^2\sigma \epsilon^{\alpha\beta} \partial_\alpha (\lambda_j \partial_\beta y^j). \qquad (14.3.35)$$

Being the integral of a total divergence, this vanishes.

However, in arriving at (14.3.35) we have used the form (14.3.34) which is valid only locally. The correct statement, in general, is not that δV vanishes, but only that δV is a topological invariant which vanishes if the map $\Sigma \to K$ given by $\sigma^\alpha \to y^i(\sigma^\alpha)$ is topologically trivial. To finite orders in α', as was described in §3.4, one describes string propagation on $M^4 \times K$ by sigma-model perturbation theory, in which $y^i(\sigma^\alpha)$ is topologically trivial. Nonperturbatively, the sigma model governing string propagation, just like other quantum field theories, will have instantons. More precisely, instantons are present if the second Betti number of K is not zero, so that there are topologically nontrivial two-dimensional submanifolds of space-time onto which the string world sheet (of suitable genus) can be mapped in a noncontractible way. This is precisely the situation in which the mode a that we are discussing exists; in other words, whenever a exists, the relevant world-sheet instantons will also exist. Since the sigma-model coupling constant is of order α', the world-sheet instantons will make contributions of order $e^{-1/\alpha'}$. Thus, the Peccei–Quinn-like symmetry of the model-dependent axions associated with $b_2(K)$ is valid to all finite orders in α', but not exactly.

The violation of the Peccei–Quinn symmetries of model-dependent axions by world-sheet instantons may be strong enough to prevent these modes from being relevant to low-energy phenomenology; this would leave us with the one model-independent mode ϕ. The Peccei–Quinn symmetry of the model-dependent modes is, however, good enough to have a striking theoretical application which we will meet in chapter 16.

14.3.3 The 'Nonzero Modes'

So far, we have been discussing only what we called the 'zero modes', which are the modes of zero action or energy. However, it is also of interest to discuss more general solutions of the equations of motion derived from (14.3.3). The zero modes were the (nontrivial) solutions of the equations of motion with $C = 0$; now it is our task to discuss the solutions with $C \neq 0$. The variational equations derived from (14.3.3) require $d^*C = 0$. This must be supplemented with the Bianchi identity $dC = 0$ which follows from the definition $C = dB$ of the field strength C. Therefore, a solution of the equations of motion with $C \neq 0$ is characterized by

$$dC = d^*C = 0. \tag{14.3.36}$$

In other words, C (if not zero) is a harmonic $p+1$ form. This is a concept with which we are already familiar from our discussion of the zero modes, which were harmonic p forms. In particular, specializing to the case of a compact manifold K, we know that the number of linearly independent solutions of (14.3.36) is a topological invariant, the $p+1$ Betti number b_{p+1}.

It is perhaps curious that while the zero modes correspond to harmonic p forms, the solutions of nonzero energy correspond to harmonic $p+1$ forms. Because of this state of affairs, the mathematical tools needed to understand the zero modes carry over directly to the solutions of nonzero energy. But there is one particular property that should be stressed. A solution C of (14.3.36) can always be written locally in terms of a gauge potential B as $C = dB$. As was discussed in §12.4, this is guaranteed by the Poincaré lemma. Can one *globally* write C as dB, or does one encounter singularities analogous to the Dirac string? The answer can be deduced easily. On a compact manifold K, a harmonic differential form C obeying (14.3.36) always represents a nonzero cohomology class and so cannot be written globally as $C = dB$. Thus, on a compact manifold the nonzero modes are always generalizations of the Dirac magnetic monopole.

In considering compactification on $M^4 \times K$, if the $p+1$ Betti number of K is nonzero, one has the option of introducing a nonzero-energy solution of (14.3.36) in specifying the vacuum state. This involves introducing a monopole-like structure in the vacuum; analysis of the Bianchi identity $dC = 0$ shows that the 'monopole' strength is a conserved quantity which cannot change in time. In the more delicate situation that we encountered in studying anomaly cancellation, with the Bianchi identity replaced by $dC = \text{tr} R \wedge R - \text{tr} F \wedge F$, more careful analysis shows that to avoid global anomalies (a loss of modular invariance after compactification) the monopole charge must obey a quantization condition. It also turns out in that case that the 'monopole' charge is not really a constant of integration but can change dynamically (in tunneling events) by integral quanta.

14.3.4 The Exterior Derivative and the Dirac Operator

In the foregoing we have considered the zero modes of the wave operator $\Delta = dd^* + d^*d$. Since $d^2 = d^{*2} = 0$, we can alternatively define this operator as

$$\Delta = S^2, \tag{14.3.37}$$

where

$$S = d + d^* \tag{14.3.38}$$

is a first-order operator.

As a first-order operator which arises naturally in Riemannian geometry, S is reminiscent of the Dirac operator. For applications to the family problem and the fermion quantum numbers, it will be important for us to understand the precise connection between the operator $S = d + d^*$ and the Dirac operator. In what follows, we suppose that K is a manifold of dimension n.

The operator S turns a p form into a linear combination of a $p-1$ form and a $p+1$ form, so the field Ψ on which S acts must be understood as a linear combination of differential forms of different degree. The general component of Ψ is a p form which is antisymmetric in p tangent vector indices, where p may have any value from 0 to n. In all (summing over p), Ψ is a field with 2^n components, since each of the n independent tangent vector indices may be either present or absent. This is very much like a system of fermions in which there are n single-particle states corresponding to the n independent values of a tangent vector index. Thus, we say that the ith single particle level is filled if and only if Ψ has an index of type i, meaning that it takes the value i. As in any system of fermions, it is convenient to introduce the fermion creation and annihilation operators. Thus, let a_i^* be an operator that creates an index of type i. In other words, a_i^* will annihilate any p form Ψ that already has an index of type i; acting on a p form Ψ that does not have such an index, a_i^* makes a $p+1$ form by adding an index of type i. The formula for this is

$$(a_i^*\Psi)_{j_1\ldots j_{p+1}} = \frac{1}{p+1}\Big\{g_{ij_1}\Psi_{j_2\ldots j_{p+1}} \pm \text{cyclic permutations}\Big\}, \tag{14.3.39}$$

where g is the metric tensor of K. The adjoint of a_i^* is the operator a_i

that removes an index of type i (and annihilates any differential form that does not contain such an index). The formula is

$$(a_i\Psi)_{j_1...j_{p-1}} = \Psi_{ij_1...j_{p-1}}. \tag{14.3.40}$$

The creation and annihilation operators obey $\{a_i, a_j{}^*\} = g_{ij}$ as usual. The utility of the creation and annihilation comes in large part from the fact that they make possible rather convenient formulas for the exterior derivative d and its adjoint d^*. The operator d creates an index of type i while differentiating the field it acts on in the i direction, so it is

$$d = g^{ij}a_i{}^*D_j. \tag{14.3.41}$$

Here D_j is the covariant derivative. On the other hand, the adjoint operator d^* removes an index of type i while differentiating in the i direction, so it is

$$d^* = g^{ij}a_iD_j. \tag{14.3.42}$$

Combining these formulas, the operator $S = d + d^*$ that appeared in the study of harmonic differential forms is

$$S = g^{ij}(a_i{}^* + a_i)D_j. \tag{14.3.43}$$

In this last formula appear the very important operators

$$\Gamma_i = (a_i + a_i{}^*). \tag{14.3.44}$$

These matrices deserve to be called gamma matrices; they obey the standard gamma matrix algebra $\{\Gamma_i, \Gamma_j\} = 2g_{ij}$. It is perhaps surprising[*] that one can also define a second set of gamma matrices,

$$\tilde{\Gamma}_i = i(a_i - a_i{}^*), \tag{14.3.45}$$

which are also easily seen to obey the standard gamma matrix algebra. In addition, the Γ_i and $\tilde{\Gamma}_i$ mutually anticommute, $\{\Gamma_i, \tilde{\Gamma}_j\} = 0$. For our purposes it is more useful to have two sets of mutually commuting rather than mutually anticommuting gamma matrices. While this can always be done, the details depend somewhat on the number of dimensions, and

---

[*] To the reader who has read the construction of the spinor representation of $SO(2N)$ in appendix 5.A, this may not seem so surprising.

we specialize now to the case of principal interest in which n is even. We introduce the two chirality operators

$$\Gamma^{(n)} = i^{n(n-1)/2}\Gamma_1\Gamma_2\ldots\Gamma_n$$
$$\tilde{\Gamma}^{(n)} = i^{n(n-1)/2}\tilde{\Gamma}_1\tilde{\Gamma}_2\ldots\tilde{\Gamma}_n,$$

(14.3.46)

where the phase $i^{n(n-1)/2}$ has been included to ensure that the square of either $\Gamma^{(n)}$ or $\tilde{\Gamma}^{(n)}$ is unity. Now we redefine the second set of gamma matrices to be

$$\overline{\Gamma}_i = i\Gamma^{(n)} \cdot \tilde{\Gamma}_i.$$

(14.3.47)

One may readily check that the $\overline{\Gamma}_i$ anticommute among themselves in the proper fashion for gamma matrices, but *commute* with the Γ_i.

Either the Γ_i or the $\overline{\Gamma}_i$ may be viewed as 'the' gamma matrices of $SO(n)$. One set of gamma matrices can be realized on a field ϕ_α with a single spinor index α, so two sets of commuting gamma matrices can be realized on a field $\Phi_{\alpha\beta}$ with two independent spinor indices α and β. On the other hand, we constructed the two sets of commuting gamma matrices starting with a field Ψ which was a linear combination of differential forms of varying degree, so it must be that a field $\Phi_{\alpha\beta}$ with two independent spinor indices is equivalent to a linear combination of differential forms. That this is so is simply an exercise in $SO(n)$ group theory. In one's first encounter with the Dirac equation one learns that because of the anticommutation laws for gamma matrices, the independent Dirac matrices are the antisymmetrized products

$$\Gamma_{i_i\ldots i_p} = (\Gamma_1\Gamma_2\ldots\Gamma_p \pm \text{permutations})/p!$$

(14.3.48)

of p gamma matrices. As a result, the tensors that one can form by multiplying two spinors η and λ are the zero form $\overline{\lambda}\eta$, the one form $\overline{\lambda}\Gamma_i\eta$, the two form $\overline{\lambda}\Gamma_{i_1 i_2}\eta$, and in general the p form $\overline{\lambda}\Gamma_{i_1 i_2\ldots i_p}\eta$ for any $p \leq n$. As far as group theory is concerned, combining two spinors η_α and λ_β is the same as combining the two different spinor indices of a single bispinor field $\Phi_{\alpha\beta}$ with two such indices, so this confirms that the field $\Phi_{\alpha\beta}$ is really equivalent to a linear combination of differential forms of various p.

Now we can use what we have learned to get a much better understanding of the operator S. Let us first recall the general structure of the Dirac operator. Typically one considers a spinor field ϕ_α^x with a spinor index α and perhaps an additional Yang–Mills index x. The Dirac operator is then

$$(\not{D}\phi)^x_\alpha = (\Gamma^i_{\alpha\sigma}D_i)\phi^x_\sigma. \tag{14.3.49}$$

With our knowledge that the field on which S acts is a bispinor $\Phi_{\alpha\beta}$, equation (14.3.43) can be rewritten in a form that is precisely analogous,

$$(S\Phi)_{\alpha\beta} = \Gamma^i_{\alpha\sigma}D_i\Phi_{\sigma\beta}. \tag{14.3.50}$$

We see that S is just a Dirac operator acting on a spinor field that has an extra spinor index, the second spinor index β of $\Phi_{\alpha\beta}$ being an 'internal' index analogous to x in (14.3.49). As a matter of fact, in a rather important application of the S operator that we will encounter towards the end of this chapter, the second spinor index of the bispinor field will originate precisely as an internal symmetry index.

Since S is a kind of Dirac operator, it is natural to ask what is the analog for S of the Dirac index that we studied in §14.2.1. To define an index problem, one introduces a 'chirality' operator T that anticommutes with S and obeys $T^2 = +1$. Letting n_+ and n_- be the number of zero modes of S with $T = +1$ and $T = -1$ respectively, we then define the 'index' as $n_+ - n_-$. It is a topological invariant for the same reasons that we discussed in §14.2.1. However, in the case of the S operator there are several possible choices of T, and they lead to different topological invariants.

The possibility which corresponds most closely to our treatment of the standard Dirac operator is to let $T = \Gamma^{(n)}$. In this case the index is a topological invariant that is known as the Hirzebruch signature and is often denoted σ. It can be shown to vanish in $4k + 2$ dimensions by an argument based on complex conjugation which is similar to arguments that were given in §14.2 to show that the standard Dirac index (without Yang–Mills fields) is zero in such dimensions. The point is that the creation and annihilation operators a_i and a^*_i are manifestly real from their definition, so (14.3.46) shows that $\Gamma^{(n)}$ and $\tilde{\Gamma}^{(n)}$ are real in $4k$ dimensions and imaginary in $4k + 2$ dimensions. The operator S is manifestly real, so, in $4k + 2$ dimensions, complex conjugation maps zero modes of S with $\Gamma^{(n)} = +1$ into zero modes of $\Gamma^{(n)} = -1$, ensuring that the index σ vanishes. On the other hand, in $4k$ dimensions, σ need not vanish.

Another choice of the chirality operator which will be of much greater importance in this book is to use $\tilde{T} = i^n\Gamma^{(n)} \cdot \tilde{\Gamma}^{(n)}$. With this choice, the index of S is a topological invariant called the Euler characteristic χ. It is perhaps the single most fundamental topological invariant of a manifold. Note that $\tilde{T}$ is real both in $4k$ and in $4k + 2$ dimensions, so that the Euler characteristic can be nonzero in any even dimension. It is not obvious that the Euler characteristic defined in this way as the index of the operator S coincides with the definition given at the end of chapter 12, where the Euler characteristic of a manifold was defined as the integral over the manifold of a certain polynomial in the curvature tensor. This equality is the Gauss–Bonnet–Chern theorem, and can be extracted as a special case of the Atiyah–Singer index theorem. We will do this explicitly in the two-dimensional case in §14.4 below.

The Euler characteristic of a manifold K has a simple expression in terms of the Betti numbers of K. Tracing back the definition of $\Gamma^{(n)}$ and $\tilde{\Gamma}^{(n)}$, one sees that

$$\tilde{T} = i^n \Gamma^{(n)} \tilde{\Gamma}^{(n)} = (-1)^n \prod_{i=1}^n (a_i + a_i^*)(a_i - a_i^*) = \prod_{i=1}^n (-1)^{N_i}, \quad (14.3.51)$$

where $N_i = a_i^* a_i$ is the number operator of the ith fermion mode. For a p form, the number of filled states is p, so (14.3.51) means that for p forms $\tilde{T} = (-1)^p$.

Now, the zero modes of S are the harmonic differential forms, and the number of p forms which are such zero modes is the Betti number b_p. The contribution of p forms to the index of S is hence $\pm b_p$, where we must take a plus or minus sign depending on whether $\tilde{T} = +1$ or $\tilde{T} = -1$ or in other words depending on whether p is even or odd. It follows then that the Euler characteristic is

$$\chi(K) = \sum_{p=0}^n (-1)^p b_p, \quad (14.3.52)$$

and this is our desired formula expressing the Euler characteristic in terms of the Betti numbers. One important property of this should be pointed out. In discussing the index of a generic Dirac operator, we noted that if n_+ and n_- are the number of positive- and negative-chirality zero modes, then in general only the index $n_+ - n_-$ is a topological invariant; and one may expect that either n_+ or n_- will vanish depending on whether the index is positive or negative. The S operator is an exception to this; the numbers n_+ and n_- are both topological invariants in the case of the S operator, being respectively the sum of the b_p for even or odd p. In ordinary Riemannian geometry (as opposed to more specialized subjects like the study of complex manifolds), the operator S is the only known operator with this property. We will see in chapter 16 that it is at least possible that this property may be related to the solution of the gauge hierarchy problem.

Along the lines of equation (14.2.9) in §14.2.2 above, the signature σ and Euler characteristic χ can be written

$$\sigma = \text{Tr}\left(\Gamma^{(n)} \exp -\beta\Delta\right),$$
$$\chi = \text{Tr}\left(\Gamma^{(n)}\tilde{\Gamma}^{(n)} \exp -\beta\Delta\right), \qquad (14.3.53)$$

where $\Delta = S^2$ plays the role of the 'Hamiltonian'. In our later work, we will encounter not quite the operator S but rather $\tilde{S} = S(1 + \tilde{\Gamma}^{(n)})/2$ – which is simply S projected onto the subspace of Hilbert space with $\tilde{\Gamma}^{(n)} = +1$. In this sector, $T = \tilde{T}$, so it does not matter which chirality operator we use in defining the index of $\tilde{S}$. Choosing $T = \Gamma^{(n)}$, the index is

$$\text{index}\,\tilde{S} = \text{Tr}\left(\Gamma^{(n)}(1 + \tilde{\Gamma}^{(n)})/2 \exp -\beta\Delta\right) = (\sigma + \chi)/2. \qquad (14.3.54)$$

Of course, in $4k + 2$ dimensions σ vanishes, and (14.3.54) reduces to

$$\text{index}\,\tilde{S} = \chi/2, \qquad (14.3.55)$$

a result that will enter in our later study of the number of fermion generations.

14.4   Index Theorems on the String World Sheet

Though our goal in this chapter is mainly to elucidate the properties of wave operators in space-time, we will now pause to discuss how some of the concepts appear in the analysis of a wave operator on a string world sheet, *i.e.*, on a compact Riemann surface Σ of genus g. This will enable us to illustrate many of the ideas we have considered earlier, and others which we will be introducing shortly, in a simple setting. The results in any case have various applications in string theory.

14.4.1 The Dirac Index

As in chapters 3 and 4, many of the essential features reflect   the fact that in two dimensions the rotation group $SO(2)$ is abelian; an irreducible representation is specified by giving the eigenvalue of the one generator W of $SO(2)$. In our applications this eigenvalue will always be an integer or half-integer. A spinor ψ has positive- and negative-chirality components ψ_+ and ψ_- with $W = 1/2$ and $W = -1/2$. The covariant derivative ∇_α

likewise has components ∇_+ and ∇_- with $W = +1$ and $W = -1$. Acting with the Dirac operator gives a spinor $\not{\nabla}\psi$ which like ψ has components with $W = \pm 1/2$. The Dirac equation $\not{\nabla}\psi = 0$ is explicitly

$$\nabla_+\psi_- = 0,$$
$$\nabla_-\psi_+ = 0. \tag{14.4.1}$$

The two equations in (14.4.1) are the equations for zero modes of negative or positive chirality, respectively. If we denote the number of linearly independent solutions of the first and second equations in (14.4.1) as n_- and n_+, respectively, then the Dirac index is

$$\text{index } \not{\nabla} = n_+ - n_-. \tag{14.4.2}$$

If the covariant derivative in (14.4.1) is constructed from the spin connection only, with no coupling to, say, a Yang–Mills field, then the two equations in (14.4.1) are complex conjugates of each other, so $n_+ = n_-$ and the Dirac index vanishes.

A nonzero index can arise if an additional gauge coupling is present. Consider a spinor field ψ^a that carries a Yang–Mills index a, labeling some representation R of a gauge group G. We will denote the gauge field as A_α. The Dirac equation is

$$\nabla_+\psi_-^a = 0,$$
$$\nabla_-\psi_+^a = 0. \tag{14.4.3}$$

The two equations are still complex conjugates of each other if the representation R is real or pseudoreal, but not otherwise. In this more general situation, the Dirac index is not necessarily zero. It is given by a formula discussed in §14.2:

$$\text{index } \not{\nabla} = \frac{1}{2\pi} \int_\Sigma \text{tr } F. \tag{14.4.4}$$

Here F is the curvature two form, and the trace is taken, of course, in the R representation.

Equation (14.4.4) evidently vanishes if the group G is semisimple. Only a $U(1)$ component of G is relevant, and we will focus on the case in which G is actually $U(1)$ (or $SO(2)$). The $U(1)$ gauge fields for which the right-hand side of (14.4.4) is nonzero have a net nonzero flux of the magnetic field integrated over Σ, and thus represent generalizations of the Dirac magnetic monopole. According to Dirac's classical analysis, the magnetic charge of the monopole or more precisely the right-hand side of (14.4.4) must be an integer in order for coupling of the monopole to a charged field such as ψ to be possible. Equation (14.4.4) actually constitutes a new proof of Dirac's quantization law, since the left-hand side of (14.4.4) is certainly an integer.

In the above, ψ might be coupled to any $U(1)$ gauge field that we wish to consider, but there is one choice that is particularly interesting. On any orientable two-dimensional manifold, the spin connection ω_α is an $SO(2)$ or $U(1)$ gauge field. The $U(1)$ gauge field A_α of the previous construction could thus be chosen to be simply ω_α. A slight generalization of this is possible; because $U(1)$ is abelian, it makes sense to write

$$A_\alpha = n\,\omega_\alpha \tag{14.4.5}$$

with an arbitrary integer or half integer n.[*] Because of the nonlinearity of the $SO(N)$ gauge transformation law for $N > 2$, in more than two dimensions (14.4.5) makes sense in a gauge-invariant way only for $n = 1$, but in two dimensions we do not have this restriction.

Equation (14.4.5) is certainly an interesting example of an abelian gauge field to study, since it introduces no arbitrary ingredients that are not present anyway whenever we study the surface Σ. We will see that the index theorem (14.4.4) in this special situation has many interesting applications.

With the choice (14.4.5), the gauge field strength $\epsilon^{\mu\nu} F_{\mu\nu}$ is simply $n/2$ times the Ricci curvature scalar R. If we denote the Dirac operator acting on spinors of 'charge' n as $\nabla_{(n)}$, then the index theorem becomes

$$\text{index } \nabla_{(n)} = \frac{n}{4\pi} \int_\Sigma \sqrt{g}\,R. \tag{14.4.6}$$

We will work out three interesting applications of (14.4.6), to the Euler characteristic, the conformal ghosts, and the superconformal ghosts.

14.4.2 The Euler Characteristic

We have already analyzed, in $2n$ dimensions for any n, an operator S whose index is the Euler characteristic. We discovered in (14.3.50) that

---

[*] n must be an integer or half integer if A_α is to respect the Dirac quantization condition on magnetic charge for any choice of the surface Σ, though other fractions would make sense on some surfaces. We will see that in the applications that arise naturally n is always an integer or half integer.

S is simply the ordinary Dirac operator acting on a field ψ_α^a with a spinor index α and an extra 'Yang–Mills' index which is merely another spinor index. In two dimensions, this construction simplifies because $SO(2)$ is abelian. The a index takes two possible values, corresponding to $W = 1/2$ and $W = -1/2$ in the terminology of the last subsection. The operator S is simply

$$S = \begin{pmatrix} \nabla_{(1/2)} & 0 \\ 0 & \nabla_{(-1/2)} \end{pmatrix}, \tag{14.4.7}$$

where $\nabla_{(\pm 1/2)}$ are the operators whose index is given in (14.4.6) above.

What is the index of S? We must recall that there were two possible choices, corresponding to $\operatorname{tr} e^{-\beta S^2}\Gamma$ and $\operatorname{tr} e^{-\beta S^2}\Gamma\tilde{\Gamma}$. They correspond, respectively, to the sum and difference of the two operators $\nabla_{(1/2)}$ and $\nabla_{(-1/2)}$ that appear in (14.4.7). The Euler characteristic was seen in (14.3.53) to arise from the latter choice, so we have

$$\chi = \operatorname{index} \nabla_{(1/2)} - \operatorname{index} \nabla_{(-1/2)} = \frac{1}{2\pi} \int_\Sigma R, \tag{14.4.8}$$

where (14.4.6) has been used. Equation (14.4.8) has already been obtained at the end of chapter 12 on the basis of a different definition of the Euler characteristic, so we see that these two definitions are equivalent, at least for a manifold of dimension two. The two definitions (previously χ was defined as the integral of a certain characteristic class, here as the index of S) are actually equivalent in any dimension. This can be established by further study of the index theorem.

We now have considerably more information about the Euler characteristic than we had when we first defined it in chapter 12. For instance, we now know (from (14.3.52)) that

$$\chi = \sum_{p=0}^{n} (-1)^p b_p. \tag{14.4.9}$$

Any compact connected manifold has $b_0 = 1$ and, for a two-dimensional manifold, Poincaré duality gives $b_2 = 1$. At the end of chapter 12 we determined the Euler characteristic of a Riemann surface of genus g to be $2 - 2g$. Combining these results, we see that the first Betti number of a genus g surface is

$$b_1(g) = 2g. \tag{14.4.10}$$

As we discussed in §12.2, $2g$ is the number of independent noncontractible loops on the genus g surface. Equation (14.4.10) is thus an illustration of the relation between the Betti number b_p of a manifold M and the number of topologically independent p-dimensional submanifolds of M.

14.4.3 Zero Modes of Conformal Ghosts

In §3.3, we discussed the zero modes of conformal ghosts and antighosts on a compact Riemann surface of genus g. The wave equations for the positive-chirality ghosts c_- and antighosts b_{++} were

$$\nabla_- b_{++} = 0,$$
$$\nabla_- c_- = 0. \qquad (14.4.11)$$

The numbers of b and c zero modes on a genus g surface were called B_g and C_g, respectively, in §3.3. We found for these numbers a somewhat irregular behavior for low genus, but for the difference $\Delta_g = C_g - B_g$ we found the simple result

$$\Delta_g = 3(1 - g). \qquad (14.4.12)$$

It is natural to suspect that the difference Δ_g might be the index of some operator, and that this is responsible for the fact that the behavior of Δ_g is simpler than that of B_g or C_g. We will now see that this is the case.

First of all, there is no harm in replacing the second equation in (14.4.11) with

$$\nabla_+ \tilde{c}_+ = 0, \qquad (14.4.13)$$

where $\tilde{c}_+$ is the complex conjugate of c_-. We will interpret the difference between the number of solutions of (14.4.13) and the number of solutions of the first equation in (14.4.11) as the index of an operator. The operator in question is simply $\nabla_{(3/2)}$ in the terminology introduced earlier. Indeed, $\nabla_{(3/2)}$ would act on a field ψ_α^a, α being a spinor index ($W = \pm 1/2$) and a a 'gauge' index of $W = 3/2$. The positive- and negative-chirality components of ψ have, respectively, $W = 1/2 + 3/2 = 2$ and $W = -1/2 + 3/2 = 1$. They can be identified with b_{++} and $\tilde{c}_+$, respectively. So

$$\Delta_g = \text{index}\,\nabla_{(3/2)} = \frac{3}{4\pi} \int_\Sigma R. \qquad (14.4.14)$$

The integral on the right-hand side of (14.4.14) is 3/2 of the Euler characteristic; using the known value of the Euler characteristic of a genus g surface, we recover (14.4.12).

14.4.4 Zero Modes of Superconformal Ghosts

In a similar spirit we can study the zero modes of superconformal ghosts, which we have not considered previously. The right-moving modes are a ghost $\gamma_{(-1/2)}$ and an antighost $\beta_{(+3/2)}$ (the subscripts are eigenvalues of W). The wave equations are

$$\nabla_- \beta_{(+3/2)} = 0$$
$$\nabla_- \gamma_{(-1/2)} = 0. \tag{14.4.15}$$

Rather as in the bosonic case considered in §3.3, the zero modes of γ correspond to generators of superconformal symmetries, while those of β correspond to what might be called 'supermoduli'. If U_g and V_g are the number of γ and β zero modes, respectively, on a genus g surface, then as in the purely bosonic case, U_g and V_g behave somewhat irregularly for small g. Indeed, $U_0 = 2$ (corresponding to the existence in the quantized bosonic sector of superstrings of an anomaly-free symmetry group with two odd generators $G_{\pm 1/2}$ in addition to L_0 and $L_{\pm 1}$), while $U_1 = 1$ (the only zero mode on a torus with flat metric being $u =$ constant), and $U_g = 0$ for $g > 1$. As for V_g, the values are $V_0 = 0$, $V_0 = 1$, $V_g = 2g$ for $g > 1$.

We will not prove all of these statements here, but we can check them to a certain extent by using an index theorem to compute the difference $\tilde{\Delta}_g = U_g - V_g$. As in the bosonic case, this difference behaves more smoothly than U_g and V_g. The same arguments as in the bosonic case show that $\tilde{\Delta}_g$ can be interpreted as the index of what we have called $D_{(1)}$. Hence

$$\tilde{\Delta}_g = 2(1 - g). \tag{14.4.16}$$

Curiously, $\tilde{\Delta}_g$ is precisely equal to the Euler characteristic of the Riemann surface.

14.5   Zero Modes of Nonlinear Fields

So far we have discussed the zero modes of spinors and antisymmetric tensor fields in background gauge and gravitational fields. These zero modes have turned out to be related to many rich geometrical and topological ideas of which we have discussed only a few aspects. The classical linear equations that we have studied are only approximations from the superstring theory point of view, but they are good enough approximations for most of the questions we have studied, because most of our conclusions have depended only on very general considerations. For example, the Dirac index only depended on the universality class of a theory, and the zero modes of differential forms will be zero modes of almost any gauge-invariant equation those fields might obey, because the gauge-invariant field strength of the zero modes vanishes.

We now turn our attention to the other massless ten-dimensional fields, which are the gauge and gravitational fields. These fields obey nonlinear equations such as the classical Einstein and Yang–Mills equations, as opposed to linear equations in prescribed backgrounds, so it is natural to call these fields nonlinear fields. In discussing the wave equations of the nonlinear fields, the first question one runs into is the question of whether a presumed vacuum state $M^4 \times K$ really obeys the nonlinear Einstein and Yang–Mills equations. This discussion has a completely different flavor from the discussion in the last few sections. Whether a given space K, with given expectation values for matter fields, obeys the nonlinear equations depends not just on the universality class of a theory but on a particular choice of action. It makes all the difference in the world whether the gravitational field, for instance, is governed by the minimal Einstein–Hilbert action $\sqrt{g}R$ or by some other action with higher-order contributions determined by string theory. For this reason, the discussion of which spaces $M^4 \times K$ might obey the nonlinear equations that arise from string theory requires a discussion of a completely different kind from the one presented here.

Here, we simply assume that some suitable vacuum state $M^4 \times K$ has been found, and we ask what spectrum of massless particles will emerge in four dimensions. In the case of gauge and gravitational fields, this means that we linearize their field equations around a presumed solution and ask what is the spectrum of zero modes. The discussion will necessarily depend only on very simple and general arguments because we will not assume that we know what are the correct gauge and gravitational field equations. We will simply assume that the gauge and gravitational fields are governed by some gauge-invariant and generally covariant equations of which $M^4 \times K$ is a solution. The important conclusions indeed depend only on this. Statements that depend on the detailed form of the gauge and gravitational equations are not terribly interesting, unless one learns to use the exact equations coming from string theory, because in this discussion almost any small error is unacceptably large. A mode that is a zero mode to within one percent corresponds to a particle whose mass may be as large as one percent of the Planck mass.

In discussing the massless modes of the nonlinear fields, we recall that the coordinates of M^4 are denoted x^μ, $\mu = 1, \ldots, 4$, while the coordinates of K are y^j, $j = 5, \ldots, 10$ and all ten coordinates together are written X^M, $M = 1, \ldots, 10$. The first general statement to be made about compactification on $M^4 \times K$ is certainly that it is covariant under Poincaré transformations of M^4. If the underlying ten-dimensional theory was generally covariant, then the four-dimensional Poincaré group is actually a local symmetry group, and we must therefore expect the appearance of a massless graviton in the four-dimensional theory. Since a constant is the unique zero form, the massless mode is simply $g_{\mu\nu}(x^\lambda)$, with all indices tangent to M^4 and no dependence on the y^j. Generically, this is the only massless field in four dimensions arising from the components $g_{\mu\nu}$ of the ten-dimensional metric g_{MN}. In a similar spirit, suppose that the underlying ten-dimensional theory has a local symmetry under gauge transformations of some gauge group G, and suppose that expectation values of matter fields on K break G down to some subgroup H. Then the K independent modes $A_\mu^a(x^\lambda)$ (with a running over the Lie algebra of H) behave in four dimensions as massless gauge fields of H, and generically no other massless modes can be expected to come from the components of the gauge field tangent to M^4.

A more subtle case arises if one asks for massless modes of the components $g_{\mu j}$ of the metric tensor with one index tangent to M^4 and one index tangent to K. Historically, this case animated the invention of Kaluza–Klein theory and much subsequent work on the subject down to the present. Massless modes of $g_{\mu j}$ will be observed in four dimensions as massless spin one particles – in other words, massless gauge bosons. Such particles appear if a suitable subgroup of the underlying ten-dimensional general covariance is left unbroken in the compactification on $M^4 \times K$. To understand how this can occur, one must first review the description of continuous symmetries in general relativity. Under a general coordinate transformation of a manifold K, $y^k \to y^k + \epsilon V^k(y^j)$ (ϵ being a small parameter and V^k a vector field), the change in the metric g_{ij} of K is easily seen from the standard formulas of general relativity to be $\delta g_{ij} = \epsilon(D_i V_j + D_j V_i)$. So the metric of K is left invariant under the coordinate transformation $y^k \to y^k + \epsilon V^k$ generated by V if V obeys the Killing vector equation

$$D_i V_j + D_j V_i = 0. \tag{14.5.1}$$

The coordinate transformation generated by a Killing vector field V leaves the metric of K invariant, so it is a symmetry of any generally covariant equation for the metric of K. If one is studying an equation not for the metric only but for a coupled system consisting of the metric of K and suitable matter fields (such as gauge fields) one gets a symmetry if V can be combined with a suitable transformation of the matter fields that leaves their expectation values invariant.

If one finds several Killing vector fields $V_a^i, a = 1, \ldots, n$, the corresponding coordinate transformations will generate a Lie algebra of some kind. In general

$$[V_a^i \partial_i, V_b^j \partial_j] = f_{abc} V_c^k \partial_k, \qquad (14.5.2)$$

where f_{abc} are the structure constants of some Lie algebra H, which will then generate a symmetry group of K.

Let us now consider compactification of a ten-dimensional generally covariant theory on $M^4 \times K$, assuming K to have a symmetry group H. A rigid (M^4-independent) rotation of K would be the coordinate transformation $(x^\mu, y^k) \to (x^\mu, y^k + \sum_a \epsilon^a V_a^k)$, the ϵ^a being infinitesimal constants. This rigid transformation leaves fixed the metric of $M^4 \times K$, if the V_a are Killing vector fields. If the original ten-dimensional theory is generally covariant, we are free to consider not only rigid rotations of K but x^μ-dependent ones. The formula is

$$(x^\mu, y^k) \to (x^\mu, y^k + \sum_a \epsilon^a(x^\nu) V_a^k), \qquad (14.5.3)$$

where now the parameters in the transformation are not constants ϵ^a but functions of the x^μ. Working at very long wavelengths compared to the Planck length so that a description just in terms of massless fields is adequate, we take the $\epsilon^a(x^\mu)$ to be very slowly varying functions. The transformation in (14.5.3) will be a symmetry of any generally covariant theory compactified on $M^4 \times K$, and it will look in the effective four-dimensional theory like M^4-dependent local gauge transformations with gauge group H. So the effective four-dimensional theory must have massless gauge bosons of this group. The ansatz exhibiting them is just (in the linearized approximation)

$$g_{\mu j} = \sum_a A_\mu(x^\nu)^a V_{ja}(y^k), \qquad (14.5.4)$$

where the $A_\mu(x^\nu)$ are the massless gauge fields that appear in M^4. To verify the correctness of (14.5.4), it is enough to note that under the gauge transformation (14.5.3), the fields A_μ^a transform in the appropriate fashion, $\delta A_\mu^a = \partial_\mu \epsilon^a + f^{abc} \epsilon_b A_{\mu c}$.

What gauge symmetries can arise in this way in the process of compactification on $M^4 \times K$? We will consider this question in some detail if only for its historical interest, though the considerations it leads to no longer seem to be of central interest in the context of superstring theory.

In discussing what symmetry groups can arise, we temporarily drop the restriction to ten dimensions and contemplate compactification of a $(4+n)$-dimensional theory for arbitrary n. Determining the gauge groups that can arise in four dimensions amounts to asking what symmetry groups an n-dimensional manifold K can have. To investigate this question, let x be a point in K. The 'orbit' of x under a symmetry group H of K is the space W_x consisting of all points in K into which x is transformed by the action of the group. K is said to be a homogeneous space under the action of H if for each x the orbit W_x comprises all of K, which is tantamount to saying that for any point y in K there is some element h of the group H such that $h(x) = y$. It follows from the definitions that whether or not K is homogeneous under the action of H, each of the W_x always has this property. The maximum possible symmetry groups for manifolds of given dimension always arise for homogeneous spaces; intuitively this is because if K is not a homogeneous space, some of the dimensions of K are 'wasted' because H is really acting on the submanifolds W_x of lower dimension. Let us suppose then that K is a homogeneous space.

The 'little group' of a point x in K is the subgroup H_0 of H that leaves x fixed. The difference between the dimension $\dim H$ of the group H and the dimension $\dim H_0$ of the group H_0 is $\dim H - \dim H_0 = n$, n being the dimension of K. The reason for this is that a generator of H that does not leave x fixed moves it in some tangent direction. As there are only n such directions, there can be at most n linearly independent generators of H that do not leave x fixed; and if K is a homogeneous space, there must be precisely n such independent generators, since there must be *some* generator of H that moves x in any prescribed direction.

Up to isomorphism, there is precisely one space homogeneous under the action of a given group H with a given little group H_0 of a point. Such a homogeneous space is conveniently defined as consisting of elements of H subject to the equivalence relation that two elements a and b of H are considered equivalent if $a = bh_0$. The space of equivalence classes is called the quotient space H/H_0 and is homogeneous under the action of H by left multiplication, $a \rightarrow ha$ for $h \in H$ and $a \in H/H_0$. Picking an element a of H/H_0, say $a = 1$, its little group consists of elements $h \in H$ such that $h \cdot 1 = 1 \cdot h_0$ for some $h_0 \in H_0$; in other words, the little group is isomorphic to H_0. The uniqueness up to isomorphism of this construction is not difficult to prove, but we will not do so here.

Rather than trying to completely determine the maximal symmetry groups for n manifolds K, let us simply ask for what n the observed gauge symmetry group of nature, namely $Q = SU(3) \times SU(2) \times U(1)$, is possible. (If one wishes a larger group than this, then n would have to be larger than we will determine.) Suppose that Q acts homogeneously on K, and let the little group of a point be a subgroup Q_0. Q_0 should not contain all of either $SU(3)$ or $SU(2)$ or $U(1)$, for if so this factor of Q would leave K invariant and would not really be a symmetry group. The maximal subgroup of $SU(3) \times SU(2) \times U(1)$ that does not contain any of the three factors is $SU(2) \times U(1) \times U(1)$, the $SU(2)$ being an 'isospin' subgroup of $SU(3)$, and the two $U(1)$'s being linear combinations of the 'hypercharge' generator of $SU(3)$ which commutes with 'isospin', an arbitrary $SU(2)$ generator, and the $U(1)$ factor of $SU(3) \times SU(2) \times U(1)$. (There are infinitely many inequivalent $SU(2) \times U(1) \times U(1)$ subgroups of Q, corresponding to inequivalent choices of the two $U(1)$'s, but we will not classify them here.) So the minimum dimension of a space with $SU(3) \times SU(2) \times U(1)$ symmetry arises if we pick the little group $Q_0 = SU(2) \times U(1) \times U(1)$. With this choice, Q_0 has dimension five, while the dimension of $Q = SU(3) \times SU(2) \times U(1)$ is twelve, so the dimension of the quotient space, which is the lowest possible dimension of a space with $SU(3) \times SU(2) \times U(1)$ symmetry, is $12 - 5 = 7$.

For superstring theory, this is at best a near miss. We have six compact dimensions, and we would have needed seven. If we cannot get $SU(3) \times SU(2) \times U(1)$ by compactifying from ten to six dimensions, can we at least get an interesting subgroup of this? There is indeed a unique six manifold with $SU(3) \times SU(2)$ symmetry (namely $CP(2) \times S^2$, corresponding to the little group $Q_0 = SU(2) \times U(1) \times U(1)$), but this space does not make sense in superstring theory because it is a manifold on which one cannot define spinors (the relevant concepts here were discussed in §12.1). If one could have obtained at least $SU(3) \times SU(2)$ via compactification, one might have thought of doing physics with type IIA superstrings, which have a $U(1)$ gauge boson in the ten-dimensional supergravity multiplet, but this fails because one cannot even get $SU(3) \times SU(2)$ from compactification.

The fact that ten dimensions is not quite enough is by no means the only problem with trying to get a realistic gauge group from compactification. The other major problem is that, regardless of how many dimensions one starts with, one cannot obtain realistic fermion quantum numbers in four dimensions if the gauge groups are assumed to come entirely from symmetries of a compact manifold in the fashion we have been discussing. This can be shown to follow from a rather subtle theorem, first proved by Atiyah and Hirzebruch, about zero modes of the Dirac operator on manifolds with continuous symmetry; however, we will not enter here into a discussion of this theorem and its application to Kaluza–Klein theory. With the discovery of anomaly-free superstring theories that already have gauge groups in ten dimensions before compactification, and which have the relatively attractive phenomenology that we will be discussing in the remainder of this book, there is no longer a compelling need to try to get gauge symmetries via compactification. On the contrary, the $SO(32)$ and $E_8 \times E_8$ theories have more than enough gauge symmetry, the problem being to break the symmetries in the right way, not to obtain more symmetry.

So far we have discussed the zero modes of the components of the gravitational and gauge fields with all indices tangent to M^4, and the mixed components $g_{\mu k}$ of the metric with one index tangent to M^4 and one index tangent to K. It remains to discuss the zero modes of the components g_{ij} of the metric and A_k^a of the gauge field with all indices tangent to K. The very complexity of the equations governing those fields will make the discussion brief.

The components g_{ij} and A_j^a of the gravitational and gauge field enter (perhaps along with other fields) in the nonlinear equations that determine the structure of K. We are here assuming nothing about those equations except that they are generally covariant and gauge invariant. It might happen that the equations that determine the structure of K determine its structure uniquely. In this case, K is rigid, with no possibility of low-frequency oscillations. On the other hand, it might happen that the equations that determine the structure of K do not determine this structure uniquely. Integration constants ϕ_i might enter in solving the relevant equations. Such integration constants are usually called moduli in mathematical discussions; they are somewhat analogous to the moduli of Riemann surfaces, discussed in §3.3. Without further information, we cannot predict how many moduli there are, if any. But if undetermined integration constants appear in finding the structure of K, they will always be manifested in the effective four-dimensional theory in the form of massless spin zero particles. The reason for this is that it will inevitably be possible to consider a situation in which the moduli ϕ_i are not quite constant but are slowly varying functions $\phi_i(x^\lambda)$, x^λ being the coordinates of M^4. As in the usual proof of Goldstone's theorem (which governs a situation in which the undetermined integration constants in the vacuum are present because of a spontaneously broken symmetry), the $\phi_i(x^\lambda)$ will be seen in the low-energy theory as massless fields.

The occurrence of moduli represents a loss of predictive power in the theory, because observable quantities such as the fine-structure constant and particle masses are likely to depend on the undetermined integration constants. On the other hand, the massless particles associated with the moduli, if really present, would be an interesting and perhaps testable prediction. They might very well have coherent couplings to matter, in which case they might give rise to detectable deviations from the usual gravitational force laws. In addition, if such a scalar ϕ has coherent couplings to matter, its wave equation in the expanding universe is something like

$$(\partial_t^2 - \partial_i^2)\phi = \rho, \tag{14.5.5}$$

where ρ is the source of the ϕ field – some operator which we suppose has an expectation value in the universe. Averaging (14.5.5) over local inhomogeneities in the universe and so discarding the space derivatives, (14.5.5) reduces on a large scale to

$$\frac{d^2}{dt^2}\phi = \langle\rho\rangle, \tag{14.5.6}$$

where $\langle\rho\rangle$ is the average value of ρ in the universe, at given cosmic time. Equation (14.5.6) forces ϕ to change in time. Since coupling constants and mass ratios will almost certainly depend in a nontrivial way on the field ϕ (particularly since we suppose it has coherent couplings to matter), the time dependence of ϕ would be seen as a time dependence of the natural 'constants'.

The connection that we have discussed between moduli and massless particles is valid both classically (in an approximation in which the vacuum is governed by the classical equations which arise from superstring theory) and quantum mechanically. (In the exact theory, the vacuum is determined by minimizing some effective potential.) In the classical theory there is one general statement that can be made; there is always at least one modulus, the dilaton expectation value, because of a scaling argument that was described in §3.4.6 and §13.2. Concerning the exact quantum theory, we are not in a position at present to make any statement, but experimental bounds on the phenomena mentioned at the end of the last paragraph suggest that such moduli do not exist. The sharpest bounds come actually from the failure to observe changes in time of the natural constants.

14.6   Models of the Fermion Quantum Numbers

In this section we finally apply the tools that we have developed to work out a class of models of the quantum numbers of massless fermions that arise in four dimensions after compactification. Of particular note is the ease with which we will obtain, in four dimensions, several fermion 'generations', identical in their gauge quantum numbers, despite starting in ten dimensions with a single irreducible multiplet. The class of models that we present here is most interesting in the case of the heterotic $E_8 \times E_8$ theory, whose construction was described in chapter 6. The requisite facts about E_8 were explained in appendix 6.A.

We consider space-time to be a ten-dimensional manifold $M^4 \times K$, with M^4 being four-dimensional Minkowski space and K being some compact six-dimensional manifold. On general grounds, discussed in §14.2.3, we know that if we do not include an expectation value of the $E_8 \times E_8$ gauge field A_m tangent to K, then the chiral asymmetry of fermion quantum numbers in four dimensions will vanish. Since the chiral asymmetry depends only on the universality class of a theory (this is why we view it as a fundamental problem!), it is clear that if we do include gauge-field expectation values (while leaving $SU(3) \times SU(2) \times U(1)$ unbroken), the chiral asymmetry of four-dimensional fermions will be invariant under continuous variations of those expectation values. The chiral asymmetry of four-dimensional fermions can only depend on topological invariants associated with the gauge fields. This is indeed manifest in the index formula (14.2.8) which vanishes if the gauge fields are topologically trivial.

To obtain an interesting model of four-dimensional fermions, we must choose on K a topologically nontrivial configuration of $E_8 \times E_8$ gauge fields – while leaving $SU(3) \times SU(2) \times U(1)$ unbroken. Depending on K, there are many possibilities that we might consider for the gauge field expectation value. The possibilities correspond essentially to the classification of vector bundles on K. The range of possibilities is so vast that at first sight it may not be clear where to start in constructing an interesting model. However, one interesting possibility presents itself by analogy with our discussion of world-sheet phenomena in §14.4.1. Whenever one considers a six-dimensional Riemannian manifold K, there is one vector bundle that is implicit in the discussion. This is the tangent bundle. The gauge field that enters in parallel transport of tangent vectors is the spin connection $\omega_m{}^a{}_b$, which in six dimensions is an $SO(6)$ gauge field. By embedding $SO(6)$ in $E_8 \times E_8$, we can view the spin connection as an $E_8 \times E_8$ gauge field. In this way, we obtain a candidate gauge field on K for use in compactification, which has the attractive property that it involves relatively little by way of arbitrary choices, using only ingredients that are present anyway in any discussion of a six-dimensional Riemannian manifold K.

What embedding of $SO(6)$ in $E_8 \times E_8$ do we wish to consider? As discussed in appendix 6.A, E_8 has a maximal subgroup $SO(16)$, under which the adjoint representation of E_8 decomposes as $\mathbf{120} \oplus \mathbf{128}$, the $\mathbf{120}$ and $\mathbf{128}$ being respectively the adjoint representation and the positive-chirality spinor of $SO(16)$. $SO(16)$ has in turn a maximal subgroup $SO(10) \times SO(6)$, and it is that latter $SO(6)$ subgroup that we will use. Thus, we take the $SO(16)$ gauge field to be

$$A_i = \begin{pmatrix} 0 & 0 \\ 0 & \omega_i \end{pmatrix} \tag{14.6.1}$$

with nonzero entries only in a 6×6 block in the lower right-hand corner, where we place the spin connection ω_i.

In some sense, the $SO(6)$ subgroup of E_8 just identified is the smallest one (the Casimir operators are as small as possible, for example), so embedding $SO(6)$ in one of the two E_8 factors in the way just stated (and not at all in the second E_8) can be viewed as the minimal choice. There actually are some good theoretical reasons for making this choice. For instance, the equation associated with anomaly cancellation

$$dH = \operatorname{tr} R \wedge R - \operatorname{tr} F \wedge F \tag{14.6.2}$$

places a topological restriction on possible compactifications. The restriction is that the right-hand side, which represents a certain combination of characteristic classes, must be zero at the level of cohomology. This restriction is obeyed with the precise embedding given in (14.6.1); in that case the right-hand side of (14.6.2) is identically zero, not just zero as a differential form. With other embeddings of $SO(6)$ in $E_8 \times E_8$, since the quadratic $SO(6)$ Casimir operator would be larger, the right-hand side of (14.6.2) would not be identically zero and would typically not be zero even in cohomology.

If we do choose the embedding in (14.6.1), then the second E_8 is left unbroken while the first is broken to the subgroup of E_8 that commutes with $SO(6)$. The decomposition of the E_8 Lie algebra under $SO(10) \times SO(6)$,

$$\mathbf{248} = (\mathbf{45}, \mathbf{1}) \oplus (\mathbf{1}, \mathbf{15}) \oplus (\mathbf{10}, \mathbf{6}) \oplus (\mathbf{16}, \mathbf{4}) \oplus (\overline{\mathbf{16}}, \overline{\mathbf{4}}) \tag{14.6.3}$$

shows that the subgroup of E_8 that commutes with $SO(6)$ is precisely $SO(10)$. Thus, our ansatz breaks $E_8 \times E_8$ to $SO(10) \times E_8$.

This is already a significant step towards a phenomenologically viable model. E_8 is not a realistic candidate for a grand unified gauge group in four dimensions (since it only has real representations), but $SO(10)$ is. In fact, $SO(10)$ is known as the only orthogonal group that is a reasonable candidate for unification in four dimensions. The $\mathbf{16}$ of $SO(10)$ neatly accommodates a single generation of quarks and leptons (together with an unseen massive left-handed neutrino). Efforts to use other orthogonal groups for unification have led to severe difficulties. In the above discussion, $SO(10)$, rather than some other orthogonal group, appeared not because we contrived this but because $SO(16)$ is a maximal subgroup of E_8, and 6 is the number of compact dimensions.

What about the chiral asymmetry? We know from (14.2.13) how to compute the chiral asymmetry of fermions that transform in some representation Q_i of $SO(6)$ and a corresponding representation L_i of $SO(10)$. The net number N_{L_i} of chiral fermions that transform in four dimensions as L_i is simply

$$N_{L_i} = \text{index}_{Q_i} \, \slashed{D}_K, \tag{14.6.4}$$

the index of the Dirac operator $\slashed{D}_K$ on K for fermions that transform in the Q_i representation of $SO(6)$. We also know from our study of the Dirac operator that this index vanishes unless (14.6.4) is a complex representation. Inspection of (14.6.3) shows that the only complex representations of $SO(6)$ which appear are the **4** and $\overline{\bf 4}$. Fermions that transform as **4** (or $\overline{\bf 4}$) of $SO(6)$ transform as **16** (or $\overline{\bf 16}$) of $SO(10)$, so any chiral asymmetry that we obtain in four dimensions will be for fermions that transform in these representations. This is to the good, since as remarked above the **16** (or $\overline{\bf 16}$; the difference between them is a matter of convention) is the appropriate fermion representation for grand-unified models based on $SO(10)$. The fact that the construction described here automatically gives not only the correct group $SO(10)$ but also the correct fermion representation, the **16**, is a very encouraging sign.

What number of fermion generations will we obtain? In nature we observe so far three generations (left-handed **16**'s) and no antigenerations (left-handed $\overline{\bf 16}$'s). An index theorem will only predict what we will call N_{gen}, the difference between the number of left-handed **16**'s and left-handed $\overline{\bf 16}$'s in four dimensions. Of course, it is not surprising if at some energy scale generations and antigenerations should pair up and gain mass to the extent possible; in this case the number of antigenerations at low energy will be zero, and the number of generations will equal what we are calling N_{gen}. This is indeed what seems to have happened in nature, since antigenerations (mirror fermions with $V + A$ weak interactions) are not observed.

According to (14.6.4) and (14.6.3), the net number of chiral generations N_{gen} equals the index of the operator $\slashed{D}_K$ in the **4** of $SO(6)$. Happily, this is an index problem which we have already analyzed. The **4** of $SO(6)$ is the positive-chirality spinor representation. The Dirac operator $\slashed{D}_K$ acting on a field $\psi_{\alpha a}$, α being a spinor index and a being a Yang–Mills index labeling the **4** of the tangent space group $SO(6)$, is precisely the operator

$\tilde{S}$ discussed at the end of §14.3.4. Its index is given by (14.3.55) as

$$\text{index } \tilde{S} = \chi(K)/2, \qquad (14.6.5)$$

where $\chi(K)$ is the Euler characteristic of K. Actually, (14.6.5) is not quite what we should identify as the number of fermion generations in four dimensions. The difference between **16** and $\overline{\mathbf{16}}$ depends on a choice of basis in the Lie algebra of $SO(10)$, and physicists have simply by convention defined the observed fermions to be **16**'s rather than $\overline{\mathbf{16}}$'s of $SO(10)$. The number of generations in nature is thus positive more or less by definition, and so should be identified as the absolute value of (14.6.5):

$$N_{gen} = |\chi(K)/2|. \qquad (14.6.6)$$

This is our final result for the number of generations.

Equation (14.6.6) represents a fascinating possibility for the answer to the question of why nature chooses to replicate structure, presenting us at low energies with several generations of fermions with identical gauge quantum numbers. Beginning with an irreducible fermion multiplet in ten dimensions, we obtain according to (14.6.6) several generations in four dimensions, the number being determined by one of the really fundamental topological invariants of the compact manifold K, namely its Euler characteristic.

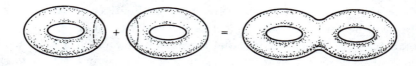

Figure 14.3. The connected sum $A + B$ of two spaces A and B is obtained by removing a hole from each and gluing them together along the boundary.

The Euler characteristics of various simple spaces can be calculated using the following tools. First, of all, we have the formula for the Euler characteristic derived in (14.3.52):

$$\chi(K) = \sum_p (-1)^p b_p. \qquad (14.6.7)$$

For a product of spaces $K = A \times B$ we have

$$\chi(A \times B) = \chi(A) \times \chi(B). \qquad (14.6.8)$$

This can be proved using the Künneth formula discussed in §12.4 in conjunction with (14.6.7), or alternatively by using the characterization of the Euler characteristic as the index of the operator S. For a connected sum of two spaces A and B (the connected sum $A + B$ is formed as in fig. 14.3 by removing hemispheres from A and B and gluing them together along the resulting boundary) we have

$$\chi(A + B) = \chi(A) + \chi(B) - 2. \qquad (14.6.9)$$

The correction term -2 in (14.6.9) is the Euler characteristic of the two hemispheres that are discarded in making the connected sum. Equation (14.6.9) is the n-dimensional generalization of the formula that was used at the end of chapter 12 to determine the Euler characteristic of a Riemann surface of any genus; and it can be proved in the same way. Equation (14.6.9) can actually be generalized to give the behavior of the Euler characteristic in an arbitrary cutting and pasting process and thereby to compute the Euler characteristic of any reasonably simple space.

The above formulas are adequate for working out many interesting examples. We know from §12.4 that the nonzero Betti numbers of a sphere S^n are $b_0 = b_n = 1$. For $K = S^6$, (14.6.7) therefore gives $\chi = 2$ and one generation. For $K = S^2 \times S^4$, (14.6.8) gives $\chi = 4$, corresponding to two generations. For $K = S^2 \times S^2 \times S^2$, we get $\chi = 8$ or four generations. The connected sum $K = A + B$, with $A = B = S^2 \times S^4$, would have $\chi = 4 + 4 - 2 = 6$, corresponding to a three generation model. Clearly, any desired number of generations can be obtained in this way. Without additional physical principles, we cannot predict the number of generations. But it is still very satisfying to see the group $SO(10)$ and the correct fermion representation emerging in a natural way. These successes are unlikely to be entirely accidental, and the relation between the number of generations and the topology of K is very possibly the seed of an eventual explanation of the origin of flavor.

In chapter 16, we will explore a refinement of the model considered here, in which the holonomy group of K is not $SO(6)$ but $SU(3)$; the unbroken $SO(10)$ of the above discussion is then enlarged to E_6, and the massless fermions are in **27**'s of E_6, not **16**'s of $SO(10)$. Many other possible approaches to making detailed models of the fermion quantum numbers have been discussed in recent literature.

14.7   Anomaly Cancellation in Four Dimensions

There is one further theoretical point that we wish to explore. A theory is anomaly free if the effective action is invariant under gauge and coordinate transformations. If such invariance holds exactly in a given theory, it must hold in any valid approximation such as reduction to a low-energy four-dimensional effective theory after compactification. Thus, the anomaly-free ten-dimensional theories must reduce in four dimensions to anomaly-free four-dimensional theories. Let us see how this works.

Let F_0 and R_0 be the vacuum expectation values of the Yang–Mills field strength and the Riemann tensor on K. It is not necessary for the analysis here that the background fields satisfy the field equations, but it is necessary that the four form

$$\text{tr}R_0^2 - \text{tr}F_0^2 \tag{14.7.1}$$

should be zero in cohomology, so that the Bianchi identity

$$dH = \text{tr}R_0^2 - \text{tr}F_0^2 \tag{14.7.2}$$

has a solution. Let G denote the ten-dimensional gauge group $SO(32)$ or $E_8 \times E_8$ and suppose that the components of G that have expectation values lie in some subgroup J of G. The unbroken subgroup H in four dimensions is the subgroup of G that commutes with J. The adjoint representation A of G decomposes under $H \times J$ into a sum of terms

$$A \approx \oplus_i L_i \otimes Q_i, \tag{14.7.3}$$

where the L_i are irreducible representations of H and the Q_i are irreducible representations of J. In particular,

$$\sum_i \dim L_i \cdot \dim Q_i = 496. \tag{14.7.4}$$

Now we consider the Dirac equation for ten-dimensional 'gluino' fields χ, which transform as (14.7.3). According to (14.1.4), the massless ten-dimensional Dirac equation $\displaystyle{\not}D_{10}\chi = 0$ can be written

$${\not}D_4\chi + {\not}D_K\chi = 0, \tag{14.7.5}$$

where ${\not}D_4$ and ${\not}D_K$ refer to Dirac operators on M_4 and K, respectively.

According to (14.2.8), the net chiral asymmetry of four-dimensional fermions that transform as L_i under the unbroken group H is

$$n_i = n_i^L - n_i^R = \frac{1}{48(2\pi)^3} \int_K [\mathrm{tr}_{Q_i} F_0^3 - \frac{1}{8}\mathrm{tr}_{Q_i} F_0 \mathrm{tr} R_0^2]. \qquad (14.7.6)$$

The symbol tr_{Q_i} means that the trace is taken in the Q_i representation of H.

Let T be a generator of H. The T^3 anomaly for four-dimensional massless fermions in the L_i representation of J is just $\mathrm{tr}_{L_i} T^3$; summing over i, with n_i multiplets of each type, the total anomaly is

$$I = \sum n_i \mathrm{tr}_{L_i} T^3. \qquad (14.7.7)$$

Only complex representations L_i can contribute in the sum since $\mathrm{tr} T^3 = 0$ for real representations.

Let X and Y be generators (or products of generators) of H and J respectively. Then it follows from (14.7.3) that

$$\sum_i \mathrm{tr}_{L_i} X \mathrm{tr}_{Q_i} Y = \mathrm{Tr}(XY), \qquad (14.7.8)$$

where Tr is the trace in the 496-dimensional adjoint representation of G. The point in (14.7.8) is that the sum over i of $\mathrm{tr}_{L_i} \cdot \mathrm{tr}_{Q_i}$ reconstructs the trace in the adjoint representation of G. We apply this with X being T^3 and Y being F_0 or F_0^3. So (14.7.7) becomes, with the aid of (14.7.6),

$$I = \int_K (\mathrm{Tr} T^3 F_0^3 - \tfrac{1}{8}\mathrm{Tr} T^3 F_0 \mathrm{tr} R_0^2). \qquad (14.7.9)$$

The next step is to recast $\mathrm{Tr} T^3 F_0^3$ using the identity

$$\mathrm{Tr} F^6 = \frac{1}{48}\mathrm{Tr} F^2 \mathrm{Tr} F^4 - \frac{1}{14,400}(\mathrm{Tr} F^2)^3 \qquad (14.7.10)$$

from §13.5.3, which plays an important role in canceling anomalies. Applying that equation for an arbitrary linear combination $\alpha T + \beta F_0$ and extracting the coefficient of $\alpha^3 \beta^3$ gives

$$\mathrm{Tr} T^3 F_0^3 = \tfrac{1}{240}\mathrm{Tr} T^3 F_0 \mathrm{Tr} F_0^2. \qquad (14.7.11)$$

In writing this we have dropped terms of the form $\mathrm{Tr} T F_0$ and $\mathrm{Tr} T F_0^3$, which vanish if H is semisimple. We comment on the more general case later. Substituting (14.7.11) in (14.7.9) gives

$$I = -\frac{1}{8} \int_K \mathrm{Tr} T^3 F_0 (\mathrm{tr} R_0^2 - \mathrm{tr} F_0^2). \qquad (14.7.12)$$

Substituting $\mathrm{tr} R_0^2 - \mathrm{tr} F_0^2 = dH$ and integrating by parts gives zero since

$$\mathrm{Tr} T^3 F_0 = \sum_i \mathrm{tr}_{L_i} T^3 \mathrm{tr}_{Q_i} F_0 \qquad (14.7.13)$$

and by the abelian Bianchi identity, $d\, \mathrm{tr}_{Q_i} F_0 = 0$.

In the above argument, we used the assumption that H is semisimple in order to drop certain terms from (14.7.11). At first sight, it might appear that otherwise, $U(1)$ components of H will be anomalous. This cannot be, since the gauge-invariant effective action of an anomaly-free ten-dimensional theory must remain gauge invariant under any valid approximation. What ingredients are we missing? When H is not semisimple, a new physical phenomenon is possible. $U(1)$ components of H can gain mass by absorbing some of the would be massless axions discussed in §14.3.2 above. Precisely when (14.7.11) seems to predict anomalies for a given $U(1)$ gauge field, that gauge field is eliminated from the low-energy spectrum by combining with an 'axion' into a massive gauge meson. We shall not explore this phenomenon, however; the interested reader is urged to tackle this as an exercise or consult the references.

Four-dimensional mixed gauge-gravitational anomalies can be analyzed in a similar way; the 'axionic' phenomenon just mentioned is important in the analysis, since in any case four-dimensional mixed anomalies do not arise if H is semisimple. Purely gravitational anomalies are not possible in four dimensions.

15. Some Algebraic Geometry

In chapters 12 and 14 we developed some simple tools in differential geometry and used them to gain some insight concerning the compactification of hidden dimensions as well as some insight concerning phenomena on the string world sheet. We now turn our attention to some more specialized mathematical tools involving complex manifolds and algebraic geometry. Again, the motivation is twofold. The world sheet of a string is a complex manifold – a Riemann surface, in fact – and as string theory develops, the deeper study of world-sheet phenomena is likely to involve deeper aspects of algebraic geometry, which have already begun to enter in recent works on multiloop diagrams. Also, algebraic geometry has been a tool in recent attempts to formulate more realistic models of string compactification.

In this chapter, we develop some of the basic concepts of complex geometry, with examples selected for their role both in world-sheet phenomena and in the study of compactification. We will unfortunately not be able to describe in this book recent work on the application of algebraic geometry to multi-loop diagrams. This subject is probably not yet ripe for synthesis; and the requisite mathematical machinery is more extensive than we will be able to present even in this moderately lengthy chapter. By laying at least some of the elementary foundations we hope to facilitate the task of the reader who wishes to delve further elsewhere. We will try to present a moderately thorough survey of applications of algebraic geometry to string compactification. This subject serves – along with world-sheet phenomena – as the motivation for our mathematical work in this chapter; it will also be the theme of the next chapter.

1̄5.1 Low-Energy Supersymmetry

Since the analysis of low-energy supersymmetry is the application of algebraic geometry that we will explore in detail, it seems appropriate to begin by discussing the motivation for this analysis.

15.1.1 Motivation

One striking motivation for thinking about low-energy supersymmetry is the gauge-hierarchy problem, the question of why the mass scale of weak-interaction symmetry breaking is so tiny compared to more fundamental scales such as the Planck mass. We do not know how to solve this problem, but one necessary ingredient is presumably that the ordinary $SU(2) \times U(1)$ Higgs doublet must remain massless at the compactification scale and indeed to within extraordinary precision.[*] This is rather puzzling. There are many mechanisms (index theorems and chiral symmetries, for example) that can yield massless charged fermions, but it is difficult to explain the existence of massless charged spin zero fields. Goldstone bosons are massless, but they are always neutral under unbroken gauge symmetries. In fact, the only known way to explain the existence of massless charged scalars is to postulate the existence of an unbroken supersymmetry of the low-energy world. Under this assumption, massless charged scalars can naturally arise as supersymmetric partners of massless charged fermions. Of course, if we suppose that the $SU(2) \times U(1)$ Higgs doublet would be exactly massless in the limit of unbroken supersymmetry, then the tiny but nonzero scale of $SU(2) \times U(1)$ breaking in the real world must be related to a small scale of supersymmetry breaking.

A second motivation for studying conditions for unbroken supersymmetry is that this is one way to find solutions of the equations of motion of the theory. A state of unbroken supersymmetry in four dimensions always obeys the equations of motion. This is most obvious in global supersymmetry, where the Hamiltonian is positive semi-definite and vanishes when and only when supersymmetry is unbroken. It is also true in supergravity. At present, states of unbroken supersymmetry are very nearly the only examples known of compactified solutions of the equations; the other known examples are related in comparatively simple ways to states of unbroken supersymmetry.[†]

The last major motivation for studying configurations with unbroken supersymmetry is that the hypothesis of unbroken supersymmetry is very restrictive, but not too restrictive, for phenomenology. More precisely, it is the hypothesis of unbroken $N = 1$ supersymmetry in four dimensions that has these virtues. A larger unbroken supersymmetry algebra could

---

[*] There is the alternative possibility that the $SU(2) \times U(1)$ Higgs doublet is not an elementary field but a composite that forms from dynamical symmetry breaking. This fascinating idea seems to be plagued with innumerable phenomenological difficulties, and has not so far been the basis for interesting developments in string theory.

[†] For example, it is possible to have spontaneously broken supersymmetry with orbifolds (see §9.5.2); it is also possible on manifolds of $SU(3)$ holonomy to break supersymmetry by working with a spin structure that does not contain the covariantly constant spinor.

not lead to a realistic model, because in four-dimensional supersymmetric theories with $N \geq 2$ supersymmetry, the massless fermions always transform in a real representation of the gauge group, in stark contrast to what is observed in nature. Thus it is really the conditions for unbroken $N = 1$ supersymmetry in four dimensions that will interest us.

We will concentrate on the $SO(32)$ and $E_8 \times E_8$ theories because they have elementary gauge fields and can generate chiral fermions in four dimensions. As we will see, their phenomenology is rather interesting, especially in the $E_8 \times E_8$ case. Many considerations that we will develop could, however, be carried out for the type II theories as well. Low-energy supersymmetry is not an issue for the $SO(16) \times SO(16)$ theory, since it does not possess unbroken supersymmetry even in ten dimensions.

Understanding the conditions for unbroken supersymmetry will lead us into the fascinating terrain of algebraic geometry, a subject which also has applications to world-sheet phenomena, as noted earlier. The fact that similar mathematical tools enter in both problems is indeed one reason for being hopeful that the approach to compactification based on algebraic geometry may bear an element of truth.

15.1.2 Conditions for Unbroken Supersymmetry

In a locally supersymmetric theory, the infinitesimal supersymmetry parameter $\eta_\alpha(X^M)$ can have an arbitrary dependence on the space-time coordinates X^M. For each choice of $\eta_\alpha(X)$, there is a corresponding conserved supercharge Q. Out of this infinity of conserved supercharges, we wish to identify those that generate unbroken supersymmetries. An unbroken supersymmetry Q is simply a conserved supercharge that annihilates the vacuum state $|\Omega\rangle$. Saying that Q annihilates $|\Omega\rangle$ is equivalent to saying that for all operators U, $\langle\Omega| \{Q, U\} |\Omega\rangle = 0$. This will certainly be so if U is a bosonic operator, since then $\{Q, U\}$ is fermionic, so the real issue is whether $\langle\Omega| \{Q, U\} |\Omega\rangle$ vanishes when U is a fermionic operator.

Now, when U is fermionic $\{Q, U\}$ is simply δU, the variation of U under the supersymmetry transformation generated by Q. Also, in the classical limit, δU and $\langle\Omega| \delta U |\Omega\rangle$ coincide. So finding an unbroken supersymmetry at tree level means finding a supersymmetry transformation such that $\delta U = 0$ for every fermionic field U. Also, in the classical limit, it is enough to check this for the elementary fermion fields.

In the low-energy effective field theory in ten dimensions, the only elementary fermions are the gravitino ψ_M, the spin one-half 'dilatino' λ, and the gluinos χ^a. As we discussed in chapter 13, their supersymmetry variations are

$$\delta\psi_M = \frac{1}{\kappa}D_M\eta + \frac{\kappa}{32g^2\phi}(\Gamma_M{}^{NPQ} - 9\delta_M^N\Gamma^{PQ})\eta H_{NPQ} + (\text{Fermi})^2,$$

$$\delta\chi^a = -\frac{1}{4g\sqrt{\phi}}\Gamma^{MN}F_{MN}^a\eta + (\text{Fermi})^2,$$

$$\delta\lambda = -\frac{1}{\sqrt{2}\phi}(\Gamma\cdot\partial\phi)\eta + \frac{\kappa}{8\sqrt{2}g^2\phi}\Gamma^{MNP}\eta H_{MNP} + (\text{Fermi})^2.$$

$$(15.1.1)$$

Here ϕ is the dilaton field, F_{MN}^a is the Yang–Mills field strength, and H is the gauge-invariant field strength of the antisymmetric tensor field B_{MN}. While we can arbitrarily specify the space-time metric and the dilaton field ϕ in trying to obey $\delta\psi_M = \delta\lambda = \delta\chi = 0$, we cannot arbitrarily specify F or H; they must obey certain Bianchi identities. In the minimal ten-dimensional field theory, the Bianchi identity for H is $dH = -\text{tr}F\wedge F$, but as we have seen in chapter 13, string theory corrects this to

$$dH = \text{tr}R\wedge R - \text{tr}F\wedge F. \qquad (15.1.2)$$

Since the stringy correction in (15.1.2) plays an important role in the discussion, one may ask whether other string-theoretic corrections to (15.1.1) can be neglected. In principle, we would like to search for vacuum states not just in the limiting low-energy field theory but in the exact string theory with all its massive degrees of freedom. To do so by studying the string-theoretic generalization of (15.1.1) directly would be rather difficult; one would either have to study the supersymmetry variations of the whole infinite tower of massive fields or, integrating them out, one would have to study an infinite series of corrections to (15.1.1) containing terms of higher and higher order in α'. In either case the appropriate equations are not known in closed form. Despite this, we will be able to show in §16.6.3 that the results that we obtain from the limiting low-energy field theory are valid to all finite orders in α'.

The analysis of (15.1.1) is rather complicated in general, and so we simplify the discussion by assuming at the outset that the gauge-invariant three form H vanishes, and that the dilaton ϕ is a constant. In §16.7, after developing some necessary tools, we will return to the problem and find a more general class of supersymmetric vacuum states with nonzero H and nonconstant ϕ. With our simplifying assumptions, the condition for finding a supersymmetry generator η that leaves the vacuum invariant reduces to

$$0 = \delta\psi_M = D_M\eta$$
$$0 = \delta\chi^a = \Gamma^{ij}F_{ij}^a\eta. \qquad (15.1.3)$$

Thus, the first of these equations says that a generator η of an unbroken supersymmetry should be covariantly constant. This is an extremely strong condition because it implies the integrability condition $[D_M, D_N]\eta = 0$ or, in other words,

$$R_{MNPQ}\Gamma^{PQ}\eta = 0, \tag{15.1.4}$$

with R_{MNPQ} being the Riemann tensor.

We will look for a vacuum state in which ten-dimensional space-time is of the form $T^4 \times K$, where T^4 is a maximally symmetric four-dimensional space (de Sitter space, anti de Sitter space or Minkowski space) and K is a compact six manifold. We will now show, however, that (15.1.4) implies the vanishing of the four-dimensional cosmological constant, thereby eliminating the de Sitter and anti de Sitter possibilities. Recall our convention that an index M, N, P is tangent to ten-dimensional space-time while an index μ, ν, λ is tangent to T^4 and an index i, j, k is tangent to K. By definition, the maximally symmetric space T^4 has curvature tensor $R_{\mu\nu\alpha\beta} = (r/12)(g_{\mu\alpha}g_{\nu\beta} - g_{\mu\beta}g_{\nu\alpha})$, r being the four-dimensional Ricci scalar. Equation (15.1.4) with $M, N = 1, \ldots, 4$ now immediately implies that $r = 0$ so that the maximally symmetric space T^4 must in fact be flat Minkowski space, as we would hope.

This is a nontrivial result, since in four-dimensional supergravity theories unbroken supersymmetry does not necessarily imply the vanishing of the cosmological constant. For instance, unbroken $N = 1$ supersymmetry in four dimensions is possible in either Minkowski space or anti de Sitter space. While unbroken supersymmetry does not automatically lead to vanishing cosmological constant, it does so in the circumstances at hand. We might consider this a phenomenological success, although it hardly touches on the real mystery of the cosmological constant, which is why it vanishes *after* supersymmetry breaking.

Returning to (15.1.3), with T^4 being flat Minkowski space M^4, the components of the first equation with $M = 1, \ldots, 4$ tell us that η must be independent of the four uncompactified coordinates. The first equation in (15.1.3) thus reduces to a statement about the compact six manifold K; it must be possible on K to find a spinor field η that is covariantly constant. In the next section, we will explore the consequences of this statement. In the process, we will begin our encounter with algebraic geometry. Eventually we will develop the tools to understand also the second equation in (15.1.3).

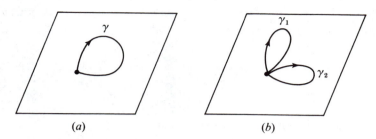

Figure 15.1. Upon parallel transport around a closed curve, tangent vectors are transformed by an orthogonal matrix obtained by integrating the spin connection around the curve, as shown in (*a*). The matrices obtained in this way always form a group, because closed curves can be composed as in (*b*).

15.1.3 Manifolds of $SU(3)$ Holonomy

On a Riemannian manifold K of dimension n, the spin connection ω is, in general, an $SO(n)$ gauge field. Upon parallel transport around a contractible closed curve γ, a physical field ψ is transformed into $U\psi$, where $U = P \exp \int_\gamma \omega \cdot dx$ is the path-ordered exponential of ω around the curve γ. The $SO(n)$ matrices U that can be obtained in this way always form a group H, called the holonomy group of the manifold. (H forms a group because if parallel transport around a curve γ_1 gives an $SO(n)$ matrix U_1, and parallel transport around a curve γ_2 gives an $SO(n)$ matrix U_2, then the product matrix $U_1 \cdot U_2$ arises from parallel transport around the composition curve $\gamma_1 \cdot \gamma_2$ indicated in fig. 15.1*b*. Since we restrict ourselves to contractible loops, what we have defined is called the local holonomy group. Global properties will enter later.)

Generically, H is all of $SO(n)$, but we will be interested in cases in which this is not so. In fact, for the 'physical' case of $n = 6$, we would like to find conditions under which K admits a spinor field η obeying $D_i \eta = 0$. As in gauge theories, a covariantly constant field η always returns to its original value upon parallel transport around a contractible closed curve. Thus, the holonomy group of a manifold that admits a covariantly constant spinor field η consists of $SO(6)$ matrices U that obey $U\eta = \eta$. What is the subgroup of $SO(6)$ that obeys this condition? The Lie algebra of $SO(6)$ is isomorphic to that of $SU(4)$, and the positive- and negative-chirality spinors of $SO(6)$ are the fundamental **4** and $\overline{\mathbf{4}}$ of $SU(4)$, respectively.[*] We may as well suppose that η has positive chirality

---

[*] These facts were explained in appendix 6.A.

and thus belongs to the **4** of $SU(4)$. The subgroup of $SU(4)$ that leaves fixed an element of the **4** is $SU(3)$. In fact, by an $SU(4)$ transformation we can always put η in the form

$$\eta = \begin{pmatrix} 0 \\ 0 \\ 0 \\ \eta_0 \end{pmatrix}, \tag{15.1.5}$$

and the unbroken $SU(3)$ group is simply the subgroup of $SU(4)$ that acts on the first three components.

Since the **4** of $SU(4)$ contains precisely one $SU(3)$ singlet, a six manifold K whose holonomy is precisely $SU(3)$, rather than a subgroup, admits precisely one covariantly constant spinor field η of positive chirality. The complex conjugate $\bar{\eta}$ of η is then the unique covariantly constant spinor field of negative chirality. If we desire unbroken $N = 1$ supersymmetry in four dimensions, then the holonomy group of K should be precisely $SU(3)$. To see this, note that if we denote the positive- and negative-chirality spinor representations of $SO(1,3)$ as **2** and **2'**, then under an $SO(1,3) \times SO(6)$ or $SO(1,3) \times SU(4)$ subgroup of $SO(1,9)$, the 16-component positive-chirality spinor of $SO(1,9)$ decomposes as $\mathbf{16} \approx (\mathbf{2,4}) \oplus (\mathbf{2',\bar{4}})$.[†] If the **4** and $\bar{\mathbf{4}}$ both contain a single covariantly constant spinor field, then the covariantly constant components of the **16** will transform under $SO(1,3)$ as $\mathbf{2} \oplus \mathbf{2'}$. But $\mathbf{2} \oplus \mathbf{2'}$ (with the **2'** being the complex conjugate of the **2**, since the **16** of $SO(1,9)$ is real or alternatively because the two spinor representations of $SO(1,3)$ are complex conjugates of one another) corresponds to a single real four-component Majorana spinor of $SO(1,3)$. So $SU(3)$ holonomy of K will yield unbroken $N = 1$ supersymmetry in four dimensions.

Our next task is to understand what sort of manifold K will admit a metric of $SU(3)$ holonomy. The basic tool is the fact that if η is covariantly constant, then so is any tensor field constructed just from products of η with itself. One of the relevant tensors is the 'Kähler form', $k_{ij} = \bar{\eta}\Gamma_{ij}\eta$. Another, even more basic in a sense, is the 'complex structure', $J^i{}_j = g^{ik}k_{kj}$. Finally, we will meet the 'holomorphic volume form', $\omega_{ijk} = \eta^T \Gamma_{ijk}\eta$. In coming to grips with what it means for a manifold to admit a metric of $SU(3)$ holonomy, we will consider in turn the consequences

---

[†] This amounts to saying that in the **16** of $SO(1,9)$, positive (or negative) $SO(1,3)$ chirality is correlated with positive or negative $SO(6)$ chirality; it is equivalent to the statement $\Gamma^{(4)} = \Gamma^{(K)}$ derived in §14.2.1.

of the existence of the covariantly constant complex structure, Kähler form and holomorphic volume form. In the process we will meet many of the foundational concepts of algebraic geometry, relevant also in other applications such as applications to world-sheet phenomena.

15.2   Complex Manifolds

The embedding of $SU(3)$ or $U(3)$ in $SO(6)$ can be thought of in the following way. There is, of course, no real number whose square is -1, but we can easily find a 2×2 real matrix whose square is -1, namely the matrix

$$I = \begin{pmatrix} 0 & 1 \\ -1 & 0 \end{pmatrix}. \tag{15.2.1}$$

A $U(3)$ matrix is a 3×3 matrix of complex numbers,

$$U = \begin{pmatrix} a_{11} & a_{12} & a_{13} \\ a_{21} & a_{22} & a_{23} \\ a_{31} & a_{32} & a_{33} \end{pmatrix} \tag{15.2.2}$$

with matrix elements that are complex numbers a_{ij}. Replacing the imaginary unit i by the 2×2 real matrix I, we can represent the complex number $a_{ij} = \mathrm{Re}\, a_{ij} + i \mathrm{Im}\, a_{ij}$ by the 2×2 real matrix $\tilde{a}_{ij} = \mathrm{Re}\, a_{ij} + I \cdot \mathrm{Im} a_{ij}$. Turning every complex number into a 2×2 real matrix in this way, the 3×3 complex matrix U becomes a 6×6 real matrix $\tilde{U}$. If U is unitary, then $\tilde{U}$ is orthogonal. This is the embedding of $U(3)$ in $SO(6)$.

In particular, the $U(3)$ matrix $U = i$ (in other words, the matrix with matrix elements $a_{ij} = i \cdot \delta_{ij}$) is represented in this way as the 6×6 real matrix

$$\overline{I} = \begin{pmatrix} 0 & 1 & 0 & 0 & 0 & 0 \\ -1 & 0 & 0 & 0 & 0 & 0 \\ 0 & 0 & 0 & 1 & 0 & 0 \\ 0 & 0 & -1 & 0 & 0 & 0 \\ 0 & 0 & 0 & 0 & 0 & 1 \\ 0 & 0 & 0 & 0 & -1 & 0 \end{pmatrix}. \tag{15.2.3}$$

Apart from the identity matrix, $\overline{I}$ is the only $U(3)$ (or $SU(3)$) invariant matrix in the fundamental **6** of $SO(6)$. After all, to commute with $U(3)$ a matrix must be a constant in each irreducible $U(3)$ representation. As the decomposition of the **6** under $U(3)$ contains only two terms ($\mathbf{6} \approx \mathbf{3} \oplus \mathbf{\overline{3}}$), there only are two matrices in the **6** of $SO(6)$ that commute with $U(3)$. They are the identity matrix and $\overline{I}$. In particular, $\overline{I}$ can be uniquely described (up to normalization) by saying that it commutes with $U(3)$ and is real and traceless.

15.2.1 Almost Complex Structure

Now let us return to our manifold K of $SU(3)$ holonomy, with its covariantly constant spinor field $\eta(y)$ of positive chirality. On this manifold we can define, as at the end of the last section, the tensor field $J^i{}_j(y) = g^{ik}(y)\overline{\eta}(y)\Gamma_{kj}\eta(y)$. For each $y \in K$, $J^i{}_j$ can be viewed as a matrix acting on tangent vectors, the action being $v^i \to J^i{}_j v^j$, for any tangent vector v^i. Viewed in this way as a matrix acting on tangent vectors, J is real, traceless and $SU(3)$ invariant, so (with proper normalization) J coincides with the matrix $\overline{I}$ defined in (15.2.3). In particular, J obeys $(J^2)^i{}_j = -\delta^i{}_j$, or more succinctly $J^2 = -1$. A tensor field J on a manifold (with one covariant and one contravariant index) that obeys $J^2 = -1$ is called an *almost complex structure*.

The matrix J or equivalently the matrix $\overline{I}$ of equation (15.2.3) certainly cannot be diagonalized over the real numbers, but it can be diagonalized over the complex numbers. J is in fact the matrix that assigns the value i or $-i$ to states in the **3** or $\overline{\mathbf{3}}$ of $U(3)$. At any point y in K, there is a suitable basis of complex coordinates z^a, $a = 1, 2, 3$ and their complex conjugates $\overline{z}^{\overline{a}}$, $\overline{a} = 1, 2, 3$, in which J takes the form $J^a{}_b = i\delta^a{}_b, J^{\overline{a}}{}_{\overline{b}} = -i\delta^{\overline{a}}{}_{\overline{b}}$, with other components zero. We will call this the canonical form of J. Relative to a basis y^i, $i = 1, \dots, 6$, in which J takes the form of (15.2.3), the z^a can be defined as $z^1 = y^1 + iy^2, z^2 = y^3 + iy^4, z^3 = y^5 + iy^6$. This choice of the z^a is, of course, only unique up to a $U(3)$ transformation.

15.2.2 The Nijenhuis Tensor

Given a manifold K with an almost complex structure J and any one point p in K, one can find local coordinates $z^a, \overline{z}^{\overline{a}}$ in which J takes the canonical form at that one point p. We would now like to ask, 'Can we choose complex coordinates to put J in the canonical form not just at the one point p but in a whole open set containing p?' Coordinates with this property may be called local holomorphic coordinates. If local holomorphic coordinates exist (in a neighborhood of each point $p \in K$), then the almost complex structure J is said to be integrable. Determining which almost complex structures are integrable is at least qualitatively similar to a problem that is familiar from general relativity.

Suppose that we are given on K not a tensor field $J^i{}_j$ with one contravariant and one covariant index, but a symmetric tensor field g_{ij} with two covariant indices. (In most applications g is the metric tensor, but this is irrelevant at the moment.) By simple linear algebra, one can show that given any point $p \in K$, there is a coordinate system such that g takes the standard form $g_{ij} = \delta_{ij}$ at p. (This is true in the 'locally inertial coordinates' of general relativity.) Now we would like to ask whether we can find coordinates which will put g in the standard form not just at the one point p but in a whole neighborhood of p. Such a coordinate system might be called a flat coordinate system.

To determine when a flat coordinate system exists requires certain fundamental concepts. One tries to define a tensor field made from derivatives of g. It turns out there is essentially only one such tensor field, the Riemann tensor R_{ijkl}. (All others can be made from R and its derivatives, together with the metric.) A necessary condition for existence of a flat coordinate system in a whole neighborhood of a point p is that R_{ijkl} should vanish in this neighborhood. (This condition is necessary, since if a coordinate system with $g_{ij} = \delta_{ij}$ exists, then evaluating R_{ijkl} in this coordinate system gives immediately $R_{ijkl} = 0$.) It is not too hard to prove the converse; if $R_{ijkl} = 0$, there is at least locally a flat coordinate system.[*]

Now we would like to carry out an analogous argument in the case of an almost complex structure $J^i{}_j$. We first try to define a new tensor field constructed from $J^i{}_j$ and its derivatives. There is essentially one such tensor field, the Nijenhuis tensor

$$N^k{}_{ij} = J^l{}_i(\partial_l J^k{}_j - \partial_j J^k{}_l) - J^l{}_j(\partial_l J^k{}_i - \partial_i J^k{}_l). \qquad (15.2.4)$$

To prove that N is a tensor is straightforward. One approach is to intro-

---

[*] Such a coordinate system might not exist globally; it does not exist globally if K is, for instance, a torus or Klein bottle. The proper global statement is as follows. If $R_{ijkl} = 0$ everywhere on K, then it is possible to cover K with open sets $O_{(\alpha)}$ and to choose on each $O_{(\alpha)}$ flat coordinates $y^i_{(\alpha)}$. On intersection regions $O_{(\alpha)} \cap O_{(\beta)}$, $y^i_{(\alpha)}$ and $y^i_{(\beta)}$ are related by $y^i_{(\alpha)} = M_{(\alpha\beta)}{}^i{}_j y^j_{(\beta)} + \phi_{(\alpha\beta)}{}^i$, where (for each α and β) $M_{(\alpha\beta)}$ is an orthogonal matrix and $\phi_{(\alpha\beta)}$ is a constant. M and ϕ represent rigid rotations and translations of the coordinates, respectively. The $y^i_{(\alpha)}$ and $y^i_{(\beta)}$ differ (at most) by a rotation plus translation, because this is the most general change of coordinates from $y^i_{(\alpha)}$ to $y^i_{(\beta)}$ that is compatible with the fact that the metric is in standard form in each coordinate system. In the case of complex manifolds, it is an arbitrary holomorphic change of coordinates that plays the role of M and ϕ.

duce an arbitrary metric g on the manifold K and define

$$\overline{N}^k{}_{ij} = J^l{}_i(D_l J^k{}_j - D_j J^k{}_l) - J^l{}_j(D_l J^k{}_i - D_i J^k{}_l), \qquad (15.2.5)$$

with D_l denoting the covariant derivative based on the affine connection calculated from g. Then $\overline{N}$ (like any other expression made from J and its covariant derivatives) is certainly a tensor. On the other hand, using the definition of the covariant derivative in general relativity and the fact that $J^2 = -1$, one easily sees that $\overline{N} = N$. Thus, N is a tensor made from J and its derivatives *without* need to use an auxiliary metric; as such, it is essentially unique.

Now, just as in the analogous problem in general relativity, if we can find local holomorphic coordinates that put J in the standard form in a whole neighborhood of a point p, then evaluating N in this coordinate system shows immediately that $N = 0$. The converse is a rather difficult theorem due to Newlander and Nirenberg. According to the Newlander–Nirenberg theorem, if $N = 0$ then given any point $p \in K$ it is possible to find local holomorphic coordinates in a whole neighborhood of p. Thus, an almost complex structure is integrable if and only if $N = 0$. An integrable almost complex structure is called a complex structure. A manifold K endowed with a complex structure J is called a complex manifold.

In general (as in the real case discussed in a previous footnote), the local holomorphic coordinates z^a of a complex manifold exist only locally. By definition, however, one can always cover a complex manifold by open sets $O_{(\alpha)}$ on each of which we have a local holomorphic coordinate system $z^a_{(\alpha)}$ which puts the complex structure J in the standard form

$$J^a{}_b = i\delta^a{}_b, \quad J^{\bar{a}}{}_{\bar{b}} = -i\delta^{\bar{a}}{}_{\bar{b}}, \qquad (15.2.6)$$

other components zero. How are the $z^a_{(\alpha)}$ and $z^a_{(\beta)}$ related on overlap regions $O_{(\alpha)} \cap O_{(\beta)}$? Evidently, they must be related by a change of coordinates that preserves the standard form of J. It is easily seen that the coordinate transformations that preserve the standard form of J are the holomorphic (analytic) coordinate transformations $z^a \to \tilde{z}^b = \tilde{z}^b(z^a)$ (that is, the $\tilde{z}^b$ are functions of the z^a, and not of their complex conjugates $\overline{z^a}$). Thus, the local holomorphic coordinates $z^a_{(\alpha)}$ and $z^a_{(\beta)}$ are holomorphic (analytic) functions of one another on overlap regions $O_{(\alpha)} \cap O_{(\beta)}$. The converse to this is also true and gives an alternative way to express what it means to have a complex manifold. Suppose that we are given a manifold K of dimension $2N$. Suppose that K can be covered by open sets $O_{(\alpha)}$ on each of which we can find local complex coordinates $z^a_{(\alpha)}$ with the property that on intersection regions $O_{(\alpha)} \cap O_{(\beta)}$, the $z^b_{(\beta)}$ are analytic functions of the $z^a_{(\alpha)}$. Then K is a complex manifold. To show this, define on each $O_{(\alpha)}$ a tensor field $J_{(\alpha)}$ with one covariant and one contravariant index by saying that its nonzero components in the $z^a_{(\alpha)}$ coordinate system are $J_{(\alpha)}{}^a{}_b = i\delta^a{}_b, J_{(\alpha)}{}^{\bar{a}}{}_{\bar{b}} = -i\delta^{\bar{a}}{}_{\bar{b}}$. On $O_{(\alpha)} \cap O_{(\beta)}$, $J_{(\alpha)} = J_{(\beta)}$, since J is invariant under holomorphic changes of coordinates. Because $J_{(\alpha)} = J_{(\beta)}$ on intersection regions, the $J_{(\alpha)}$ fit together into a tensor field defined globally on all of K. This tensor field is manifestly an integrable complex structure, since we are presented in the construction with the requisite local holomorphic coordinates $z^a_{(\alpha)}$.

Part of the motivation for the concept of a complex manifold is that on such a manifold it is possible to define a concept of holomorphic (or analytic) functions. Let K be a complex manifold. A complex-valued function f on K is said to be a holomorphic function if at each point p, in any system of local holomorphic coordinates, f obeys $\partial f/\partial z^{\bar{a}} = 0$. This definition is independent of the choice of local holomorphic coordinates, since any two systems of local holomorphic coordinates are related by a holomorphic change of coordinates. Many familiar theorems of ordinary complex analysis carry over in this more general situation. For instance, it is not possible to have a local maximum in the modulus of a nonconstant holomorphic function. (This is a local statement, so it is true on an arbitrary complex manifold as a consequence of the familiar theorem on the complex plane.) It follows from this that on a compact complex manifold (on which any nonsingular function has a maximum), any holomorphic function without singularities anywhere is a constant.

Now let us return to our original problem. Consider a $2N$-dimensional manifold K whose holonomy group is not $SO(2N)$ but $U(N)$ or a subgroup thereof. At any point $x \in K$, tangent vectors, which transform as the vector of $SO(2N)$, decompose as $N \oplus \overline{N}$ under $U(N)$. There is a unique matrix $J^i{}_j(x)$ acting on tangent vectors at x which takes the value $+i$ or $-i$, respectively, for vectors in the N or $\overline{N}$ of $U(N)$. J defines an almost complex structure. Since J is invariant under the holonomy group, it is covariantly constant. Thus, all of the covariant derivatives of J vanish, and in particular the Nijenhuis tensor vanishes. Thus, a manifold of $U(N)$ holonomy is a complex manifold.

15.2.3 Examples of Complex Manifolds

Not every real manifold can be given the structure of a complex manifold; for instance, the sphere S^n admits a complex structure if $n = 2$ (as we will see shortly) but not otherwise. On the other hand, if a manifold does admit a complex structure, it may admit complex structures that are inequivalent to each other. Two manifolds M and N have the same topological type if it is possible to find a continuous and one-to-one map $f : M \to N$. In this case M and N are considered equivalent as real manifolds. On the other hand, two complex manifolds M and N are equivalent as complex manifolds only if it is possible to find an invertible map $f : M \to N$ which is *holomorphic* in the sense that local holomorphic coordinates of M are mapped by f into local holomorphic coordinates of N. It is not unusual to find manifolds that are equivalent as real manifolds but not as complex manifolds; in this case we say that it is possible to put inequivalent complex structures on the same real manifold. Let us give some examples of complex manifolds.

(1) *'Flat Space' and the Complex Torus*

The most basic example of a complex manifold of complex dimension n is C^n, the Cartesian product of n copies of the complex plane C. It is the same as R^{2n}, Euclidean space of $2n$ dimensions, endowed with a complex structure.

It is now easy to construct our first example of a complex manifold with nontrivial topology. Let Γ be a lattice in C^n (again, this is exactly the same thing as a lattice in R^{2n}). Then the quotient $E = C^n/\Gamma$ is a complex manifold which is known as a complex torus. A holomorphic function on E is the same thing as a holomorphic function on C^n which is invariant under the lattice translations. By continuously varying the choice of lattice Γ, one obtains manifolds E with the same topological type, but different complex structure.

(2) *Riemann Surfaces*

The next example is one that is extremely important in string theory: an arbitrary oriented Riemannian manifold of real dimension two is a complex manifold. (For genus one, this is a special case of our previous example.) Thus (in the Euclidean formulation) the world sheet of an oriented closed string is a complex manifold. To establish this result, let Σ be a Riemannian manifold of real dimension two, with metric tensor g_{ij}. If Σ is oriented, it possesses a covariantly constant antisymmetric tensor ϵ_{ij}, which we may normalize so that $\epsilon_{12} = +1$ in locally inertial frames.[*] This tensor obeys the familiar identity $\epsilon^{ik}\epsilon_{kj} = -\delta^i{}_j$. By virtue of that identity, the tensor field $J^i{}_j = g^{ik}\epsilon_{kj}$ is an almost complex structure, obeying $J^2 = -1$. Moreover, it follows immediately from the definition

---

[*] In previous chapters the antisymmetric symbol ϵ was defined to be a tensor density. For the purposes of this chapter it is more convenient to include an additional factor of $\sqrt{g}$ in the definition so that it becomes a true tensor.

of the Nijenhuis tensor that this tensor vanishes in the case of two real dimensions, or one complex dimension.[†] Hence Σ is a complex manifold, and it is possible to choose a local complex coordinate z to put J in the standard form. (The nonzero components of J are $J^z{}_z = -J^{\bar z}{}_{\bar z} = i$.) We are using here the Newlander–Nirenberg theorem, which is a fairly difficult theorem even in the case of real dimension two, where it was first proved by Korn and Lichtenstein.

These considerations allow us to extract a further corollary, which has actually been used extensively in the first volume of this book. In the coordinate system defined by z and $\bar z$, ϵ, being antisymmetric, obeys $\epsilon_{zz} = \epsilon_{\bar z \bar z} = 0$, and also $\epsilon_{z\bar z} = -\epsilon_{\bar z z} = if(z, \bar z)$ for some f. As ϵ is real and nonzero, $f = e^{2\rho}$ with some real function ρ. Using ϵ_{ij} and $J^i{}_j$, we can reconstruct the metric tensor as $g_{ij} = -\epsilon_{ik}J^k{}_j$. With the known form of J and ϵ in the $z, \bar z$ coordinates, we see that the metric takes the form $g_{zz} = g_{\bar z \bar z} = 0$, $g_{z\bar z} = g_{\bar z z} = \frac{1}{2}e^{2\rho}$. Equivalently, in terms of real coordinates x^i defined by $z = x^1 + ix^2$, the metric takes the form $g_{ij} = e^{2\rho}\delta_{ij}$. The existence of local coordinates on an arbitrary string world sheet with $g_{ij} = e^{2\rho}\delta_{ij}$ is something that we used extensively in volume one of this book; we now see that the existence of such coordinates is equivalent to the assertion that every oriented manifold of real dimension two is a complex manifold.

We have shown that every Riemannian metric g on an oriented string world sheet Σ induces an integrable complex structure J. Moreover, under a Weyl rescaling of the metric $g \to e^{2\psi}g$, the antisymmetric tensor transforms as $\epsilon \to e^{2\psi}\epsilon$, so the complex structure $J^i{}_j = g^{ik}\epsilon_{kj}$ is invariant under Weyl rescaling. Thus, the induced complex structure J is the same for any two metrics g and $g' = e^{2\psi}g$ that differ only by Weyl rescaling; this is usually described by saying that the complex structure of a Riemann surface depends only on the conformal class of the metric, or on the conformal structure determined by the metric. If, however, two metrics on a Riemann surface Σ do not differ merely by conformal rescaling, they determine two different complex structures on Σ. The possible complex structures on a Riemann surface make up the Teichmuller parameters that appear (in different guises, depending on the formalism) in loop integrals in string theory. Our discussion in §3.3 of the conformal structures on a

---

[†] To prove this, note that in two dimensions $\partial_l J^k{}_j - \partial_j J^k{}_l$, being antisymmetric in j and l, is necessarily of the general form $\epsilon_{jl}V^k$, where ϵ is the antisymmetric tensor and V^k is some vector field. With $J^l{}_i = g^{lm}\epsilon_{mi}$, it follows that $W^k_{ij} = J^l{}_i(\partial_l J^k{}_j - \partial_j J^k{}_l)$ is symmetric in i and j, so that the Nijenhuis tensor $N^k{}_{ij} = W^k_{ij} - W^k_{ji}$ vanishes.

Riemann surface of given genus can be reinterpreted as a discussion of the complex structures on such a surface.

(3) *Complex Projective Space*

Our next example of a complex manifold is very different but equally fundamental. Let Z^k, $k = 1, \ldots, N+1$ be $N+1$ complex variables, not all zero. For any nonzero complex number λ, we consider the family $\{Z^k\}$ to be equivalent to the family $\{\lambda Z^k\}$. The equivalence classes depend on N independent complex coordinates, and form a manifold called complex projective space or CP^N. CP^N is a compact manifold; to see this note that by scaling we can assume $\sum_{i=1}^{N+1} |Z^k|^2 = 1$. If this were the only condition, $\{Z^k\}$ could be viewed as defining a point on the sphere S^{2N+1}, which is certainly compact. Since in addition we wish to remove a phase degree of freedom, $\{Z^k\} \approx \{e^{i\alpha} Z^k\}$, CP^N is actually the quotient of the compact space S^{2N+1} by the compact group $U(1)$ of complex numbers of modulus one; hence CP^N is compact. Compactness of CP^N could also be proved by noting that CP^N is equivalent to the homogeneous space $SU(N+1)/U(N)$.[*]

To prove that CP^N is a complex manifold, note first that if $Z^1 \neq 0$, we can use the scaling symmetry to set $Z^1 = 1$; the remaining Z^i, $i = 2, \ldots, N+1$ are then a set of N independent complex coordinates. This coordinate system, however, breaks down when $Z^1 = 0$. More generally, for $i = 1, \ldots, N+1$, let $O_{(i)}$ be the subset of CP^N with $Z^i \neq 0$. The $O_{(i)}, i = 1, \ldots, N+1$, form an open cover of CP^N. In each $O_{(i)}$, we choose local complex coordinates $Z^j_{(i)} = Z^j/Z^i, 1 \leq j \leq N+1, j \neq i$. (The idea behind this definition is that the $Z^j_{(i)}$ are invariant under $\{Z^k\} \to \{\lambda Z^k\}$ and are well defined where $Z^i \neq 0$.) To show that CP^N is a complex manifold, we must show that in $O_{(i)} \cap O_{(l)}$, the $Z^k_{(i)}$ are analytic functions of the $Z^k_{(l)}$. This is so, since in fact $Z^k_{(i)} = (Z^l/Z^i) Z^k_{(l)}$.

In one case, complex projective space is already familiar: CP^1 is the Riemann sphere, the ordinary two sphere S^2. Indeed, CP^1 is described by two homogeneous coordinates (Z, W). If $W \neq 0$, we set $W = 1$ by scaling, and we are left with the complex Z plane. If $W = 0$ we can set $Z = 1$ by scaling (the definition of CP^1 requires that Z and W are not both zero), so this is one additional point, the 'point at infinity' that must be added to the complex plane to make the Riemann sphere. The fact that there

---

[*] Except when strict adherence to this convention would be awkward, we will denote local complex coordinates by lower case letters z^a, w^b, etc. Homogenous 'coordinates' of CP^N (which are not true coordinates since they are defined only up to scaling) will be denoted by upper case letters Z_i, W_j.

are no Teichmuller parameters in string theory at tree level corresponds to the statement that CP^1 has a unique complex structure; the same is true for CP^N for any N.

(4) *Hypersurfaces in CP^N*

Many additional examples of complex manifolds can be obtained as submanifolds of CP^N. The simplest way to do this is to pick a polynomial $P(Z^1, Z^2, \ldots, Z^{N+1})$ which is homogeneous of some degree k in the sense that $P(\lambda Z^1, \lambda Z^2, \ldots, \lambda Z^{N+1}) = \lambda^k P(Z^1, Z^2, \ldots, Z^{N+1})$. The homogeneity of P means that the equation $P = 0$ makes sense in CP^N. For generic P this equation defines a compact complex submanifold of CP^N of dimension $N - 1$ called a degree k hypersurface. The necessary condition on P will be stated shortly. More generally, we could consider n polynomials P_i, $i = 1, \ldots, n$, homogeneous of degrees $k_1, k_2, \ldots, k_n$. Because of the homogeneity of the P_i, the equation $P_1 = P_2 = \ldots = P_n = 0$ makes sense in CP^N. For $n \leq N$, this defines (for generic P_i) a compact complex submanifold of CP^N of dimension $N - n$, called a complete intersection of hypersurfaces.

To understand the requirement that must be imposed on the polynomials P_i in the above construction, consider the analogous problem for real manifolds. For simplicity, let us start with the $x - y$ plane X and attempt to define a smooth submanifold of dimension one or in other words a curve by an equation $f(x, y) = 0$, where $f(x, y)$ is some smooth function. A generic smooth function, such as $f(x, y) = y - x^2$, defines a smooth curve (a parabola, in this case). A typical example that leads to a curve with singularity is $\tilde{f}(x, y) = y^2 - x^2$. In this case the curve $\tilde{f}(x, y) = 0$ consists of two lines, and is singular at $x = y = 0$ where the two lines meet. Analytically, the signal for the fact that the curve $\tilde{f} = 0$ is singular at $x = y = 0$ is that at that point $\partial \tilde{f} / \partial x = \partial \tilde{f} / \partial y = 0$ or in other words $d\tilde{f} = 0$ (here d is the exterior derivative, defined in chapter 12; the statement that $d\tilde{f} = 0$ amounts to saying that all of the partial derivatives of f vanish). Near any point in the plane X where $d\tilde{f} \neq 0$, $\tilde{f}$ can be chosen as one of the coordinates. Hence near any point on the curve $\tilde{f} = 0$ at which $d\tilde{f} \neq 0$, the equation $\tilde{f} = 0$ is equivalent to vanishing of one of the coordinates and defines a smooth curve.

These considerations carry over to the complex situation. The hypersurface in CP^N defined by vanishing of a homogeneous polynomial P is nonsingular if at no point on this hypersurface do the partial derivatives $\partial P / \partial Z^i$ all vanish or in other words if the equations $P = dP = 0$ have no solution except $Z^i = 0$ (which is not a point on CP^N). The generalization of this to a complete intersection of hypersurfaces defined by the vanishing of homogeneous polynomials $P_1, P_2, \ldots, P_n$ is that the surface

$P_1 = P_2 = \ldots = P_n = 0$ is nonsingular if on this surface there is no point at which the n form $dP_1 \wedge dP_2 \wedge \ldots \wedge dP_n$ vanishes. The idea here is again that locally, near any point in CP^N with $dP_1 \wedge dP_2 \wedge \ldots \wedge dP_n \neq 0$, the P_i can be chosen as n of the coordinates of CP^N.

As an illustration of these ideas, consider the curve in the (real) x–y plane defined by $f = 0$ where f is a polynomial, say $f = x^n + y^n - 1$. It is easy to give a qualitative description of this curve; for instance, it is compact if n is even. One might be curious to analytically continue this curve to complex values of x and y. Thus, one can consider the equation $x^n + y^n - 1 = 0$ where now x and y are *complex* variables. Over the complex numbers this curve is of course not compact, regardless of n. One might wonder whether it is possible to add 'points at infinity' to compactify it. The way to do this is to replace $x^n + y^n - 1$ with a homogeneous polynomial in CP^2. Thus, we introduce a third variable z and consider the homogeneous equation $x^n + y^n - z^n = 0$, where x, y, z are homogeneous coordinates of CP^2 (and so the triplet x, y, z is identified with $\lambda x, \lambda y, \lambda z$ for nonzero complex λ). If z is not zero, then by scaling we can take $z = 1$, so in this way we get back to the equation $x^n + y^n - 1 = 0$ in the complex plane. However, by working in CP^2 we have also the 'points at infinity', where $z = 0$ and $x^n + y^n = 0$. Notice that there are precisely n such points.

The equation $x^n + y^n - z^n = 0$ in CP^2 defines a compact complex manifold of complex dimension one, which must be a Riemann surface of some genus. One may wonder what is the genus of this surface. The answer is that it has genus $(n-1)(n-2)/2$, as we will see in the last section of this chapter.

Although the complex structure of CP^N is unique, that is not so (except in a few cases) for a hypersurface in CP^N. Let P be a polynomial, homogeneous of degree n in the $N+1$ homogeneous coordinates $Z^1, \ldots, Z^{N+1}$, and let P' be a second polynomial homogeneous of the same degree n in $N+1$ complex coordinates $W^1, \ldots, W^{N+1}$. The equations $P = 0$ and $P' = 0$ define complex manifolds Q and Q' which have the same topological type, since we could interpolate smoothly between the polynomials P and P'.[*] However, Q and Q' do not necessarily have the same complex structure. For them to have the same complex structure, there must be a holomorphic map from the Z^i to the W^j which turns P into P'. To be invertible and free of poles, and invariant under rescaling of the Z^i and W^j, the map $Z^i \to W^j$ must actually be linear.

---

[*] Consider a one-parameter family of polynomials $P_t = tP + (1-t)P'$, $0 \leq t \leq 1$; as t goes from zero to one the zeros of P_t define a one-parameter family of manifolds starting at Q and ending at Q', proving that Q and Q' have the same topology.

For instance, an example that will interest us later is the case of a quintic hypersurface in CP^4, determined by the zeros of a fifth-order polynomial P in five variables $Z^1, \ldots, Z^5$. A little arithmetic shows that there are $5 \cdot 6 \cdot 7 \cdot 8 \cdot 9/1 \cdot 2 \cdot 3 \cdot 4 \cdot 5 = 126$ independent quintic polynomials in five variables. On the other hand, a linear change of coordinates would depend on an invertible 5×5 matrix with 25 independent matrix elements. The quintic polynomial P thus has $126 - 25 = 101$ independent degrees of freedom that cannot be absorbed in redefining the coordinates. Hence, the complex structure of a quintic hypersurface in CP^4 depends on at least 101 complex parameters. For $N > 2$, all of the complex structures of a hypersurface in CP^N can be found in this way, so the complex structure of the quintic hypersurface in CP^4 depends on precisely 101 parameters.

15.3   Kähler Manifolds

On a manifold K of $U(N)$ (or $SU(N)$) holonomy, it is possible to define a natural complex structure $J^i{}_j$ which was the subject of the previous section. Our discussion in that section merely focused on the implications of the vanishing of the Nijenhuis tensor $N^i{}_{jk}$ constructed from J. However, $U(N)$ holonomy means not just that the Nijenhuis tensor vanishes but also that $J^i{}_j$ is covariantly constant. We will devote this section to analyzing what it means for $J^i{}_j$ to be covariantly constant. Actually, we will find it somewhat more convenient to study not the complex structure J but the two form $k_{ij} = g_{ik}J^k{}_j$, with g being the metric tensor. Covariant constancy of J or k are equivalent statements, since g is covariantly constant. A metric of $U(N)$ holonomy is called a Kähler metric; a manifold that admits such a metric is called a Kähler manifold; the two form k is called the Kähler form. In this section we will work out the general local expression for a Kähler metric and describe a few of the basic examples.

15.3.1 The Kähler Metric

In a system of local holomorphic coordinates z^a, and their complex conjugates $\bar{z}^{\bar{a}}$, a vector v^i has holomorphic components v^a and antiholomorphic components $v^{\bar{a}}$. The splitting of v^i into v^a and $v^{\bar{a}}$ of course corresponds to the splitting of the fundamental vector of $SO(2N)$ as $N \oplus \bar{N}$ of $U(N)$. The distinction between holomorphic and antiholomorphic indices is invariant under holomorphic changes of coordinates and does not depend on the choice of a particular holomorphic coordinate system. It therefore is very natural to decompose tensor fields into pieces with definite numbers of holomorphic and antiholomorphic indices.

For example, the metric tensor g_{ij} at a point $p \in K$ is invariant under the action of $U(N)$ on the tangent space at p. Since it is impossible to make a $U(N)$ singlet from $N \otimes N$ or $\overline{N} \otimes \overline{N}$ for $N > 2$, g obeys $g_{ab} = g_{\overline{a}\overline{b}} = 0$. Also, being symmetric, g obeys $g_{a\overline{b}} = g_{\overline{b}a}$. Using $k_{ij} = g_{ik}J^k{}_j$ and the explicit form of J in local holomorphic coordinates, we find that $k_{ab} = k_{\overline{a}\overline{b}} = 0$ and $k_{a\overline{b}} = -ig_{a\overline{b}} = -k_{\overline{b}a}$.

The study of differential forms is another example in which it is important to decompose tangent-vector indices into holomorphic and antiholomorphic indices. Thus, consider a p form $\psi_{i_1 i_2 ... i_p}$; we recall that this is a pth rank covariant tensor that is completely antisymmetric in all indices. On a complex manifold it is natural to ask how many of the indices of the p form ψ are holomorphic indices and how many are antiholomorphic indices. For many purposes, the basic concept is the (p, q) form $\psi_{a_1 a_2 ... a_p \overline{a}_1 \overline{a}_2 ... \overline{a}_q}$ which is completely antisymmetric in p holomorphic and q antiholomorphic indices. It is convenient to adopt a convention about ordering of indices; we will always write the holomorphic indices of a (p, q) form first.

15.3.2 Exterior Derivatives

On a real manifold the 'exterior derivative' of a p form ψ is defined to be the $p + 1$ form

$$(d\psi)_{i_1 i_2 ... i_{p+1}} = \frac{1}{p+1} \partial_{i_1} \psi_{i_2 ... i_{p+1}}$$

$$\pm \text{ cyclic permutations.}$$

(15.3.1)

On a complex manifold, these notions can be refined. We define a 'holomorphic' exterior derivative ∂ which maps (p, q) forms to $(p + 1, q)$ forms by saying that if ψ is a (p, q) form then $\partial\psi$ is the $(p + 1, q)$ form

$$(\partial\psi)_{a_1 a_2 ... a_{p+1} \overline{a}_1 \overline{a}_2 ... \overline{a}_q} = \frac{1}{p+1} \partial_{a_1} \psi_{a_2 ... a_{p+1} \overline{a}_1 ... \overline{a}_q} .$$

$$\pm \text{ cyclic permutations}$$

(15.3.2)

Just like d, ∂ obeys $\partial^2 = 0$. We define an 'antiholomorphic' exterior derivative $\overline{\partial}$ which maps (p, q) forms to $(p, q + 1)$ forms by saying that if ψ is a (p, q) form, then $\overline{\partial}\psi$ is the $(p, q + 1)$ form

$$(\overline{\partial}\psi)_{a_1 ... a_p \overline{a}_1 ... \overline{a}_{q+1}} = \frac{1}{q+1} (-1)^p \partial_{\overline{a}_1} \psi_{a_1 ... a_p \overline{a}_2 ... \overline{a}_{q+1}} .$$

$$\pm \text{ cyclic permutations.}$$

(15.3.3)

The factor $(-1)^p$ has been included to ensure that $d = \partial + \overline{\partial}$. Granted this, the identity $d^2 = 0$ implies that

$$\partial^2 = \overline{\partial}^2 = \partial\overline{\partial} + \overline{\partial}\partial = 0. \tag{15.3.4}$$

The operators ∂ and $\overline{\partial}$ are often called the Dolbeault operators. Their essential property is that they can be defined just in terms of the complex structure of the manifold K, with no choice of a metric tensor. Despite the use of ordinary derivatives, rather than some sort of covariant derivatives, in defining them, the objects $\partial\psi$ and $\overline{\partial}\psi$ transform homogeneously under holomorphic changes of coordinates.

Because $d^2 = 0$, a form ψ that can be written $\psi = d\chi$ automatically obeys $d\psi = 0$. The Poincaré lemma gives a local inverse to this; if ψ is a p form with $p > 0$ and $d\psi = 0$ then locally ψ can be written $\psi = d\chi$ for some $p - 1$ form χ. This statement has an analog for complex manifolds. If ψ is a (p, q) form with $p > 0$ (or $q > 0$) and $\partial\psi = 0$ (or $\overline{\partial}\psi = 0$), then locally ψ can be written $\psi = \partial\chi$ (or $\psi = \overline{\partial}\chi$) where χ is a $(p - 1, q)$ form (or $(p, q - 1)$ form). For complex manifolds there is a further statement that does not have a direct analog for real manifolds. If ψ is a (p, q) form with $p > 0, q > 0$, and $\partial\psi = \overline{\partial}\psi = 0$, then locally ψ can be written $\psi = \partial\overline{\partial}\chi$, with χ being a $(p - 1, q - 1)$ form.

15.3.3 The Affine Connection and the Riemann Tensor

We are now in a position to give the general local description of a metric of $U(N)$ holonomy. The covariantly constant tensor k_{ij} is not just a two form but a $(1, 1)$ form, since we noted earlier that $k_{ab} = k_{\overline{a}\overline{b}} = 0$. Also, being covariantly constant, k obeys $\partial k = \overline{\partial} k = 0$. It follows then that k can be expressed locally as

$$k = -i\partial\overline{\partial}\phi, \tag{15.3.5}$$

where ϕ is a $(0, 0)$ form or in other words a scalar function. Here ϕ is known as a Kähler potential. The Kähler potential is not uniquely determined; if F is an arbitrary holomorphic function, then $\tilde{\phi} = \phi + F + \overline{F}$ obeys $\partial\overline{\partial}\tilde{\phi} = \partial\overline{\partial}\phi$. Locally, this is the general form of the indeterminacy in ϕ.

Bearing in mind the relation $k_{a\overline{b}} = -ig_{a\overline{b}}$ between the metric tensor g and the two form k, (15.3.5) is equivalent to a general local expression

$$g_{a\overline{b}} = g_{\overline{b}a} = \frac{\partial^2\phi}{\partial z^a \partial \overline{z^b}} \tag{15.3.6}$$

for a metric of $U(N)$ holonomy. Actually, so far we have only sketched why it is always possible to locally write a metric of $U(N)$ holonomy in the form (15.3.6). The converse, however, is also true; any metric that locally can be put in the form of (15.3.6) (with $g_{ab} = g_{\bar{a}\bar{b}} = 0$) has $U(N)$ holonomy.

To prove this, it is useful to first work out the form of the affine connection for Kähler manifolds. One easily finds that the only nonzero components of the affine connection $\Gamma^i_{jk} = g^{il}(\partial_j g_{lk} + \partial_k g_{lj} - \partial_l g_{jk})/2$ are

$$\Gamma^a_{bc} = g^{a\bar{d}}\partial_b g_{c\bar{d}}, \qquad \Gamma^{\bar{a}}_{\bar{b}\bar{c}} = g^{\bar{a}d}\partial_{\bar{b}} g_{\bar{c}d}. \qquad (15.3.7)$$

It follows from this form of the affine connection (as the reader should check!) that the standard complex structure J (with nonzero components $J^a{}_b = i\delta^a{}_b, J^{\bar{a}}{}_{\bar{b}} = -i\delta^{\bar{a}}{}_{\bar{b}}$) is covariantly constant. Since the subgroup of $SO(2N)$ under which J is invariant is $U(N)$, covariant constancy of J means that the holonomy group of a Kähler manifold is (at most) $U(N)$, completing the demonstration that metrics of $U(N)$ holonomy are precisely Kähler metrics.

Like the affine connection, the Riemann tensor greatly simplifies on a Kähler manifold. Using the definition

$$R^i{}_{jkl} = \partial_k \Gamma^i_{jl} - \partial_l \Gamma^i_{kj} + \Gamma^i_{km}\Gamma^m_{lj} - \Gamma^i_{lm}\Gamma^m_{kj} \qquad (15.3.8)$$

and the form (15.3.7) of the affine connection, it is immediate that $R^i{}_{jkl}$ vanishes unless the indices i and j are both of the same type (both holomorphic or both antiholomorphic). It follows then (since the metric g has $g_{ab} = g_{\bar{a}\bar{b}} = 0$) that $R_{mjkl} = g_{mi}R^i{}_{jkl}$ vanishes unless the indices m and j are of *opposite* type. In view of the symmetry relation $R_{mjkl} = R_{klmj}$ of Riemannian geometry, R_{mjkl} also vanishes unless the indices k and l are of opposite type. The nonzero components of the Riemann tensor are thus of the form $R_{a\bar{b}c\bar{d}}$. The cyclic identity of Riemannian geometry $R_{ijkl} + R_{jkil} + R_{kijl} = 0$ simplifies for Kähler manifolds to the statement that

$$R_{a\bar{b}c\bar{d}} = R_{c\bar{b}a\bar{d}} = R_{a\bar{d}c\bar{b}}. \qquad (15.3.9)$$

The nonzero components $R_{a\bar{b}c\bar{d}}$ or equivalently $R^{\bar{a}}{}_{\bar{b}c\bar{d}} = g^{a\bar{a}}R_{a\bar{b}c\bar{d}}$ of the curvature tensor of a Kähler manifold are easily calculated. One of the resulting formulas is

$$R^{\bar{a}}{}_{\bar{b}c\bar{d}} = \partial_c \Gamma^{\bar{a}}_{\bar{b}d}. \qquad (15.3.10)$$

A particularly useful simplification occurs for the Ricci tensor, $R_{\bar{b}c} = R^{\bar{a}}{}_{\bar{b}\bar{a}c}$ which in view of (15.3.10) is

$$R_{\bar{b}c} = -\partial_c \Gamma^{\bar{a}}_{\bar{b}a}. \qquad (15.3.11)$$

It follows from (15.3.7) that (by analogy with a similar formula for real manifolds) $\Gamma^{\bar{a}}_{b\bar{a}} = \partial_{\bar{b}} \ln \det g$. Hence, the formula for the Ricci tensor reduces to

$$R_{\bar{b}c} = -\partial_{\bar{b}}\partial_c \ln \det g, \qquad (15.3.12)$$

which we will find useful later.

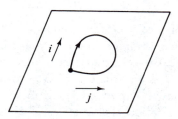

Figure 15.2. Under parallel transport of a vector V^k around a small loop in the ij plane, it changes by an amount $\delta V^k \sim R_{ij}{}^k{}_l V^l$.

The above restrictions on the Riemann tensor of a Kähler manifold can be understood intuitively. Recall that on a general Riemannian manifold M of dimension $2N$, the Riemann tensor R_{ijkl} is antisymmetric in k and l. Keeping i and j fixed, it is a $2N \times 2N$ antisymmetric matrix or in other words a generator of $SO(2N)$ which we may call $R_{(ij)}$. (Explicitly, $R_{(ij)}$ is the matrix whose kl matrix element is R_{ijkl}.) The well-known relation $[D_i, D_j]V^k = R_{ij}{}^k{}_l V^l$ shows that in parallel transport around a small loop (which can be expressed by a commutator of covariant derivatives $[D_i, D_j]$, as in fig. 15.15.2) tangent vectors undergo rotation by the $SO(2N)$ matrices $R_{(ij)}$. If M has $U(N)$ holonomy, the matrices $R_{(ij)}$ are not general $SO(2N)$ matrices; they are $U(N)$ matrices, embedded in $SO(2N)$ in the fashion that we described above in §15.2. Now, it is easy to describe which $SO(2N)$ matrices M_{kl} are $U(N)$ generators. In our familiar complex basis, the requirement is $M_{ab} = M_{\bar{a}\bar{b}} = 0$. Hence to say that for fixed i and j the matrix $R_{(ij)}$ is a $U(N)$ matrix amounts to saying that $R_{ijab} = R_{ij\bar{a}\bar{b}} = 0$. Under these conditions, parallel transport around an infinitesimal loop gives a $U(N)$ transformation; since finite loops can be made by piecing together infinitesimal ones, this is the necessary and sufficient condition for $U(N)$ holonomy.

15.3.4 Examples of Kähler Manifolds

We will now describe some of the basic examples of Kähler manifolds.

The most basic example is certainly C^n, complex n space with complex coordinates $z^k, k = 1, \ldots, n$. The standard flat metric $ds^2 = \sum_k |dz^k|^2$ is a Kähler metric, since the holonomy (being trivial) is a subgroup of $U(n)$. The Kähler potential can be chosen to be

$$\phi = \sum_k |z^k|^2. \tag{15.3.13}$$

Notice that this Kähler potential is globally defined throughout C^n. A less trivial example is the complex torus $E = C^n/\Gamma$, with Γ as before a lattice in C^n. The flat metric on E is still a Kähler metric (the holonomy is still trivial!), but the Kähler potential (15.3.13) is now satisfactory only locally, as it is not invariant under the lattice translations.

Our second example is a basic one in string theory. Every oriented real two-dimensional Riemannian manifold is a Kähler manifold. In fact, an oriented real two-dimensional manifold has holonomy $SO(2)$ (or a subgroup). As $SO(2)$ is the same as $U(1)$, every metric on an oriented two-dimensional manifold is a Kähler metric. The fact that the metric of a two-dimensional manifold can locally be put in the form of (15.3.6) is another form of the statement that locally the line element of such a manifold can be put in the form $ds^2 = e^{2\rho}dzd\bar{z}$. (Implicit in (15.3.6) is the additional, not very deep, assertion that locally one can always find a solution ϕ of the linear equation $\partial^2\phi/\partial z\partial\bar{z} = e^{2\rho}$.)

Another very important example of a Kähler manifold is CP^N. Up to normalization, there is a unique metric on CP^N that is invariant under unitary transformations of the $N + 1$ homogeneous coordinates $Z^1, \ldots, Z^{N+1}$. This metric is

$$ds^2 = (\overline{Z}^f Z^f)^{-1}\left(dZ^a - Z^a\frac{\overline{Z}^b dZ^b}{Z^c\overline{Z}^{\bar{c}}}\right)\left(d\overline{Z}^{\bar{a}} - \overline{Z}^{\bar{a}}\frac{Z^d dZ^{\bar{d}}}{Z^e\overline{Z}^{\bar{e}}}\right). \tag{15.3.14}$$

This metric has been adjusted to be invariant under rescaling of the Z^a, since in CP^N the Z^a are only defined up to scaling. It has also been adjusted to vanish if dZ^a (or $d\overline{Z}^{\bar{a}}$) is a multiple of Z^a (or $\overline{Z}^{\bar{a}}$), since points that differ by $\delta Z^a \sim Z^a$ are equivalent in CP^N and must have zero separation. The metric of (15.3.14), which is known as the Fubini–Study metric, is manifestly invariant under $SU(N + 1)$ transformations of the Z^a. To see that (15.3.14) is a Kähler metric, it is enough to note that in the region $Z^{N+1} \neq 0$, we can set $Z^{N+1} = 1$ by scaling, and then we find that the Fubini–Study metric can be derived in this region from the Kähler potential $\phi = \ln(1 + \sum_{a=1}^{N} Z^a\overline{Z}^{\bar{a}})$.

Having found that CP^N is a Kähler manifold, it is natural to ask whether a hypersurface in CP^N or an intersection of hypersurfaces is likewise a Kähler manifold. This question can be answered very simply. A complex submanifold of a Kähler manifold is always a Kähler manifold in its own right. Indeed, this follows rather easily from the local characterization (15.3.6) of Kähler metrics. If K is a manifold and Q is a submanifold, any metric on K always induces a metric on Q; one simply uses the metric of K to measure distances on Q. Differently put, the induced metric g_{ij} on Q is the same as the metric g_{ij} of K except that in viewing g_{ij} as a metric on Q, one is only interested in the components with i and j tangent to Q. Now, if the metric of K can be written as in (15.3.6) in terms of second derivatives of a Kähler potential, then the same is true for the induced metric on a complex submanifold Q; working on Q just means evaluating $\partial^2\phi/\partial Z^a\partial\overline{Z}^{\overline{b}}$ only for values of a and $\overline{b}$ tangent to Q.

Thus, every complex submanifold of CP^N is a Kähler manifold. There is also a partial converse to this statement. Every compact Kähler manifold (with a mild restriction on the topological class of the Kähler form, which we will discuss shortly) can be embedded in CP^N. However, such an embedding of a given Kähler manifold in CP^N may be difficult to describe; embedding in CP^N is certainly not always the most economical way to describe a given Kähler manifold. The complete intersection of hypersurfaces which we discussed at the end of the last section is a particularly simple example of a complex submanifold of CP^N, but the 'typical' complex submanifold of CP^N cannot be obtained in this way. A relatively simple example of a complex submanifold of CP^N that cannot be obtained as a complete intersection of hypersurfaces is the 'twisted cubic curve' in CP^3. It is the subspace of CP^3 consisting of points whose homogeneous coordinates are of the form (u^3, u^2v, uv^2, v^3), where u and v are complex numbers not both zero. Since u and v can be considered to determine a point on CP^1, the twisted cubic curve can be regarded as an exotic embedding of CP^1 in CP^3. It is not too hard to convince oneself that it cannot be realized as a complete intersection.

Returning now to general theory, let K be a Kähler manifold on which we have chosen a particular Kähler metric g. The corresponding Kähler form k, since it is a closed two form, defines an element of $H^2(K;R)$ (the second de Rham cohomology group of K with real coefficients). This element is called the Kähler class of the Kähler metric g. Just as a complex manifold may admit a range of choices of complex structure, a Kähler manifold may admit Kähler metrics with a range of possible Kähler classes. However, the Kähler class of a compact Kähler manifold is always nonzero; in other words, k cannot be written globally as $k = d\alpha$ with α a one form. To prove this, one shows that on a Kähler manifold of complex dimension N or real dimension $2N$, the N-fold wedge product of the Kähler form k with itself is always a multiple of the completely antisymmetric Levi–Civita tensor $\epsilon_{i_1 \ldots i_{2N}}$. Recall that at any point p, one can choose standard complex coordinates z^a with $k = dz^1 \wedge d\overline{z^1} + \ldots + dz^n \wedge d\overline{z^n}$. From this, it is easy to deduce that $k \wedge k \wedge \ldots \wedge k = N! \epsilon$. The volume of a compact manifold is $V = \int \epsilon$, and in particular this does not vanish. If on the other hand $k = d\alpha$ globally, we would have

$$
\begin{aligned}
N! \cdot V &= \int k \wedge k \wedge \ldots \wedge k = \int d\alpha \wedge k \wedge \ldots \wedge k \\
&= \int d(\alpha \wedge k \wedge \ldots \wedge k) = 0.
\end{aligned}
\tag{15.3.15}
$$

So the Kähler class of a compact Kähler manifold is always nonzero.

Not every complex manifold admits a Kähler metric; there are strong topological restrictions. For example, for the compact manifold M to be a Kähler manifold, the second cohomology group $H^2(M; R)$ must be nonzero, or in other words the the second Betti number of M must be positive. Otherwise, the Kähler class of a hypothetical Kähler metric on M would automatically vanish, but we have just seen that the Kähler class of an arbitrary Kähler metric on a compact Kähler manifold is not zero. In fact, the even Betti numbers of a compact Kähler manifold must all be positive, since the cohomology class of the n-fold wedge product $\{k \wedge k \wedge \ldots \wedge k\}_n$ is always nonzero for all $n \leq N$. In fact if we could write the n-fold wedge product as $\{k \wedge \ldots \wedge k\}_n = d\lambda$ then the volume of K would be $N! \cdot V = \int_K (d\lambda) \wedge \{k \wedge \ldots \wedge k\}_{N-n} = \int_K d(\lambda \wedge \{k \wedge \ldots \wedge k\}_{N-n}) = 0$, which is impossible. There are many examples of compact complex manifolds which cannot be endowed with Kähler metrics since some of their even Betti numbers vanish.

Locally, nothing can prevent us from defining a Kähler metric as in (15.3.6) on an arbitrary complex manifold, so the obstruction to finding a Kähler metric on a given complex manifold is purely topological. In fact, locally any two complex manifolds of the same dimension are indistinguishable from one another.

15.4   Ricci-Flat Kähler Manifolds and $SU(N)$ Holonomy

If a manifold K does admit a Kähler metric g, this Kähler metric is not unique. We are free to pick an arbitrary scalar field ψ, and to define a new Kähler metric g' (which will have the same Kähler class) by saying that $g'_{a\bar{b}} = g_{a\bar{b}} + \partial_a \partial_{\bar{b}} \psi$. Any two Kähler metrics with the same Kähler class are related in this way.

Having understood the basic content of $U(N)$ holonomy, we would now like to understand the additional restrictions associated with $SU(N)$ holonomy. Given a Kähler manifold K – which by our above remarks admits an infinity of Kähler metrics – is it possible to find on K a Kähler metric that has not $U(N)$ holonomy but $SU(N)$ holonomy? It is relatively easy to see that there is a topological obstruction to finding such a metric. In fact, the spin connection of a Kähler manifold is a $U(N)$ or $SU(N) \times U(1)$ gauge field. The $U(1)$ part of the spin connection is an abelian gauge field, which we may call A. A metric on K of $SU(N)$ holonomy would be a Kähler metric such that A is a pure gauge or, in other words, such that the gauge-invariant field strength $F = dA$ vanishes. Such a metric exists only if the $U(1)$ part of the spin connection is topologically trivial. If instead the spin connection describes a $U(1)$ field with a nontrivial quantized magnetic flux, this cannot be removed by continuous variation of the Kähler metric. As we discussed in chapter 14, the closed two form F defines an element of $H^2(K; R)$, the second de Rham cohomology group of K with real coefficients. This element is called the first Chern class of K, $c_1(K)$. (It is, in fact, the first Chern class of the so-called canonical line bundle of K, which will be defined in example (2) in §15.6.3.) Only if $c_1(K) = 0$ can K admit a metric of $SU(N)$ holonomy.

15.4.1 The Calabi–Yau Metric

In 1957 E. Calabi conjectured that a Kähler manifold K of vanishing first Chern class always admits a Kähler metric of $SU(N)$ holonomy. More specifically, he conjectured that there would be up to scaling precisely one metric of $SU(N)$ holonomy for any given complex structure on K and any given Kähler class. Calabi proved the uniqueness of this hypothetical metric. Its existence was proved by S.-T. Yau twenty years later. By virtue of this rather difficult theorem, metrics of $SU(N)$ holonomy correspond precisely to Kähler manifolds of vanishing first Chern class. This is a crucial simplification in our search for vacuum states of unbroken supersymmetry, because metrics of $SU(N)$ holonomy are extremely difficult to describe (none are known explicitly except in certain singular limits), but Kähler manifolds of zero first Chern class can be found by qualitative methods, as we will see.

It would hardly be possible to prove here the existence of metrics of $SU(N)$ holonomy on Kähler manifolds of $c_1 = 0$, but the following facts may make this seem more plausible. We already know that a manifold of $SU(N)$ holonomy admits a spinor field η that is covariantly constant, $D_i\eta = 0$.[*] Such a spinor field necessarily obeys $[D_i, D_j]\eta = 0$ or

$$R_{ijkl}\Gamma^{kl}\eta = 0 \qquad (15.4.1)$$

so that

$$0 = \Gamma^j\Gamma^{kl}R_{ijkl}\eta = 0. \qquad (15.4.2)$$

Using the gamma matrix identity $\Gamma^j\Gamma^{kl} = \Gamma^{jkl} + g^{jk}\Gamma^l - g^{jl}\Gamma^k$, and the curvature identity $R_{ijkl} + R_{iklj} + R_{iljk} = 0$, (15.4.2) implies that

$$\Gamma^k R_{ik}\eta = 0. \qquad (15.4.3)$$

This in fact implies that $R_{ik} = 0$; thus, a metric of $SU(N)$ holonomy is necessarily Ricci-flat.

We will now establish this in another way and at the same time prove the converse. In fact, we will show that (up to a factor) the Ricci tensor of a Kähler manifold is the field strength of the $U(1)$ part of the spin connection. Recall the nature of the embedding of $U(1)$ in $SO(2N)$. The $U(1)$ generator is precisely the complex structure $J^i{}_j$, whose nonzero components are $J^a{}_b = i\delta^a{}_b$, $J^{\bar{a}}{}_{\bar{b}} = -i\delta^{\bar{a}}{}_{\bar{b}}$. Given an $SO(2N)$ generator or in other words an antisymmetric matrix M_{kl}, its $U(1)$ part is $\mathrm{tr}JM = M^k{}_l J^l{}_k$. Now, as we discussed in connection with fig. 15.2, under parallel transport around a small loop in the ij plane, vectors are rotated by a matrix $R_{(ij)}$ whose kl matrix element is R_{ijkl}. The $U(1)$ part of this matrix is $F_{ij} = \mathrm{tr}JR_{(ij)} = R_{ijkl}J^{lk}$. This is the $U(1)$ part of the rotation undergone by tangent vectors that are transported around a small loop in the ij plane; it is the field strength of the $U(1)$ part of the spin connection. The nonzero components of F_{ij} are hence

$$F_{a\bar{b}} = -F_{\bar{b}a} = R_{a\bar{b}}{}^k{}_l J^l{}_k = iR_{a\bar{b}}{}^c{}_c - iR_{a\bar{b}}{}^{\bar{c}}{}_{\bar{c}}, \qquad (15.4.4)$$

but

$$R_{a\bar{b}}{}^c{}_c = R_{a\bar{b}d c}g^{\bar{d}c} = -R_{a\bar{b}cd}g^{c\bar{d}} = -R_{a\bar{b}}{}^{\bar{d}}{}_{\bar{d}}. \qquad (15.4.5)$$

So, in fact, $F_{a\bar{b}} = 2iR_{a\bar{b}}{}^c{}_c$. Comparing this to the Ricci tensor $R_{a\bar{b}} = R_a{}^c{}_{\bar{b}c}$ and the Kähler identity $R_a{}^c{}_{\bar{b}c} = -R_{a\bar{b}}{}^c{}_c$, we get finally the relation

---

[*] We established this earlier for $SU(3)$ holonomy, but it is true more generally for $SU(N)$ holonomy. With the embedding of $SU(N)$ in $SO(2N)$ that arises in Kähler geometry the spinor representation of $SO(2N)$ always contains an $SU(N)$ singlet.

between the Ricci tensor $R_{a\bar{b}}$ and the $U(1)$ field strength $F_{a\bar{b}}$:

$$F_{a\bar{b}} = -2iR_{a\bar{b}} = -F_{\bar{b}a}. \qquad (15.4.6)$$

Thus, a metric of $SU(N)$ holonomy is the same thing as a Ricci-flat Kähler metric. This satisfying result is our first taste of the fact that in finding states of unbroken supersymmetry we actually are finding vacuum states that obey the equations of motion of string field theory. After all, Ricci-flatness is certainly the leading long-distance approximation to the equations of motion of the theories we are discussing!

While we cannot here prove the existence of such metrics on general Kähler manifolds of $c_1 = 0$, we can point out a few facts that tend to make this plausible. We saw in equation (15.3.12) above that the Ricci tensor is $R_{a\bar{b}} = -\partial_a\partial_{\bar{b}}\ln\det g$. The condition that $R_{a\bar{b}} = 0$ is consequently that $\ln\det g = F(z^a) + \overline{F}(\overline{z^a})$, where $F(z^a)$ is an arbitrary holomorphic function of the z^a. Locally, by a holomorphic change of coordinates $z^a \to \tilde{z}^a(z^b)$, we can put F in an arbitrary form, say $F = 1/2$. The Ricci flat condition is hence locally

$$\ln\det g = 1. \qquad (15.4.7)$$

Here, of course, g must be a Kähler metric, described locally as $g_{a\bar{b}} = \partial_a\partial_{\bar{b}}\phi$, with some scalar function ϕ. So locally, in trying to find a Ricci-flat Kähler metric, we are trying to adjust one scalar function ϕ to obey one equation (15.4.7). This can be done locally. (Flat space, described by $\phi = \sum|z^a|^2$, is one solution.) Whether the one desired equation for one unknown can be solved globally is another question. We have already noted a topological obstruction – $c_1(K)$ must vanish. Yau's proof of the Calabi conjecture means that this is the only obstruction; a Kähler manifold of $c_1 = 0$ admits a unique Ricci-flat Kähler metric for any choice of the complex structure and Kähler class.

15.4.2 Covariantly Constant Forms

For future use, let us state the criterion for $c_1 = 0$ in another way. We know that the fundamental representation of $SO(2N)$ splits as $N \oplus \overline{N}$ under $SU(N)$. Out of the antisymmetric combination of N copies of the N of $SU(N)$ we can make an $SU(N)$-invariant $\epsilon_{a_1 a_2 \ldots a_N}$, which is not a $U(N)$ invariant of course. If we consider then an N form ω on a manifold of $SU(N)$ holonomy, such a field has a component that is neutral under the holonomy group and so is invariant under parallel transport in the case of a manifold of $SU(N)$ holonomy, but not in the case of a manifold of $U(N)$ holonomy. Actually, the covariantly constant N forms on a manifold of $SU(N)$ holonomy are forms of type $(N,0)$ or $(0,N)$, since it is from $N \otimes N \otimes \ldots \otimes N$ or from $\overline{N} \otimes \overline{N} \otimes \ldots \otimes \overline{N}$ that one can form an $SU(N)$ singlet. One of the essential properties of a manifold of $SU(N)$ holonomy is thus the existence of a covariantly constant $(N,0)$ form ω; its complex conjugate is a covariantly constant $(0,N)$ form $\overline{\omega}$. In terms of a positive-chirality covariantly constant spinor η, ω can be expressed as $\omega_{i_1 i_2 \ldots i_N} = \eta^T \Gamma_{i_1 i_2 \ldots i_N} \eta$. Being covariantly constant, ω nowhere vanishes (it would vanish everywhere if it vanishes at one point). The converse to this is also true. A charged field interacting with a topologically nontrivial $U(1)$ gauge field always has to have zeros somewhere, so if a complex manifold K admits an everywhere nonzero N form, the $U(1)$ part of the spin connection of K must be topologically trivial and the first Chern class of K must be zero.

A differential form α of type $(p,0)$ that obeys $\overline{\partial}\alpha = 0$ is called a holomorphic p form, since in local holomorphic coordinates all of the components of α are holomorphic functions. The covariantly constant $(N,0)$ form of a manifold of $SU(N)$ holonomy is certainly holomorphic. In practice, one often tries to prove that a complex manifold K has $c_1 = 0$ by searching for an everywhere nonzero holomorphic N form, though it would be adequate merely to find an everywhere nonzero $(N,0)$ form, holomorphic or not.

15.4.3 Some Manifolds of SU(N) Holonomy

We will now give some examples of manifolds K that admit metrics of $SU(N)$ holonomy. For our first example, we pick K to be a compact Riemann surface Σ. In this case, the whole spin connection is a $U(1)$ gauge field. For the spin connection to be a topologically trivial $U(1)$ gauge field, the requirement is that $\int_\Sigma F = 0$, F being the $U(1)$ field strength. We know from the discussion in chapter 14 that, among compact Riemann surfaces, $\int_\Sigma F$ vanishes only in the case of the torus. Thus it is only the torus that might admit a metric of $SU(1)$ holonomy. Actually, a metric of $SU(1)$ holonomy is the same thing as a flat metric, since the group $SU(1)$ is trivial. For real dimension two, Yau's proof of the Calabi conjecture just implies that the torus admits a flat metric, and we certainly know that that is true.

In more than one complex dimension, how can we find examples of complex manifolds of $c_1 = 0$? As we discussed in §15.2.3, some of the simplest examples of complex manifolds are the complete intersections of hypersurfaces in CP^N. We would like to determine which of these manifolds have $c_1 = 0$ and admit metrics of $SU(N)$ holonomy. To begin with, we consider the case in which we are given a single homogeneous polynomial P of degree k in the $N+1$ complex variables $Z^1, Z^2, \ldots, Z^{N+1}$. The equation $P = 0$ defines a hypersurface Q in CP^N. For what values of k and N does the hypersurface Q have $c_1 = 0$? We will answer this at least in part by exhibiting for certain k and N an everywhere nonzero holomorphic N form.

First let us work in the region of Q in which $Z^{N+1} \neq 0$. In this region we can parametrize Q with coordinates $x^a = Z^a/Z^{N+1}, a = 1, \ldots, N$. Now, consider the holomorphic $N-1$ form

$$\omega = dx^1 \wedge dx^2 \wedge \ldots \wedge dx^{N-1}/(\partial P/\partial x^N). \qquad (15.4.8)$$

Here by P we mean $P(x_1, x_2, \ldots, x_N, 1)$. At first sight it seems that the definition of ω depends on arbitrarily singling out x^N. Actually, however, we should remember that on Q the polynomial P is a constant (in fact, $P = 0$ on Q), so on Q we have $dP = 0$. Explicitly, $dP = \sum_{a=1}^{N}(\partial P/\partial x^a)dx^a$. So on Q we have (for any m)

$$dx^m/(\partial P/\partial x^N) = -dx^N/(\partial P/\partial x^m) + \sum_{i \neq m, N} f_i dx^i, \qquad (15.4.9)$$

where the form of the f_i can be worked out but will not matter. Substituting (15.4.9) in (15.4.8) (the $f_i dx^i$ terms drop out, since $dx^i \wedge dx^i = 0$), we find that for any m

$$\omega = (-1)^{N-m} dx^1 \wedge dx^2 \wedge \ldots \widehat{dx}^m \ldots \wedge dx^N/(\partial P/\partial x^m). \qquad (15.4.10)$$

The symbol $\widehat{dx}^m$ means that dx^m is to be omitted. Now, the importance of (15.4.10) is not merely that we wish to treat all of the x^m democratically. The original definition (15.4.8) of ω had an apparent singularity at $\partial P/\partial x^N = 0$. The alternative form (15.4.10) shows that this apparent singularity is harmless; in fact, the ability to write ω as in (15.4.10) for arbitrary m shows that as long as $Z^{N+1} \neq 0$, ω can be singular only if the polynomials $\partial P/\partial x^m$ are all zero for $m = 1, \ldots, N$. But, as we sketched in example (4) of §15.2.3, the $\partial P/\partial x^m$ never simultaneously vanish on the hypersurface $P = 0$ if that hypersurface is nonsingular. It is also clear from the definition that ω is nowhere zero except perhaps when $Z^{N+1} = 0$.

To investigate the behavior at $Z^{N+1} = 0$, let us assume, say, that $Z^1 \neq 0$, and define coordinates $y^m = Z^m/Z^1, m = 2, \ldots, N + 1$ or in other words $y^m = x^m \cdot (Z^{N+1}/Z^1)$. If P is homogeneous of degree k, then $P(1, y^2, \ldots, y^{N+1}) = P(x^1, x^2, \ldots, x^N, 1) \cdot (Z^{N+1}/Z^1)^k$. In terms of the

y^k, ω becomes

$$\omega = (-1)^{1-m}(Z^1/Z^{N+1})^{N+1-k} dy^2 \wedge dy^3 \\ \wedge \ldots d\hat{y}^m \ldots \wedge dy^{N+1}/(\partial P/\partial y^m). \tag{15.4.11}$$

In deriving (15.4.11) one must discard terms that vanish at $P = 0$ or that are proportional to, say, $dy^i \wedge dy^i$. Using (15.4.11), we see that if $N + 1 - k = 0$, then ω has neither a zero nor a pole at $Z^{N+1} = 0$, so under this condition we have found an everywhere nonzero (and nonsingular) holomorphic N form. For $k \neq N + 1$, we see from (15.4.11) that ω has only zeros and no poles or only poles and no zeros. With some slight additional work, the existence of a meromorphic $N - 1$ form on Q with only poles and no zeros or only zeros and no poles can be used to show that $c_1(Q) \neq 0$ if $k \neq N + 1$.

A hypersurface of degree $N + 1$ in CP^N is thus a relatively simple example of a Kähler manifold of $c_1 = 0$. To construct additional examples, consider n polynomials $P_1, \ldots, P_n$ in CP^N homogeneous of degrees $k_1, \ldots, k_n$, and let Q be the locus of $P_1 = \ldots = P_n = 0$. Let

$$\omega = dx^1 \wedge dx^2 \wedge \ldots \wedge dx^{N-n}/(\det M), \tag{15.4.12}$$

where M is the $n \times n$ matrix whose matrix elements are

$$M_{ab} = \partial P^a/\partial x^{N-n+b}. \tag{15.4.13}$$

As before, $x^k = Z^k/Z^{N+1}$. A reasoning similar to what we gave before shows that ω is everywhere nonzero and nonsingular if and only if

$$\sum_{i=1}^{n} k_i = N + 1. \tag{15.4.14}$$

In this way, we get additional examples of Kähler manifolds with $c_1 = 0$.

As an example of this, we see that a cubic equation in CP^2, such as Fermat's equation $X^3 + Y^3 + Z^3 = 0$, defines a Riemann surface of $c_1 = 0$. By our previous discussion, this must be topologically a torus. A quartic equation in CP^3, such as $X^4 + Y^4 + Z^4 + W^4 = 0$, defines a manifold that admits a metric of $SU(2)$ holonomy. This manifold is known as the $K3$ surface. It can be shown (by methods that go beyond the scope of our exposition here) to be *topologically* the only example of a manifold that admits a metric of $SU(2)$ holonomy. As for manifolds of complex dimension three, there are many topological types that admit metrics of

$SU(3)$ holonomy. In view of the above, an example of such a manifold would be a quintic hypersurface in CP^4, described by an equation such as

$$\sum_{a=1}^{5} Z_a^5 = 0 \tag{15.4.15}$$

or a more general homogeneous quintic polynomial in five variables. In CP^5 we could consider two cubic equations or a quartic equation and a quadric (quadratic) one. In CP^6 we could consider two quadrics and a cubic. Finally, in CP^7 the intersection of four quadrics defines a manifold that admits a metric of $SU(3)$ holonomy. These are the only examples of this nature.[*] There are, however, many other constructions of manifolds of $SU(3)$ holonomy. We will describe one in §16.10. In contrast to the situation in complex dimension two, there are innumerable examples (perhaps infinitely many) of Kähler manifolds of complex dimension three and $c_1 = 0$. The examples just cited are only some of the simplest.

15.5   Wave Operators on Kähler Manifolds

As we have seen extensively in chapter 14, physical properties of Kaluza–Klein theories depend very much on the zero modes of wave operators such as the Dirac operator and the de Rham operators. It turns out that these wave operators have very special properties on Kähler manifolds; it will be crucial to understand these properties.

15.5.1 The Dirac Operator

The Dirac operator is a good place to start. Standard Dirac gamma matrices Γ^i obey $\{\Gamma^i, \Gamma^j\} = 2g^{ij}$. On an N-dimensional Kähler manifold K, Dirac algebra simplifies if we work in local holomorphic coordinates z^a, $a = 1, \ldots, N$, with $g^{ab} = g^{\bar{a}\bar{b}} = 0$. The gamma matrix algebra is then

$$\begin{aligned}
\{\Gamma^a, \Gamma^b\} &= \{\Gamma^{\bar{a}}, \Gamma^{\bar{b}}\} = 0, \\
\{\Gamma^a, \Gamma^{\bar{b}}\} &= 2g^{a\bar{b}}.
\end{aligned} \tag{15.5.1}$$

Up to a normalization factor this is the usual algebra of fermion creation

---

[*] One could formally find other examples obeying (15.4.15) by using linear equations, but these are not really new, since a linear equation in CP^N merely sets to zero one of the coordinates, reducing the problem to a problem in CP^{N-1} with one less equation.

and annihilation operators; one can think of $\Gamma^{\overline{b}}$ as creation operators and Γ^a as the hermitian-conjugate annihilation operators.

The spinors, on which the gamma matrices act, can be constructed in the fashion that is familiar from the study of free fermions. We begin with a 'Fock vacuum' $|\Omega\rangle$ which is annihilated by the annihilation operators, $\Gamma^a |\Omega\rangle = 0$, $a = 1, \ldots, N$. The other states are the 'one-particle state' $|\Omega^{\overline{a}}\rangle = \Gamma^{\overline{a}} |\Omega\rangle$, the 'two-particle state' $|\Omega^{\overline{a}\overline{b}}\rangle = -|\Omega^{\overline{b}\overline{a}}\rangle = \Gamma^{\overline{a}}\Gamma^{\overline{b}} |\Omega\rangle$ and so on all the way up to the completely filled state $|\overline{\Omega}\rangle = \Gamma^{\overline{1}}\Gamma^{\overline{2}} \ldots \Gamma^{\overline{N}} |\Omega\rangle$.

We wish to study not just the spinor representation of $SO(2N)$, which is what we have just constructed, but spinor fields that depend in some way on the coordinates $z^a, \overline{z}^{\overline{a}}$. A general spinor field on K has an expansion

$$\psi(z^a, \overline{z}^{\overline{a}}) = \phi(z^a, \overline{z}^{\overline{a}}) |\Omega\rangle + \phi_{\overline{b}}(z^a, \overline{z}^{\overline{a}})\Gamma^{\overline{b}} |\Omega\rangle + \phi_{\overline{b}\overline{c}}(z^a, \overline{z}^{\overline{a}})\Gamma^{\overline{b}}\Gamma^{\overline{c}} |\Omega\rangle + \cdots .$$
(15.5.2)

At first sight, one might think that ϕ, which has no indices, is the same as a spinless field, but this is not necessarily true. ϕ may interact with the $U(1)$ part of the spin connection. In fact, if we normalize the $U(1)$ generator of $U(N)$ so that the N and $\overline{N}$ of $U(N)$ have $U(1)$ charge $+1$ and -1, respectively, then the creation operators $\Gamma^{\overline{a}}$, which transform as $\overline{N}$ of $U(N)$, have charge -1. The filled state $|\overline{\Omega}\rangle$ is obtained from the Fock vacuum $|\Omega\rangle$ by acting with N creation operators, so if q is the $U(1)$ charge of $|\Omega\rangle$, then the $U(1)$ charge of $|\overline{\Omega}\rangle$ is $q - N$. On the other hand, we can relate $|\Omega\rangle$ and $|\overline{\Omega}\rangle$ by an operation of complex conjugation. Such an operation exchanges creation operators Γ^a with annihilation operators $\Gamma^{\overline{a}}$ and so exchanges the empty state $|\Omega\rangle$ with the filled state $|\overline{\Omega}\rangle$. At the same time, complex conjugation changes the sign of the $U(1)$ generator. We conclude that the filled and empty states have equal and opposite $U(1)$ charge, so $q = N/2$.

It is therefore not true on a general Kähler manifold that ϕ in (15.5.2) is the same as a neutral spinless field. In general ϕ is not neutral with regard to the $U(1)$ part of the spin connection. The only case in which ϕ is the same as a spinless field is a case of particular interest to us: the case of a Ricci-flat Kähler manifold, on which the $U(1)$ part of the spin connection is trivial anyway. In that situation ϕ is completely decoupled from the spin connection, like a neutral scalar. Hence we can choose ϕ to be a constant. The spinor $\psi = |\Omega\rangle$ is then one of the covariantly constant spinor fields of a manifold of $SU(N)$ holonomy, the other being $\tilde{\psi} = |\overline{\Omega}\rangle$.

As for $\phi_{\overline{a}}$ in (15.5.2), it has a single (covariant) antiholomorphic index, so we can think of it as a $(0, 1)$ form. (This is only appropriate for $SU(N)$ holonomy; otherwise $\phi_{\overline{a}}$ has different $U(1)$ charge from a conventional $(0, 1)$ form.) Since $\phi_{\overline{a}\overline{b}}$ is antisymmetric in its two antiholomorphic indices, it is equivalent to a $(0,2)$ form. Moving on in this way, it should be clear that states obtained by acting on the 'Fock vacuum' with k creation operators are equivalent to $(0,k)$ forms. Thus, the spinors on a manifold of $SU(N)$ holonomy are the same as $(0,k)$ forms for $k = 0,\ldots,N$.[†] The creation operators that turn $(0,k)$ forms into $(0,k+1)$ forms are gamma matrices that reverse chirality, so a $(0,k)$ form is a spinor whose chirality is $(-1)^k$.

Now, we wish to study the Dirac operator $\not{D} = \Gamma^i D_i$. Two of its eigenvalues are the two covariantly constant spinors. Are there others? A standard computation shows that

$$\begin{aligned}\not{D}^2 &= \Gamma^i\Gamma^j D_i D_j = (1/2)\{\Gamma^i,\Gamma^j\}D_i D_j + (1/4)[\Gamma^i,\Gamma^j][D_i,D_j]\\ &= D_i D^i + [\Gamma^i,\Gamma^j][\Gamma^k,\Gamma^l]R_{ijkl}/32.\end{aligned} \tag{15.5.3}$$

Using the identity

$$\begin{aligned}\Gamma^i\Gamma^j\Gamma^k\Gamma^l &= \Gamma^{ijkl} - (\Gamma^{ik}g^{jl} \pm \text{permutations})\\ &\quad + g^{ij}g^{kl} - g^{ik}g^{jl} + g^{il}g^{jk},\end{aligned} \tag{15.5.4}$$

(15.5.3) reduces to

$$-\not{D}^2 = -D_i D^i + R/4 \tag{15.5.5}$$

with R the Ricci scalar. For a manifold of $SU(3)$ holonomy, $R = 0$, so $\not{D}^2 = D_i D^i$. The operator $-D_i D^i$ is positive semidefinite, and its zero eigenvalues must be covariantly constant. The proof is as follows: if $-D_i D^i \psi = 0$, then

$$0 = \int_K \langle\bar{\psi}|(-D_i D^i)|\psi\rangle = +\int_K \langle\overline{D_i\psi}|D^i\psi\rangle. \tag{15.5.6}$$

This is possible only if $D_i\psi = 0$. The spinor of $SO(2N)$ contains no $SU(N)$ singlets other than the Fock vacuum and its charge conjugate. Covariantly constant spinors must be singlets of the holonomy group, so a manifold whose holonomy is $SU(N)$ rather than a subgroup has precisely two covariantly constant spinors. Therefore, on a manifold of $SU(N)$ holonomy, the only zero eigenvalues of the Dirac operator are the covariantly constant $(0,0)$ and $(0,N)$ forms.

[†] It is possible to define (p,q) forms on any complex manifold, so the equivalence of spinors to $(0,k)$ forms for manifolds of $SU(N)$ holonomy means, in particular, that such manifolds are always manifolds on which spinors can be defined, a nontrivial assertion as we have noted in §12.1.

We have not yet formulated the most far-reaching aspect of the connection between spinors and $(0, k)$ forms on manifolds of $SU(N)$ holonomy. Let us look more carefully at the Dirac operator:

$$\not{D} = \Gamma^{\bar{a}} D_{\bar{a}} + \Gamma^a D_a. \tag{15.5.7}$$

Let us study the first term, $\not{D}_+ = \Gamma^{\bar{a}} D_{\bar{a}}$. Acting on a $(0, k)$ form $\psi = \phi_{\bar{a}_1 \ldots \bar{a}_k} \Gamma^{\bar{a}_1} \ldots \Gamma^{\bar{a}_k} |\Omega\rangle$, $\not{D}_+$ gives the $(0, k+1)$ form

$$\not{D}_+ \psi = \frac{1}{k+1}(D_{\bar{a}_0} \phi_{\bar{a}_1 \ldots \bar{a}_k} \pm \text{cyclic permutations})\Gamma^{\bar{a}_0} \ldots \Gamma^{\bar{a}_k} |\Omega\rangle. \tag{15.5.8}$$

Comparing (15.5.8) with (15.3.3), we see that in fact $\not{D}_+$ is equivalent to the Dolbeault operator $\bar{\partial}$. As the Dirac operator is hermitian, it is $\not{D} = \bar{\partial} + \bar{\partial}^*$, where $\bar{\partial}^*$ is the adjoint of $\bar{\partial}$.

15.5.2 Dolbeault Cohomology

By analogy with our treatment of de Rham cohomology in chapter 14, there is a more 'topological' way to think about these matters. We refer to a form ψ that obeys $\bar{\partial}\psi = 0$ as a $\bar{\partial}$ closed form. We also refer to a form ψ that can be written $\psi = \bar{\partial}\lambda$ as $\bar{\partial}$ exact. Since $\bar{\partial}^2 = 0$, every $\bar{\partial}$ exact form is $\bar{\partial}$ closed. There may in addition be nontrivial examples of $\bar{\partial}$ closed forms. That is, there may be $(0, n)$ forms ψ that obey $\bar{\partial}\psi = 0$ but cannot be written as $\psi = \bar{\partial}\lambda$. Just as in the definition of de Rham cohomology, it is natural to think of two $(0, n)$ forms ψ and ψ' that are $\bar{\partial}$ closed as being equivalent if their difference is trivial in the sense that $\psi - \psi' = \bar{\partial}\lambda$ for some λ. The equivalence classes make up a vector space that is known as the Dolbeault cohomology group $H^{(0,n)}(K)$ (the group structure being the additive structure as a vector space).

Some standard arguments from de Rham cohomology carry over easily to the Dolbeault case. For instance, if ψ is $\bar{\partial}$ closed but not exact, then we cannot choose λ to make $I = \langle \psi - \bar{\partial}\lambda | \psi - \bar{\partial}\lambda \rangle$ vanish, but we can choose λ to minimize I. The variational equation

$$\frac{\delta I}{\delta \lambda} = 0 \tag{15.5.9}$$

tells us that $\psi' = \psi - \bar{\partial}\lambda$ obeys $\bar{\partial}^* \psi' = 0$ as well as $\bar{\partial}\psi' = 0$. This shows us that in every Dolbeault cohomology class there is a solution ψ' of $\bar{\partial}\psi' = \bar{\partial}^* \psi' = 0$. Conversely, the same arguments as in the de Rham case show that a solution of $\bar{\partial}\psi = \bar{\partial}^* \psi = 0$ cannot possibly be written as $\psi = \overline{\partial}\chi$, so in fact Dolbeault cohomology classes are in one-to-one correspondence with forms ψ obeying $\overline{\partial}\psi = \overline{\partial}^{*}\psi = 0$. A form that obeys $\overline{\partial}\psi = \overline{\partial}^{*}\psi = 0$ is obviously a zero eigenvalue of the 'Laplacian'

$$\Delta_{\overline{\partial}} = \overline{\partial}\,\overline{\partial}^{*} + \overline{\partial}^{*}\overline{\partial}. \qquad (15.5.10)$$

Just as in the de Rham case, an eigenfunction ψ of the Laplacian with zero eigenvalue obeys

$$0 = \langle\psi|\Delta_{\overline{\partial}}|\psi\rangle = \langle\overline{\partial}\psi|\overline{\partial}\psi\rangle + \langle\overline{\partial}^{*}\psi|\overline{\partial}^{*}\psi\rangle = 0, \qquad (15.5.11)$$

so it must obey $\overline{\partial}\psi = \overline{\partial}^{*}\psi = 0$.

Putting all of these facts together, Dolbeault cohomology classes are (just as in the de Rham case) in one-to-one correspondence with zero eigenvalues of the Laplacian. Moreover, with $\overline{\partial}^{2} = \overline{\partial}^{*2} = 0$, the Laplacian $\Delta_{\overline{\partial}}$ is up to normalization one and the same thing as the square of the Dirac operator. Hence the Dolbeault cohomology groups $H^{(0,n)}$ are precisely the spaces of $(0,n)$ forms annihilated by the Dirac operator. In particular, the discussion surrounding (15.5.6) determines the spaces $H^{(0,n)}$ for manifolds of $SU(N)$ holonomy; they are one dimensional for $n = 0, N$ and zero otherwise.

In studying the Dirac operator in real differential geometry, we found only one tool for predicting zero eigenvalues – the Dirac index. In Kähler geometry, Dolbeault cohomology gives a much richer framework for predicting zero eigenvalues of the Dirac operator and hence massless particles. Instead of one number, the index, which may predict zero eigenvalues, we have $N+1$ of them, namely the dimensions of the Dolbeault cohomology groups $H^{(0,n)}$, $n = 0, \ldots, N$. These depend only on the complex structure of the manifold K, not on a choice of metric. In complex geometry properties that depend only on the topology and complex structure of a complex manifold K behave very much like ordinary topological properties on real manifolds, even though the complex structure of a complex manifold may depend in general on freely adjustable parameters. Thus, the Dolbeault cohomology groups should be viewed as quasi-topological.

As a first illustration of the significance of this, let us note that the ordinary index theorem can never predict the occurrence of zero eigenvalues of the Dirac operator of *both* positive and negative chirality; in real differential geometry such a behavior seemed implausible.[*] Yet in Kähler

---

[*] Even in real differential geometry, there is in $8k+2$ dimensions a Z_2 index theorem that can sometimes predict that the Dirac operator has a zero eigenvalue of each chirality, but there is nothing to match the phenomena that arise in Kähler geometry.

geometry such behavior is quite natural. For example, on a manifold of $SU(N)$ holonomy we have found that the minimal Dirac equation has two zero eigenvalues, a $(0,0)$ form and a $(0,N)$ form. For odd N (which happens to be a case of interest!), these states have opposite chirality. What makes this behavior 'natural' is really the fact that zero eigenvalues of the Dirac operator correspond to Dolbeault cohomology classes, which are quasi-topological. In later generalizations, the role of Dolbeault cohomology will become even more significant.

15.5.3 The Hodge Decomposition

So far we have studied only the $(0,n)$ forms in Kähler geometry. We would now like to study the properties of (p,q) forms for arbitrary p and q.

The definition of the $\overline{\partial}$ cohomology groups carries over immediately to (p,q) forms. We consider (p,q) forms which are $\overline{\partial}$ closed but not exact. Considering two such forms ψ and ψ' to be equivalent if their difference is exact, the equivalence classes make up the Dolbeault cohomology group $H^{(p,q)}$. Precisely the same arguments as before show that elements of $H^{(p,q)}$ are in one-to-one correspondence with (p,q) forms that are annihilated by $\Delta_{\overline{\partial}} = \overline{\partial}\,\overline{\partial}^{*} + \overline{\partial}^{*}\overline{\partial}$.

This is not the first time that we have encountered 'cohomology groups' that consist of 'harmonic differential forms'. In chapter 14 we defined the Laplacian $\Delta = dd^{*} + d^{*}d$, and noted that the de Rham cohomology group H_D^n (in this section only we use the subscript D to specify de Rham cohomology) can be identified with the harmonic n forms, *i.e.*, the n forms annihilated by Δ. One may wonder what is the connection between the 'real' Laplacian Δ and the 'complex' Laplacian $\Delta_{\overline{\partial}}$. The answer is that on Kähler manifolds (but not on more general complex manifolds) they are related by $\Delta_{\overline{\partial}} = \Delta/2$. In particular they have the same zero eigenvalues, so a de Rham cohomology class is also a Dolbeault cohomology class and vice versa. In real differential geometry de Rham cohomology classes can always be taken to be n forms of definite n, but the relation of de Rham and Dolbeault cohomology makes possible a much sharper statement in the case of Kähler geometry. The de Rham cohomology classes of a Kähler manifold can be taken as (p,q) forms of definite p and q. Mathematically, the relation between de Rham and Dolbeault cohomology groups is that

$$H_D^n = \oplus_{p+q=n} H^{p,q}. \tag{15.5.12}$$

This assertion is known as the Hodge decomposition.

One way to establish the above statements is to recall from §14.3.4 that the de Rham complex of differential forms of arbitrary degree is conveniently described in terms of a field $\psi_{\alpha\beta}$ with two spinor indices. We introduced two sets of anticommuting gamma matrices $\Gamma^i, \tilde\Gamma^j$, $i, j = 1, \ldots, 2N$, with $\{\Gamma^i, \Gamma^j\} = \{\tilde\Gamma^i, \tilde\Gamma^j\} = 2g^{ij}$, $\{\Gamma^i, \tilde\Gamma^j\} = 0$. In the Kähler case, the two sets of gamma matrices become two anticommuting sets of fermion creation and annihilation operators, $\{\Gamma^a, \Gamma^{\bar b}\} = \{\tilde\Gamma^a, \tilde\Gamma^{\bar b}\} = 2g^{a\bar b}$. It is convenient to treat holomorphic and antiholomorphic indices symmetrically in a way that was not possible with only one set of gamma matrices. Thus, we will regard the

$$\Gamma^{\bar a} \text{ and } \tilde\Gamma^b \tag{15.5.13}$$

as creation operators. This choice is symmetric under exchange of holomorphic and antiholomorphic indices a and $\bar a$ if accompanied by $\Gamma \leftrightarrow \tilde\Gamma$. We introduce a 'Fock vacuum' $|\Omega\rangle$ annihilated by the annihilation operators

$$\Gamma^a \text{ and } \tilde\Gamma^{\bar a}. \tag{15.5.14}$$

A general state

$$\Gamma^{\bar b_1} \ldots \Gamma^{\bar b_q} \tilde\Gamma^{a_1} \ldots \tilde\Gamma^{a_p} \cdot |\Omega\rangle \tag{15.5.15}$$

is antisymmetric in p holomorphic and q antiholomorphic indices, so we identify it as a (p, q) form.

In §14.3.4 we defined an operator S acting on the bispinor field $\psi_{\alpha\beta}$ by the formula

$$(S\psi)_{\alpha\beta} = \Gamma^i_{\alpha\sigma} D_i \psi_{\sigma\beta}. \tag{15.5.16}$$

We organized the Dirac algebra in the previous paragraph in such a way that S is the operator $\bar\partial + \bar\partial^*$ whose square is $\Delta_{\bar\partial}$ and whose zero eigenvalues are therefore the Dolbeault cohomology classes. On the other hand, in chapter 14, with a slightly different way of organizing Dirac algebra, we interpreted the same operator S as the operator $d + d^*$ whose square is Δ and whose zero eigenvalues are de Rham cohomology classes. Therefore, we have already established the relation between Dolbeault and de Rham cohomology. However, it is more satisfying to go on and explicitly demonstrate that S^2 maps (p, q) forms to (p, q) forms, so that its zero eigenvalues are forms of definite p and q. The required formulas are useful in any case.

The bispinor field $\psi_{\alpha\beta}$ is a special case of a more general thing – a spinor field $\psi_\alpha{}^x$ with an extra index x acted on by some Yang–Mills field.

The bispinor $\psi_{\alpha\beta}$ is the special case in which x happens to be an extra spinor index. The operator S acting on $\psi_{\alpha\beta}$ is just a minimally coupled Dirac operator with the Yang–Mills field acting on the β index being the spin connection.

Considering the general case of a field $\psi_\alpha{}^x$ with an arbitrary Yang–Mills index x, we would like to calculate the square of the minimally coupled Dirac operator $\Gamma^i D_i$ acting on this field. The derivation of (15.5.3) and (15.5.5) is modified because there is an extra term in the commutator of covariant derivatives. The extra term is $[D_i, D_j] \sim F_{ij}^x T^x$, where F_{ij}^x is the Yang–Mills field strength and T^x are the group generators. The generalization of (15.5.5) is

$$(i\Gamma^i D_i)^2 = -D_i D^i + R/4 - \Gamma^{ij} F_{ij}^x T^x / 4. \tag{15.5.17}$$

We wish to specialize this general formula to the case in which the index x is an extra spinor index, acted on by the spin connection. In this case, the T^x are just $[\tilde{\Gamma}^k, \tilde{\Gamma}^l]/4$, and F_{ij}^x is just R_{ijkl}. Also, in this case, as we know from chapter 14, $(i\Gamma^i D_i)^2$ is the Laplacian Δ associated with de Rham cohomology. Hence (15.5.17) reduces in this situation to a formula for the Laplacian

$$\Delta = -D_i D^i + R/4 - R_{ijkl}[\Gamma^i, \Gamma^j][\tilde{\Gamma}^k, \tilde{\Gamma}^l]/16. \tag{15.5.18}$$

In the case of a Kähler manifold (not necessarily of $SU(N)$ holonomy!), we know that the only nonzero components of the Riemann tensor are $R_{a\bar{b}c\bar{d}}$, so the last term in (15.5.18) becomes $R_{a\bar{b}c\bar{d}}[\Gamma^a, \Gamma^{\bar{b}}][\tilde{\Gamma}^c, \tilde{\Gamma}^{\bar{d}}]$. This has one creation and one annihilation operator of each type, so it maps (p, q) forms to (p, q) forms. The same is true of the covariant derivatives D_i (because the $U(N)$ representation of a (p, q) form depends on p and q, and D_i, which contains the $U(N)$ spin connection only, cannot change the $U(N)$ representation). So Δ maps (p, q) forms to (p, q) forms, and the harmonic forms can be chosen as forms of definite (p, q). This completes our discussion of the Hodge decomposition.

15.5.4 Hodge Numbers

In real manifolds, the dimension of the de Rham cohomology group H_D^n is called the Betti number b_n. Similarly, in Kähler geometry the dimension of the Dolbeault cohomology group $H^{p,q}$ is called the Hodge number $h^{p,q}$.

The Hodge decomposition implies that

$$b_n = \sum_{p+q=n} h^{p,q}. \tag{15.5.19}$$

This implies that the Euler characteristic of a Kähler manifold is

$$\chi = \sum_n (-1)^n b_n = \sum_{p,q} (-1)^{p+q} h^{p,q}. \tag{15.5.20}$$

The Hodge numbers are governed by certain simple restrictions that are easily deduced from (15.5.18). First of all, symmetry under $\Gamma^i \leftrightarrow \tilde{\Gamma}^i$ means that

$$h^{p,q} = h^{q,p}. \tag{15.5.21}$$

This is a reflection of the fact that it is a matter of convention what we consider holomorphic and what we consider antiholomorphic. Second, (15.5.18) is invariant under exchange of all creation operators with annihilation operators (thus exchanging the N and $\overline{N}$ of $U(N)$). This implies that

$$h^{p,q} = h^{N-p,N-q}. \tag{15.5.22}$$

This is Poincaré duality, specialized to Kähler manifolds.

Let us now discuss the determination of the Hodge numbers in some simple examples. First, we consider a compact Riemann surface of genus g. We know that the Betti numbers are $b_0 = b_2 = 1$, $b_1 = 2g$. Equations (15.5.19) and (15.5.21) are enough to determine that $h^{0,0} = h^{1,1} = 1$, and $h^{1,0} = h^{0,1} = g$. Here $h^{1,0}$ is the number of globally defined holomorphic $(1,0)$ forms, or in other words globally defined holomorphic differentials $a(z)dz$. The fact that this number is g is the original Riemann–Roch theorem.

As a second example, consider a manifold of complex dimension three whose holonomy is precisely $SU(3)$, not a subgroup. At the beginning of this section we showed that fermion zero modes, which are harmonic $(0,n)$ forms, must be covariantly constant. This means that $h^{0,0} = h^{3,0} = 1$ and $h^{1,0} = h^{2,0} = 0$. The remaining Hodge numbers that are not determined by (15.5.21) and (15.5.22) are $h^{1,1}$ and $h^{2,1}$. Here (since $h^{2,0} = h^{0,2} = 0$) $h^{1,1}$ is the same as b_2, the number of harmonic two forms. As for $h^{2,1}$, it also has a qualitative significance, which we will encounter after some further adventures with Dolbeault cohomology. The Euler characteristic of a manifold whose holonomy is precisely $SU(3)$ is $\chi = 2(h^{1,1} - h^{2,1})$.

15.6   Yang–Mills Equations and Holomorphic Vector Bundles

Equation (15.1.3), under the simplified assumption that $H = d\phi = 0$, gave two conditions for unbroken supersymmetry. One of them, $0 = D_i\eta$, has been our subject until now. We now turn to the second equation,

$$0 = \Gamma^{ij} F_{ij}\eta, \tag{15.6.1}$$

with F_{ij} being the field strength derived from some Yang–Mills gauge field A, which is a connection on some vector bundle X. It is now time to investigate the implications of this equation. Written out explicitly in terms of creation and annihilation operators, (15.6.1) becomes

$$0 = (F_{\bar{a}\bar{b}}\Gamma^{\bar{a}\bar{b}} + F_{ab}\Gamma^{ab} + 2F_{a\bar{b}}\Gamma^{a\bar{b}})\eta. \tag{15.6.2}$$

All three terms in (15.6.2) must vanish separately. Requiring that the first term in (15.6.2) should vanish if η is the Fock vacuum, or that the second term should vanish if η is the completely filled state, we learn that

$$F_{\bar{a}\bar{b}} = F_{ab} = 0. \tag{15.6.3}$$

The analysis of the last term in (15.6.2) is slightly more subtle. $\Gamma^{a\bar{b}}$ contains one creation operator and one annihilation operator. Acting on a completely empty or completely filled state η, the annihilation operator must annihilate what the creation operator has created, or vice versa, so $\Gamma^{a\bar{b}}\eta = \pm g^{a\bar{b}}\eta$. Hence the vanishing of the last term in (15.6.2) amounts to the condition that

$$g^{a\bar{b}} F_{a\bar{b}} = 0. \tag{15.6.4}$$

We will assume that the fields are real in the sense that the $(1,0)$ and $(0,1)$ parts of the gauge field A are hermitian conjugates of one another. In this case, to obey (15.6.3) it is enough to require $F_{\bar{a}\bar{b}} = 0$, since the condition $F_{ab} = 0$ is related to this by complex conjugation. Locally, there is no difficulty in finding the general solution of $F_{\bar{a}\bar{b}} = 0$. This equation says that $0 = [D_{\bar{a}}, D_{\bar{b}}] = [\partial_{\bar{a}} + iA_{\bar{a}}, \partial_{\bar{b}} + iA_{\bar{b}}]$, so

$$A_{\bar{b}} = i\partial_{\bar{b}} V \cdot V^{-1}, \tag{15.6.5}$$

where V is some matrix-valued function of the coordinates $z^a, \bar{z}^b$. With A_a the hermitian conjugate of $A_{\bar{a}}$, this implies that

$$A_a = i\partial_a V^{*-1} \cdot V^*. \tag{15.6.6}$$

Here V^* is the adjoint of V, so $V^* = V^{-1}$ if and only if V is unitary. Equations (15.6.5) and (15.6.6) show that either $A_{\bar{a}}$ or A_a may be set to zero by a gauge transformation, but in general not simultaneously. They can be set to zero simultaneously only if V is unitary.

Let f be a charged field interacting with the gauge field A. Then f is said to be 'holomorphic' if

$$D_{\bar{a}}f = 0. \tag{15.6.7}$$

The integrability condition for this equation is $[D_{\bar{a}}, D_{\bar{b}}] = 0$. Locally, (15.6.7) is equivalent, in view of (15.6.5), to $f = Vg$, where g is a holomorphic function in the usual sense, $\partial_{\bar{a}}g = 0$. Globally, (15.6.7) is a significant generalization of the elementary definition of an analytic or holomorphic function. Equation (15.6.3) is fundamental in complex geometry because it leads to this generalization of ordinary complex analysis.

15.6.1 Holomorphic Vector Bundles

We now will analyze the global import of (15.6.3). First we note that in general the vector bundle X on which the gauge field is a connection may be topologically nontrivial. Thus, in general we must, as in §12.3, cover our manifold K with open sets $O_{(\alpha)}$, on each of which we have a gauge field $A_{(\alpha)}$. On overlap regions $O_{(\alpha\beta)} = O_{(\alpha)} \cap O_{(\beta)}$, $A_{(\alpha)}$ is related to $A_{(\beta)}$ by a gauge transformation $U_{(\alpha\beta)}$. Explicitly,

$$A_{i(\alpha)} = U_{(\alpha\beta)}A_{i(\beta)}U_{(\alpha\beta)}^{-1} + i\partial_i U_{(\alpha\beta)} \cdot U_{(\alpha\beta)}^{-1} \tag{15.6.8}$$

on $O_{(\alpha\beta)}$. Equation (15.6.8) implies that $U_{(\alpha\beta)} = U_{(\beta\alpha)}^{-1}$. The $U_{(\alpha\beta)}$ are called transition functions for the bundle X. Consistency of (15.6.8) requires that the transition functions on triple overlap regions $O_{(\alpha\beta\gamma)} = O_{(\alpha)} \cap O_{(\beta)} \cap O_{(\gamma)}$ should obey the consistency condition

$$U_{(\alpha\beta)}U_{(\beta\gamma)}U_{(\gamma\alpha)} = 1. \tag{15.6.9}$$

The transition functions are, of course, not uniquely defined. Making a gauge transformation on each $O_{(\alpha)}$ by a gauge transformation $U_{(\alpha)}$, the $U_{(\alpha\beta)}$ are transformed into equally good transition functions

$$\tilde{U}_{(\alpha\beta)} = U_{(\alpha)}U_{(\alpha\beta)}U_{(\beta)}^{-1}, \tag{15.6.10}$$

which equally well obey (15.6.9).

Now, in a situation in which $F_{\bar{a}\bar{b}} = 0$, we can write $A_{(\alpha)}$ on each $O_{(\alpha)}$ as in (15.6.6),

$$A_{\bar{a}(\alpha)} = i\partial_{\bar{a}}V_{(\alpha)} \cdot V_{(\alpha)}^{-1}, \tag{15.6.11}$$

with now a different $V_{(\alpha)}$ on each open set $O_{(\alpha)}$. Substituting (15.6.11) in (15.6.8), we find that on each $O_{(\alpha\beta)}$, the quantity

$$U'_{(\alpha\beta)} = V_{(\alpha)}^{-1}U_{(\alpha\beta)}V_{(\beta)} \tag{15.6.12}$$

obeys

$$\partial_{\bar{a}}U'_{(\alpha\beta)} \cdot \left[U'_{(\alpha\beta)}\right]^{-1} = 0. \tag{15.6.13}$$

Comparing (15.6.10) to (15.6.12), we see that (taking $U_{(\alpha)} = V_{(\alpha)}^{-1}$) the $U'_{(\alpha\beta)}$ can be viewed as new transition functions for the bundle X. Equation (15.6.13) means that these new transition functions are holomorphic.

A vector bundle on which the transition functions can be chosen to be holomorphic is called a holomorphic vector bundle. What we have found is that a solution of (15.6.3) is a holomorphic vector bundle. The converse is also true (though we will not prove it here). Given a vector bundle X over a complex manifold K, if we can choose the transition functions $U_{(\alpha\beta)}$ of X to be holomorphic functions, then it is possible to find on X a gauge field that obeys (15.6.3). Such a gauge field is called a holomorphic gauge field or holomorphic connection.

Topologically, two vector bundles X and $\tilde{X}$ with transition functions $U_{(\alpha\beta)}$ and $\tilde{U}_{(\alpha\beta)}$ are considered equivalent if the U and $\tilde{U}$ are related as in (15.6.12). Holomorphically, the situation is more delicate. The importance of the notion of a holomorphic vector bundle is largely that on such a bundle we can (as in (15.6.7) above) generalize the concept of an analytic or holomorphic function. The holomorphic vector bundles with transition functions U and $\tilde{U}$ define equivalent notions of holomorphic function only if the $V_{(\alpha)}$ in (15.6.12) can be chosen to be holomorphic; only in this case are X and $\tilde{X}$ considered to be equivalent holomorphically.

Given a holomorphic vector bundle, the holomorphic connection A_i on this bundle is not unique. Equally good (in the sense that it also obeys (15.6.3)) would be the new gauge field A'_i whose $(0, 1)$ part is

$$A'_{\bar{a}} = GA_{\bar{a}}G^{-1} + i\partial_{\bar{a}}G \cdot G^{-1}. \tag{15.6.14}$$

It is not too hard to prove that any two holomorphic connections A and A' on the same holomorphic vector bundle are related as in (15.6.14). G is globally defined, since A and A', as connections on the same bundle, should have the same Dirac string singularities. We need not specify the $(1,0)$ part of A', since it is related to the $(0,1)$ part by complex conjugation.

If G is unitary, then A' is gauge equivalent to A, but not otherwise. In general, we may write for G a polar decomposition $G = UP$, where U is unitary and P is hermitian. The unitary part is irrelevant, so we may simple assume that G is hermitian.

15.6.2 The Donaldson–Uhlenbeck–Yau Equation

Now, we wish to ask the following question. Given a holomorphic vector bundle X, can the connection A on this bundle be chosen to obey not just (15.6.3) but also (15.6.4)? This amounts to asking whether the hermitian matrix G in (15.6.14) can be chosen so that the hermitian matrix $H = g^{a\bar{b}} F_{a\bar{b}}$ in (15.6.4) vanishes at each point in K. Locally, we can expect to be able to adjust one hermitian matrix G to set another hermitian matrix H to zero. The question of whether this can be done globally is another story. As usual, it involves topological questions.

If A is a $U(1)$ gauge field, we can answer the question rather easily. In this case, the $(1,1)$ form F is gauge invariant and represents the first Chern class of the line bundle X. Let k be the Kähler form of the Kähler manifold K, which we suppose has complex dimension N. Then one can verify that in this case H is equal to

$$H = \epsilon^{a_1 a_2 ... a_N \bar{b}_1 \bar{b}_2 ... \bar{b}_N} F_{a_1 \bar{b}_1} k_{a_2 \bar{b}_2} ... k_{a_N \bar{b}_N} / (N-1)!^2. \qquad (15.6.15)$$

Hence, $H = 0$ if and only if the (N,N) form $\lambda = F \wedge k \wedge ... \wedge k$ vanishes. On the other hand,

$$f = \int_K \lambda \qquad (15.6.16)$$

is a topological invariant in the sense that for fixed choice of the bundle X and the Kähler class k, f is independent of the choice of a gauge field A on X. The crucial ingredient here is the fact that Chern classes are topological invariants or more specifically the fact that the cohomology class of the $(1,1)$ form F depends only on the topology of X.

Thus, a necessary condition for finding a holomorphic connection A with $H = 0$ is that the topological invariant

$$f = \int_K \text{tr} F \wedge k \wedge ... \wedge k = (N-1)!^2 \int_K H \qquad (15.6.17)$$

should vanish. This is easily seen to be (in the abelian case) also a sufficient condition. In the abelian case the hermitian matrix G of (15.6.14) is just $G = e^\sigma$ for some scalar function σ. Also, if F' is the gauge-invariant field strength derived from A', then in the abelian case $H' = g^{a\bar{b}}F'_{a\bar{b}}$ is related to $H = g^{a\bar{b}}F_{a\bar{b}}$ by $H' = H + \Delta\sigma$ (with $\Delta = g^{a\bar{b}}\partial_a\partial_{\bar{b}}$ being the ordinary Laplacian). Setting $H' = 0$ means finding σ such that $\Delta\sigma = H$; this is possible if and only if $\int_K H = 0$. If indeed the eigenfunctions of Δ are ψ_k with eigenvalues λ_k, the solution of $\Delta\sigma = H$ would have to be

$$\sigma(x) = \sum_k \frac{1}{\lambda_k}\psi_k(x)\langle\psi_k|H\rangle. \tag{15.6.18}$$

The trouble with (15.6.18) is that λ_k might vanish for some k. Equation (15.6.18) is a valid solution of $\Delta\sigma = H$ if and only if $\langle\psi_k|H\rangle = 0$ whenever $\lambda_k = 0$. Actually, the only zero eigenvalue of Δ corresponds to the constant function 1, so the necessary and sufficient condition for validity of (15.6.18) is

$$0 = \langle 1|H\rangle = \int_K H, \tag{15.6.19}$$

as we claimed.

In the nonabelian case, it is of course less straightforward to determine the global conditions for existence of a solution of (15.6.4). If A is a $U(N)$ gauge field, the first necessary condition is that the $U(1)$ part of A should obey the conditions that we have just stated. The additional requirement has been determined in recent theorems by Donaldson and by Uhlenbeck and Yau. The requirement is that the holomorphic vector bundle X must be a 'stable' bundle. This is a relatively mild topological condition whose meaning we will not try to elucidate. In our applications, the necessary condition will be evident on physical grounds.

Since the conditions we have been discussing are related to unbroken supersymmetry, it is natural to think that they will imply the equations of motion, in this case the Yang–Mills equations. Indeed, with $F_{ab} = F_{\bar{b}\bar{c}} = 0$, the Yang–Mills equations $D^i F_{ij} = 0$ reduce to

$$g^{a\bar{b}}D_a F_{\bar{b}c} = 0. \tag{15.6.20}$$

The Bianchi identity $D_i F_{jk} + D_j F_{ki} + D_k F_{ij} = 0$ implies in this situation that $D_a F_{\bar{b}c} = D_c F_{\bar{b}a}$, so that the Kähler–Yang–Mills equations (15.6.20) amount to $D_c g^{a\bar{b}} F_{a\bar{b}} = 0$. This certainly follows from the equation $g^{a\bar{b}} F_{a\bar{b}} = 0$ that we have just been discussing.

15.6.3 Examples

The above concepts may seem somewhat abstract, and therefore we will now give a number of concrete examples, some of which will be important in later applications.

(1) *Holomorphic Vector Bundles Over Riemann Surfaces*

As usual, it is convenient to begin with the case in which the complex manifold considered is a Riemann surface Σ, of complex dimension one. Let A be *any* gauge field over Σ. Then inevitably $F_{\bar{a}\bar{b}} = 0$ (since on a Riemann surface $\bar{a}$ and $\bar{b}$ have only one possible value, namely $\bar{z}$), so A defines a holomorphic vector bundle over Σ.

The above may possibly give a somewhat exaggerated impression of how many holomorphic vector bundles there are over a Riemann surface. Two gauge fields A and A' may define equivalent holomorphic vector bundles if they differ by a transformation of the type (15.6.14), and in fact holomorphic vector bundles of a given topological type over a given Riemann surface are determined by a finite number of parameters. We will address this question in the next section.

(2) *The Tangent Bundle and Its Cousins*

For our next example of a holomorphic vector bundle, let K be an arbitrary complex manifold. The 'holomorphic tangent bundle' consists of tangent vectors with a $(1, 0)$ part only; that is, a section of the holomorphic tangent bundle would be a vector field v^a with only a holomorphic index.

To show that this is a holomorphic vector bundle, we construct explicitly the transition functions. The complex manifold K can be covered with open sets $O_{(\alpha)}$, on each of which we have complex coordinates $z^a_{(\alpha)}$. On overlap regions $O_{(\alpha\beta)}$ the $z^a_{(\alpha)}$ are analytic functions of the $z^a_{(\beta)}$, this being one way to express the definition of a complex manifold; hence quantities such as $\partial z^a_{(\alpha)}/\partial z^b_{(\beta)}$ are analytic functions of $z^a_{(\alpha)}$ (or of $z^b_{(\beta)}$). Consider a general vector field v on K. On each open set $O_{(\alpha)}$, v has components $v^a_{(\alpha)}$. In overlap regions $O_{(\alpha\beta)}$, in changing from $z^a_{(\beta)}$ coordinates to $z^a_{(\alpha)}$ coordinates, the vector field v transforms in a way that is familiar from general relativity:

$$v^a_{(\alpha)} = \frac{\partial z^a_{(\alpha)}}{\partial z^b_{(\beta)}} v^b_{(\beta)}. \tag{15.6.21}$$

We see from this that the transition functions of the tangent bundle are

$$V_{(\alpha\beta)}{}^a{}_b = \partial z^a_{(\alpha)}/\partial z^b_{(\beta)}. \tag{15.6.22}$$

As these are holomorphic functions, the tangent bundle is a holomorphic vector bundle.

Instead of considering a vector field v^a with a holomorphic tangent vector index, we could consider, for instance, a differential form w_a of type $(1,0)$. These transform under coordinate transformations as $w_{a(\alpha)} = (\partial z^b_{(\beta)}/\partial z^a_{(\alpha)})w_{b(\beta)}$. Again, the transition functions are holomorphic functions, so the differential forms of type $(1,0)$ take values in a holomorphic vector bundle which is usually considered the holomorphic cotangent bundle Ω. More generally, we could consider differential forms $w_{a_1 a_2 \ldots a_p}$ of type $(p,0)$. Their transformation law under a change of coordinates from $z^a_{(\beta)}$ to $z^b_{(\alpha)}$ is

$$w_{a_1 a_2 \ldots a_p(\alpha)} = \left(\frac{\partial z^{b_1}_{(\beta)}}{\partial z^{a_1}_{(\alpha)}}\right) \cdots \left(\frac{\partial z^{b_p}_{(\beta)}}{\partial z^{a_p}_{(\alpha)}}\right) w_{b_1 b_2 \ldots b_p(\beta)}. \tag{15.6.23}$$

The transition functions are again holomorphic, so the differential forms of type $(p,0)$ take values in a holomorphic vector bundle which we will call Ω^p.

On a complex manifold K of complex dimension N, a differential form of type $(N,0)$ must be a multiple of $dz^1 dz^2 \ldots dz^N$. So the differential forms of type $(N,0)$ form a one-dimensional vector bundle called the canonical line bundle L of K. Its first Chern class is called the first Chern class of K. When this vanishes, L is a trivial line bundle, and if K is compact there is a unique everywhere nonzero and nonsingular holomorphic section ω of L. (A holomorphic section of a trivial line bundle is equivalent to a holomorphic function, and on a compact complex manifold K there is a unique nonzero and nonsingular holomorphic function, namely the constant function 1.) This everywhere nonzero holomorphic N form ω was one of the key ingredients in our discussion of manifolds of $SU(N)$ holonomy.

(3) *New Vector Bundles From Old Ones*

Generalizing our above remarks leads to a general method of making new holomorphic vector bundles out of old ones. Let X be a holomorphic vector bundle over a complex manifold K. Let $U_{(\alpha\beta)}$ be transition functions for X. If X has complex dimension n, then the $U_{(\alpha\beta)}$ are invertible $n \times n$ complex matrices or in other words elements of $GL(n, C)$. Let R be any representation of $GL(n, C)$ and let $\tilde{U}_{(\alpha\beta)}$ be $U_{(\alpha\beta)}$ written in the R representation. Then the $\tilde{U}_{(\alpha\beta)}$ are holomorphic functions which obey the appropriate identities to be transition functions for a new holomorphic vector bundle X_R.

For instance, we can let $\tilde{U}_{(\alpha\beta)}$ be the inverse transpose matrix of $U_{(\alpha\beta)}$. With this choice the $\tilde{U}$ are transition functions for a bundle $\tilde{X}$ known as the dual or complex conjugate of X. Another case of special importance is that in which R is the adjoint representation of $GL(n, C)$. In that case, X_R is usually called End X, the bundle of endomorphisms of X. The reason for this name is that if X_p is the fiber of X at a point $p \in K$, then the fiber of X_R at p is the space of all endomorphisms or linear transformations of the vector space X_p.

(4) *The Direct Sum*

Another fundamental operation is the direct sum of bundles. Let X and Y be any holomorphic vector bundles over K, whose fibers at a point p are X_p and Y_p. Then we define a new bundle $X \oplus Y$, called the direct sum of X and Y, by saying that the fiber of $X \oplus Y$ at p is $X_p \oplus Y_p$, the direct sum of the vector spaces X_p and Y_p. Equivalently, if $U_{(\alpha\beta)}$ and $V_{(\alpha\beta)}$ are transition functions for X and Y, then $X \oplus Y$ is the vector bundle whose transition functions are

$$W_{(\alpha\beta)} = \begin{pmatrix} U_{(\alpha\beta)} & 0 \\ 0 & V_{(\alpha\beta)} \end{pmatrix}. \tag{15.6.24}$$

In a similar way one can define the tensor product of X and Y, denoted $X \otimes Y$, as the bundle whose fiber at p is the tensor product of the vector spaces X_p and Y_p.

(5) *Line Bundles Over CP^N*

For our next example, we consider line bundles or $U(1)$ bundles over complex projective space CP^N. We recall that CP^N may be described in terms of $N + 1$ homogeneous complex coordinates $Z^1, Z^2, \ldots, Z^{N+1}$. We cover CP^N with open sets $O_{(a)}$, $a = 1, \ldots, N + 1$, with $O_{(a)}$ being the region in which $Z^a \neq 0$.

Now, we define one-dimensional complex bundles or line bundles over CP^N very explicitly by specifying the transition functions $V_{(ab)}$ in the overlap region $O_{(ab)} = O_{(a)} \cap O_{(b)}$. We pick an arbitrary integer n and define

$$V_{(ab)} = (Z^b / Z^a)^n. \tag{15.6.25}$$

The $V_{(ab)}$ are clearly holomorphic functions in $O_{(ab)}$, and in triple intersection regions $O_{(abc)} = O_{(a)} \cap O_{(b)} \cap O_{(c)}$ they obey $V_{(ab)} \cdot V_{(bc)} \cdot V_{(ca)} = 1$. So they are transition functions on a line bundle $L(n)$ over CP^N. These are all of the line bundles over CP^N, though we will not try to prove that here.

Let us now determine explicitly what is a section of $L(n)$ or in other words a function that takes values in this line bundle. Concretely, such a function is described by giving on each $O_{(a)}$, $a = 1, \ldots, N + 1$ a function

$f_{(a)}(Z^b, \overline{Z}^{\bar{c}})$ with two basic properties. Since the Z^b are homogeneous coordinates we require

$$f_{(a)}(\lambda Z^b, \overline{\lambda}\, \overline{Z}^{\bar{c}}) = f_{(a)}(Z^b, \overline{Z}^{\bar{c}}). \qquad (15.6.26)$$

Also, on overlap regions the $f_{(a)}$ must obey $f_{(a)} = V_{(ab)} f_{(b)}$. In view of the explicit form of the $V_{(ab)}$, this equation amounts to the statement that $(Z^a)^n f_{(a)} = (Z^b)^n f_{(b)}$, so $f = (Z^a)^n f_{(a)}$ is independent of a. On the other hand, f does not have the conventional homogeneity of (15.6.26). Rather, it obeys

$$f(\lambda Z^b, \overline{\lambda}\, \overline{Z}^{\bar{c}}) = \lambda^n f(Z^b, \overline{Z}^{\bar{c}}). \qquad (15.6.27)$$

Thus, to describe a section of $L(n)$, we can dispense with the machinery of open coverings and transition functions, and work with a single function f, which obeys not the usual homogeneity condition of (15.6.26) but the generalized condition of (15.6.27).

(6) *Vector Bundles Over Hypersurfaces*

In our phenomenological work in the next chapter, we will find that vector bundles over manifolds of $SU(3)$ holonomy can play an important role in the construction of more or less realistic models. We would therefore like to construct here some examples of such bundles. We will content ourselves with describing some examples of vector bundles over one of the simplest examples of a manifold of $SU(3)$ holonomy, namely a hypersurface Q in CP^4 described by the vanishing of a quintic polynomial P.

We begin by describing explicitly the tangent bundles of CP^4 and of Q. A tangent vector field v^a can be thought of as the generator of an infinitesimal coordinate transformation $\delta z^a = v^a$. Here in general the z^a might be any complex coordinates on any complex manifold K. In the case of CP^4 the simplest description involves choosing the z^a to be the standard homogeneous coordinates $Z^1, Z^2, \ldots, Z^5$, which are defined only modulo a scaling $Z^a \approx \lambda Z^a$. When we write in the case of CP^4 the standard formula $\delta Z^a = V^a$, $a = 1, \ldots, 5$, the $V^a(Z^b, \overline{Z}^{\bar{c}})$ must be homogeneous of degree one (*i.e.*, $V^a(\lambda Z^b, \overline{\lambda}\, \overline{Z}^{\bar{c}}) = \lambda V^a(Z^b, \overline{Z}^{\bar{c}})$). This is required because the formula $\delta Z^a = V^a$ does not make sense unless the Z^a and the V^a scale in the same way. Also, one of the five V^a is redundant since the complex dimension of CP^4 is only four. The redundancy is removed by noting that we should consider V^a and $V^a + \lambda Z^a$ to be equivalent in CP^4, since the transformation $\delta Z^a = \lambda Z^a$ is trivial in CP^4.

If we wish to discuss the tangent bundle not of CP^4 but of the hypersurface Q in CP^4, then we should impose the condition that the vector field V^a should be tangent to the hypersurface. This means that the equation $P = 0$, which defines the hypersurface, must be invariant under $\delta Z^a = V^a$. The change in P under that coordinate transformation is $\delta P = (\partial P/\partial Z^a)V^a$, and this must vanish for V^a to be tangent to the hypersurface. In summary then, a tangent vector field to Q is a set of five functions $V^a(Z^b, \overline{Z^c})$ that are homogeneous of degree one and are subject to an equivalence relation

$$V^a \approx V^a + \lambda Z^a \tag{15.6.28}$$

and a constraint

$$\frac{\partial P}{\partial Z^a} V^a = 0. \tag{15.6.29}$$

The constraint is compatible with the equivalence relation, since for any homogeneous quintic polynomial P, $Z^a(\partial P/\partial Z^a) = 5P$, and this vanishes on the hypersurface Q.

Equations (15.6.28) and (15.6.29) describe a particular holomorphic vector bundle over Q, namely the tangent bundle T. We now will describe some more general complex vector bundles over T that have the same topological type but a different complex structure. To do this, we perturb the equations that entered in defining T. There is no useful way to modify (15.6.28), but (15.6.29) is another story. In (15.6.29), $\partial P/\partial Z^a$ is, for each a, a quartic polynomial in the Z^b. Let P_a, $a = 1, \ldots, 5$ be *any* five quartic polynomials in the Z^b that obey

$$P_a Z^a = 0 \tag{15.6.30}$$

(on Q). Then we can modify (15.6.29) to

$$P_a V^a = 0. \tag{15.6.31}$$

This is still compatible with (15.6.28) because of (15.6.30). For any choice of the five polynomials P_a (subject to (15.6.30)), (15.6.28) and (15.6.31) together define a holomorphic vector bundle $\tilde{T}$ over Q. It is equivalent to T topologically but has a different complex structure. The general form of the P_a compatible with (15.6.30) is

$$P_a = \partial P/\partial Z^a + p_{abcde} Z^b Z^c Z^d Z^e, \tag{15.6.32}$$

where p_{abcde} is completely symmetric in its last four indices but vanishes if symmetrized in all five indices. A little arithmetic shows that there are 224 parameters in the choice of the p_{abcde}, so the holomorphic deformation of T that we have constructed depends on 224 parameters.

Now, $\tilde{T}$, like T, is a three-dimensional complex vector bundle or $SU(3)$ bundle, since at each point in Q, the V^a, in view of (15.6.28) and (15.6.31), have three independent complex components. In our search in the next chapter for models with unbroken supersymmetry in four dimensions and gauge group $SO(10)$, we will be interested in *four*-dimensional bundles over Q. In the case at hand, we can obtain some four-dimensional bundles over Q by dropping either (15.6.28) or (15.6.31). It turns out to be far more interesting to drop (15.6.28), so we define an $SU(4)$ bundle X over Q by saying that a section of X is a set of five functions V^a, homogeneous of degree one, that obey (15.6.31). More explicitly, for each point p in Q there are four independent solutions of (15.6.31); these form a vector space that is to be the fiber of X at p.

Upon dropping (15.6.28), there is no reason to require (15.6.30), so we no longer require that p_{abcde} vanishes if symmetrized in all five indices. Of course, p_{abcde} is still completely symmetric in the last four indices. At first sight, there seem to be $5 \cdot (5 \cdot 6 \cdot 7 \cdot 8 / 1 \cdot 2 \cdot 3 \cdot 4) = 350$ parameters in the p_{abcde}, but this is not the number of parameters that enters in determining the complex structure of X. After dropping (15.6.28), the index 'a' in V^a or P_a no longer has any intrinsic meaning. If we take any 5×5 matrix Λ then $\tilde{P}_a = \Lambda_{ab} P_b$, inserted in (15.6.31), gives a bundle equivalent to the one that we would have made from the P_a. As Λ has $5 \times 5 = 25$ components, the number of independent parameters that enter in defining X is $350 - 25 = 325$.

Topologically, X is the same as $T \oplus L$ (the direct sum of the tangent bundle T and a trivial line bundle L). To see this, note that the topological type of X cannot depend on the continuously adjustable parameters p_{abcde}. Thus, we can determine the topological type of X by considering the case $p_{abcde} = 0$. In this case, we note that X contains a trivial line bundle L (the degree of freedom that we could have removed by the equivalence relation (15.6.28) if we had chosen to do so), while the orthogonal complement of L is manifestly the same as T. While X is equivalent to $T \oplus L$ topologically, this is not so holomorphically. Instead, X is a 'holomorphic deformation' of $T \oplus L$, a structure that can be reached from $T \oplus L$ by continuously varying the complex structure.

15.7   Dolbeault Cohomology and Some Applications

Let K be a Kähler manifold of complex dimension N. We have defined the $\bar{\partial}$ operator which maps $(0, q)$ forms to $(0, q + 1)$ forms by saying that if $\psi_{\bar{a}_1\bar{a}_2...\bar{a}_q}$ is a $(0,q)$ form, then $\bar{\partial}\psi$ is the $(0,q+1)$ form

$$(\bar{\partial}\psi)_{\bar{a}_1\bar{a}_2...\bar{a}_{q+1}} = \frac{1}{q+1}\partial_{\bar{a}_1}\psi_{\bar{a}_2...\bar{a}_{q+1}} \pm \text{cyclic permutations.} \qquad (15.7.1)$$

In many ways, the $\bar{\partial}$ operator is quite similar to the ordinary exterior derivative d of a real manifold, but there are crucial differences. One of the main differences is that, unlike d, $\bar{\partial}$ has a generalization in the presence of Yang–Mills fields that preserves the essential property $\bar{\partial}^2 = 0$.

To formulate this generalization, let X be a holomorphic vector bundle, endowed with a holomorphic connection, so that the Yang–Mills field strength obeys $F_{\bar{a}\bar{b}} = 0$. Let ψ be a $(0,q)$ form with values in X. This means that $\psi^x_{\bar{a}_1\bar{a}_2...\bar{a}_q}$ is antisymmetric in the q antiholomorphic indices $\bar{a}_1, \bar{a}_2, \ldots, \bar{a}_q$ and in addition carries an index x appropriate to the bundle X. We define a gauge covariant exterior derivative $\overline{D}$ that maps $(0,q)$ forms with values in X to $(0,q+1)$ forms with values in X by the formula

$$(\overline{D}\psi)_{\bar{a}_1\bar{a}_2...\bar{a}_{q+1}} = \frac{1}{q+1}\Big\{ D_{\bar{a}_1}\psi_{\bar{a}_2...\bar{a}_{q+1}} \\ \pm \text{cyclic permutations.}\Big\} \qquad (15.7.2)$$

Thus, we have merely replaced the ordinary derivative $\partial_{\bar{a}}$ in the exterior derivative with the covariant derivative $D_{\bar{a}}$. Since for a holomorphic connection on a holomorphic vector bundle $[D_{\bar{a}}, D_{\bar{b}}] = 0$, $D_{\bar{a}}$ serves just as well as $\partial_{\bar{a}}$ in ensuring that $\overline{D}^2 = 0$.

Since $\overline{D}^2 = 0$, we can define cohomology groups in the usual way. A $(0,q)$ form ψ with values in X is considered $\overline{D}$ closed if $\overline{D}\psi = 0$ and $\overline{D}$ exact if $\psi = \overline{D}\lambda$ for some λ. The equivalence classes of $(0,q)$ forms that are closed but not exact form what is called $H^q(X)$, the qth Dolbeault cohomology group with values in X. In the usual way, $H^q(X)$ can alternatively be characterized as consisting of the zero eigenvalues of the hermitian operator $\overline{D} + \overline{D}^*$ (with $\overline{D}^*$ being here the adjoint of $\overline{D}$).

To understand the connection of this with our previous definition of Dolbeault cohomology, note that the extra index x of the form $\psi^x_{\bar{a}_1\bar{a}_2...\bar{a}_q}$ may label an arbitrary holomorphic vector bundle X. In particular, x may label the bundle Ω^p of differential forms of type $(p,0)$, which was described in example (2) in §15.6.3. In that case $\psi^x_{\bar{a}_1\bar{a}_2...\bar{a}_n}$ is the same thing as a differential form $\psi_{a_1a_2...a_p\bar{a}_1\bar{a}_2...\bar{a}_q}$ with p holomorphic and q antiholomorphic indices. The gauge-covariant operator $\overline{D}$ in this situation is the same as the $\bar{\partial}$ operator acting on (p,q) forms, so what we called $H^{(p,q)}$ in §15.5.3 is what we would now call $H^q(\Omega^p)$.

As an example, given a holomorphic vector bundle X over a complex manifold K, the cohomology group $H^0(X)$ has a particularly simple interpretation. An element of $H^0(X)$ is a scalar field ψ^x with values in X which obeys $0 = \overline{D}\psi^x$. More explicitly, it must obey $0 = D_{\overline{a}}\psi^x$ for all a. We have already discussed this equation in (15.6.7) above; if locally we gauge away the gauge fields that appear in $D_{\overline{a}}$, the equation reduces to $\partial_{\overline{a}}\psi^x = 0$, and means that the ψ^x locally are analytic functions in the usual sense. If X is actually the trivial line bundle L, an element of $H^0(X)$ is an everywhere holomorphic function which – if K is compact – must be a constant. For more general X, $H^0(X)$ generalizes the notion of an everywhere holomorphic function. For instance, let K be CP^N and let X be the line bundle $L(n)$ defined in the last section. We showed there that a section of $L(n)$ is a function f homogeneous of degree n in the $N+1$ homogeneous coordinates $Z^1, \ldots, Z^{N+1}$. f is a holomorphic section of $L(n)$ if it is holomorphic in the Z^i. A global holomorphic function homogeneous of degree n in the Z^i must be an nth-order polynomial in the Z^i. (Any other homogeneous holomorphic function would have singularities somewhere.) For negative n there are, of course, no nth-order polynomials. For nonnegative n there are $(N+n)!/N!n!$ independent nth-order polynomials in $N+1$ variables, and this is the dimension of $H^0(L(n))$.

We will now sketch some applications of the generalized notion of Dolbeault cohomology.

15.7.1 Zero Modes of the Dirac Operator

Let K be a manifold of $SU(N)$ holonomy. Then – as we have learned in §15.5.1 – a spinor field on K is the same as a collection of $(0, q)$ forms, $q = 1, \ldots, N$. A spinor field ψ^x_α with values in some holomorphic vector bundle X is the same as a collection of $(0, q)$ forms $\psi^x_{\overline{a}_1 \ldots \overline{a}_q}$ with values in X. The Dirac operator $\Gamma^i D_i$ for such a field is the same as the Dolbeault operator $\overline{D} + \overline{D}^*$, a matter that we have already discussed in the case in which X is trivial. The zero modes of the Dirac operator are the zero modes of the Dolbeault operator; they are in other words the Dolbeault cohomology classes, the elements of $H^q(X)$.

On the other hand, the Dirac zero modes give rise to massless fermions in four dimensions. This is indeed the principal reason for our interest in Dolbeault cohomology. A few general statements about Dolbeault cohomology on manifolds of $SU(N)$ holonomy will be very helpful in the next chapter.

The first general statement is that $H^q(X)$ has the same dimension as $H^{N-q}(\tilde{X})$, with $\tilde{X}$ being the dual or complex conjugate bundle of X. This is proved by complex conjugation. The complex conjugate of a $(0, q)$ form $\psi^x_{\bar{a}_1 \bar{a}_2 \ldots \bar{a}_q}$ with values in X would be a $(q, 0)$ form $\psi^{*\bar{x}}_{a_1 a_2 \ldots a_q}$ with values in $\tilde{X}$. By contracting with the covariantly constant $(0, N)$ form ω of the manifold of $SU(N)$ holonomy,

$$\psi^{*\bar{x}}_{a_1 a_2 \ldots a_q} \rightarrow \phi^{\bar{x}}_{\bar{b}_{q+1} \ldots \bar{b}_N} = \psi^{*\bar{x}}_{a_1 a_2 \ldots a_q} g^{a_1 \bar{b}_1} g^{a_2 \bar{b}_2} \ldots g^{a_q \bar{b}_q} \omega_{\bar{b}_1 \bar{b}_2 \ldots \bar{b}_N}, \quad (15.7.3)$$

we can relate ψ^* to a $(0, N - q)$ form ϕ with values in $\tilde{X}$. ϕ obeys the Dirac equation if and only if ψ does. This establishes that $H^q(X)$ and $H^{N-q}(\tilde{X})$ have the same dimension on a manifold of $SU(N)$ holonomy, a statement that is a special case of Serre's generalization of Poincaré duality to complex manifolds.

Physically, in compactification on manifolds of $SU(3)$ holonomy, the relation between H^q and H^{3-q} means that massless fermions coming from H^{3-q} are always the CPT conjugates of massless fermions coming from H^q. It will therefore be adequate to understand the modes coming from, say, H^0 and H^1. Actually, the theory of H^0 is particularly simple in the case that will interest us most – the case in which the gauge connection can be chosen to obey (15.6.4), an equation that is related to unbroken supersymmetry. To see this, note that an element of H^0 is annihilated by the Dirac operator $\Gamma^i D_i$ and hence also by its square, for which we have given a formula in (15.5.17) which we here repeat for convenience:

$$(i\Gamma^i D_i)^2 = -D_i D^i + R/4 - \Gamma^{ij} F^x_{ij} T^x / 4. \quad (15.7.4)$$

Here we may drop the $R/4$ term, since this vanishes on manifolds of $SU(N)$ holonomy. Also, in studying H^0, we may drop the last term since the relation of spinors to $(0, q)$ forms is such that (if the Yang–Mills field obeys (15.6.3) and (15.6.4)) $\Gamma^{ij} F_{ij}$ annihilates $(0,0)$ forms. Hence on $(0,0)$ forms, $(i\Gamma^i D_i)^2 = -D_i D^i$. A zero mode ψ of the Dirac operator must hence be annihilated by $-D_i D^i$, and hence it must be covariantly constant. In practice, this means that the only massless modes in four dimensions coming from H^0 (or H^3) are modes such as gluinos, which are related to supersymmetry and unbroken gauge symmetries in obvious ways. The real subtlety in studying fermion zero modes on a manifold of $SU(3)$ holonomy involves studying H^1 (or equivalently H^2).

As a check on some of the foregoing, recall that the relation of forms to spinors on a manifold of $SU(3)$ holonomy is such that a $(0, q)$ form has chirality $(-1)^q$ in the sense of the six uncompactified dimensions. In view of the correlation between six-dimensional and four-dimensional chirality described in chapter 14, a $(0, q)$ form also has chirality $(-1)^q$ in the four-dimensional sense. Thus, we can consider elements of H^1 as left-handed massless fermions in four dimensions; their CPT conjugates, which arise from H^2 (of the complex-conjugate bundle) are right-handed.

15.7.2 Deformations of Complex Manifolds

In this subsection and the following one we turn to a discussion of the relation between Dolbeault cohomology and deformations of complex structure.

Let K be a complex manifold. Then K is endowed with a tensor field $J^i{}_k$ that obeys

$$J^i{}_k J^k{}_m = -\delta^i_m \qquad (15.7.5)$$

along with the vanishing of the Nijenhuis tensor

$$0 = N^k{}_{ij} \qquad (15.7.6)$$

constructed from J. Locally, it is possible to find holomorphic coordinates z^a on K; in the coordinate system defined by the z^a and $\overline{z}^b$, the nonzero components of J are

$$J^a{}_b = i\delta^a_b, \qquad J^{\overline{a}}{}_{\overline{b}} = -i\delta^{\overline{a}}_{\overline{b}}. \qquad (15.7.7)$$

Now, we would like to ask whether it is possible to deform the complex structure of K to a 'nearby' but inequivalent complex structure. (In the special case or a Riemann surface, this is a restatement of a question that we first asked in §3.3, where we explored the conformally inequivalent metrics on such a surface.) Thus, we ask whether it is possible to perturb the complex structure J to a new one

$$J^i{}_k \rightarrow \tilde{J}^i{}_k = J^i{}_k + \tau^i{}_k \qquad (15.7.8)$$

in such a way that the key equations (15.7.5) and (15.7.6) are still satisfied. Of course, if $\tilde{J}$ is obtained from J by a diffeomorphism or coordinate transformation on K, it will obey (15.7.5) and (15.7.6) just as J does; but in this case $\tilde{J}$ will define a complex structure which (up to diffeomorphism) is equivalent to that defined by J. We are interested in characterizing the perturbations τ that obey (15.7.5) and (15.7.6) but *cannot* be obtained by a diffeomorphism applied to J. Using (15.7.7), it is not too difficult to see that (15.7.5) implies that $\tau^a{}_b = \tau^{\overline{a}}{}_{\overline{b}} = 0$, but it does not constrain $\tau^a{}_{\overline{b}}$ or its complex conjugate $\tau^{\overline{a}}{}_b$. What about (15.7.6)? Using the definition

(15.2.4) of the Nijenhuis tensor, the reader should be able to see that all components of $N^k{}_{ij}$ vanish automatically except

$$N^a{}_{\overline{bc}} = \overline{\partial}_{\overline{b}}\tau^a{}_{\overline{c}} - \overline{\partial}_{\overline{c}}\tau^a{}_{\overline{b}} \tag{15.7.9}$$

and its complex conjugate $N^{\overline{a}}{}_{bc}$, which will of course vanish if (15.7.9) does. The vanishing of $N^a{}_{\overline{bc}}$ has a simple interpretation. If we view $\tau^a{}_{\overline{b}}$ as a $(0,1)$ form with values in the holomorphic tangent bundle T, then $N^a{}_{\overline{bc}} = 0$ amounts to

$$\overline{\partial}\tau^a = 0. \tag{15.7.10}$$

This says that the deformation τ of the complex structure defines an element of $H^1(T)$.

Of course, deformations $J \to J + \tau$ are to be considered trivial if they can be obtained by an infinitesimal change of coordinates

$$z^a \to \tilde{z}^a = z^a + \epsilon v^a(z^b, \overline{z^c}), \tag{15.7.11}$$

with ϵ being a small parameter. The transformation law of J under such a transformation is determined by the transformation law of tensors in general relativity

$$J^i{}_j \to \tilde{J}^i{}_j = \frac{\partial \tilde{x}^i}{\partial x^{i'}} \frac{\partial x^{j'}}{\partial \tilde{x}^j} J^{i'}{}_{j'}. \tag{15.7.12}$$

One finds that the infinitesimal transformation law of τ^a is

$$\tau^a \to \tau^a + \epsilon \overline{\partial} v^a. \tag{15.7.13}$$

This shows that if two different deformations of J, say τ and τ', have the same cohomology class in $H^1(T)$, then they define (up to diffeomorphism) the same deformation of the complex structure. Combining our results, we see that the possible deformations of the complex structure of a complex manifold K are in one-to-one correspondence with elements of $H^1(T)$, the first Dolbeault cohomology group of K with values in the holomorphic tangent bundle T.

As an example, let K be a Riemann surface Σ. An element of $H^1(T)$ is a harmonic form $\psi^z_{\overline{z}}$ with a holomorphic one form index which has only one possible value, namely z, and an antiholomorphic $(0,1)$ form index which must be $\overline{z}$. Contracting with the covariantly constant metric tensor we can relate ψ to a field

$$b_{\overline{z}\,\overline{z}} = g_{z\overline{z}}\psi^z_{\overline{z}}, \tag{15.7.14}$$

which transforms like the antighosts that arise in covariant quantization of string theory. An element of $H^1(T)$ would be equivalent to a zero mode of the Dirac equation acting on the field $\psi_{\bar{z}}^{z}$ or $b_{\bar{z}\bar{z}}$. Thus, we learn that infinitesimal deformations of the complex structure of Σ are in one-to-one correspondence with zero modes of the antighost field b. It is pleasing to recover in this way a result that we first obtained in §3.3.

For another important application, let K be a manifold of $SU(3)$ holonomy. In this case, by use of the holomorphic three form ω, an element $\psi^{a}{}_{\bar{b}}$ of $H^{1}(T)$ is equivalent to a $(2,1)$ form

$$\chi_{a_1 a_2 \bar{b}} = \omega_{a_1 a_2 a_3} \psi^{a_3}{}_{\bar{b}}. \tag{15.7.15}$$

As ω is covariantly constant, χ is a harmonic form if and only if ψ is. Thus, on a manifold of $SU(3)$ holonomy, $H^{1}(T)$ coincides with $H^{(2,1)}$. In particular, this means that the Hodge number $h^{2,1}$, defined as the dimension of $H^{(2,1)}$, is the same as the number of complex parameters that enter in defining the complex structure of K.

For example, let K be a quintic hypersurface in CP^4. Then we saw in example (4) of §15.2.3 that the complex structure of K is determined by the 101 independent complex parameters of a quintic polynomial. Hence, K has $h^{2,1} = 101$. We know from §15.5.4 that $h^{p,q} = h^{q,p} = h^{3-p,3-q}$ and that $h^{0,0} = h^{3,0} = 1$, $h^{1,0} = h^{2,0} = 0$. This leaves only $h^{1,1}$ to be determined. $h^{1,1}$ is the number of harmonic $(1,1)$ forms. On any Kähler manifold there is always at least one such form, the Kähler form. A hypersurface in CP^{N*} always has precisely one harmonic $(1,1)$ form, so $h^{1,1} = 1$ for such a space. Hence we can evaluate the Euler characteristic of a quintic hypersurface in CP^4. It is

$$\chi = \sum_{p,q}(-1)^{p+q} h^{p,q} = -200. \tag{15.7.16}$$

There is of course a gap in our derivation of this result, since we have not proved that $h^{1,1} = 1$. In the next and last section of this chapter, we will give an independent derivation that $\chi = -200$ (and hence $h^{1,1} = 1$) for a quintic hypersurface in CP^4.

15.7.3 Deformations of Holomorphic Vector Bundles

In the last section, we showed that deformations of the complex structure of a complex manifold K are related to $H^{1}(T)$, with T being the holomorphic tangent bundle of K. Here we will consider a holomorphic vector bundle X over K, and give an analogous description of the deformations of the complex structure of X in terms of Dolbeault cohomology.

---

* Except for one exceptional case $N = 3$.

We can describe the bundle X by specifying a gauge field A_i with $F_{\overline{ab}} = 0$. In the latter equation only the $(0,1)$ part of the gauge field is relevant. If we wish to perturb the bundle, we can do this by perturbing the $(0,1)$ part of A. We write $\tilde{A}_{\overline{a}} = A_{\overline{a}} + \delta A_{\overline{a}}$, with $\delta A_{\overline{a}}$ an infinitesimal disturbance, and we ask whether it is still true that $F_{\overline{ab}} = 0$ to lowest order in δA. The requisite equation is $D_{\overline{a}}\delta A_{\overline{b}} - D_{\overline{b}}\delta A_{\overline{a}} = 0$, or more succinctly $\overline{D}\delta A = 0$. Thus, δA must be $\overline{D}$ closed, and so defines a Dolbeault cohomology class. As δA is a $(0,1)$ form in the adjoint representation of the gauge group, the relevant cohomology group is $H^1(\text{End }X)$. (Recall that we defined End X as the bundle made by writing the structure functions in the adjoint representation.)

Figure 15.3. The four generators of the fundamental group of a Riemann surface of genus two are sketched here.

As an example, let X be the trivial line bundle over a Riemann surface Σ of genus g. Then End X is trivial, like X, so $H^1(\text{End }X)$ is the same as $H^{(0,1)}$. This we know to be g dimensional for a surface of genus g, so holomorphic deformations of the complex structure of X depend on g complex parameters. An explicit description of them can be given. As sketched in fig. 15.3 for the case of genus two, the fundamental group of a surface of genus g has $2g$ generators. We can require that a complex field change by an arbitrarily prescribed phase in circumnavigating any independent noncontractible loop in Σ. As there are $2g$ independent loops, the possible boundary conditions depend on $2g$ real parameters or on g complex parameters. Since the boundary conditions can be adjusted continuously, they do not affect the topology of the bundle X, but they do affect its holomorphic structure (as we have proved by computing $H^1(\text{End }X)$).

As another example, let T be the holomorphic tangent bundle of a quintic hypersurface in CP^4. In example (6) of §15.6.2 we explicitly constructed a family of holomorphic deformations of T depending on 224 complex parameters. Though we will not prove it here, these are all of the possible continuous deformations of T, so $H^1(\operatorname{End} T)$ is 224 dimensional. Likewise, we constructed in the same discussion a 325-dimensional family of deformations of $X \sim T \oplus L$, and this corresponds to the fact that $H^1(\operatorname{End} X)$ is 325 dimensional.

15.8   Branched Coverings of Complex Manifolds

In this concluding section we will briefly and somewhat heuristically describe the notion of a branched covering of a complex manifold and the behavior of the Euler characteristic under such a branched covering. The goal is to enrich our understanding of Riemann surfaces and complex manifolds and to develop a new tool for determining the Euler characteristic of a manifold of $SU(3)$ holonomy.

Let k be the standard Kähler form of CP^N. What harmonic forms exist on CP^N? Apart from the zero form 1 and the two form k, the harmonic forms we know of are the wedge products $k \wedge k$, $k \wedge k \wedge k$, and so on up to the N-fold wedge product $(k \wedge k \wedge \ldots \wedge k)_N$. Though we will not try to prove this here, these are all of the harmonic forms on CP^N, and in particular the Euler characteristic of CP^N is $N + 1$. This fact will be useful presently.

Let X, Y, and Z be homogeneous coordinates for CP^2. Consider the Riemann surface Σ defined by the homogeneous equation

$$Z^n = X^n + Y^n \qquad (15.8.1)$$

in CP^2. We would like to determine the genus of this surface; we will do so by computing the Euler characteristic.

By themselves, X and Y would be homogeneous coordinates for a Riemann sphere $R \approx CP^1$. If $X^n + Y^n$ is not zero, then (15.8.1) permits n possible values of Z for given X and Y. Thus, if there were no roots of the equation

$$X^n + Y^n = 0, \qquad (15.8.2)$$

the surface Σ would simply consist of n copies of R. As the Euler characteristic of R is 2, the Euler characteristic of n copies of R would be $2n$.

Actually, the equation (15.8.2) has n roots on the Riemann sphere. For $X^n + Y^n \neq 0$, (15.8.1) allows n values of Z, but for $X^n + Y^n = 0$, (15.8.1) requires $Z = 0$. Because of this, the Euler characteristic of Σ is not simply n times that of R. What is the necessary correction? Let P be the collection of n points on R on which (15.8.2) is obeyed. Let $R - P$ be R with those points removed, and let $\Sigma - P$ be Σ with the corresponding points removed. Then $\Sigma - P$ is indeed n copies of $R - P$, so the Euler characteristic of $\Sigma - P$ is n times that of $R - P$:

$$\chi(\Sigma - P) = n \cdot \chi(R - P). \tag{15.8.3}$$

The Euler characteristic of R is 2. The Euler characteristic of a point is 1, so that of P is n. The Euler characteristic behaves in an additive fashion in gluing spaces together or in removing pieces,[*] so

$$\chi(R - P) = \chi(R) - \chi(P) = 2 - n. \tag{15.8.4}$$

Thus $\chi(\Sigma - P) = n(2 - n)$. Gluing in n points to recover Σ from $\Sigma - P$, we get

$$\chi(\Sigma) = \chi(\Sigma - P) + \chi(P) = n(3 - n), \tag{15.8.5}$$

which determines the genus of Σ. In particular, for $n = 3$ the Euler characteristic is zero; Σ is a surface of genus one, an ordinary torus. This agrees with the fact that in §15.4.3, we found an everywhere nonzero holomorphic one form on Σ precisely for $n = 3$.

The above construction is usually described by saying that Σ is an n-fold cover of the Riemann sphere R, branched over the n roots of (15.8.2). Σ is made by taking n copies of X everywhere except on the branch locus, where we take only one copy.

Let us try this again, considering a hypersurface H in CP^3:

$$W^n = X^n + Y^n + Z^n. \tag{15.8.6}$$

By themselves X, Y, and Z would be homogeneous coordinates for a manifold M, isomorphic to CP^2, with Euler characteristic 3. If the equation

$$X^n + Y^n + Z^n = 0 \tag{15.8.7}$$

had no roots, (15.8.6) would simply determine n possible values of W for given (X, Y, Z). Actually, (15.8.7) vanishes on a Riemann surface Σ' isomorphic to the surface Σ considered earlier (one relates (15.8.7) to the equation for Σ by multiplying Z by an nth root of -1). H can be described as an n-fold cover of $M \approx CP^2$ branched over Σ'. The Euler

---

[*] This additivity was proved and used in a special case at the end of chapter 12 to determine the Euler characteristic of a Riemann surface. It can be proved in the general case in a similar way, by using the fact that the Euler characteristic is the integral of a polynomial in the Riemann tensor.

characteristic of M is 3, so that of $M - \Sigma'$ is $3 - n(3 - n)$, and that of $H - \Sigma'$ is $3n - n^2(3 - n)$. Finally, the Euler characteristic of H is

$$\chi(H) = \chi(H - \Sigma') + \chi(\Sigma') = n(n^2 - 4n + 6). \tag{15.8.8}$$

For instance, the $K3$ surface $(n = 4)$ has $\chi = 24$.

We will push this one step further and consider the hypersurface Q in CP^4 defined by

$$U^n = W^n + X^n + Y^n + Z^n. \tag{15.8.9}$$

By themselves $W, X, Y,$ and Z would be homogeneous coordinates for a manifold $N \approx CP^3$, of Euler characteristic 4. Q is an n-fold cover of N, branched over the locus H' of

$$W^n + X^n + Y^n + Z^n = 0. \tag{15.8.10}$$

The Euler characteristic of H' is given in (15.8.8). Applying the same reasoning as in the previous examples, the Euler characteristic of Q is

$$\begin{aligned}\chi(Q) &= \chi(Q - H') + \chi(H') = n\chi(N - H') + \chi(H') \\ &= n\chi(N) + (1 - n)\chi(H') = -n(n^3 - 5n^2 + 10n - 10).\end{aligned} \tag{15.8.11}$$

Setting $n = 5$, we learn that the quintic hypersurface in CP^4, which admits a metric of $SU(3)$ holonomy, has $\chi = -200$. This, finally, is in agreement with our discussion in §15.7.2.

16. Models of Low-Energy Supersymmetry

In the last chapter we developed some of the basic tools of algebraic geometry, and in particular we developed the machinery that is needed to understand compactifications that lead to low-energy supersymmetry in four dimensions. In this chapter, we will apply what we have learned to explore in considerable detail the properties of the resulting models.

It should be noted that in our discussion of orbifolds in chapter 9, we have already discussed special examples of models with low-energy supersymmetry. The precise connection of orbifolds with the models investigated in this chapter will be explored in §16.10 below.

16.1   A Simple Ansatz

We formulated in the last chapter the condition for unbroken supersymmetry to survive after compactification from ten dimensions on $M^4 \times K$. The ten-dimensional spinor η_α generates an unbroken supersymmetry in four dimensions if a supersymmetry transformation with parameter η_α leaves the elementary Fermi fields invariant. Restricting the equations of §15.1.2 to the internal space K and dropping Fermi fields gives

$$
\begin{aligned}
(a) \qquad & 0 = \delta\psi_i = \frac{1}{\kappa}D_i\eta + \frac{\kappa}{32g^2\phi}(\Gamma_i{}^{jkl} - 9\delta_i^j\Gamma^{kl})\eta H_{jkl} \\[2mm]
(b) \qquad & 0 = \delta\lambda = -\frac{1}{\sqrt{2\phi}}(\Gamma\cdot\partial\phi)\eta + \frac{\kappa}{8\sqrt{2}g^2\phi}\Gamma^{ijk}\eta H_{ijk} \qquad (16.1.1)\\[2mm]
(c) \qquad & 0 = \delta\chi = -\frac{1}{4g\sqrt{\phi}}\Gamma^{ij}F_{ij}\eta.
\end{aligned}
$$

Here ψ, λ, and χ are respectively the gravitino, dilatino, and gluino; ϕ is the dilaton, and H is the gauge-invariant field strength of the two form B. The above equations must be supplemented by the Bianchi identity

$$
dH = \mathrm{tr}\, R \wedge R - \mathrm{tr}\, F \wedge F. \qquad (16.1.2)
$$

The above equations simplify considerably if we assume that $d\phi = H = 0$. As we will see, a good deal of interesting physics is possible under

this simplifying assumption, and we will analyze this situation in detail
before trying to formulate generalizations. Indeed, to understand the
generalizations requires a more powerful approach, which we will develop
later.

With $d\phi = H = 0$, the first equation in (16.1.1) tells us that K is a
Kähler manifold of vanishing first Chern class endowed with a metric of
$SU(3)$ holonomy predicted by Yau's proof of the Calabi conjecture. The
second condition in (16.1.1) is vacuous when $d\phi = H = 0$. The last tells
us that the gauge field A is a holomorphic connection on a holomorphic
vector bundle X and obeys the Donaldson–Uhlenbeck–Yau equation.

Once the topology of K has been chosen, a metric on K of $SU(3)$
holonomy is determined by a finite number of parameters. These are the
parameters that determine the topological class of the Kähler form on
K and the complex structure of K. Likewise, once a choice is made of
the topological class of the bundle X, the solution of (16.1.1)(c) depends
on only a finite number of parameters, namely the parameters that enter
in determining the complex structure of X. Therefore, once the discrete
topological choices have been made, there are only a finite number of continuously adjustable parameters at our disposal in trying to obey (16.1.2).
Since we have assumed that $H = 0$ in order to simplify (16.1.1), (16.1.2)
tells us that

$$0 = \operatorname{tr} F \wedge F - \operatorname{tr} R \wedge R. \tag{16.1.3}$$

At first sight, (16.1.3) may not appear promising. Equation (16.1.3) is
really an infinite set of conditions, since the four form $\operatorname{tr} F \wedge F - \operatorname{tr} R \wedge R$
must vanish at each point in K. How can we adjust the finite set of
parameters left over from (16.1.1) to obey (16.1.3)? We can only expect
an overdetermined set of conditions like (16.1.3) to be obeyed if it is
obviously obeyed, and this requires that there should be some relation
between F and R.

How can we achieve this? We will use a special case of an *ansatz* that
was introduced in §14.6 to obtain a reasonable model of the fermion quantum numbers. As we noted in our discussion there, the spin connection
ω_i of a Riemannian manifold K can be regarded as a gauge field. The
gauge group is the holonomy group H of K, which for our present purposes is $SU(3)$. If the ten-dimensional gauge group were $SU(3)$, we could
simply set the gauge fields A_i of the ten-dimensional theory equal to the
spin connection ω_i. The Yang–Mills field strength F would then equal
the Riemann curvature tensor R, and (16.1.3) would be obeyed (up to a
normalization factor that we will discuss shortly). In practice, the ten-
dimensional gauge group is not $SU(3)$ but much larger; in fact, $E_8 \times E_8$ is

the most interesting possibility. To carry out the above construction, we must pick an $SU(3)$ subgroup H' of $E_8 \times E_8$ and set the H' gauge fields equal to the spin connection. Picking an $SU(3)$ subgroup of $E_8 \times E_8$ can be interpreted as embedding the holonomy group H in $E_8 \times E_8$, so we will describe the whole construction as 'embedding the spin connection in the gauge group'.

For reasons that will become clear, we wish to pick a minimal embedding. Thus, we select a maximal subgroup $SU(3) \times E_6$ of (say) the first E_8, and we identify the holonomy group H with the first factor.[*] With the $SU(3)$ gauge fields being the same as the spin connection, (16.1.3) will be obeyed up to a possible normalization factor coming from the fact that the traces $\mathrm{tr} F \wedge F$ and $\mathrm{tr} R \wedge R$ are being evaluated in different representations. In fact, $\mathrm{tr} R \wedge R$ is being evaluated in the vector representation of $SO(1,9)$, which under $SU(3)$ transforms as $\mathbf{3} \oplus \overline{\mathbf{3}} \oplus$ singlets. For $\mathrm{tr} F \wedge F$ the situation is less straightforward. We must recall that in chapter 13 we defined 'tr' in the E_8 case to denote one thirtieth of the trace in the adjoint representation of E_8. Also, we recall the decomposition of the adjoint representation of E_8 under $SU(3) \times E_6$:

$$\mathbf{248} = (\mathbf{3}, \mathbf{27}) \oplus (\overline{\mathbf{3}}, \overline{\mathbf{27}}) \oplus (\mathbf{8}, \mathbf{1}) \oplus (\mathbf{1}, \mathbf{78}). \qquad (16.1.4)$$

Here $(\mathbf{8}, \mathbf{1})$ and $(\mathbf{1}, \mathbf{78})$ are the adjoint representations of $SU(3)$ and E_6, respectively; $(\mathbf{3}, \mathbf{27})$ is the tensor product of the fundamental $\mathbf{3}$ of $SU(3)$ with the fundamental $\mathbf{27}$ of E_6; and $(\overline{\mathbf{3}}, \overline{\mathbf{27}})$ is the complex conjugate of this. It is straightforward to show that the trace of the square of an $SU(3)$ matrix in the adjoint representation $\mathbf{8}$ is three times what it is in the $\mathbf{3} \oplus \overline{\mathbf{3}}$. Adding this to the 27 copies of $\mathbf{3} \oplus \overline{\mathbf{3}}$ explicitly visible in (16.1.4), we see that the trace of the square of an $SU(3)$ generator in the adjoint representation of E_8 is $3 + 27 = 30$ times what it is in the $\mathbf{3} \oplus \overline{\mathbf{3}}$ of $SU(3)$. This factor of 30 cancels the fact that $\mathrm{tr} F \wedge F$ is defined to be one thirtieth of the trace of $F \wedge F$ in the adjoint representation of E_8. So in (16.1.3) we may interpret $\mathrm{tr} F \wedge F$ as a trace in the $\mathbf{3} \oplus \overline{\mathbf{3}}$ of $SU(3)$. Since $\mathrm{tr} R \wedge R$ is a trace in the same representation, (16.1.3) is satisfied. This would not work if we were to choose a nonminimal embedding of $SU(3)$ in $E_8 \times E_8$, so the minimal embedding is forced upon us.

Embedding the spin connection in the gauge group was already considered in chapter 14, merely in order to obtain a realistic model of the fermion quantum numbers. In chapter 14 we considered a generical Riemannian manifold of $SO(6)$ holonomy. Embedding the holonomy group in

---

[*] The facts about E_8 group theory required in the present chapter were described in appendix 6.A.

E_8 by the chain $SO(6) \times SO(10) \subset SO(16) \subset E_8$, we broke E_8 to $SO(10)$. This led to the existence of chiral fermions in the **16** and $\overline{\mathbf{16}}$ of $SO(10)$, the net number of generations being one-half the Euler characteristic of K. We are here considering the special case in which the holonomy group H is not $SO(6)$ but an $SU(3)$ subgroup. The unbroken subgroup of E_8 is correspondingly larger – E_6 rather than $SO(10)$. At the E_6 level, the number of generations is defined as the net number of chiral **27**'s, since the **27** of E_6 decomposes under $S0(10)$ as $\mathbf{27} \approx \mathbf{16} \oplus \mathbf{10} \oplus \mathbf{1}$. The number of generations in an E_6 model obtained by picking K to be a manifold of $SU(3)$ holonomy must be one half of the Euler characteristic, just as in the more general $SO(10)$ models, since the construction considered here is just a special case of the one in chapter 14. Indeed, we will in the next section recover the statement that the number of generations is one-half of the Euler characteristic in the course of a systematic survey of the spectrum of massless particles.

The construction considered here, since it is claimed to give vacuum states with unbroken supersymmetry, should also lead to solutions of the equations of motion. To investigate this point, we consider the form of the effective low-energy field theory of the ten-dimensional theory. The effective Lagrangian, up to corrections of order $(\alpha')^2$, is

$$e^{-1}\mathcal{L}_{eff} = -\frac{1}{2\kappa^2}R - \frac{1}{4g^2\phi}\mathrm{tr}(F^2_{MN}) - \frac{1}{\kappa^2}(\partial_M \phi/\phi)^2 - \frac{3\kappa^2}{8g^4\phi^2}H^2_{MNP}$$
$$+ \frac{1}{4g^2\phi}(R_{MNPQ}R^{MNPQ} - 4R_{MN}R^{MN} + R^2)$$
$$+ \text{Fermi terms.}$$

$$(16.1.5)$$

All of the terms in (16.1.5) can be deduced just from space-time supersymmetry, and were described in chapter 13, with two exceptions. The first exception is that the Lorentz Chern–Simons term in

$$H = dB + \omega_L - \omega_{YM} \qquad (16.1.6)$$

is not needed in the minimal supergravity Lagrangian. The second exception is the $R_{MNPQ}R^{MNPQ} + \ldots$ interaction in (16.1.5). Although these two terms are not required by space-time supersymmetry, it can be shown (with considerable effort) that they are related to each other by supersymmetry, so that both must be present if one is. Alternatively, it is not very difficult to compute on-shell three-point couplings in the heterotic string theory (the key steps were explained in §7.4.4) and show that both correction terms are generated with the stated coefficients. (The analogous calculation in the Type I theory is trickier.) Further examination of string scattering amplitudes reveals that the next correction to the effective Lagrangian arises in order $(\alpha')^3$, where one meets a plethora of terms such as an R^4 interaction.

It is easy to see that, to this order, the equations of motion are obeyed by our ansatz. For example, the Einstein equations are obeyed since the Ricci tensor vanishes for metrics of $SU(3)$ holonomy (the correction to the Einstein equations due to the R^2 terms in (16.1.5) is harmless since $D^i R_{ijkl} = 0$ for a Ricci-flat metric). The Yang–Mills equations are likewise obeyed since (as we know from the end of §15.6.2) they follow from (16.1.1). The slightly delicate case is the dilaton equation, which (with our assumption that $d\phi = H = 0$) becomes

$$0 = \mathrm{tr} F_{ij} F^{ij} - \mathrm{tr} R_{ij} R^{ij}, \qquad (16.1.7)$$

where $\mathrm{tr} R_{ij} R^{ij} \equiv R_{ijkl} R^{ijkl}$, the 'trace' in $\mathrm{tr} R_{ij} R^{ij}$ being over the indices of R_{ijkl} that have been suppressed. Just like (16.1.3), (16.1.7) is an implausible equation unless there is some special relation between F and R. Embedding the spin connection in the gauge group gives such a relation, and indeed (16.1.7) is obeyed in this case.

This is in agreement with our expectation that states of unbroken supersymmetry should obey the equations of motion. However, (16.1.5) is only valid to within corrections of order $(\alpha')^2$, and we would like to extend the analysis to higher orders. After developing suitable tools, we will return to this question and prove that solutions of the equations of motion can be constructed to all finite orders in α' with manifolds of $SU(3)$ holonomy as the starting point. In the process we will also learn how to relax the assumption of embedding the spin connection in the gauge group and to find more general solutions.

Since α' is a constant of nature if string theory is correct, what is the meaning of expanding in powers of α'? A manifold of $SU(3)$ holonomy has a radius r that is not determined by the Einstein equations. (The condition for $SU(3)$ holonomy is invariant under rescaling of the metric $g_{ij} \to t g_{ij}$.) All curvatures and field strengths vanish for large r, the dimensionless expansion parameter being α'/r^2. Our preceding discussion amounted to verifying the equations of motion up to and including terms of order $(\alpha'/r^2)^2$.

Embedding the spin connection in the gauge group gives a number of immediate phenomenological dividends. E_8 is not a suitable gauge group for grand unification in four dimensions, since it does not have complex representations. Indeed, the only exceptional group that has complex representations and so is suitable for grand unification in four dimensions is

E_6. The group E_6 is also rather special from a supersymmetric view-point. It is the only candidate grand-unified gauge group in which the natural representation for fermions – the **27** – coincides with the natural representation for Higgs bosons, as one would wish for supersymmetric model building. In $SO(10)$, for example, fermions and Higgs bosons are naturally in the **16** and **10**, respectively; these are unified in the **27** of E_6.

In our construction E_6 appeared not because it is the four-dimensional group that we would like to get, but because embedding the spin connection in the gauge group is the simplest way to obey the requisite equations, and automatically breaks $E_8 \times E_8$ down to $E_6 \times E_8$. Also, the chiral supermultiplets appear in the **27** and $\overline{\textbf{27}}$ of E_6, which are the only really suitable representations, not because we have contrived this, but because they are the only complex representations of E_6 that appear in the decomposition (16.1.4) of the Lie algebra of E_8.

In this chapter, when we wish to cite a particular example of a manifold of $SU(3)$ holonomy, it will usually be a quintic hypersurface Q in CP^4, defined by the vanishing of a homogeneous quintic polynomial P in five complex variables $Z_1, Z_2, \ldots, Z_5$. The polynomial P may be the polynomial

$$P = Z_1^5 + Z_2^5 + Z_3^5 + Z_4^5 + Z_5^5 \qquad (16.1.8)$$

or some more general polynomial with less symmetry, such as $P' = \sum_i Z_i^5 + \epsilon Z_1 Z_2 Z_3 Z_4 Z_5$. As we noted in chapter 15, after removing 25 degrees of freedom that can be absorbed in a linear redefinition of the Z_i, there are 101 independent complex parameters that enter in the choice of P. The description of this manifold by polynomials will permit a very extensive study of the properties of the model. The complete intersection of k hypersurfaces in CP^{3+k}, which was also described in chapter 17, can be studied in a similar way, though we will not do so. At the end of this chapter we will describe another and very different class of simple examples of manifolds of $SU(3)$ holonomy.

16.2 The Spectrum of Massless Particles

In this section we survey the spectrum of massless particles in the effective four-dimensional theory that arises by compactification of a ten-dimensional world on a manifold of $SU(3)$ holonomy. We will assume that the spin connection has been embedded in the gauge group, though much of the discussion remains valid after later generalization. We will first consider massless particles that arise as zero modes of the fields that carry $E_8 \times E_8$ charges, and then we will turn our attention to the neutral fields.

16.2.1 Zero Modes of Charged Fields

The gluino field χ_α^x has a spinor index α and an index x labeling the adjoint representation of E_8 (or $E_8 \times E_8$; for brevity we concentrate on one E_8 factor in which the gauge field has been embedded). Assuming that only $SU(3)$ gauge fields of a maximal subgroup $SU(3) \times E_6 \subset E_8$ are present in the vacuum, the Dirac equation decouples the components of χ that transform as different representations of $SU(3)$.

The $SU(3)$ singlet modes are not affected by the embedding of the spin connection in the gauge group, so they are spinor fields of K. A spinor field ψ_α on K is equivalent to a collection of differential forms $\psi_{\bar{a}_1 \ldots \bar{a}_q}$ of type $(0, q)$. The zero modes of the Dirac operator form the Dolbeault cohomology groups $H^{0,q}$, and give rise in four dimensions to massless fermions of chirality $(-1)^q$. We know from §15.5.4 that $H^{0,1} = H^{0,2} = 0$, while $H^{0,0}$ and $H^{0,3}$ are one dimensional, consisting of the covariantly constant spinor fields. Hence for $SU(3)$ singlets there is precisely one zero mode of positive chirality and one of negative chirality. The decomposition (16.1.4) of the Lie algebra of E_8 shows that the $SU(3)$ singlet states transform in the adjoint representation of E_6, so the modes that we have just identified are the four-dimensional gluinos of positive and negative chirality.[*]

We now turn to $SU(3)$ nonsinglets. Fermions ψ_α^x transforming in some representation R of $SU(3)$ can be regarded as spinor fields with values in some holomorphic vector bundle X_R determined by R. The Dirac zero modes are the Dolbeault cohomology groups $H^q(X_R)$. As was shown in §15.7.1, H^{3-q} is related to H^q by CPT, and H^0 will vanish if R is nontrivial. So the interesting modes come from $H^1(X_R)$. We are thus dealing with a $(0, 1)$ form $\psi_{\bar{b}}^x$ with an index x labeling some representation of $SU(3)$.

With the spin connection embedded in the gauge group, we can be very specific about the resulting structure. For modes that transform as **3** of $SU(3)$, the Yang–Mills index is equivalent to a holomorphic tangent vector index. Thus $\psi_{\bar{b}}^x$ is equivalent in this case to a $(0, 1)$ form $\psi_{\bar{b}}^a$ with an extra holomorphic tangent vector index. Equivalently, we can relate ψ to a $(2, 1)$ form $\tilde{\psi}_{a_1 a_2 \bar{b}} = \omega_{a_1 a_2 a_3} \psi_{\bar{b}}^{a_3}$, $\omega_{a_1 a_2 a_3}$ being here the holomorphic three form. For the $(2, 1)$ form $\tilde{\psi}$, the relevant cohomology

---

[*] The ten-dimensional gluinos of the second E_8 likewise have only four-dimensional gluinos among their zero eigenvalues.

group is $H^{2,1}$, and the number of zero eigenvalues is the Hodge number $h^{2,1}$. The decomposition (16.1.4) of the Lie algebra of E_8 shows that the modes that transform as **3** of $SU(3)$ transform as **27** of E_6, so we have shown that the number of left-handed massless **27**'s is $N_{27} = h^{2,1}$.

The alternative description of the same modes in terms of $\psi_{\bar{b}}{}^a$ is also useful. As a is a tangent vector index with values in the tangent bundle T, the relevant cohomology group in this description is $H^1(T)$. We know from §15.7.2 that $H^1(T)$ has a qualitative significance – it parametrizes infinitesimal deformations in the complex structure of K.

Now we move on to fields that transform as $\bar{\mathbf{3}}$ of $SU(3)$. In this case, the Yang–Mills index is an antiholomorphic tangent vector index, so we are dealing with a $(0,1)$ form $\phi_{\bar{a}}{}^{\bar{b}}$ with an antiholomorphic tangent vector index $\bar{b}$. Such a field is equivalent to a $(1,1)$ form $\tilde{\phi}_{b\bar{a}} = g_{b\bar{b}}\phi_{\bar{a}}{}^{\bar{b}}$. The relevant cohomology group is $H^{1,1}$, and the number of Dirac zero modes is the Hodge number $h^{1,1}$. The decomposition of the E_8 Lie algebra shows that modes that transform as $\bar{\mathbf{3}}$ of $SU(3)$ transform as $\overline{\mathbf{27}}$ of E_6, so we have determined that the number of massless left-handed $\overline{\mathbf{27}}$'s in four dimensions is $N_{\overline{27}} = h^{1,1}$.

The number of generations is $N_{gen} = |N_{27} - N_{\overline{27}}| = |h^{2,1} - h^{1,1}|$. As this coincides with one-half of the Euler characteristic, we recover our old result from §14.6.

Finally, we consider modes that transform in the adjoint representation of $SU(3)$. The decomposition of the E_8 Lie algebra shows that these modes are E_6 singlets, so they do not carry any currently known gauge interactions. Nonetheless, it will turn out that they can play a significant role in phenomenology. The $SU(3)$ bundle X that is made from the tangent bundle T by writing the spin connection in the adjoint representation of $SU(3)$ was called $End\,T$ in example (3) in §15.6.3; it is the bundle whose fiber at a given point $x \in K$ consists of (traceless) linear transformations of the holomorphic tangent vectors at p. The Dirac zero modes form the Dolbeault cohomology group $H^1(End\,T)$. The qualitative significance of this group was explored in §15.7.3. Viewing T as a holomorphic vector bundle over the manifold K (and ignoring the fact that T happens to be the holomorphic tangent bundle of K), the cohomology group $H^1(End\,T)$ labels the possible deformations of the holomorphic structure of T.

As an example, consider the quintic hypersurface Q in CP^4, defined by the zeros of a homogeneous quintic polynomial P. In this case, a very explicit description of $H^{2,1}$ can be given because, as explained in §15.7.2 and §15.2.3, elements of $H^{2,1}$ correspond to perturbations of the polynomial P that cannot be absorbed in linear changes of coordinates. The connection arises because the choice of P determines the complex structure of Q. We will here explicitly count the relevant perturbations of P. Taking for P the minimal quintic polynomial $P = \sum Z_i^5$, we would like to count perturbations δP of P that cannot be absorbed in linear changes of coordinates $\delta Z_i = a_{ij} Z_j$. It is easy to see that a perturbation of the form $\delta P = Z_i^4 Z_j$ can, for any i and j, be absorbed in the change of coordinates $\delta Z_i = -Z_j/5$, so the relevant polynomials δP are the ones that are at most cubic in any one Z_k. These come in several patterns:

$$
\begin{aligned}
(20) \quad & Z_1^3 Z_2^2 \\
(30) \quad & Z_1^3 Z_2 Z_3 \\
(30) \quad & Z_1^2 Z_2^2 Z_3 \\
(20) \quad & Z_1^2 Z_2 Z_3 Z_4 \\
(1) \quad & Z_1 Z_2 Z_3 Z_4 Z_5.
\end{aligned}
\qquad (16.2.1)
$$

The number in front of each monomial in (16.2.1) is the number of monomials of similar structure that can be obtained from it by permutations of the Z_j. Adding the numbers in (16.2.1), we see explicitly that deformation of the complex structure of Q depends on 101 complex parameters, so that the model under discussion has 101 generations. The relation of the fermion generations to polynomials is useful not just for counting the generations but, as we will see, for many other purposes as well.

The facts required for counting the other fields were also described in chapter 15. On Q there is precisely one harmonic $(1,1)$ form, namely the Kähler form, so there is precisely one antigeneration. The net number of generations is thus $101 - 1 = 100$. Finally, the E_6 singlet modes arising from $H^1(End\, T)$ are 224 in number, corresponding to the counting of certain polynomials discussed in example (6) of §15.6.3.

The appearance of massless **27**'s and $\overline{\mathbf{27}}$'s in the above discussion would be comprehensible in terms of real differential geometry; upon embedding the spin connection in the gauge group, massless generations and antigenerations arise in real differential geometry from harmonic p forms of even or odd p. However, the appearance in the above discussion of massless E_6 singlets for arbitrary choice of the quintic polynomial P has no analog in real differential geometry; no index theorem or other topological invariant of real differential geometry would predict such modes. Their appearance is one of the wonders of algebraic geometry.

16.2.2 Fluctuations of the Gravitational Field

For ten-dimensional fields that carry $E_8 \times E_8$ charges, we have found it convenient to count the zero modes of fermions. Boson zero modes are of course related to these by unbroken supersymmetry. In discussing the zero modes of other fields, we will find it more convenient to discuss directly the boson zero modes. We begin here with the zero modes that arise in the fluctuations of the gravitational field. The ten-dimensional metric tensor g_{MN} has components $g_{\mu\nu}$, with all indices tangent to four-dimensional Minkowski space M^4; $g_{\mu j}$, with one index tangent to M^4 and one tangent to the compact space K; and g_{ij}, with both indices tangent to K. The first case is universal in Kaluza–Klein theory: from $g_{\mu\nu}$ the only massless mode is the four-dimensional graviton.

Massless modes coming from $g_{\mu j}$ are massless gauge bosons as seen in four dimensions, and are in one-to-one correspondence with continuous symmetries of K. However, a manifold whose holonomy is $SU(3)$ (and not a proper subgroup thereof)* never has continuous symmetries, so one does not get massless modes from $g_{\mu j}$. To prove this, note that a vector field V^i that generates a continuous symmetry must obey the Killing vector equation $D_i V_j + D_j V_i = 0$. This implies that $0 = D^i(D_i V_j + D_j V_i)$. But $D^i D_j V_i = D_j D^i V_i + R_{jk} V^k$. The Killing vector equation implies that $D^i V_i = 0$, and $SU(3)$ holonomy implies that $R_{jk} = 0$. So a Killing vector field on a Ricci-flat space obeys $0 = D^i D_i V_j$. This implies that $0 = \int_K V^j D^i D_i V_j$. For compact K, integration by parts gives

$$0 = \int_K (D^i V^j)(D_i V_j). \qquad (16.2.2)$$

Hence a Killing vector field V on a compact Ricci-flat manifold K must be covariantly constant. But $SU(3)$ holonomy is incompatible with the existence of a covariantly constant vector field, since under $SU(3)$ the vector of $SO(6)$ decomposes as $\mathbf{3} \oplus \bar{\mathbf{3}}$, which does not contain an $SU(3)$ singlet.

We move on now to zero modes of the metric tensor g_{ij} of K. Such modes represent degeneracies in the vacuum state, and therefore we can find them easily. In our construction of vacuum states, the starting point was a metric on K of $SU(3)$ holonomy. Any deformation of the metric of K that preserves the fact that the holonomy is $SU(3)$ must correspond to a zero mode of the wave operator that governs metric disturbances, and hence to a massless particle in four dimensions. But Yau's proof of the

---

* A six-dimensional manifold whose holonomy group is a proper subgroup of $SU(3)$ must have a very simple structure; it must have a covering space that is either a torus or a product of a torus and the unique space $K3$ of $SU(2)$ holonomy. Such manifolds do not seem suitable for phenomenology since, for instance, their Euler characteristics vanish.

Calabi conjecture gives a precise description of the metric disturbances that preserve $SU(3)$ holonomy. They correspond to deformations of the complex structure or the Kähler class of K.

To verify the above statements, and also to show that there are no zero modes that do not preserve $SU(3)$ holonomy, it is useful to write down an explicit equation for the metric disturbances. To lowest order in α'/r^2, the equation for the metric g_{ij} of K is the Einstein equation $R_{ij} = 0$. Expanding $g_{ij} = g^0_{ij} + h_{ij}$, where g^0 is a Ricci-flat background and h_{ij} is the metric disturbance, the equation for h is $\Delta h = 0$, where Δ is a certain linear operator which can be obtained by linearizing the Einstein equations around a classical solution; it is often called the Lichnerowicz Laplacian. In the gauge $D^i h_{ij} - \frac{1}{2} D_j h^i{}_i = 0$, the explicit form of the equation is

$$0 = (\Delta h)_{ij} = -D_k D^k h_{ij} - R_{isjt} h^{st}. \tag{16.2.3}$$

With $SU(3)$ holonomy, the equations for the components $h_{a\bar{b}}$ and $h_{\bar{a}\bar{b}}$ are decoupled in (16.2.3). The equation for $h_{a\bar{b}}$ is more or less easily seen to coincide with the equation for a harmonic $(1,1)$ form that is given by the Laplacian in §15.5.3. In fact a zero mode of $h_{a\bar{b}}$ represents a variation in the Kähler class of the Kähler metric on K. With somewhat more work, the equation for $h_{\bar{a}\bar{b}}$ obtained from (16.2.3) can be seen to imply that $\tilde{h}^a_{\bar{b}} = g^{a\bar{a}} h_{\bar{a}\bar{b}}$ is an element of $H^1(T)$ and so represents a deformation of the complex structure of K.

Zero modes of $h_{\bar{a}\bar{b}}$ (or equivalently deformations of the complex structure of K) are complex and give rise to complex massless scalars in four dimensions. If the spin connection is embedded in the gauge group, these massless scalars are in one-to-one correspondence with massless **27**'s, since both are controlled by elements of $H^1(T)$. For instance, in the case of the quintic hypersurface Q in CP^4, the deformations of complex structure are in one-to-one correspondence with the same 101 monomials discussed above. With unbroken supersymmetry, the massless scalars coming from deformations of the complex structure of K are necessarily the spin zero parts of massless chiral superfields, which we will call $X_{(\alpha)}$.

What about zero modes of $h_{a\bar{b}}$ or harmonic $(1,1)$ forms? In the case of the quintic hypersurface Q in CP^4, there is a unique harmonic $(1,1)$ form, the Kähler form. This mode is of special significance; it is described by the ansatz $h_{a\bar{b}} = g^0_{a\bar{b}}$ and so represents an overall dilation of K. Thus, for the quintic hypersurface the choice of Kähler class is just a choice of radius or volume. The other gravitational zero modes determine the 'shape' of K.

Harmonic $(1,1)$ forms are naturally real. With unbroken supersymmetry, the real massless scalars that arise in four dimensions from zero modes of $h_{a\bar{b}}$ must have pseudoscalar partners to fill out a complex supermultiplet; we will find those partners shortly.

16.2.3 The Other Bose Fields

The other Bose fields that we must consider in $N = 1$ supergravity are the dilaton ϕ and the two form B_{MN} with its components $B_{\mu\nu}$, $B_{\mu j}$, and B_{ij}.

For ϕ and $B_{\mu\nu}$, little needs be added here to the general remarks in §14.3.1. In compactification on $M^4 \times K$, they invariably give rise to a single scalar and pseudoscalar mode, respectively, with a wave function that is independent of position on K. The pseudoscalar mode has axion-like couplings. These modes combine into the spin zero part of a chiral supermultiplet.

As for $B_{\mu j}$, on general grounds its zero modes are massless spin one fields as seen in four dimensions, and are in one-to-one correspondence with harmonic one forms on K. But a manifold of $SU(3)$ holonomy has no harmonic one forms (the cohomology groups $H^{0,1}$ and $H^{1,0}$ vanish), so there are no massless fields originating from $B_{\mu j}$.

Finally, we consider B_{ij}. Since $H^{2,0} = H^{0,2} = 0$ for a manifold of $SU(3)$ holonomy, the only nontrivial result arises for the $(1,1)$ components $B_{a\bar{b}}$. The zero modes are harmonic $(1,1)$ forms and give rise to real massless scalars in four dimensions. These are indeed the missing modes that complete the incomplete supermultiplets that we encountered in our discussion of the gravitational fluctuations. Every harmonic $(1,1)$ form on K gives rise to a gravitational zero mode and to a zero mode of B; the two types of modes are related by supersymmetry and combine into the spin zero parts of complex chiral supermultiplets $Y_{(\beta)}$ in four dimensions.

16.3   Symmetry Breaking by Wilson Lines

Embedding the spin connection in the gauge group is a simple way to obey the necessary equations for unbroken supersymmetry and simultaneously obtain a semblance of phenomenology. The principal successes are the automatic appearance of a reasonable gauge group, E_6, with fermions and Higgs bosons in the proper representation, the **27**. Also, the fact that one can obtain a multitude of fermion generations from a unified underlying structure means that in principle the problem of flavor may be soluble. At the same time there are some glaring faults. Simple constructions of manifolds of $SU(3)$ holonomy give far too many fermion generations. For instance, the quintic hypersurface in CP^4 gives 100 generations; it can be shown by counting polynomials that a complete intersection of hypersurfaces in CP^{3+N} always gives at least 64 generations. Also, while E_6 is an interesting candidate for grand unification of a four-dimensional theory, phenomenology would certainly not permit us to leave E_6 unbroken to low energies. We must find a satisfactory way to break E_6 at some very high energy down to an acceptable low-energy subgroup.

To break E_6, we must give an expectation value to fields that carry E_6 charges. In the field theory limit, the only massless charged bosons are the E_6 gauge fields. (The 'stringy' modes have positive mass squared in the field-theory limit, and present tools would not permit us to investigate states in which they have expectation values.) Of course, any nontrivial expectation value of E_6 gauge bosons will break E_6 down to a subgroup. Generically, however, supersymmetry will be broken at the same time. Since unbroken supersymmetry is the rationale for manifolds of $SU(3)$ holonomy, an E_6-breaking mechanism that spoils supersymmetry would ruin the rationale for the whole construction. If manifolds of $SU(3)$ holonomy are to be useful, we must find a way to break E_6 without breaking supersymmetry.

At first sight, it would appear hopeless to do so without relaxing the simplifying ansatz $d\phi = H = 0$ that enabled us to find solutions of the requisite equations. In §16.1, we encountered equations such as $0 = \text{tr} F \wedge F - \text{tr} R \wedge R$ and $0 = \text{tr} F_{ij} F^{ij} - \text{tr} R_{ij} R^{ij}$ that place extremely strong restrictions on the Yang–Mills field strength F. Indeed, these equations presumably require the E_6 field strength to vanish. How then can we break E_6?

Saying that the E_6 field strength F_{ij} vanishes indeed means locally that the E_6 gauge field A_i is a pure gauge, $A_i = \partial_i U \cdot U^{-1}$. Whether this is also true globally is another story. If the manifold K is simply connected, then $F_{ij} = 0$ ensures that we can set the gauge field to zero by a gauge transformation. When, however, K is not simply connected, $\pi_1(K) \neq 0$, there is a more general possibility that appears in electrodynamics as the Bohm–Aharonov effect. Let γ be a noncontractible loop in K, beginning and ending at some point x. Then the 'Wilson line'

$$U_\gamma = P \exp \oint_\gamma A \cdot dx \qquad (16.3.1)$$

is gauge covariant, and if $U_\gamma \neq 1$ it cannot be set to one by a gauge transformation. As long as $F_{ij} = 0$, U_γ depends only on the topological class of the loop γ in $\pi_1(K)$. If the U_γ are not in the center of E_6, then E_6 is broken down to the subgroup that commutes with all of the U_γ. Yet an E_6 gauge field with $F_{ij} = 0$ does not contribute to any local equations such as the classical equations of motion or the conditions for unbroken supersymmetry, so these are obeyed if they were obeyed at $U = 1$.

16.3.1 Symmetry Breaking Patterns

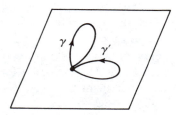

Figure 16.1. Two loops γ and γ' are 'multiplied' as shown here. This is the multiplication law in the definition of the fundamental group $\pi_1(K)$.

In a typical situation, there are many topological classes of noncontractible loops γ in space-time. For each γ, we define U_γ as in (16.3.1). The U_γ obey an important general constraint. Let γ and γ' be two different noncontractible loops in K starting and ending at the same point x. The group structure of the fundamental group $\pi_1(K)$ is defined by saying that the product loop $\gamma\gamma'$ shown in fig. 16.1 is the loop that traverses first γ' and then γ. Using this definition, we see that

$$U_{\gamma\gamma'} = P \exp \int_{\gamma\gamma'} A \cdot dx = \left(P \exp \int_\gamma A \cdot dx \right) \left(P \exp \int_{\gamma'} A \cdot dx \right) \quad (16.3.2)$$

so that in fact

$$U_{\gamma\gamma'} = U_\gamma \cdot U_{\gamma'}. \quad (16.3.3)$$

Equation (16.3.3) may be succinctly described by saying that the map $\gamma \to U_\gamma$ is a homomorphism of the fundamental group into E_6. Once such a homomorphism is chosen, E_6 is broken down to the subgroup that commutes with all of the gauge-covariant composite fields U_γ.

Now, we can easily determine explicitly the possible E_6-breaking patterns that can be achieved by this mechanism. E_6 contains a maximal subgroup $SU(3)_C \times SU(3)_L \times SU(3)_R$, where $SU(3)_C$ is the color group of strong interactions, and $SU(3)_L$ and $SU(3)_R$ describe, respectively, the weak interactions of left and right-handed quarks. The weak interaction group $SU(2)_L$ is embedded in $SU(3)_C \times SU(3)_L \times SU(3)_R$ in the form

$$(1) \otimes \begin{pmatrix} SU(2)_L & \\ & 1 \end{pmatrix} \otimes (1). \qquad (16.3.4)$$

Weak hypercharge $U(1)_Y$ is generated by

$$Y = (0) \otimes \begin{pmatrix} 1/3 & & \\ & 1/3 & \\ & & -2/3 \end{pmatrix} \otimes \begin{pmatrix} 4/3 & & \\ & -2/3 & \\ & & -2/3 \end{pmatrix}. \qquad (16.3.5)$$

It can be written as the sum of $SU(3)_L$ and $SU(3)_R$ pieces, $Y = Y_L + Y_R$, with

$$Y_L = (0) \otimes \begin{pmatrix} 1/3 & & \\ & 1/3 & \\ & & -2/3 \end{pmatrix} \otimes (0)$$

$$(16.3.6)$$

$$Y_R = (0) \otimes (0) \otimes \begin{pmatrix} 4/3 & & \\ & -2/3 & \\ & & -2/3 \end{pmatrix}.$$

We consider first the simplest situation in which the fundamental group of K is a cyclic group Z_n, generated by some element γ that obeys $\gamma^n = 1$. In this case, the symmetry breaking depends on a single E_6 element $U = U_\gamma$. It follows from (16.3.3) that $U^n = 1$, since (16.3.3) implies that $(U_\gamma)^n = U_{\gamma^n} = U_1 = 1$. Assuming that it commutes with $SU(3) \times SU(2) \times U(1)$, U can – after diagonalizing its $SU(3)_R$ part – be put in the form

$$U = (\alpha) \otimes \begin{pmatrix} \beta & & \\ & \beta & \\ & & \beta^{-2} \end{pmatrix} \otimes \begin{pmatrix} \gamma & & \\ & \delta & \\ & & \epsilon \end{pmatrix}. \qquad (16.3.7)$$

Here $\alpha, \beta, \gamma, \delta$, and ϵ are nth roots of unity, to ensure that $U^n = 1$. In order that U belongs to $SU(3)_C \times SU(3)_L \times SU(3)_R$, we require that

$\alpha^3 = 1$ (α then defines an element of the center of $SU(3)$) and $\gamma\delta\epsilon = 1$. The subgroup of E_6 that commutes with U (or any single E_6 element) has rank six. For generical values of the phases in (16.3.7), the unbroken group is $SU(3)_C \times SU(2)_L \times U(1) \times U(1) \times U(1)$, the three $U(1)$'s being Y_L and the diagonal $SU(3)_R$ matrices. Thus, in this case the low-energy theory consists of the standard model plus two additional abelian gauge interactions. While this is the group left unbroken by a generical matrix of type (16.3.7), the unbroken groups are larger in special cases. For instance, with $\gamma = \delta$, and no other special restrictions, the unbroken group is $SU(3)_C \times SU(2)_L \times SU(2)_R \times U(1) \times U(1)$. This corresponds to a weak interaction model with a generalized left–right symmetry. We will not classify here all of the possibilities.

Though we have assumed that the fundamental group of K is generated by a single U, a similar result will emerge as long as the fundamental group of K is abelian. If $\pi_1(K)$ is abelian, then the various U_γ commute with one another since (16.3.3) implies that $U_\gamma U_{\gamma'} = U_{\gamma\gamma'} = U_{\gamma'\gamma} = U_{\gamma'}U_\gamma$. In such a situation (assuming $SU(3)_C \times SU(2)_L \times U(1)_Y$ is to be unbroken), the U_γ can be put simultaneously in the form of (16.3.7). If, however, $\pi_1(K)$ is nonabelian, an essentially new possibility appears. If $\gamma \to V_\gamma$ is an irreducible two-dimensional representation of $\pi_1(K)$, then we can set

$$
U_\gamma = (\alpha_\gamma) \otimes \begin{pmatrix} \beta_\gamma & & \\ & \beta_\gamma & \\ & & \beta_\gamma^{-2} \end{pmatrix} \otimes \begin{pmatrix} \phi_\gamma & \\ & V_\gamma \end{pmatrix}.
\tag{16.3.8}
$$

Here $\phi_\gamma = \det V_\gamma^{-1}$; $\gamma \to \alpha_\gamma$ and $\gamma \to \beta_\gamma$ are one-dimensional representations of $\pi_1(K)$; and α_γ is a cube root of unity (defining an element of the center of $SU(3)$) for each γ. This structure leaves unbroken a rank-five group, which, in the absence of additional restrictions on the matrix elements that appear in (16.3.8), is $SU(3)_C \times SU(2)_L \times U(1) \times U(1)$. The two $U(1)$'s are Y_L and Y_R, defined in (16.3.6) above. This indeed is the smallest group to which E_6 can be broken by Wilson lines alone while keeping the standard model unbroken, since every E_6 matrix that commutes with the standard model commutes with both Y_L and Y_R.

Thus, if E_6 is broken by Wilson lines only, there must be at least one new gauge interaction, and there is a definite prediction of what it must be. On the other hand, it is quite possible that other mechanisms for E_6 breaking also play a role. For instance, we will later see that upon relaxing the assumption $d\phi = H = 0$, it is possible to use flat directions in the superpotential to break E_6 to $SO(10)$ or $SU(5)$ while keeping unbroken supersymmetry.

For this and other reasons, it is very interesting to ask to what low-energy groups $SU(5)$ or $SO(10)$ can be broken by use of Wilson lines. The question is easily answered. In the case of $SU(5)$, a Wilson line U that leaves the standard model unbroken must be an element of the $U(1)$ group generated by hypercharge. Such an element would break $SU(5)$ precisely down to the standard model. For $SO(10)$, the discussion is a bit different. $SO(10)$ contains a subgroup $SU(5) \times U(1)$, the extra $U(1)$ being $B - L$, the difference between baryon and lepton number. Any $SO(10)$ element that commutes with the standard model also commutes with $B - L$. Thus, in $SO(10)$ models, symmetry breaking by Wilson lines that leaves the standard model unbroken would leave unbroken at least $SU(3)_C \times SU(2)_L \times U(1)_Y \times U(1)_{B-L}$. Just as in the E_6 case, symmetry breaking by Wilson lines predicts at least one new gauge interaction in the $SO(10)$ case, but curiously the new interaction that is predicted is different.

This discussion should make it clear that symmetry breaking by Wilson lines can lead to a more or less realistic low-energy gauge group. Our next order of business is to find examples of models of $SU(3)$ holonomy with reasonably large fundamental group, to which the above discussion can be applied.

16.3.2 A Four Generation Model

If we are given a manifold K_0 of $SU(3)$ holonomy that is simply connected, it is not possible to break E_6 in the manner just described. But very frequently K_0 can be modified to give a manifold of $SU(3)$ holonomy whose fundamental group is nontrivial. Suppose that we can find a symmetry group F of K_0. F will automatically be a discrete symmetry group, since we have proved that a manifold of $SU(3)$ holonomy does not have continuous symmetries. In this situation, it is natural to introduce an equivalence relation with two points x and y considered equivalent if $y = fx$ for some $f \in F$. The equivalence classes make up a new space that we will denote as $K = K_0/F$.

This quotient space K is a manifold if F acts freely in the sense that for $f \in F$ and $x \in K_0$, $fx \neq x$ if $f \neq 1$. The idea here is that a manifold is a topological space that locally looks like Euclidean space. Local measurements at a point $x \in K_0$ are unaffected by the fact that x may have an image point fx that is considered equivalent to x. More exactly, they are unaffected by this – and so $K = K_0/F$ is a manifold – as long as F acts freely and x and fx are distinct points. By local measurements, an observer at x cannot discover that he is considered equivalent to another observer at fx, as long as $x \neq fx$. As an example, let K_0 be ordinary three-dimensional Euclidean space, with Euclidean coordinates x_1, x_2, x_3. The translation T defined by $T(x_i) = x_i + a_i, i = 1, 2, 3$ acts freely, and if F is the group generated by this translation, then K_0/F is a manifold. On the other hand, let P be the parity transformation $P(x_i) = -x_i$. Then P acts freely except at $x_i = 0$, where it has a fixed point. If F is the Z_2 group generated by P, then K_0/F has a singularity at $x_i = 0$ and is not a manifold.

In this section, we will assume that F acts freely so that $K = K_0/F$ is a manifold.[*] Assuming that K_0 has $SU(3)$ holonomy, the same is always true also for K, as we will prove later. However, K will not be simply connected. Let γ be a path on K_0 that starts at a point $x \in K$ and ends at fx, for some $f \in F$. Then γ is a closed loop in K, even though (if $f \neq 1$) it is not a closed loop in K_0. In this construction, we may choose for f an arbitrary element of F. Moreover, if K_0 is simply connected, then the topological class of γ depends only on the choice of f. Thus, if K_0 is simply connected, the fundamental group of K is $\pi_1(K) \approx F$.

Thus, we can construct manifolds K of $SU(3)$ holonomy that are not simply connected by finding the free action of a discrete group F on a simply connected space of $SU(3)$ holonomy K_0. Moreover, any manifold K of $SU(3)$ holonomy that is not simply connected can be constructed in this way, with K_0 being the 'universal covering space' of K, and with $F \approx \pi_1(K)$.

Most simple constructions of manifolds of $SU(3)$ holonomy give simply connected examples. For instance, this is true for the hypersurface Q in CP^4 defined by the zeros of a homogeneous quintic polynomial such as

$$P = Z_1^5 + Z_2^5 + \ldots + Z_5^5. \tag{16.3.9}$$

How can we find a discrete group that acts freely on this hypersurface? A metric of $SU(3)$ holonomy on the quintic hypersurface is uniquely determined (after fixing the Kähler class or volume) by the complex structure. The complex structure is invariant under any holomorphic change of coordinates. To find a holomorphic map of Q into itself, one usually looks for a holomorphic map of CP^4 into itself that leaves fixed the hypersurface Q. A holomorphic map of CP^4 into itself (without poles or other singularities) is always a linear transformation $Z_i \to A_{ij}Z_j$, with some matrix

---

[*] However, the space K_0/F seems to make sense in string theory even if F does not act freely, as we learned in our discussion of orbifolds in §9.5.2. We will explore the connection between orbifolds and manifolds of $SU(3)$ holonomy in §16.10 below.

A_{ij}. The linear transformations of the Z_i are certainly the holomorphic maps of Q into itself which are easy to find; with more effort they can be shown to be the only ones. Moreover, linear transformations of the Z_i always leave fixed the Kähler class of CP^4 and hence also of Q.

Linear transformations of the Z_i that leave P fixed and act freely on the hypersurface $P = 0$ can be found without too much difficulty. Let A be the transformation

$$A(Z_i) = Z_{i+1}, \qquad (16.3.10)$$

where we define $Z_{i+5} = Z_i$. Then A manifestly leaves fixed the polynomial P. Also, $A^5 = 1$, so A generates a group isomorphic to Z_5.

That A acts freely on the hypersurface is slightly less obvious. A fixed point of A on CP^4 must have coordinates Z_i that obey $Z_{i+1} = \lambda Z_i$ with some complex λ. (The possibility to have $\lambda \neq 1$ comes from the fact that scaling of the coordinates is irrelevant in CP^4.) Since $Z_{i+5} = Z_i$, this implies that $\lambda^5 = 1$, from which one can easily see that points on CP^4 with $Z_{i+1} = \lambda Z_i$ do not obey $P = 0$. Thus, while A does not act freely on CP^4, it does act freely on the hypersurface $P = 0$. The same argument applies to A^k as long as k is not divisible by 5, so the Z_5 group generated by A acts freely on the hypersurface.

To get a larger group that acts freely, let $\alpha = e^{2i\pi/5}$ be a fifth root of unity, and define a transformation B by

$$B(Z_k) = \alpha^k Z_k. \qquad (16.3.11)$$

Then $B^5 = 1$, and $BA = AB$ (as transformations of the Z_i, A and B obey $BA = AB\alpha$, but the phase α is irrelevant in CP^4). So A and B together generate a group F isomorphic to $Z_5 \times Z_5$. By analogy with our previous comments, it is easily seen that $A^k B^l$ has no fixed points on the hypersurface unless k and l are divisible by 5, so F acts freely.

We now wish to study the manifold $K = Q/F$ whose fundamental group (since Q is simply connected) is $Z_5 \times Z_5$. The first question to ask is what is the Euler characteristic of K? There are many ways to answer this question. One way is to note that the Euler characteristic may be defined as the integral of a certain curvature polynomial, discussed in §12.5.2. For a six manifold M the form is

$$\chi = \frac{1}{48\pi^3} \cdot \int_M \epsilon^{i_1 i_2 \ldots i_6} \epsilon^{j_1 j_2 \ldots j_6} R_{i_1 i_2 j_1 j_2} R_{i_3 i_4 j_3 j_4} R_{i_5 i_6 j_5 j_6}. \qquad (16.3.12)$$

Suppose now that a manifold K_0 (not necessarily simply connected) admits the action of a discrete symmetry group F with $N(F)$ elements. If F

acts freely, then K_0 consists of $N(F)$ 'fundamental regions' of the action of F. $K = K_0/F$ consists of a single 'fundamental region', with some boundaries identified. To compute the Euler characteristic of K_0 from (16.3.12), we would carry out the integration over all $N(F)$ fundamental regions. To compute the Euler characteristic of K from (16.3.12), we would carry out the same integral over only one fundamental region. So the Euler characteristics of K and K_0 are related by

$$\chi(K) = \chi(K_0)/N(F). \tag{16.3.13}$$

Returning to the hypersurface Q in CP^4, which admits free action of $F = Z_5 \times Z_5$, we can now determine the Euler characteristic of $K = Q/F$. With $\chi(Q) = -200$ and $N(F) = 5 \times 5 = 25$, we have

$$\chi(K) = -8. \tag{16.3.14}$$

The physical model compactified on K thus has $|\chi|/2 = 4$ generations. This rather simple four generation model, on which there is a natural possibility for E_6 breaking since $\pi_1(K) = Z_5 \times Z_5$, will be used for illustrative purposes in much of what follows.

It is instructive to verify the result that $\chi(K) = -8$ more explicitly. To do this, we will calculate the Hodge numbers $h^{p,q}$ of K. A harmonic (p,q) form on K is the same as a harmonic (p,q) form on the covering space Q that is invariant under F. It follows, for instance, that K has $h^{0,1} = h^{0,2} = 0$, since Q has no harmonic $(0,1)$ forms or harmonic $(0,2)$ forms. On the other hand, on Q there is a unique harmonic $(1,1)$ form, the Kähler form. It is invariant under F (or $K = Q/F$ would not be a Kähler manifold), so K has $h^{1,1} = 1$. K also has $h^{0,0} = 1$, like any connected compact complex manifold, since the constant function is the unique harmonic $(0,0)$ form.

To compute $h^{2,1}$, we will count the deformations in the complex structure of K. The definition of K involves the choice of a quintic polynomial P that is $Z_5 \times Z_5$ invariant, so that one can take the quotient of the hypersurface $P = 0$ by $Z_5 \times Z_5$. Deformations in the complex structure of P correspond to $(Z_5 \times Z_5)$-invariant perturbations in P that cannot be absorbed in change of variables. It is easily seen that apart from P itself there are five independent $(Z_5 \times Z_5)$-invariant quintic polynomials:

$$
\begin{array}{lll}
(a) & Z_1^3 Z_2 Z_5 + \dots & \\
(b) & Z_1^3 Z_3 Z_4 + \dots & \\
(c) & Z_1^2 Z_2 Z_3^2 + \dots & (16.3.15) \\
(d) & Z_1^2 Z_2^2 Z_4 + \dots & \\
(e) & Z_1 Z_2 Z_3 Z_4 Z_5. &
\end{array}
$$

Here '+...' refers to the addition of terms obtained by cyclic permutations of the Z_i. Addition of any of the five polynomials in (16.3.15) cannot be absorbed in linear transformation of the Z_i, so the complex structure of $K = Q/F$ depends on five complex parameters, and K has $h^{2,1} = 5$.

We have now determined all of the independent Hodge numbers of K except $h^{3,0}$. Compatibility with the value $\chi(K) = -8$ requires that $h^{3,0} = 1$. In fact, on Q there is a unique holomorphic $(3,0)$ form – namely the covariantly constant $(3,0)$ form ω that is the hallmark of a manifold of $SU(3)$ holonomy. On these grounds alone, K has $h^{3,0}$ equal to 1 or 0. To show that $h^{3,0}(K) = 1$, we must show that ω is invariant under $F = Z_5 \times Z_5$. Otherwise, K would not have a holomorphic three form and would not have $SU(3)$ holonomy.

Instead of using methods special to this one example, we will prove a general statement, which was made earlier. Let K_0 be *any* manifold of $SU(3)$ holonomy, and F any discrete group that acts freely on K_0. Then $K = K_0/F$ has $SU(3)$ holonomy. We will prove this by showing that $h^{0,3}(K) = 1$, so that on K there is a holomorphic three form.

Consider the $\bar{\partial}$ operator acting on $(0, k)$ forms. The index of the hermitian operator $D = \bar{\partial} + \bar{\partial}^*$ is known as the arithmetic genus, A. Since zero modes of D are harmonic $(0, k)$ forms, whose chirality is $(-1)^k$, the arithmetic genus is $A = \sum_k (-1)^k h^{0,k}$. Like the index of any operator, the arithmetic genus can be expressed by an integral formula analogous to (16.3.12). The same argument that we used for the Euler characteristic can be used to prove that if K_0 is any complex manifold (no restriction on its first Chern class) and F acts freely, then the arithmetic genus of K is related to that of K_0 by $A(K) = A(K_0)/N(F)$.

Let us apply this to the case in which K_0 has $SU(3)$ holonomy. Then from our knowledge of the Hodge numbers we can see that $A(K_0) = 0$. The relation $A(K) = A(K_0)/N(F)$ then tells us that $A(K) = 0$. Since K has $h^{0,0} = 1$ (like any connected complex manifold) and $h^{0,1} = h^{0,2} = 0$ (since there are no harmonic $(0, 1)$ forms or $(0, 2)$ forms on K_0), vanishing of $A(K)$ implies that K has $h^{0,3} = 1$. Hence K has $SU(3)$ holonomy.

16.4 Relation to Conventional Grand Unification

In many ways, the Wilson lines U that we have introduced to describe E_6 breaking are similar to ordinary Higgs fields. The relation would be $U = e^{i\phi}$, with ϕ being a Higgs boson in the adjoint representation of E_6. The natural appearance of the adjoint representation is indeed attractive, since this is certainly known to be a suitable representation for the initial stage of grand-unified symmetry breaking. There are, however, a number of

differences from conventional grand unification. Many of these differences
spring from the fact that the eigenvalues of U are quantized. For instance,
if the fundamental group of K is $\pi_1(K) = Z_n$, then U obeys $U^n = 1$ and
its eigenvalues are nth roots of unity. By contrast, the eigenvalues of
an ordinary Higgs boson ϕ in the adjoint representation of E_6 depend
continuously on freely adjustable coupling constants. We will explore
the relation to conventional grand unification in what follows, and draw
conclusions at the end on this section.

16.4.1 Alternative Description of Symmetry Breaking

We begin with an alternative description of the symmetry breaking mechanism. Let K_0 be a simply connected manifold, and let F be a discrete
symmetry group that acts freely. Let $K = K_0/F$. An ordinary field $\psi(x)$
on K_0/F, transforming in some representation of E_6, is equivalent to a
field $\psi(x)$ on K_0 that obeys

$$\psi(fx) = \psi(x) \tag{16.4.1}$$

for all $f \in F$. Now, consider a theory with an (otherwise unbroken)
symmetry group E_6. Then we can generalize (16.4.1) as follows. For each
$f \in F$, we pick an element $U_f \in E_6$, and we require that ψ obey not
(16.4.1) but

$$\psi(fx) = U_f\psi(x). \tag{16.4.2}$$

From this equation, we can deduce that for any $f, f' \in F$,

$$U_f U_{f'}\psi(x) = U_f\psi(f'x) = \psi(ff'x) = U_{ff'}\psi(x). \tag{16.4.3}$$

Hence, (16.4.2) only makes sense if

$$U_f U_{f'} = U_{ff'}, \tag{16.4.4}$$

or in other words if the map $f \to U_f$ is a homomorphism of F into E_6.
    This condition also appeared in our discussion of symmetry breaking
by Wilson lines, and suggests a relation of (16.4.2) to that mechanism.
Indeed, the precise connection is easily described. In the last section we
started with an E_6 gauge field A_i with $F_{ij} = 0$. Despite the vanishing field strength, the nontrivial Wilson lines can make it impossible to
gauge A_i away with a single-valued gauge transformation U. Vanishing
field strength ensures, however, that we can gauge A_i to zero by a gauge
transformation that is *not* single valued. Such a non-single-valued gauge

transformation will introduce a 'twist' in the boundary conditions obeyed by charged fields, and this is what appears in (16.4.2). Symmetry breaking by Wilson lines is exactly equivalent to a framework in which the E_6 gauge field vanishes in vacuum, but charged fields obey (16.4.2).

This alternative description makes it clear that we may pick an *arbitrary* homomorphism $f \to U_f$ in describing the vacuum state; this may not have been altogether clear in the last section. The alternative description is also very convenient for understanding other physical properties, as we will see.

16.4.2 E_6 Relations among Coupling Constants

Even though one must assume that the underlying unified group is spontaneously broken, four-dimensional grand unification has significant implications for the tree-level coupling constants. We would like to understand the analogous issues in the case in which unification occurs not in four dimensions but in ten dimensions. Consider, then, a simply connected manifold K_0 of $SU(3)$ holonomy. Finding a freely acting discrete symmetry group F, we formulate physics on the quotient $K = K_0/F$; E_6 breaking is described by choosing a homomorphism $f \to U_f$ of F into E_6. This homomorphism maps F into a discrete subgroup of E_6, which we will call $\overline{F}$. The unbroken gauge group in four dimensions is the subgroup G of E_6 that commutes with $\overline{F}$.

Under $G \times \overline{F}$, the **27** of E_6 has a decomposition

$$\mathbf{27} = \oplus_i \, R_i \otimes T_i, \tag{16.4.5}$$

where R_i and T_i are certain representations of G and $\overline{F}$ respectively. T_i is of interest to four-dimensional physicists only indirectly, but R_i is of utmost interest. Different choices of R_i represent, for example, up quarks or charged leptons.

To find massless fermions, we look for zero modes of the Dirac operator on K. (The relevant Dirac operator of course includes the effect of embedding the spin connection in the gauge group.) A zero mode of the Dirac operator on K is the same as a zero mode on the covering space K_0 that obeys the appropriate restriction

$$\psi(fx) = U_f \psi(x) \tag{16.4.6}$$

that was formulated in (16.4.2) above. The substance of this condition is that allowed modes transform under $\overline{F}$ the same way that they transform under F. Thus, modes that transform as R_i under G transform (according to (16.4.5)) as T_i under $\overline{F}$, so according to (16.4.6) they must transform as T_i under F as well.

Before trying to extract the consequences of this statement, let us practice with the easier case in which E_6 is *not* broken. Then (16.4.6) reduces to the more elementary statement

$$\psi(fx) = \psi(x) \tag{16.4.7}$$

with which we began this section. Equation (16.4.7) means that the modes on K_0, which are kept in doing physics on $K = K_0/F$ instead of K, are the modes that are invariant under F. Let ψ be a zero mode of the Dirac equation on K_0, transforming as **27** of E_6. According to (16.4.7), if ψ is not invariant under F, it is to be discarded in formulating physics on K; if it is invariant under F, it is to be kept. If ψ is kept it contributes in four dimensions an entire **27** of (positive or negative chirality) massless fermions, since E_6 is unbroken.

Now let us discuss in an analogous way (16.4.6), whose implications are at first less transparent. A zero mode ψ of the Dirac operator on K_0 will in general transform in some representation T_j of F. Equation (16.4.7) would tell us that ψ gives rise in four dimensions to either a complete **27** of E_6 or nothing, depending on whether it is E_6 invariant. Equation (16.4.6) gives a more complicated answer. Given that ψ transforms as T_j under F, (16.4.6) says that we are to keep those components of the **27** that also transform as T_j under $\overline{F}$. This means in view of (16.4.5) that we are to keep those components that transform as R_j under G.

Thus, a given zero mode ψ on K_0 does not give rise to a complete E_6 multiplet when the theory is formulated on K instead of K_0. Running this backwards, suppose that we are interested in massless fermions that transform as R_i under G. According to (16.4.5) these transform as T_i under $\overline{F}$. Consequently, (16.4.6) asserts that these modes originate from Dirac zero modes on K_0, which transform as T_i under F. This means, for instance, that massless up quarks and massless charged leptons, if they transform in distinct representations of G, originate from different zero modes of the Dirac operator on K_0. In turn, this means that the physical up quark is related by E_6 not to physical charged leptons, but to modes on K_0 that violate (16.4.6) and have nothing to do with physics formulated on K.

In four-dimensional unified theories, it is often asked, 'Which charged lepton is the $SU(5)$ (or E_6) partner of the up quark?' In the present context, there is no way to answer this question; since there is no natural way to assign the physical fermions to $SU(5)$ or E_6 multiplets. The of which lepton is related to the up quark by $SU(5)$ is usually raised because of its implications for the branching ratios in proton decay. When an up quark emits an X or Y boson (color triplet $SU(5)$ partners of the photon), which lepton is created? In the present discussion, we need not answer the latter question, since the color triplet $SU(5)$ partners of the photon violate (16.4.6) and do not exist. (This is how E_6 was broken!) The proton will still decay, of course. There are an infinity of massive spin one bosons with the same quantum numbers and the same order of magnitude of mass and coupling as the usual X and Y bosons. But there is no natural way to say which of these is the $SU(5)$ partner of the photon. The question about branching ratios in proton decay still makes sense, but the ingredients required to answer it are different.

When a classical field theory is formulated on K_0, it leads to various four-dimensional fields ϕ^i transforming in various representations of E_6, with a variety of E_6-invariant couplings such as cubic couplings $\lambda_{ijk}\phi^i\phi^j\phi^k$. When it is formulated on $K = K_0/F$, the classical couplings are the same λ_{ijk} (since the starting point is exactly the same classical Lagrangian!), but some of the fields ϕ^i (those that violate (16.4.6)) must be set to zero. States ϕ and ϕ' on K that came from the same E_6 multiplet on K_0 have couplings that are related by E_6. States ϕ and ϕ' on K that came from different E_6 multiplets on K_0 have couplings that are not related by E_6 even if the theory is formulated on K_0, and certainly not if it is formulated on K.

Now we can apply this to draw a very significant conclusion. There is on K_0 only a single multiplet of massless E_6 gauge bosons in the adjoint representation, so the strong, weak, and electromagnetic gauge couplings of the four-dimensional world obey (at tree level) the standard E_6 relations. Hence the conventional Georgi–Quinn–Weinberg calculation of $\sin^2\theta$ and the unification scale is valid in this context. This calculation of $\sin^2\theta$, which will be reviewed in §16.9, is of course one of the significant successes of grand unification.

However, as the physical quarks allowed on K did not originate on K_0 as E_6 partners of one another, their Yukawa couplings do not obey any relations due to E_6. There are no E_6 relations between quark and lepton masses or between up and down quark masses. This is just as well, since such relations are notoriously problematic, especially in models such as this one with simple Higgs content. (The only Higgs bosons with couplings to quarks and leptons originate as **27**'s of E_6.)

The above comments were stated in field-theoretic terms, but they are also valid in string theory. Indeed, the argument concerning the validity of E_6 relations on gauge couplings holds in an arbitrary classical field theory formulated on $K = K_0/F$, and therefore it holds, in particular, for the infinite component classical field theory that can be extracted from string theory. There is an analogous statement that can only be understood with the aid of string-theoretic insights. Before breaking $E_8 \times E_8$, the gauge couplings of the two E_8's are equal. After embedding the spin connection in, say, the first E_8 and thereby breaking $E_8 \times E_8$ to $E_6 \times E_8$, are the E_6 and E_8 couplings equal? This would not be so, in general, in a classical field theory. If, for instance, $F_{(1)}$ and $F_{(2)}$ are the field strengths of the two E_8's, we could contemplate an interaction of the type $\Delta L = \mathrm{tr} F_{(1)}^2 \cdot \mathrm{tr} F_{(2)}^2$. After embedding the spin connection in the first E_8, this would give a contribution $\Delta L = \langle \mathrm{tr} F_{(1)}^2 \rangle \cdot \mathrm{tr} F_{(2)}^2$, which contributes to the coupling constant of the unbroken E_8 without affecting the E_6 coupling. In string theory, such contributions do not arise; the E_6 and E_8 couplings are equal. This can be proved using the σ-model formulation of string propagation in background fields, which was introduced in §3.4; but this is something that we will not explore in this book.

16.4.3 Counting Massless Particles

In counting the massless particles that arise upon embedding the spin connection in the gauge group, we encountered an interesting property that we did not stress. The net number of chiral **27**'s was determined by an index theorem as one-half of the Euler characteristic of K. But upon more careful study, we discovered that we could predict separately the net number of positive- and negative-chirality **27**'s. These were given as the Hodge numbers $h^{1,1}$ and $h^{2,1}$, respectively. For instance, for the quintic hypersurface in CP^4, we found one positive-chirality **27** and 101 negative-chirality **27**'s. It is rather puzzling from the point of view of low-energy physics that the positive-chirality **27** does not pair up with one of the negative-chirality **27**'s to gain a mass. This could be understood only by study of the microscopic structure. The unexpected appearance of massless particles whose existence is difficult to explain from the low-energy point of view is clearly something of great interest. A conspicuous example of a light particle in nature whose lightness is hard to understand is the electroweak symmetry-breaking Higgs doublet. It is therefore natural to ask whether after breaking of E_6 by Wilson lines there are still massless particles whose masslessness is difficult to account for in four-dimensional terms. We will first address this question in general, and then discuss the extent to which we can actually gain some useful insights about the Higgs doublet.

Given a representation R_i of the unbroken gauge group G, let n_i^+ and

n_i^- be the number of massless fermion multiplets in four dimensions with positive or negative chirality that transform as R_i. Since the modes that transform as R_i are in no way related by E_6 to the modes that transform in some other representation R_j, the numbers $n_i^\pm$ and $n_j^\pm$ need not be equal for $i \neq j$. At first sight, it may appear that proceeding in this way we will arrive in four dimensions with a completely random collection of massless fermions. What saves the day is that the chiral asymmetry $N_i = n_i^+ - n_i^-$ is the *index* of the Dirac operator $\not{D}_{(i)}$, which acts on fermions that transform as R_i under G. The index N_i is independent of i, as long as E_6 is only broken by Wilson lines. The reason for this is that the index theorem expresses the index of the Dirac operator in terms of a polynomial in the Yang–Mills field strength and the Riemann tensor; there is no contribution from global phases like those in (16.4.6) or equivalently from Wilson lines. The fact that $N_i = N_j$ means that as far as the chiral asymmetry is concerned, the fermions come in complete E_6 multiplets. Differences in the $n_i^\pm$ that cancel out of the N_i correspond to the possible existence of extra massless particles in real representations of G – particles whose existence is puzzling from the low-energy point of view.

Whether such particles are present depends on the details of a particular model. As an example of what can happen, recall that any compact Kähler manifold K_0 has $h^{1,1} \geq 1$, since the Kähler form is always a harmonic $(1,1)$ form. In many simple cases, $h^{1,1}$ is precisely one; this is so, for instance, for the quintic hypersurface in CP^4 and actually for any complete intersection of hypersurfaces in CP^{3+N}. Let us examine what happens in such cases.

Let F be a freely acting symmetry group of K_0 (that preserves the complex structure), and let $K = K_0/F$. A harmonic form on K is a harmonic form on K_0 that is invariant under F. The Kähler form of K_0 always has this property. It is the only harmonic $(1,1)$ form on K, since by our hypothesis there was only one harmonic $(1,1)$ form on K_0. Hence $h^{1,1}(K) = h^{1,1}(K_0) = 1$. As for $h^{2,1}(K)$, it must equal $N_{gen} + 1$, with N_{gen} the net number of generations of the theory formulated on K.

Now we introduce Wilson lines, mapping F into a discrete subgroup $\overline{F}$ of E_6. Denoting the unbroken subgroup of E_6 as G, the **27** of E_6 decomposes as in (16.4.5) into

$$27 = \oplus_i R_i \otimes T_i, \qquad (16.4.8)$$

where R_i and T_i are certain representations of G and $\overline{F}$. We now wish to calculate the numbers $n_i^\pm$ of massless positive- and negative-chirality fermion multiplets transforming as R_i under G. Here n_i^+ and n_i^- will equal respectively the number of harmonic $(1,1)$ forms and $(2,1)$ forms that transform as T_i under F. Since an index theorem ensures that $n_i^+ - n_i^- = -N_{gen}$ is independent of i, it is enough to determine n_i^+. What makes this easy is the assumption that the Kähler form is the only harmonic $(1,1)$ form on K_0. Hence, for each i, n_i^+ is either 1 or 0 depending on whether the Kähler form contributes to n_i^+. But we know that the Kähler form transforms as the trivial representation T_0 of F. T_0 may or not be one of the $\overline{F}$ representations appearing in (16.4.8). If it is, let R_0 be the representation of G for fermions that transform as T_0 under $\overline{F}$. For fermions that transform as R_0 we have $n_0^+ = 1$, since the Kähler form contributes, and therefore $n_0^- = N_{gen} + 1$. For fermions that transform as R_i for $i \neq 0$, we have $n_i^+ = 0$, and hence $n_i^- = N_{gen}$.

Let us check this for our familiar toy model based on a hypersurface Q in CP^4 defined by vanishing of a quintic polynomial P such as $P = Z_1^5 + Z_2^5 + \ldots + Z_5^5$. We defined

$$A : Z_k \rightarrow Z_{k+1} \qquad\qquad (16.4.9)$$

and

$$B : Z_k \rightarrow \alpha^k Z_k \qquad\qquad (16.4.10)$$

with $\alpha = e^{2i\pi/5}$. These obey $A^5 = B^5 = 1$, and $AB = BA$, so they generate a group $F \approx Z_5 \times Z_5$. The model based on $K = Q/F$ has $N_{gen} = 4$.

Since F is abelian, its representations are easily described. A representation of F is one dimensional, obtained by assigning the values of A and B, which may be arbitrary fifth roots of unity:

$$A = \alpha^s, \quad B = \alpha^t \qquad\qquad (16.4.11)$$

with any s, t such that $0 \leq s, t \leq 4$. In all F has 25 representations. Depending on the choice of Wilson lines, an arbitrary representation T_i of F may appear in (16.4.8), so we may as well discuss the general case. If T_i is the trivial representation, we expect from the previous discussion that $n_i^- = N_{gen} + 1 = 5$. If T_i is a nontrivial representation, we expect $n_i^- = 4$. Let us verify these claims explicitly.

The key is the connection between elements of $H^{2,1}(Q)$ and quintic polynomials. In fact, as was seen in §16.2.1, the 101 elements of $H^{2,1}$ correspond to quintic polynomials that are at most cubic in any one of the Z_i. Let us call these the allowed polynomials. It is quite easy to count massless fermions on Q that transform in a given representation T_i

of F. One simply studies the allowed polynomials and asks how many of them transform in a given representation of Q. The claim that $n_0^- = 5$ for the trivial representation while $n_i^- = 4$ for nontrivial representations amounts to the statement that the 101 allowed polynomials consist of five that are invariant under F, and four each that transform in each of the 24 nontrivial representations of F. We actually already exhibited the five invariant allowed polynomials in §16.3.2 in the course of computing explicitly the Hodge numbers of $K = Q/F$. It remains then to show that there are precisely 4 allowed polynomials in each of the 24 nontrivial sectors. We will leave the general case to the reader, and concentrate on an interesting example.

Since F is generated by A and B, choosing a homomorphism of F into E_6 amounts to choosing two Wilson lines U_A and U_B. The example we will consider is $U_A = 1$ and

$$U_B = (1) \otimes \begin{pmatrix} \alpha & & \\ & \alpha & \\ & & \alpha^{-2} \end{pmatrix} \otimes \begin{pmatrix} \alpha & & \\ & \alpha & \\ & & \alpha^{-2} \end{pmatrix}. \tag{16.4.12}$$

This breaks E_6 to a left–right symmetric extension of the standard model $SU(3)_C \times SU(2)_L \times SU(2)_R \times U(1) \times U(1)$. Recalling that the **27** of E_6 decomposes under $SU(3)_C \times SU(3)_L \times SU(3)_R$ as $(\mathbf{3}, \mathbf{\bar{3}}, \mathbf{1}) \oplus (\mathbf{1}, \mathbf{3}, \mathbf{\bar{3}}) \oplus (\mathbf{\bar{3}}, \mathbf{1}, \mathbf{3})$, one can readily determine how the various quarks and leptons transform under U_B. Indeed, all fifth roots of unity appear. To understand in detail the four-dimensional massless fermions that arise in this model, it is necessary to study the allowed polynomials that are invariant under A and transform as α^t under B, with $t = 0, 1, \ldots, 4$. We already listed in (16.3.15) the five such polynomials that occur for $t = 0$, so our task is really to find those that appear for $t \neq 0$, and in particular to show that there are always precisely four. Explicitly, the four polynomials that arise in the four possible sectors are as follows:

$$t = 1 : (Z_1^3 Z_3 Z_5 + \ldots) \qquad\qquad t = 2 : (Z_1^3 Z_2^2 + \ldots)$$
$$(Z_1^3 Z_4^2 + \ldots) \qquad\qquad\qquad (Z_1^3 Z_4 Z_5 + \ldots)$$
$$(Z_1^2 Z_2^2 Z_5 + \ldots) \qquad\qquad\quad (Z_1^2 Z_2^2 Z_4 + \ldots)$$
$$(Z_1^2 Z_2 Z_3 Z_4 + \ldots) \qquad\qquad (Z_1^2 Z_2 Z_3 Z_5 + \ldots)$$

$$\text{(16.4.13)}$$

$$t = 3 : (Z_1^3 Z_5^2 + \ldots) \qquad\qquad t = 4 : (Z_1^3 Z_3^2 + \ldots)$$
$$(Z_1^3 Z_2 Z_3 + \ldots) \qquad\qquad\quad (Z_1^3 Z_2 Z_4 + \ldots)$$
$$(Z_1^2 Z_3^2 Z_5 + \ldots) \qquad\qquad\quad (Z_1^2 Z_2^2 Z_3 + \ldots)$$
$$(Z_1^2 Z_2 Z_4 Z_5 + \ldots) \qquad\qquad (Z_1^2 Z_3 Z_4 Z_5 + \ldots).$$

In each case the symbol '$+\ldots$' represents the addition of terms obtained by adding cyclic permutations of the variables. This confirms that in each sector with $t \neq 0$, there are four left-handed massless multiplets and no right-handed multiplets, while for $t = 0$ there are five left-handed massless multiplets and one right-handed massless multiplet.

Let us now discuss the physical implications of this. The choice of U_B was made to ensure that color nonsinglets have $t \neq 0$. However, some of the color singlets have $t = 0$. Consequently, above and beyond the four E_6 generations that are predicted in this toy model by an index theorem, we get additional massless particles that are color singlets. In fact, these extra states transform under the weak-interaction group $SU(2)_L \times SU(2)_R$ as $(1/2, 1/2) \oplus (0, 0)$, since those are the representations that with our choice of U_B correspond to $t = 0$. The $(1/2, 1/2)$ modes are particularly interesting, since they have the quantum numbers (and couplings) of conventional weak doublets.

Now, let us discuss the physical problem to which we would like to apply these results. One of the classic problems of grand-unified theories is the so-called fine-tuning problem. Why is the weak doublet so light (its mass is at most of order 1 TeV) compared to the fundamental mass scales of grand unification or gravity? Its lightness, unlike the lightness of the observed quarks and leptons, cannot be interpreted as a consequence of gauge invariance. What makes this problem particularly worrisome is the fact that the color triplet scalars that are related by the grand-unified group to the usual Higgs bosons must be extremely heavy – or they would mediate proton decay at an unacceptable rate.

Without supersymmetry, no reasonable answer is known to these questions in the context of four-dimensional grand unification.[*] In supersym-

---

[*] Models with dynamical breaking of weak interaction symmetries are a possible

metric grand-unified theories in four dimensions, the first part of the problem is greatly alleviated; it is not so difficult to understand the lightness of the Higgs doublet, since it is in a multiplet with fermions that can be light for various reasons, and since nonrenormalization theorems will keep the Higgs doublet massless to all orders if it is massless at tree level. However, it is quite awkward in supersymmetric grand unification to explain why the Higgs doublet is light while at the same time its color triplet partners are superheavy.

In one important respect, supersymmetry makes the problem worse. In supersymmetric theories, the quarks have color triplet partners, which cannot be heavier than the scale of supersymmetry breaking. If supersymmetry survives to low energies, the scalar partners of quarks will mediate proton decay at a hopelessly unacceptable rate unless this is somehow prevented. In conventional supersymmetric grand unification, this is prevented by postulating a discrete symmetry that forbids the unwanted couplings. This may be, for instance, a Z_2 symmetry X under which quark and lepton superfields Q and L are odd, while Higgs superfields H are even. Such a symmetry permits the desired superfield couplings $H\overline{L}L$ and $H\overline{Q}Q$, while forbidding unwanted couplings QQQ.

Let us now investigate these issues in the context of the superstring models we have been discussing. Without any special contrivance, with only a favorable choice of the discrete parameters that enter in defining the symmetry breaking pattern, we have obtained massless 'Higgs bosons' whose color triplet partners are superheavy. These are the extra massless modes, not predicted by index theorems, that appear at $t = 0$ in the above construction. This is the part of the problem that is difficult in supersymmetric grand unification in four dimensions. It remains to explain why color triplet partners of quarks do not mediate proton decay. This is conventionally the easy part of the problem; in the context of four-dimensional field theory, one may arbitrarily postulate the desired symmetries. Such arbitrary choices are not available here. We must instead find the desired symmetries by, for instance, finding that they automatically appear in a suitable manifold of $SU(3)$ holonomy. With this and other motivations, we will investigate the global symmetries that arise in string compactification in §16.5 below. We will see that our four generation toy model comes relatively close to having the right global symmetries but just misses. A study of the Georgi–Quinn–Weinberg calculation in §16.9 will reveal another problem with the toy model.

---

approach, but are afflicted with severe phenomenological difficulties and seem well-nigh incompatible with grand unification.

It might be useful to explain in another way 'why' symmetry breaking by Wilson lines alleviates the usual problem in obtaining light Higgs doublets whose color triplet partners are superheavy. The problem arises conventionally because the desired hierarchy of scalar masses would arise only if Higgs bosons that break the grand-unified symmetry have just the right expectation values. Typically, the standard electroweak doublet would be massless only if a Higgs boson ϕ in the adjoint representation of the grand-unified group has just the right eigenvalues; and this is an artificial requirement, since those eigenvalues depend on freely adjustable coupling constants in the underlying Lagrangian. Symmetry breaking by Wilson lines imposes a sort of quantization condition on the ϕ eigenvalues, since ϕ is replaced by a group element U that obeys a condition such as $U^n = 1$. It is not surprising that this natural quantization of Higgs eigenvalues gives possibilities for solving the fine-tuning problem.

16.4.4 Fractional Electric Charges

One of the themes of this section has been to describe the relation of the superstring models under discussion to conventional grand unification. We have so far discussed the analogs of most of the characteristic predictions of grand unification, but one remains to be considered. One of the characteristic features of grand unification is that the quantization of electric charge is explained; unification of a four-dimensional theory in a unified group G predicts that the quantum of electric charge is the smallest charge that appears in any representation of G. The real reason for this prediction is that conventional gauge symmetry breaking by Higgs bosons could be turned off continuously by adjusting coupling constants to make the Higgs expectation value go to zero. It must therefore be possible for the electric charges to go over smoothly to charges that would be allowed with unbroken G or in other words charges that occur in representations of G. When the fundamental group of a manifold is finite, however, the Wilson lines U obey quantization conditions such as $U^n = 1$, and cannot be turned off continuously. Consequently, the question of quantization of electric charge must be reconsidered.

In field theory, as opposed to string theory, symmetry breaking by Wilson lines does not disturb the standard quantization of electric charge. The reason for this is that in field theory, passing from a simply connected manifold K_0 to a quotient $K = K_0/F$ simply means throwing away some of the states – those that do not obey (16.4.2). Since the states on K_0 fit into G multiplets, they all obey the usual quantization of electric charge, and the same will therefore be true for the subset of states that is kept in

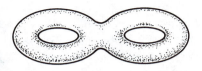

Figure 16.2. On a manifold that is not simply connected, a closed string may wrap around a noncontractible loop.

working on K and imposing (16.4.2).

In string theory, however, it is not true that the states on $K = K_0/F$ are just a subset of the states that would arise if the theory were formulated on K_0. We have considered the noncontractible loops in K as technical ingredients in understanding symmetry breaking, but they also have an important physical role. A closed string could wrap around such a noncontractible loop, as shown in fig. 16.2, giving a new branch of the physical spectrum that arises only when the theory is formulated on K instead of K_0. In a theory that contains closed strings only, a closed string wrapping around a noncontractible loop cannot break, so the lightest state in this sector must be stable. If the radius of K is of order r, the mass of the lightest state in the winding sector will be of order r/α', so if the expansion parameter α'/r^2 in a field-theoretic description is small, these states have masses at least of order $1/\sqrt{\alpha'}$. This means that in the heterotic string theory, in which closed strings carry electric charges, we must re-examine the question of electric charge quantization. In the type I case, the usual quantization of electric charge seems to be valid.

Before attempting a calculation, let us point out some general facts that constrain the answer that may emerge. Let γ be the homotopy class of a closed string in $\pi_1(K)$. Any two closed-string states in the same homotopy class can be related to one another by emission and absorption of 'ordinary' closed strings, so their electric charges must differ by integers. The difference between the electric charge of a color singlet closed string and an integer multiple of the charge of the electron must depend only on the homotopy class of the the closed string. Let μ_γ be this charge defect of a closed string of homotopy class γ, that is, its difference from being an integer. When two closed strings of homotopy classes γ and γ' join, they form the homotopy class $\gamma\gamma'$. Conservation of electric charge means that the charge defect must add in this process, so

$$\mu_\gamma + \mu_{\gamma'} = \mu_{\gamma\gamma'} \ (\mathrm{mod}\,1). \qquad (16.4.14)$$

The mod 1 in this equation reflects the fact that μ is only defined modulo an integer. Suppose now that the fundamental group of K is finite. Then given any homotopy class γ, there is some integer n such that wrapping n times around gamma would give a contractible loop, $\gamma^n = 1$. Equation (16.4.14) tells us then that $n \cdot \mu_\gamma = 0 \bmod 1$ or in other words that $\mu_\gamma = k/n$ for some integer k. We would like to calculate k.

To do so, we will work in the fermionic description of the $E_8 \times E_8$ heterotic superstring in which E_8 is realized on sixteen left-moving fermions λ^A, $A = 1, \ldots, 16$. We can diagonalize the Wilson line U_γ by passing to a complex basis of eight λ_i, $i = 1, \ldots, 8$ and their complex conjugates λ_i^*, with

$$U_\gamma \lambda_i = e^{2i\pi m_i/n} \lambda_i. \tag{16.4.15}$$

for some integers m_i. As the electric charge operator Q commutes with U_γ (otherwise electromagnetism is spontaneously broken!) we can simultaneously diagonalize Q along with U:

$$Q\lambda_i = q_i \lambda_i, \tag{16.4.16}$$

where the numbers q_i are the electric charges of the λ_i. For simplicity we will assume that U_γ and Q are embedded in just one E_8.

Since the electric charge defect μ_γ is a topological invariant, to compute it we may as well consider a string that is interacting with the Wilson line U only. Consider the λ's in the NS-like sector in which they obey antiperiodic boundary conditions. For $U = 1$, the boundary condition is $\lambda_i(\sigma + \pi) = -\lambda_i(\pi)$, and the energy levels are proportional to $\epsilon = 2k + 1$, with k being an arbitrary integer. For more general U, the boundary condition is $\lambda_i(\sigma + \pi) = -U\lambda_i(\sigma)$ or – in terms of (16.4.15)– it is

$$\lambda_i(\sigma + \pi) = -e^{2i\pi m_i/n} \lambda_i. \tag{16.4.17}$$

The energy levels of λ_i are then

$$E_k(m_i) = 2(k - m_i/n) + 1, \tag{16.4.18}$$

where k is still an arbitrary integer. The charge defect μ_γ is simply the electric charge of the 'filled Dirac sea', normal ordered by subtracting the value at $m_i = 0$. The charge Q_i of the Dirac sea made from negative energy fermions of type λ_i is simply the charge q_i times the 'number of negative-energy states'. The latter, of course, requires some regularization, so we write tentatively

$$Q_i = q_i \lim_{\epsilon \to 0} \sum_{k<0} e^{-\epsilon E_k(m_i)}, \tag{16.4.19}$$

where the factor of $e^{-\epsilon E_k}$ is included to regulate the contribution of the states of very large negative energy. Upon explicitly subtracting the value at $m_i = 0$, (16.4.19) is replaced by

$$Q_i = q_i \lim_{\epsilon \to 0} \sum_{k<0} \left(e^{-\epsilon E_k(m_i)} - e^{-\epsilon E_k(0)} \right). \qquad (16.4.20)$$

With E_k in (16.4.18), one can easily compute the sum and limit in (16.4.20) to get

$$Q_i = q_i m_i / n. \qquad (16.4.21)$$

Summing (16.4.21) over i, we finally find the value of the charge defect. It is

$$\mu_\gamma = \frac{1}{n} \sum_i q_i m_i. \qquad (16.4.22)$$

In particular, we see that the usual quantization law of electric charge is in general not obeyed. $\sum_i q_i m_i$ is an integer multiple of the usual charge quantum[*], but the factor of $1/n$ in (16.4.22) means that strings that wind around a noncontractible loop can have an electric charge that is n times smaller than the charges of other unconfined color singlet particles.

This completes our discussion of quantization of electric charge. However, the realization that the quantum of electric charge can be smaller than the charge of the electron raises an immediate question, 'What happened to the monopoles?' In any $U(1)$ gauge theory one can define at infinity the Dirac monopole field. Quantum-mechanical coupling of a monopole of charge g to a charged particle of charge q is consistent if the Dirac condition $g = 2\pi/q \cdot$ integer is obeyed. If q is n times smaller than the charge e of the electron, g must be n times larger than the conventional Dirac quantum $2\pi/e$. How can this come about?

In a $U(1)$ gauge theory the monopole gauge field defined at infinity cannot be extended throughout all space without encountering a singularity. In the early 1970's, Polyakov and 't Hooft showed that in four-dimensional unified theories one can always 'unwrap' the monopole field and obtain a nonsingular configuration whose magnetic charge equals the Dirac quantum. If unification occurs not in four dimensions but only in ten dimensions, then the discussion is modified. In this case we cannot unwrap the monopole in the four-dimensional unified gauge group, because there is none. Symmetry breaking by Wilson lines means that there is no approximation in which the theory is a four-dimensional theory with

---

[*] With the standard embedding of color $SU(3)$ in $SO(16)$ or E_8, three colors of quarks with $q_i = 1/3$ and the same value of m make an integer contribution to $\sum q_i m_i$.

E_6 symmetry. E_6 is valid only in ten dimensions, and one must try to unwrap the monopole in the ten-dimensional gauge group. To be precise, in a ten-dimensional theory one must start with the Dirac monopole not on the sphere at infinity S^2 but on $S^2 \times K$, with K being the Kaluza–Klein space; and one tries to extend it without singularity from $S^2 \times K$ to $R^3 \times K$, with R^3 being ordinary three-dimensional space. The topological problem is completely different, and there is no reason to expect the answer to be the same. Indeed, the solution of the topological problem, which involves techniques that we will not be able to develop here, shows that the minimum magnetic charge is always $n \cdot 2\pi/e$ when the minimum electric charge is e/n.

16.4.5 Discussion

We have seen that symmetry breaking by Wilson lines preserves the successes of conventional grand unification. At least as regards counting of states, the states fit into representations of the unified group, plus possible real representations. The conventional renormalization group calculation of $\sin^2 \theta$ is valid. On the other hand, there are some interesting differences. Yukawa couplings do not obey group-theory relations (but other relations that we will begin to explore presently). There are new possibilities for a solution of the fine-tuning problem. Quantization of electric charge and magnetic charge does not take its usual form. Most of the differences between conventional grand unification and symmetry breaking by Wilson lines stem from the fact that the Wilson lines obey a natural quantization law, in contrast to the Higgs bosons of conventional grand-unified symmetry breaking.

One principal attraction of symmetry breaking by Wilson lines is that one does not introduce Higgs bosons or a Higgs potential but uses ingredients that are present anyway (if K is not simply connected). The other principal attraction is that symmetry breaking by Wilson lines depends only on discrete choices and so is quasi-topological in character. This opens the possibility that it may one day be possible to understand symmetry breaking by Wilson lines on general, qualitative grounds rather than by trying to solve a difficult dynamical problem.

We have discussed symmetry breaking by Wilson lines as if it is the only mechanism for grand-unified symmetry breaking. It is possible that other mechanisms are at work as well. Later, we will discuss another mechanism, involving flat directions in superpotentials, for breaking E_6. This latter mechanism can very naturally break E_6 to $SO(10)$ or $SU(5)$. It may be that only the latter group should be broken by Wilson lines.

Most of our discussion would carry over to those cases without too much modification.

16.5 Global Symmetries

In our discussion of the fine-tuning problem, we found that we could readily obtain massless Higgs doublets with superheavy color triplet partners, but that a satisfactory understanding of the longevity of the proton would require a mechanism to suppress cubic couplings of quark superfields. Global symmetries are one possible mechanism for doing this; they are interesting in many other respects as well. The present section will be devoted to a discussion of the global symmetries that arise in manifolds of $SU(3)$ holonomy. We begin with some generalities.

16.5.1 CP Conservation in Superstring Models

A manifold K of $SU(3)$ holonomy contains certain covariantly constant tensor fields. Prominent among these are the Kähler form k, which is a $(1,1)$ form, and the holomorphic volume form ω, which is a $(3,0)$ form. If K has holonomy $SU(3)$ (rather than a subgroup) k is up to normalization the only covariantly constant $(1,1)$ form. Therefore, any isometry f of K must map k into a multiple of itself. Since k is real, the multiple is ± 1.

If $f \cdot k = +k$, then f preserves the complex structure $J^i{}_j = g^{im}k_{mj}$ and is called a holomorphic mapping of K into itself. It maps local holomorphic coordinates into local holomorphic coordinates. If, on the other hand, $f \cdot k = -k$, then f is an antiholomorphic mapping that changes the sign of the complex structure and exchanges holomorphic and antiholomorphic coordinates. A manifold of $SU(3)$ holonomy may or may not admit an antiholomorphic isometry. For instance, the hypersurface in CP^4 defined by zeros of the polynomial

$$P = \sum_{i=1}^{5} Z_i^5 + \mu \, Z_1 Z_2 Z_3 Z_4 Z_5 \tag{16.5.1}$$

is invariant under the antiholomorphic transformation $Z_i \to Z_i^*$ if μ is real. But it does not possess this or any other antiholomorphic symmetry for general complex μ.

The qualitative significance of an antiholomorphic symmetry springs from the fact that the Levi–Civita six form ϵ, which expresses the orientation of K, may be written $\epsilon = k \wedge k \wedge k$. Thus, in three complex dimensions, a transformation that reverses the sign of k is orientation-reversing. Physically, this corresponds to the fact that antiholomorphic symmetries are related to CP invariance, as we will now see.

Let us think about string theory compactified on $M^4 \times K$. With the exception of the type IIA theory, ten-dimensional superstring theories do not conserve parity in the ten-dimensional sense; they are not invariant under transformations (even isometries) that reverse the orientation of $M^4 \times K$. Such transformations would exchange massless fermions in the **16** of the tangent space group $SO(1,9)$, which exist in the theory, with fermions in the **16'**, which do not exist. An orientation reversing isometry $\hat{C}$ of K reverses the orientation of $M^4 \times K$, and so is not a symmetry of the theory. Likewise, the ordinary parity transformation P of the uncompactified coordinates, namely $(t, x, y, x) \rightarrow (t, -x, -y, -z)$, reverses the orientation of M^4 and also of $M^4 \times K$, so it is not a symmetry. However, the combination $\hat{C}P$ reverses the orientation of both M^4 and K, so it preserves the orientation of $M^4 \times K$, and is a symmetry. As it is the combination of P with an internal symmetry related to complex conjugation, one may suspect that $\hat{C}P$ corresponds to the conventional notion of CP, and that is indeed true. To see this precisely, note that the **16** of $SO(1,9)$ decomposes under $SO(1,3) \times SO(6)$ as $(\mathbf{2}, \mathbf{4}) \oplus (\mathbf{2'}, \overline{\mathbf{4}})$, with **2** and **2'** being left- and right-handed spinors of $SO(1,3)$, and **4** and $\overline{\mathbf{4}}$ being positive and negative-chirality spinors of $SO(6)$. P exchanges the **2** and **2'** of $SO(1,3)$, while an orientation reversing transformation $\hat{C}$ of K exchanges the **4** and $\overline{\mathbf{4}}$. Hence $\hat{C}P$ exchanges the $(\mathbf{2}, \mathbf{4})$ with the $(\mathbf{2'}, \overline{\mathbf{4}})$. We have obtained **27**'s and $\overline{\mathbf{27}}$'s of E_6 from the **4** and $\overline{\mathbf{4}}$ of $SO(6)$, so $\hat{C}P$ exchanges left-handed **27**'s with right-handed $\overline{\mathbf{27}}$'s, and deserves to be called CP.

16.5.2 R Transformations in Superstring Models

We now consider holomorphic symmetries that leave the Kähler form k invariant. A holomorphic transformation f maps $(3, 0)$ forms into $(3, 0)$ forms, so it must map the covariantly constant $(3, 0)$ form ω into a multiple of itself:

$$f \cdot \omega = e^{i\beta} \, \omega \qquad (16.5.2)$$

for some β. In §16.3.2, we showed that $\beta = 0$ if f acts freely, but this is not true in general. The value of β has a qualitative significance. The three form ω is related to a covariantly constant positive-chirality spinor η by $\omega_{ijk} = \eta^T \Gamma_{ijk} \eta$. Evidently, $f \cdot \eta = \pm e^{i\beta/2} \eta$. (The sign is a matter of convention since f is uniquely defined only modulo a 2π rotation.) The

covariantly constant spinor η generates unbroken supersymmetry transformations. The unbroken supersymmetries therefore transform the same
way that η does. If $\beta \neq 0$, then f does not commute with the unbroken
supersymmetries in four dimensions, but rather rotates them by a phase of
$e^{\pm i\beta/2}$ for positive or negative chirality. Such a symmetry is usually called
an R symmetry in discussions of four-dimensional supersymmetry. Under
an R symmetry, the superpotential W is not invariant but is rotated by
a phase. In the next subsection, we will give examples of holomorphic
transformations f that are or are not R symmetries.

16.5.3 Global Symmetries of the Toy Model

Analysis of global symmetries in manifolds of $SU(3)$ holonomy leads to
a surprisingly rich structure. To get some experience, let us consider our
familiar practice case, the quintic hypersurface Q in CP^4, defined by zeros
of a quintic polynomial. To begin with, we take the minimal form for this
polynomial:

$$P = Z_1^5 + Z_2^5 + \ldots + Z_5^5. \tag{16.5.3}$$

To find symmetries of the hypersurface, we look merely for linear transformations of the Z_i that map P into itself, up to phase. There are many
of these. P is invariant under the 120 permutations of the Z_k. It is also
invariant under phase transformations $Z_k \to \alpha^{m_k} Z_k$, with $\alpha = e^{2i\pi/5}$ and
m_k arbitrary elements of Z_5 (arbitrary integers modulo five). Since an
overall scaling of the Z_k is irrelevant in CP^4, the transformations with
exponents $\{m_k\}$ and $\{m_k + 1\}$ are equivalent; only four of the five m_k
are independent. This leaves a symmetry group with $120 \times 5^4 = 75,000$
elements. This is the symmetry group of the 100 generation model that
arises from the polynomial in (16.5.3).

We would like to know which of these symmetries are R symmetries
and more generally how the holomorphic three form ω transforms under
them. In §15.4.3, we derived several explicit expressions for ω. One of
these, specialized to the present choice of P, is

$$\omega = dZ_1 \wedge dZ_2 \wedge dZ_3/Z_4^4. \tag{16.5.4}$$

Using this form of ω we can read off the fact that ω changes sign under
the transposition $Z_1 \leftrightarrow Z_2$ (with Z_k fixed for $k > 2$). It follows from this
because of the underlying permutation symmetry (which is not manifest
in (16.5.4)) that ω is odd under any transposition. Hence ω is odd under
odd permutations of the Z_k and even under even permutations. Thus,
the odd permutations of the Z_k are R symmetries in the four-dimensional

sense. Likewise, we can see from (16.5.4) that under $Z_1 \to \alpha^m Z_1$ (with other Z_k invariant), ω transforms as $\omega \to \alpha^m \omega$. Permutation symmetry implies that more generally under

$$Z_k \to \alpha^{m_k} Z_k \tag{16.5.5}$$

the transformation of ω is

$$\omega \to \alpha^{\sum m_k} \omega. \tag{16.5.6}$$

Thus, a transformation of this type is an R symmetry in four dimensions unless $\sum m_k = 0$.

We are actually more interested in the symmetries of the four generation model based on Q/F, where F is the $Z_5 \times Z_5$ group generated by $A(Z_k) = Z_{k+1}$ and $B(Z_k) = \alpha^k Z_k$. If we are given a linear transformation U of the Z_i that leaves P invariant, under what conditions does this yield a symmetry not only of Q but also of Q/F? At first sight one might believe that the requirement is that U should commute with A and B. If so, Q/F would have no symmetries, since a 5×5 matrix that commutes with A and B must be a multiple of the identity, and this is irrelevant in CP^4.

It is not true, however, that U must commute with A and B to give a symmetry of Q/F. Let us consider the general situation of a manifold K_0 and a symmetry U. We would like to know whether U is a symmetry not just of K_0 but also of $K = K_0/F$, where F is some group that acts freely on K_0. Let x be a point in K_0. In K, x is equivalent to fx, for any $f \in F$. So the action of U makes sense in K only if Ux is equal (in K) to $U \cdot fx$. In other words, U makes sense as a transformation of K only if for every $x \in K$ and $f \in F$, there is some $f' \in F$ such that $U \cdot fx = f' \cdot Ux$. This means, in other words, that for every $f \in F$, UfU^{-1} must be an element f' of F.

Going back to our example, with K_0 being the hypersurface and F the group generated by A and B, a symmetry U of the four generation model must be a linear transformation of the Z_k that obeys

$$UAU^{-1} = A^k B^l$$
$$UBU^{-1} = A^m B^n. \tag{16.5.7}$$

Here k, l, m, and n are integers that are defined only modulo five, so we may think of them as elements of the group Z_5. Because A and B obey the commutation relation $BA = AB\alpha$, it follows that k, l, m and n will necessarily obey $kn - lm = 1$. This means that the matrix

$$\begin{pmatrix} k & l \\ m & n \end{pmatrix} \qquad (16.5.8)$$

has determinant one, and so belongs to the group $SL(2, Z_5)$ of 2×2 matrices with entries in Z_5 and determinant one. By analogy with $SL(2, Z)$, we might call this 'the modular group over Z_5'. Simple counting shows that there are 120 elements of this group.

For any choice of $k, l, m,$ and n, there can be (up to normalization) at most one matrix that obeys (16.5.7). Indeed, if U and $\tilde{U}$ both obey (16.5.7) (with the same $k, l, m,$ and n) then $U^{-1}\tilde{U}$ commutes with A and B and so must be a multiple of the identity. It is also true that a suitable 5×5 matrix exists for any choice of $k, l, m,$ and n. In fact, like the ordinary modular group, the modular group over Z_5 is generated by the matrices

$$S = \begin{pmatrix} 0 & -1 \\ 1 & 0 \end{pmatrix}, \quad T = \begin{pmatrix} 1 & 1 \\ 0 & 1 \end{pmatrix}. \qquad (16.5.9)$$

To show that there is a linear transformation of the Z_k corresponding to every element of $SL(2, Z_5)$, it is enough to find linear transformations corresponding to S and T. These are

$$\hat{S} : Z_k \to \sum_m \alpha^{km} Z_m / \sqrt{5} \qquad (16.5.10)$$

and

$$\hat{T} : Z_k \to \alpha^{3k^2} Z_k, \qquad (16.5.11)$$

as the reader can verify. Henceforth we will drop the hat and refer to $\hat{S}$ and $\hat{T}$ as S and T.

So far, we have merely established that the group $H = SL(2, Z_5)$ transforms the fundamental group F of Q/F in a sensible way. We still must understand the condition under which an element of H is actually a symmetry of Q/F – and the physical role of elements of H that are not symmetries. To this end, it is useful to consider not just the particular polynomial P, but the general homogeneous quintic polynomial that is invariant under F. The invariant polynomials were already listed in (16.3.15), in the process of computing Hodge numbers of Q/F. The general invariant quintic is

$$
\begin{aligned}
\tilde{P}(Z_1,\ldots,Z_5) =& C_0 \sum Z_k^5 + C_1(Z_1^3 Z_2 Z_5 + \ldots) + C_2(Z_1^3 Z_3 Z_4 + \ldots) \\
& + C_3(Z_1^2 Z_2 Z_3^2 + \ldots) + C_4(Z_1^2 Z_2^2 Z_4 + \ldots) \\
& + C_5 Z_1 Z_2 Z_3 Z_4 Z_5,
\end{aligned}
$$

$$(16.5.12)$$

where the C_i are complex parameters. By scaling of the Z_k one may set, say, $C_0 = 1$, but the other five parameters in $\tilde{P}$ are really independent, corresponding to the fact that the model has five generations (and one antigeneration).

For any choice of the C_i, the equation $\tilde{P} = 0$ defines a manifold that admits a unique metric of $SU(3)$ holonomy. As the C_i vary, one obtains a five-complex-parameter family of such metrics. The existence of this multiparameter family of metrics corresponds, physically, as we have discussed in §16.2.2 above, to the existence of five massless superfields in the low-energy theory. We can think of the five independent C_i as physical fields. Let N be the parameter space of the C_i. Long wavelength oscillations in the shape of the compact manifold Q/F are governed by a supersymmetric nonlinear sigma model of maps of four-dimensional space-time into N. Any point in N corresponds to a possible vacuum state. Within the context of our toy model, an important element of understanding low-energy physics is to determine the expectation values of the C_i or in other words to understand what point in N corresponds to the physical vacuum.

If U belongs to H and the quintic polynomial $\tilde{P}$ is invariant under A and B, then it follows from (16.5.7) that $U\tilde{P}$ is also invariant under A and B. Thus U induces a linear transformation of the C_i, and hence a mapping of N into itself. (The reader may wish to determine which transformation of the C_i is brought about by S and T; it is particularly simple for T.) This means that H maps the parameter space N into itself, and consequently that H is a symmetry group of the nonlinear sigma model based on N and consequently of the physical theory based on reducing ten dimensions to $M^4 \times (Q/F)$. Thus, our four generation toy model has the nonabelian global symmetry group H with 120 elements!

There is certainly a dramatic contrast between the sort of models under discussion here and conventional grand unification. Conventionally, global symmetries and especially elaborate global symmetry groups seem somewhat artificial and *ad hoc*, but here the simplest model we could find with a reasonably small number of generations automatically possesses the rather intricate discrete symmetry group H. However, most of the H symmetry is spontaneously broken. If a particular vacuum state in four dimensions is picked, corresponding to a particular choice of F-invariant polynomial (or to a particular point in N), then H is broken down to a subgroup that leaves fixed this polynomial.

There is a Z_2 subgroup of H that is always unbroken (in the absence of Wilson lines). It is generated by $\phi : Z_k \to Z_{-k}$. This corresponds, in the sense of (16.5.8), to the matrix

$$\begin{pmatrix} -1 & 0 \\ 0 & -1 \end{pmatrix}, \qquad (16.5.13)$$

which is the one nontrivial element of the center of H. It is easily seen to leave the general polynomial in (16.5.12)invariant. At first sight, this looks rather like what we wanted in our discussion of the fine-tuning problem in §16.4. A Z_2 symmetry has made its appearance, not because we wanted it, but because the quotient of a quintic hypersurface in CP^4 by $Z_5 \times Z_5$ always has Z_2 symmetry.

Particular choices of polynomial give larger unbroken subgroups of H. For instance, let us return finally to the minimal polynomial $P = \sum Z_k^5$. This is invariant under a nonabelian subgroup of H with 20 elements. This group is generated by T and the transformation $W : Z_k \to Z_{2k}$. (Notice that $W^2 = \phi$.) The group H_0 generated by T and W is characterized by $T^5 = W^4 = 1$, $WT = T^{-1}W$. In view of (16.5.6) and the preceding discussion, it follows that W is an R symmetry (the holomorphic three form is odd under W) but ϕ and T are not.

The most general four generation model with H_0 unbroken corresponds not to the minimal polynomial P but to

$$P' = \sum Z_k^5 + \mu\, Z_1 Z_2 Z_3 Z_4 Z_5. \qquad (16.5.14)$$

In the usual modern language, it would not be natural to assume that the vacuum is described by the polynomial P, since this is not singled out by any symmetry (of a four generation model). It *is* natural to assume that the vacuum is determined by a polynomial of the form P', since this is the most general form (of a four generation model) that leaves H_0 unbroken. The latter statement amounts to the observation that the polynomials multiplying C_1, C_2, C_3, and C_4 in (16.5.12) are not invariant under H_0. They are, in fact, not invariant under T, so (16.5.14) can be selected by asking for an unbroken global Z_5 symmetry generated by T.

If Wilson lines are included, an unbroken symmetry must be required to leave them invariant, modulo a possible gauge transformation. For instance, the Wilson lines introduced in (16.4.12) are invariant under T, so with this symmetry breaking choice there is still an unbroken Z_5 symmetry that leaves (16.5.14) a technically natural choice of vacuum state. As for W, it does not leave invariant the Wilson lines in (16.4.12) (even modulo a gauge transformation), and so is not a symmetry after E_6 is broken in this way. But $\phi = W^2$ is a symmetry, if accompanied by a suitable gauge transformation V_ϕ. The requisite gauge transformation may be described by saying that the $SU(3)_C \times SU(3)_L \times SU(3)_R$ element $U_1 \otimes U_2 \otimes U_3$ is mapped into $U_1^{-1} \otimes U_3 \otimes U_2$; there is an element of E_6 that does this. It may be noted that V_ϕ can be interpreted in four dimensions as a generalized charge conjugation that exchanges quarks with antiquarks and exchanges $SU(2)_L$ with $SU(2)_R$, so ϕ is a kind of charge conjugation symmetry. Its appearance is a fairly natural, though not particularly desirable, feature of models in which the $SU(2)_L$ symmetry of weak interactions is extended to $SU(2)_L \times SU(2)_R$.

16.5.4 Transformation Laws of Matter Fields

We will next determine how the massless charged fields of the toy model transform under the global symmetries we have described. The goal is to obtain constraints on the Yukawa couplings. The basic tool is the relation between massless chiral fields and polynomials. There is indeed a one-to-one correspondence between massless **27**'s in four dimensions and allowed perturbations δP of the polynomial P that enters in defining the toy model. We will consider the case in which P is of the form (16.5.14), so that there is an interesting unbroken discrete symmetry group.

In the absence of Wilson lines that break E_6, the allowed perturbations of P are the quintic polynomials invariant under $Z_5 \times Z_5$. These have already entered in the above discussion, and are

$$
\begin{aligned}
\psi_2 &= Z_1^2 Z_2 Z_3^2 + \ldots \\
\psi_{-2} &= Z_1^2 Z_2^2 Z_4 + \ldots \\
\psi_1 &= Z_1^3 Z_2 Z_5 + \ldots \\
\psi_{-1} &= Z_1^3 Z_3 Z_4 + \ldots \\
\psi_0 &= Z_1 Z_2 Z_3 Z_4 Z_5.
\end{aligned}
\tag{16.5.15}
$$

Here '$+\ldots$' refers to the addition of terms obtained by cyclic permutation of variables. For each polynomial in (16.5.15), there is a corresponding massless **27** in the toy model, which transforms under the unbroken symmetry group H_0 in the way described in the previous section. We can therefore readily read off the transformation law of the **27**'s.[*] The ψ_k have been named so that under T, ψ_k transforms to $\alpha^k \psi_k$. At the same time, W maps ψ_k to ψ_{-k}, and the ψ_k are all invariant under $\phi = W^2$.

This completes the description of the transformation law of the **27**'s under H_0. The toy model contains in addition a single $\overline{\bf 27}$, which we will call χ, corresponding to the Kähler form. As the Kähler form is even under holomorphic transformations, χ is invariant under H_0.

Now let us discuss what happens after E_6 breaking. An interesting example to discuss is the E_6 breaking model in equation (16.4.12) above. With this choice, color singlet states that transform as $(1/2, 1/2) \oplus (0,0)$ under $SU(2)_L \times SU(2)_R$ are unaffected by the Wilson lines, and are governed by our above comments. These are the massless Higgs bosons. As for other components of the **27**, they are described not by the polynomials in (16.5.15) but by the more general polynomials in (16.4.13). For each $SU(3)_C \times SU(2)_L \times SU(2)_R \times U(1) \times U(1)$ multiplet other than the Higgs bosons, there are four relevant polynomials instead of the five in (16.5.15) (this is the statement that there are extra massless Higgs bosons whose color triplet partners are superheavy). We will leave it to the reader to show from (16.4.13) that the four polynomials in any sector transform under T as $\alpha^2, \alpha, \alpha^{-1}$, and α^{-2}.

We recall now from our discussion of the fine-tuning problem that to prevent baryon-number-violating cubic couplings of quark and lepton superfields, while allowing quark and lepton masses, it would be desirable to find a global symmetry X that would transform the quarks and leptons nontrivially while leaving some of the Higgs bosons invariant. In supersymmetric grand unification in four dimensions, X can be arbitrarily postulated, but in the superstring models under discussion, X must be found. We now see that the global symmetry T of the toy model has many, but not all, of the requisite features. It is true that the quarks and leptons all transform nontrivially under T, and it is true that there are Higgs bosons (coming from the polynomial ψ_0 in (16.5.15), which does not have an analog in the other sectors) that are neutral under T. The T assignments of the quarks and leptons are such as to forbid many, but not all, of the baryon-number-violating couplings. A cubic coupling of three

---

[*] Things would not be so straightforward if we were dealing with R symmetries, since the scalar and spinor components of a **27** transform differently under an R symmetry, and we would have to decide which if either transforms like the corresponding polynomial. To understand constraints on Yukawa couplings due to R symmetries is somewhat subtle, and is best done using the formula for Yukawa couplings given in the next section.

quark superfields that transform as α^2, α^2, and α, or as α, α, and α^{-2}, is permitted by T.

One might wonder whether the Z_2 symmetry ϕ, which is an unavoidable symmetry of the four generation toy model (before E_6 breaking), might prevent proton decay. After all, it is a Z_2 symmetry that is usually postulated for this purpose phenomenologically. However, before E_6 breaking, the fact that ϕ is always unbroken means that all allowed perturbations of the defining polynomial P and hence all massless 27's are even under ϕ, which therefore gives no constraints on Yukawa couplings. E_6 breaking turns ϕ into a C symmetry, as was described in the last section; a C symmetry can at most relate decays of protons to decays of antiprotons. (Other patterns of E_6 breaking would spoil ϕ conservation altogether.)

In field theory, one would expect that all Yukawa couplings allowed by the unbroken symmetries are nonzero. Is this also true in string theory? We will see in the next section that there are additional restrictions on Yukawa couplings.

16.6   Topological Formulas for Yukawa Couplings

While four-dimensional grand-unified theories with minimal matter content predict certain relations among Yukawa couplings, we have learned that analogous formulas do not hold in the case of superstring models. This is just as well, since the relations among Yukawa couplings that hold in certain four-dimensional grand-unified theories have always been troublesome. Yet it is natural to ask whether there are some more or less simple statements that can be made about Yukawa couplings in superstring models to replace the group-theory relations that are not valid.

A brief review of some facts about supersymmetry will be useful in both this section and the next one. In four-dimensional supersymmetric theories, the interactions of chiral superfields Φ_k are governed by two types of terms. There are terms that are written as integrals over all of superspace,

$$L_1 = \int d^4\theta\, K(\Phi^k, \overline{\Phi}^{\bar{l}}). \tag{16.6.1}$$

Such terms are often called D terms. There are terms that involve integration over the fermionic coordinates of one chirality only,

$$L_2 = \int d^2\theta\, W(\Phi^k) + \text{c.c.} \tag{16.6.2}$$

These are called F terms. W is called the superpotential and must (for supersymmetry) be a holomorphic function of the complex superfields Φ^k.

By contrast, K in (16.6.1) need not be a holomorphic function of the fields; indeed, a holomorphic or antiholomorphic piece in (16.6.1) would vanish upon doing the $d^4\theta$ integral. (There can be derivative terms in K and W, with some restrictions in the latter case, but these play only a limited role in low-energy physics and we will not need to consider them.) Upon doing the θ integrals, (16.6.1) gives rise to the kinetic energy of the fields,

$$L_{kin} = g_{i\bar{j}}\partial_\mu\phi^i\partial^\mu\overline{\phi}^{\bar{j}} + g_{i\bar{j}}\psi^i\partial\overline{\psi}^{\bar{j}} \qquad (16.6.3)$$

with ϕ^k and ψ^k being scalar and spinor components of Φ^k. Here $g_{i\bar{j}} = \partial^2 K/\partial\Phi^i\overline{\Phi}^{\bar{j}}$ is the 'metric' in 'scalar field space'; thus, one finds by doing the $d^4\theta$ integral in (16.6.1) that the scalar field space is a Kähler manifold with Kähler potential K. Upon doing the $d^2\theta$ integral in (16.6.2), one gets Yukawa terms and a scalar potential:

$$L_{int} = \sum_k \left|\frac{\partial W}{\partial\phi^k}\right|^2 + \left(\frac{\partial^2 W}{\partial\phi^i\partial\phi^k}\psi^i\psi^k + h.c.\right). \qquad (16.6.4)$$

Equation (16.6.4) exhibits the significance of the superpotential. In the absence of other fields or interactions, the condition for unbroken supersymmetry is $\partial W/\partial\phi^k = 0$. Upon coupling to gauge fields, the potential term acquires an extra term

$$\Delta V = \sum_a (D_a)^2. \qquad (16.6.5)$$

Here the sum runs over gauge generators. The D_a are functions of chiral matter fields that transform in the adjoint representation of the gauge group. In renormalizable theories, $D_a = \langle\phi|T_a|\phi\rangle$ with T_a being gauge generators acting on the matter fields ϕ. The general form of the D_a can be much more complicated. Unbroken supersymmetry now requires $D_a = 0$ as well as $\partial W/\partial\phi^k = 0$. Upon coupling to supergravity, (16.6.4) becomes more complicated; the conditions for unbroken supersymmetry and zero cosmological constant become $\partial W/\partial\phi^k = W = D_a = 0$.

We see from (16.6.4) that the cubic terms in W give rise directly to Yukawa-like $\phi\psi\psi$ couplings. We cannot, however, relate these to physical Yukawa couplings without correctly normalizing the Bose and Fermi fields ϕ and ψ. The normalization of ϕ and ψ is determined by the kinetic energy, or, in view of (16.6.3), by the quadratic terms in K. Thus, to predict the Yukawa couplings in four dimensions, we must learn to extract both cubic terms $\phi\psi\psi$ and quadratic terms $\overline{\psi}\partial\psi$ from the underlying ten-dimensional theory.

16.6.1 A Topological Formula for the Superpotential

We will first extract a formula for the cubic $\phi\psi\psi$ terms in the effective four-dimensional Lagrangian, with ϕ and ψ being Bose and Fermi components of chiral superfields. We will work out this formula treating the underlying theory as a ten-dimensional field theory, and later discuss the string-theoretic corrections. The formula that we will extract is valid not only if the spin connection is embedded in the gauge group, but more generally if the ten-dimensional gauge field A_i is any holomorphic connection on a holomorphic vector bundle X that obeys the Donaldson–Uhlenbeck–Yau equation.

A massless fermion ψ in four dimensions arises as a zero mode of the underlying ten-dimensional gluino field χ. Likewise, ϕ originates as a zero mode in the small fluctuations of the ten-dimensional gauge field A_i around its vacuum expectation value. We must therefore look in ten dimensions for an $A \cdot \chi \cdot \chi$ coupling. Such a coupling is not hard to find. It is just the minimal gauge coupling

$$L_{gauge} = \Gamma^{i\alpha\beta} A_i^x \chi_\alpha^y \chi_\beta^z f_{xyz}, \tag{16.6.6}$$

where indices $x, y,$ and z label the adjoint representation of E_8, and f_{xyz} are the E_8 structure constants. ((16.6.6) arises by expanding out the covariant derivative in the Dirac action $\overline{\chi}\slashed{D}\chi$.) Now, the Lagrangian density in (16.6.6) is to be integrated over all ten coordinates to compute the Lagrangian or action functional,

$$S = \int_{M^4 \times K} L_{gauge}. \tag{16.6.7}$$

If we wish to find a four-dimensional Lagrangian that must be integrated over M^4 only to give S, we simply rewrite (16.6.7) as $S = \int_{M^4} \cdot \int_K L_{gauge}$. This shows that the four-dimensional Lagrangian from which Yukawa couplings are to be extracted is

$$L_{Yuk}^{(4)} = \int_K \Gamma^{i\alpha\beta} A_i^x \chi_\alpha^y \chi_\beta^z f_{xyz}. \tag{16.6.8}$$

No assumption of $SU(3)$ holonomy is needed for (16.6.8), but now we wish to specialize to this case.

In the decomposition $\mathbf{16} = (\mathbf{2}, \mathbf{4}) \oplus (\mathbf{2'}, \overline{\mathbf{4}})$ of the spinor of $SO(1, 9)$ in terms of $SO(1, 3) \times SO(6)$, the left-handed spinors in the four-dimensional sense are the $(2', \overline{4})$ component. Under the assumed holonomy group $SU(3)$ of the compact manifold K, the $\overline{4}$ transforms as $1 \oplus \overline{3}$. As we have seen in §16.2, the only zero modes that arise from the singlets of $SU(3)$ are gluinos whose Yukawa couplings are determined by supersymmetry and gauge invariance. We will therefore concentrate on the $\overline{3}$ part, which is equivalent to a $(0,1)$ form $\chi_{\overline{a}}^x$ with an extra index x that labels the adjoint representation of E_8. The ten-dimensional gauge field A_i has $(1,0)$ and $(0,1)$ pieces, transforming as 3 and $\overline{3}$ of $SU(3)$. As it is not possible to make an $SU(3)$ singlet from $3 \times \overline{3} \times \overline{3}$, it is the $(0,1)$ part of A_i that we must consider.

In $SU(3)$ group theory, the unique singlet that can be made from $\overline{3} \times \overline{3} \times \overline{3}$ is the completely antisymmetric combination. Thus, even without detailed study of the Dirac algebra, we can read off from (16.6.8) the form

$$L_{Yuk}^{(4)} = \int_K \epsilon^{\overline{abc}} f_{xyz} A_{\overline{a}}^x \chi_{\overline{b}}^y \chi_{\overline{c}}^z, \qquad (16.6.9)$$

where ϵ is the antisymmetric tensor of $SU(3)$.

Equation (16.6.9) is an answer to our problem. If we actually know the Ricci-flat Kähler metric of K and are able to solve the Dirac equation for zero modes, then inserting the zero mode wave functions in (16.6.9) and integrating over K gives Yukawa couplings in four dimensions. Expressed this way, it appears that evaluating (16.6.9) would require extensive knowledge about K. Actually, it is possible to rewrite (16.6.9) in a quasi-topological form that can be evaluated using only qualitative information.

The completely antisymmetric combination of the $SU(3)$ indices in (16.6.9) amounts to a wedge product of the $(0,1)$ forms A and χ. Indeed, the three $(0,1)$ forms in (16.6.9) are effectively being combined into a $(0,3)$ form. In a 'topological' expression, the integrand is most naturally a six form or $(3,3)$ form. We can make a $(3,3)$ form from the $(0,3)$ form that implicitly appears in (16.6.9) by multiplying by the covariantly constant $(3,0)$ form of the manifold of $SU(3)$ holonomy. Thus, (16.6.9) is equivalent to

$$L_{Yuk}^{(4)} = \int_K \omega \wedge A^x \wedge \chi^y \wedge \chi^z f_{xyz}, \qquad (16.6.10)$$

which is more revealing.

Suppose that we are given chiral superfields Φ_k in four dimensions. Each Φ_k corresponds to a harmonic $(0,1)$ form $u_{(k)}^x$ on K. We wish to calculate the cubic term in a superpotential $W = \lambda_{klm} \Phi_k \Phi_l \Phi_m + \cdots$. To do so, it is enough to set the $(0,1)$ forms A and χ in (16.6.10) equal to $u_{(k)}, u_{(l)},$ and $u_{(m)}$. We get our formula for Yukawa couplings

$$\lambda_{klm} = \int_K \omega \wedge u^x_{(k)} \wedge u^y_{(l)} \wedge u^z_{(m)} f_{xyz}. \tag{16.6.11}$$

As a check, note that (16.6.11) is completely symmetric in $k, l,$ and m, something that was not transparent in our starting point.

Now, in our derivation, the $u_{(k)}$ are harmonic $(0,1)$ forms. As such they define elements of the Dolbeault cohomology group $H^1(X)$, X being the holomorphic vector bundle that enters in the choice of vacuum state. In fact, (16.6.11) is invariant under

$$u_{(m)} \rightarrow u_{(m)} + \overline{D}\epsilon_{(m)}, \tag{16.6.12}$$

with $\overline{D}$ being the gauge covariant $\overline{\partial}$ operator. To verify that (16.6.11) is invariant under (16.6.12), one needs to show that

$$0 = \int_K \omega \wedge u^x_{(k)} \wedge u^y_{(l)} \wedge \overline{D}\epsilon^z_{(m)} f_{xyz}. \tag{16.6.13}$$

This can be proved by integrating by parts, and using the fact that $\overline{D}\omega = \overline{D}u_{(k)} = 0$, since ω and the $u_{(k)}$ are harmonic. The invariance of (16.6.11) under (16.6.12) means that (16.6.11) depends only on the cohomology classes of the $u_{(k)}$ in the Dolbeault cohomology group $H^1(X)$ (X being the holomorphic vector bundle that enters in defining the vacuum state). This in turn means that (16.6.11) is a quasi-topological formula. To evaluate it, we do not need to know the actual Ricci-flat metric of K; we only need the topology and complex structure of K. Indeed, in (16.6.11) the only ingredients are the holomorphic three form ω and the wave functions $u_{(k)}$. The holomorphic three form depends only on the complex structure of K, not any particular choice of metric. It frequently can be described explicitly, as we did in some examples in §16.5.3. As for the wave functions, since we do not need explicit zero-mode wave functions but only wave functions in the correct $\overline{D}$ cohomology class, finding the suitable wave functions likewise depends only on the topology and complex structure of K. Being quasi-topological, (16.6.11) is actually a quite tractable formula for the superpotential, though to evaluate it (except in special cases) requires methods that go beyond the scope of our exposition here.

Apart from esthetic criteria, why is it desirable to have a topological formula for the superpotential? There are many mysteries in the low-energy world that are very possibly related to the vanishing, at least at tree level, of some Yukawa couplings. Examples are the extreme lightness of the electron and first-generation quarks and the smallness of the Cabibbo angle. In most conventional approaches to the question of why some Yukawa couplings might vanish, one attempts to postulate global symmetries that would require this. (In the last section, we tried to find, rather than postulate, such symmetries.) This approach has not led to really compelling results; it seems that global symmetries that explain vanishing of couplings that are indeed small typically also predict vanishing of couplings that are not small. The fact that the superpotential is given by a topological formula in the context of ten-dimensional superstrings raises the possibility that some Yukawa couplings vanish because they are forbidden on topological grounds, rather than because of symmetries. Topological reasoning that would force the vanishing of some Yukawa couplings would not necessarily have unwanted consequences in other areas.

The general formulas (16.6.9) and (16.6.11) simplify somewhat in the case in which the spin connection is embedded in the gauge group. In this case, the massless particles in four dimensions that originate from the adjoint representation of E_8 transform as $\mathbf{27}, \overline{\mathbf{27}}$, or $\mathbf{1}$ under the unbroken E_6. The cubic superfield couplings allowed by E_6 invariance are $\mathbf{27}^3, \overline{\mathbf{27}}^3, \mathbf{1}^3$, and $\mathbf{1} \cdot \mathbf{27} \cdot \overline{\mathbf{27}}$. All of these are present in general and can be extracted from (16.6.11). For example, massless superfields $\Phi^p_{(k)}$ that transform as $\mathbf{27}$ of E_6 (here p is an E_6 index and (k) a generation index) originate microscopically from elements $u_{(k)}{}^a{}_{\bar b}$ of $H^1(T)$.[*] Because of E_6 invariance the $\mathbf{27}^3$ terms in the superpotential must be of the form

$$W = \lambda_{klm} \Phi^p_{(k)} \Phi^q_{(l)} \Phi^r_{(m)} d_{pqr}, \qquad (16.6.14)$$

where d_{pqr} is the cubic symmetric tensor characteristic of E_6, and we wish to determine the couplings λ_{klm}. Inserting the $u_{(k)}{}^a{}_{\bar b}$ in (16.6.9), $SU(3)$ invariance implies that the tangent-vector indices of the three wave functions $u_{(k)}{}^a{}_{\bar b}$ are contracted in a completely antisymmetric fashion, so

---

[*] Since the indices may seem bewildering, let us note that an element of $H^1(T)$ would be denoted $u^a_{\bar b}$. Since there is in general more than one element of $H^1(T)$, u carries in addition the generation index (k).

(16.6.9) reduces in this case to

$$\lambda_{klm} = \int_K \omega_{def}\omega_{abc}g^{a\bar{a}}g^{b\bar{b}}g^{c\bar{c}}u_{(k)}{}^d{}_{\bar{a}}u_{(l)}{}^e{}_{\bar{b}}u_{(m)}{}^f{}_{\bar{c}} \qquad (16.6.15)$$

We will find this formula useful in §16.6.4.

Likewise, superfields $\tilde{\Phi}_{(f)p}$ that transform as $\overline{\mathbf{27}}$ of E_6 can have cubic superfield couplings of the general form

$$W = \tilde{\lambda}_{fgh}\tilde{\Phi}_{(f)p}\tilde{\Phi}_{(g)q}\tilde{\Phi}_{(h)r}d^{pqr}. \qquad (16.6.16)$$

with d^{pqr} being again the E_6 invariant. To determine the $\tilde{\lambda}_{fgh}$, recall that the $\tilde{\Phi}_{(f)}$ originate microscopically as harmonic $(1,1)$ forms $u_{(f)a\bar{b}}$. $SU(3)$ invariance of (16.6.9) again means that complete antisymmetry on the $(1,0)$ indices must be implicit in the E_8 structure constants, so (16.6.9) reduces in this situation to

$$\tilde{\lambda}_{fgh} = \int_K \overline{\omega}_{\bar{d}\bar{e}\bar{f}}\omega_{abc}g^{a\bar{a}}g^{b\bar{b}}g^{c\bar{c}}g^{d\bar{d}}g^{e\bar{e}}g^{f\bar{f}}u_{(f)d\bar{a}}u_{(g)e\bar{b}}u_{(h)f\bar{c}}. \qquad (16.6.17)$$

Happily, some simple $SU(3)$ group theory shows that (16.6.17) is equivalent to a much more natural looking expression, namely the wedge product of the three $(1,1)$ forms:

$$\tilde{\lambda}_{fgh} = \int_K u_{(f)} \wedge u_{(g)} \wedge u_{(h)}. \qquad (16.6.18)$$

The great merit of this formula compared to our previous ones is that it depends only on the de Rham cohomology classes of the u's in $H^2(K)$. (This is so since (16.6.18) is invariant under $u \to u + d\alpha$, as we discussed in §12.4.) De Rham cohomology is a topological invariant, so unlike our previous formulas, (16.6.18) is a strictly topological formula that does not depend on the choice of a complex structure on K.

As an example of a consequence of topological formulas for the superpotential that would be puzzling from the low-energy point of view, suppose that K does not admit an antiholomorphic symmetry, so that CP is spontaneously broken. In this case, one expects Yukawa couplings to be complex, and that will be true for the λ_{klm} of (16.6.15), since the ingredients in that formula are complex. However, the $\tilde{\lambda}_{fgh}$ of (16.6.18) are real even if CP is violated, since $(1,1)$ forms, which are the only ingredients in (16.6.18), are naturally real.

We have only discussed Yukawa couplings of superfields that originate from the adjoint representation of E_8. There are other massless chiral superfields corresponding to deformations of the complex structure or of the Kähler form called $X_{(\alpha)}$ and $Y_{(\beta)}$, respectively. Though we have not discussed these explicitly, they actually enter the above discussion because, for instance, the λ_{klm} in (16.6.15) depend on the complex structure and thus on the expectation values of the $X_{(\alpha)}$. Thus, (16.6.14), for instance, would be more accurately written

$$W = \lambda_{klm}(X_{(\alpha)})\Phi^p_{(k)}\Phi^q_{(l)}\Phi^r_{(m)}d_{pqr}. \qquad (16.6.19)$$

Of course, upon expanding around an expectation value of the $X_{(\alpha)}$, which must somehow be determined, the $X_{(\alpha)}$ dependence of the λ_{klm} will contribute only tiny corrections to low-energy physics. While the superfield couplings derived in this section depend on the complex structure of K, they do not depend on the choice of Kähler metric and hence are independent of the $Y_{(\beta)}$. We will consider the significance of this fact in §16.8.

The reader may wonder whether, apart from the terms we have derived, there are terms in the superpotential that couple the $X_{(\alpha)}$ and $Y_{(\beta)}$ terms to themselves only. At least with the spin connection embedded in the gauge group, such terms cannot arise. Indeed, an X^3 term in the superpotential (for instance) would yield a term $|\partial W/\partial X|^2 = |X|^4$ in the ordinary potential energy V of equation (16.6.4). This would prevent X from getting a vacuum expectation value. But we know from the Calabi–Yau classification of metrics of $SU(3)$ holonomy that the complex structure and Kähler class can have arbitrary values, or in other words that the vacuum expectation value of X can be arbitrary. (We are here tacitly using the result of §16.7 below, where it will be shown that solutions of the equations of the theory can be constructed starting with general metrics of $SU(3)$ holonomy, at least to all finite orders in α'.) A similar argument, using the fact that the number of massless **27**'s is a topological invariant independent of the vacuum expectation value of X, shows that there are no $X \cdot \mathbf{27} \cdot \overline{\mathbf{27}}$ couplings, even though there may be no global symmetries forbidding such terms.

16.6.2 The Kinetic Terms

A knowledge of the superpotential is the key to predicting which Yukawa couplings vanish, and there is much physics in this. The extreme lightness of some fermions, the smallness of mixing angles, and even the stability of the proton may possibly be determined by the vanishing of some

Yukawa couplings. To obtain quantitative predictions for the Yukawa couplings that do not vanish, however, it is necessary to know the D terms of (16.6.1), as well as the superpotential. The general strategy to obtain formulas for the D terms is the same as for the superpotential – one begins with the exact ten-dimensional action, and, after integrating over the compact manifold K, one obtains an effective four-dimensional action. In this way one can obtain formulas for the kinetic terms in the effective four-dimensional theory. With the spin connection embedded in the gauge group, the resulting formulas have a topological character. Partly for brevity and partly because the nonrenormalization theorem discussed presently does not seem to have an analog for the kinetic terms, we will not discuss the detailed formula for the kinetic terms in four dimensions that arises after compactification.

One simple fact about the normalization of the Yukawa couplings should certainly be pointed out, however. If r is the radius of K, then upon integrating over K the coefficients of terms in the effective four-dimensional theory are multiplied by appropriate powers of r. It is not too difficult to see that both gauge couplings and Yukawa couplings are formally of order $1/r$. Four-dimensional gauge couplings that are not too small correspond to r being not too much larger than the Planck length; nonzero Yukawa couplings are then formally of the same order of magnitude.

16.6.3 A Nonrenormalization Theorem and Its Consequences

In analyzing the superpotential of the matter fields in the above sections, we worked exclusively in terms of the ten-dimensional field theory. This means that if r is the radius of the compact manifold K, our results are valid only to lowest order in α'/r^2. We will now show that the results concerning the superpotential are actually valid to all finite orders in α'/r^2. (In certain cases, there seem to be corrections that are nonperturbatively small as $\alpha' \to 0$.)

The argument is surprisingly simple. The radius r of K actually is one of the moduli that enter in finding a metric on K of $SU(3)$ holonomy. As such, it corresponds in four dimensions to a massless field, which we will also call r. The superpartner of r is a pseudoscalar p that originates as a zero mode of the ten-dimensional two form B_{MN}. This pseudoscalar decouples at zero momentum because of the same abelian gauge invariance $\delta B = d\Lambda$ that causes B to decouple at zero momentum in the original ten-dimensional theory. Saying that p decouples at zero momentum is tantamount to the statement that p is an axion-like mode governed by the global symmetry

$$\delta p = \text{constant}. \tag{16.6.20}$$

This symmetry was discussed in §14.3.2, and was shown to be valid to all finite orders in α'; it was also indicated that in the heterotic string theory it is *not* valid nonperturbatively. Supersymmetry unites r and p in a complex superfield Y.

Now, the superpotential W is a holomorphic function of complex variables, so if it depends on r at all, it depends on r and p in the combination $Y = r + ip$. But W may not depend on p, since then p would enter in non-derivative couplings, and the axionic symmetry would be violated. Hence, W cannot depend on r either. So W cannot depend on the dimensionless parameter α'/r^2. W may therefore be evaluated at $\alpha' = 0$, and is given by the field-theory formula. (There may be nonperturbative corrections, since (16.6.20) is only valid to finite orders in α'.) As a check on the argument, note that it implies that the field theory formula for W cannot depend on the Kähler class of the metric; this was already observed in §16.6.1.

This surprisingly simple argument dramatically increases the significance of the topological formula for the superpotential that was derived earlier. If α'/r^2 is even moderately small, the nonperturbative (worldsheet instanton) effects, which according to our discussion in §14.3.2 may spoil the symmetry (16.6.20), will be very tiny, so the topological formula for the superpotential will be extremely accurate. As of this writing, recent investigations have indicated that on many (not all) manifolds of $SU(3)$ holonomy, nonperturbative effects do correct the topological formula by an amount proportional to $e^{-r^2/\alpha'}$, but the analysis is too lengthy to be presented here.

The nonrenormalization of the superpotential also has surprisingly strong implications. We have already encountered tantalizing indications that Ricci-flat Kähler manifolds, with the spin connection embedded in the gauge group, give solutions of the equations of motion of string theory. This was certainly true to the low order in α' to which we checked the equations in §16.1. We can now easily show that, possibly with a suitable readjustment of the vacuum state, this remains true to all finite orders in α'.

Imagine compactifying the ten-dimensional theory on $T^4 \times K$, with T^4 being some four manifold, which we do not assume *a priori* to be four-dimensional Minkowski space. Ten-dimensional string theory will give some sort of effective theory governing physics on T^4. We wish to show that starting with K being a Ricci-flat Kähler manifold (with the spin connection embedded in the gauge group), we can find, at least to all finite orders in α', a solution of the underlying equations. This solution has unbroken supersymmetry, with T^4 being four-dimensional Minkowski space. It is enough to show that these statements hold in terms of the effective theory on T^4, rather than the underlying ten-dimensional field theory.

Order by order in α', string theory gives corrections to the equations of the effective theory on T^4. To begin with, we know that these corrections preserve the fact that the effective theory on T^4 is a supersymmetric theory,[*] but we do not know that Minkowski space is a solution with unbroken supersymmetry. The discussion following (16.6.5) amounted to saying that what we must show is that the effective theory has $W = \partial W/\partial \phi^k = D_a = 0$. The relevant D_a are those that correspond to massless $E_6 \times E_8$ gauge bosons, since massive ones can be integrated out in determining what we mean by the effective low-energy theory. Unbroken $E_6 \times E_8$ will ensure that the D_a terms vanish, so we must concern ourselves with the conditions $W = \partial W/\partial \phi^k = 0$. Since W receives no α'-dependent contributions (at least to finite orders), these statements will be true if they hold at $\alpha' = 0$. This is the case since we have already verified the field equations in the low-energy field-theory limit. Thus, at least to all finite orders in α', we obtain in this way solutions of the equations of motion of the theory. In the next section, we will generalize the argument in an interesting way.

It is important to note that the above argument is macroscopic in character and says nothing about what is happening microscopically. In the above we integrated out the massive degrees of freedom to discuss the effective theory of the massless ones. The argument does not guarantee that α'-dependent corrections do not modify the equations or bring about a shift in the expectation values of the massive fields. The argument guarantees only that whatever shift in the equations or the expectation values of the massive fields comes about because of string-theory corrections, the vacuum configuration of the light fields in four dimensions will not be affected in this process.

Perhaps it is worthwhile to state this more explicitly. Consider a compactified theory that in four dimensions has massive fields X_j and massless fields ϕ_m. The mass terms in the Lagrangian are

$$L_M = -\frac{1}{2} \sum_i M_i^2 X_i^2. \tag{16.6.21}$$

---

[*] Otherwise this theory, which at $\alpha' = 0$ has a massless spin 3/2 particle, would not be consistent.

Of course, we have not included mass terms for the massless fields ϕ_m. We now turn on string-theoretic corrections of order $(\alpha')^p$ for some p. They may contribute, among other things, linear terms in some of the fields:

$$\Delta L = \sum_i \epsilon_i X_i + \sum_j \rho_j \phi_j. \tag{16.6.22}$$

What are the consequences of this modification of the theory? Solving the equations

$$\frac{\partial}{\partial X_i}(L_M + \Delta L) = 0 \tag{16.6.23}$$

for massive fields, we find that the vacuum expectation values of of the X_i are shifted by

$$\delta X_i = \epsilon_i/M_i^2. \tag{16.6.24}$$

In constructing a perturbative expansion of the vacuum state, (16.6.24) just represent a small harmless shift in the expectation values of the X_i. Quite different are the ρ_j of (16.6.22). Since the ϕ_j are massless, a nonzero ρ_j would formally shift ϕ_j by

$$\delta\phi_j \sim \rho_j/0. \tag{16.6.25}$$

In actuality, the shift would be large but, presumably, not infinite. But the nonsensical form of (16.6.25) means that a perturbative expansion of the vacuum around a starting point that gives $\rho_j \neq 0$ is not possible. Perturbative corrections that give nonzero ρ_j would destabilize the original vacuum state, shifting it by a large amount in a way that invalidates (and cannot be predicted from) perturbation theory.

What we have shown using the nonrenormalization theorem for the superpotential is that order by order in α' in an expansion in which the starting point is a Ricci-flat Kähler manifold, the four-dimensional superpotential is unmodified and therefore the expectation values of the massless fields are unshifted. In the above language, the ρ_j vanish and the vacuum is stable, although possibly subject to small, finite shifts in the vacuum expectation values of the massive fields. On the other hand, the nonrenormalization theorem allows the possibility that in constructing a vacuum solution order by order in α', shifts in the expectation values of the massive fields may arise. We will explicitly exhibit the need for such shifts, in a slightly different context, in the next section.

It should be pointed out that shifts in the massive fields can include corrections to the form of the metric, so our argument permits the possibility that starting with a Ricci-flat Kähler metric and constructing a solution of the field equations order by order in α', the resulting solution may not be Ricci-flat Kähler. Recent calculations indicate that this is so.

Our analysis is perturbative in nature and raises the possibility that nonperturbatively (with the breakdown of (16.6.20)) the massless modes will be excited and the vacuum will be destabilized. Recent analyses, using relationships between the various string theories, indicate that this does not occur, but this is a matter that we will not try to elucidate here.

The fact that finite-order string theory corrections do not modify the superpotential is reminiscent of the fact that in four-dimensional field theory, the superpotential is unrenormalized to all finite orders of perturbation theory. One may ask whether the loop expansion of string theory likewise leaves W unmodified. At one-loop order, this can be proved in close analogy with the proof in chapter 9 that Newton's constant is unrenormalized at one-loop order (in the uncompactified theory). Nonrenormalization of W to all finite orders in the loop expansion seems quite plausible, but has not been completely proved.

Our arguments in this section do not apply to the kinetic terms in the Lagrangian, which come from a superspace integral $\int d^4\theta\, K$. The difference is that K is *not* an analytic function of the superfields; in fact, K does depend on the radius of the compact manifold; this is how Yukawa couplings come to depend on that radius. It is likewise true in four-dimensional field theory that K is renormalized in perturbation theory.

16.6.4 Application to the Toy Model

One of the principal motivations for investigating the topological relations for the superpotential was to understand whether it is possible to predict zeros of Yukawa couplings that do not follow from symmetries. This is indeed possible, but since the analysis is rather technical we will content ourselves with a rather special example in which the full force of topological reasoning is not required.

We return to our usual toy model, the quintic hypersurface Q in CP^4. As usual, we take the quotient of Q by $F \approx Z_5 \times Z_5$ to obtain a four-generation model. Imposing the discrete symmetry T defined in (16.5.11), the quintic polynomial defining Q must be of the form (16.5.14). Ignoring E_6 breaking for simplicity, the five massless **27**'s in four dimensions correspond to the five polynomials in (16.5.15), and have the transformation law under the global symmetry T that is given there.

Clearly, the global symmetry T restricts the Yukawa couplings; uncovering such restrictions was the purpose of working out the T quantum numbers in (16.5.15). Can we find additional restrictions on the Yukawa couplings? The Yukawa couplings are represented by the quasi-topological integral in (16.6.15). While we could subject this integral to topological reasoning, it is much faster to note that instead of evaluating the integral on the space Q/F of Euler characteristic -8, which is used to get a four generation model, we could evaluate the integral on the simply connected covering space Q. The advantage of this is that Q has a gigantic symmetry group, including the phase transformations (16.5.5), most of which is lost if we work on Q/F. By evaluating the integral (16.6.15) on Q instead of Q/F, we can predict the vanishing of some terms that respect all of the symmetries of Q/F but violate some of the symmetries of Q.

Symmetries of Q that are not symmetries of Q/F might be called 'pseudosymmetries'. The use of pseudosymmetries to restrict Yukawa couplings on Q/F is a particularly simple case of topological reasoning.

Let us now see how this works. In (16.6.15) there are two explicit factors of ω, and three wave function $u_{(k)}, u_{(l)}, u_{(m)}$. To get a nonzero integral, the product $\omega^2 u_{(k)} u_{(l)} u_{(m)}$ must be invariant under (16.5.5). In view of (16.5.6), this means that the product of the three u's must transform like

$$u_{(k)} u_{(l)} u_{(m)} \to \alpha^3 \sum m_k \cdot u_{(k)} u_{(l)} u_{(m)}. \qquad (16.6.26)$$

This condition is easy to implement because the $u_{(m)}$ are in one-to-one correspondence with suitable quintic polynomials and transform the same way under (16.5.5). Working with the ψ_m of (16.5.15) instead of the $u_{(m)}$, the product

$$\psi_k \psi_l \psi_m \qquad (16.6.27)$$

must transform like (16.6.26) if the corresponding cubic superfield coupling is not to vanish. On the other hand, the product (16.6.27) of three quintic polynomials is a polynomial of fifteenth order. The only fifteenth order polynomial that transforms like (16.6.26) is

$$Z_1^3 Z_2^3 Z_3^3 Z_4^3 Z_5^3 \qquad (16.6.28)$$

and our conclusion is that a cubic superfield coupling of three families vanishes unless (16.6.28) appears in the product of the corresponding three polynomials ψ_k, ψ_l, ψ_m. It is easy to see now that the allowed Yukawa couplings are

$$\psi_0^3, \ \psi_0 \psi_2 \psi_{-2}, \ \psi_2 \psi_{-1} \psi_{-1}, \ \psi_{-2} \psi_1 \psi_1, \ \psi_2 \psi_2 \psi_1, \ \psi_{-2} \psi_{-2} \psi_{-1}, \qquad (16.6.29)$$

but that

$$\psi_0 \psi_{-1} \psi_1 \qquad (16.6.30)$$

is forbidden. What is fascinating about this result is that (16.6.30) respects every global symmetry that is conserved by (16.6.29). To a low-energy observer, the failure to observe the coupling (16.6.30) in a theory in which all of the couplings in (16.6.29) are present would be rather puzzling. Nature present us with some very similar puzzles in the form of surprisingly small quark and lepton masses and mixing angles. We have seen how such a puzzle can have a simple and uncontrived microscopic explanation.

16.7   Another Approach to Symmetry Breaking

Throughout this chapter, we have studied the consequences of equations (16.1.1) for unbroken supersymmetry in four dimensions, which we repeat here for convenience:

$$(a)   0 = \delta\psi_i = \frac{1}{\kappa}D_i\eta + \frac{\kappa}{32g^2\phi}(\Gamma_i{}^{jkl} - 9\delta_i^j\Gamma^{kl})\eta H_{jkl}$$

$$(b)   0 = \delta\lambda = -\frac{1}{\sqrt{2}\phi}(\Gamma\cdot\partial\phi)\eta + \frac{\kappa}{8\sqrt{2}g^2\phi}\Gamma^{ijk}\eta H_{ijk} \qquad (16.7.1)$$

$$(c)   0 = \delta\chi = -\frac{1}{4g\sqrt{\phi}}\Gamma^{ij}F_{ij}\eta.$$

Our discussion has been entirely based on a simple ansatz for obeying these equations, namely embedding the spin connection in the gauge group and setting $H = d\phi = 0$. While this obeys (16.7.1), those equations are only approximations to an exact structure that includes terms of higher order in α'. The nonrenormalization theorem for the superpotential was used in the last section to show that the higher-order corrections do not excite massless fields and destabilize the vacuum, but instead excite massive fields only and can be absorbed in small, perturbative corrections to the vacuum.

It is unsatisfying to be limited to the simple ansatz that we have considered without evidence that more general possibilities are not viable. We will now use the nonrenormalization theorem for the superpotential to find more general vacuum states with unbroken $N = 1$ supersymmetry in four dimensions. We will give only a rather brief introduction to a vast and as yet incompletely explored subject.

We consider compactification on $M^4 \times K$ where K is a compact six-dimensional manifold whose radius will be denoted as r. We will expand in powers of $1/r$ or more precisely in powers of α'/r^2. The goal will be to formulate statements that are valid at least to all finite orders in α'/r^2. It is important to keep in mind that r is a dynamical variable and is in fact one of the four-dimensional fields of the theory. Indeed, r is related by supersymmetry to the axion-like mode that entered in the proof of the nonrenormalization theorem.

Let us first try to set $H = d\phi = 0$ in (16.7.1) but *without* embedding the spin connection in the gauge group. Equation (16.7.1)(a) tells us that K should be a manifold of $SU(3)$ holonomy. Notice that the right-hand side of equation (16.7.1)(a) contains one derivative and so is of order $1/r$. Equation (16.7.1)(b) is obeyed if we set $H = d\phi = 0$. As for (16.7.1)(c), we have come to grips with this equation in §15.6. This equation says that the gauge field A is a holomorphic connection on a holomorphic vector bundle X over K, and obeys the equation

$$g^{a\bar{b}} F_{a\bar{b}} = 0. \tag{16.7.2}$$

We will refer to (16.7.2) together with the holomorphic condition $F_{\bar{a}\bar{b}} = 0$ as the Kähler–Yang–Mills equations. The ten-dimensional gauge group – $E_8 \times E_8$ in the most interesting case – will be broken down to the subgroup that commutes with the gauge fields present in the vacuum state. Embedding the spin connection in the gauge group is one solution of (16.7.2); it breaks $E_8 \times E_8$ down to $E_6 \times E_8$ (or perhaps a subgroup if Wilson lines are included). Equation (16.7.2) has many other solutions, however. According to theorems of Donaldson (in complex dimension two) and Uhlenbeck and Yau (in complex dimension three or more), discussed in §15.6.2, there is a unique solution of (16.7.2) on any stable holomorphic vector bundle X over K.

There are many examples of such bundles. For instance, in §15.6.3 we constructed irreducible four-dimensional holomorphic vector bundles over a quintic hypersurface Q in CP^4. We found that such rank-four bundles X can be found by deforming $T \oplus L$, T being the tangent bundle of Q and L being a trivial line bundle. On X, the unique holomorphic connection A obeying (16.7.2) is an $SU(4)$ gauge field. As E_8 contains the maximal subgroup $SO(10) \times SO(6) \approx SO(10) \times SU(4)$, the use of this solution of (16.7.2) in the vacuum state will give a model with $E_8 \times E_8$ broken down to $SO(10) \times E_8$ (or a subgroup thereof if Wilson lines are included). It is good to keep this example in mind in what follows, though (16.7.2) has many other solutions; for instance, on a Kähler manifold with $h^{1,1} > 1$, it has abelian solutions, which were discussed in chapter 15.

Expanding in powers of $1/r$, it is natural to consider the gauge field A to be of order $1/r$, since it appears in the covariant derivative $D = \partial + iA$. While the right-hand side of equation (16.7.1)(a) is of order $1/r$, the right-hand side of (16.7.1)(c) is of order $1/r^2$. String-theoretic corrections omitted in (16.7.1) are of higher order.

By picking K to have $SU(3)$ holonomy, with $H = d\phi = 0$ and A a solution of (16.7.2), we have obeyed (16.7.1). So what is still missing? Equation (16.7.1) must be supplemented with the Bianchi identity

(16.1.2), which we repeat here for convenience:

$$dH = \text{tr}\,R \wedge R - \text{tr}\,F \wedge F. \qquad (16.7.3)$$

This is obeyed with $H = 0$ if the spin connection is embedded in the gauge group, but not otherwise. It was precisely because of this point that in our earlier discussion in §16.1 we limited ourselves to embedding the spin connection in the gauge group.

Since $\text{tr}\,F \wedge F$ and $\text{tr}\,R \wedge R$ represent respectively, the second Chern classes of X and of the tangent bundle T of K, and since dH is trivial in cohomology, (16.7.3) has a solution only if

$$c_2(X) = c_2(T). \qquad (16.7.4)$$

This requirement was already noted in §13.5.3 and §14.7. If (16.7.4) is valid, then (16.7.3) can be obeyed.

The right-hand side of (16.7.3) is of order $1/r^4$, so (16.7.3) is compatible with choosing H to be of order $1/r^3$. With this choice, (16.7.1) is not actually obeyed, but the error is only of order $1/r^3$. Our task is to determine whether it is possible to modify the configuration so that (16.7.1) will be obeyed to order $1/r^3$ and in fact to all finite orders. In the process it is necessary to take account of α'-dependent corrections to (16.7.1).

In fact, the $1/r^3$ error in (16.7.1) is a stringy effect of order α'. One may guess this simply because the expansion parameter is really α'/r^2, but let us present a more detailed discussion. The $\text{tr}\,R \wedge R$ term in (16.7.3) is not present in the limiting supersymmetric field theory formulated in §13.1.3; it is of order α', though α' has been set to one in the formulas. If one deletes the $\text{tr}\,R \wedge R$ term from (16.7.3), then (16.7.3) would imply that $c_2(X) = 0$; in complex dimension three, a solution of (16.7.2) with $c_2(X) = 0$ must have $F = 0,^*$ as we will see below. In this case (16.7.3) would allow $H = 0$. Thus, it is really because of effects of order α' that in attempting to obey (16.7.1) we are encountering an error of order $1/r^3$. The error should be analyzed using the general nonrenormalization theorem for α'-dependent corrections, which was discussed in the last section.

---

* This, of course, is not very interesting for phenomenology since it leaves unbroken $E_8 \times E_8$. The inclusion of the $\text{tr}\,R \wedge R$ term in (16.7.3), which makes it possible to use nontrivial solutions of (16.7.3) (like embedding the spin connection in the gauge group and the more general options explored in this section) is a major advantage of string theory relative to the limiting low-energy field theory.

In a Kaluza–Klein theory with a compact manifold of radius r, the natural mass scale is $1/r$, and an essentially infinite number of four-dimensional particles will appear with masses of this order. The fact that (16.7.1) is obeyed in order $1/r$ means that the four-dimensional gravitino is massless in this order. Its mass, if any, will arise in order $1/r^3$ or in whatever order we fail to obey (16.7.1) (or its stringy generalization).

A four-dimensional theory with a massless spin 3/2 particle (which does not decouple at zero momentum) must be supersymmetric. The fact that for extremely large r the gravitino is essentially massless compared to all of the other modes means that compactification from ten to four dimensions on $M^4 \times K$ must give a four-dimensional theory that is supersymmetric. Whether supersymmetry is spontaneously broken or not is another story. If the gravitino acquires a mass in order $1/r^3$ (or higher orders), this must be interpretable as spontaneous supersymmetry breaking.

How do we explore spontaneous supersymmetry breaking? The condition for unbroken supersymmetry can be succinctly stated in terms of the superpotential W; it is

$$W = \frac{\partial W}{\partial \phi^k} = 0 \qquad (16.7.5)$$

with ϕ^k ranging over all chiral superfields.[†] Equation (16.7.5) shows immediately that to determine whether effects of order $1/r^3$ and higher (or in other words effects of order α' and higher) can trigger spontaneous supersymmetry breaking, we must investigate the superpotential. If the r-dependent corrections to the leading large r behavior do not include corrections to W, they cannot trigger spontaneous supersymmetry breaking. This question is, however, precisely the one that is addressed by the nonrenormalization theorems. At least in finite orders, there are no α'-dependent (or r-dependent) corrections to the leading large r behavior of the superpotential. Since the configuration we are considering has unbroken supersymmetry in the extreme large r limit, supersymmetry must be unbroken to all orders in $1/r$. In our above discussion, (16.7.1) was obeyed only to within an error of order $1/r^3$, but the facts just cited mean that it must be possible to adjust the metric and the various other fields so as to eliminate this error.

Let us ask just what this argument proves and what it does not prove. The nature of the nonrenormalization theorem for the superpotential is

---

[†] There also is a condition $D^a = 0$, with a ranging over the Lie algebra. The structure of this condition is such that the ability to obey it is stable under slight modification of the D^a, as long as the unbroken gauge group is semisimple. In this case we need not consider it in discussing the corrections to the large r behavior.

that one integrates out all of the massive modes and analyzes the effective superpotential for the massless degrees of freedom. A simple argument involving the axionic symmetry then shows that this superpotential receives no r-dependent or α'-dependent contributions. The argument guarantees that order by order in $1/r$, a vacuum solution can be obtained in which the superpotential is not modified and therefore expectation values of the massless fields are not shifted. The argument does not guarantee what this vacuum solution looks like from the point of view of the massive degrees of freedom. If one does not integrate them out, but leaves them in the discussion, then the r-dependent corrections may well shift the vacuum expectation values of the massive modes. All of this was discussed from a slightly different point of view in the last section.

Let us now discuss the explicit perturbative construction of the vacuum state order by order in $1/r$ for low orders. This will also permit us to illustrate explicitly the considerations in the discussion in the previous section. We expand the ten-dimensional equations in powers of $1/r$, treating space-time derivatives and gauge fields as being of order $1/r$. In order $1/r^2$, the only equation that we meet is the Einstein equation

$$R_{ij} = 0. \tag{16.7.6}$$

This is obeyed by taking K to be a Ricci-flat Kähler manifold. In order $1/r^3$ we meet the Yang–Mills equations

$$D^i F_{ij} = 0, \tag{16.7.7}$$

which we obey by taking the gauge field to be a holomorphic connection on a holomorphic vector bundle X that obeys the Donaldson–Uhlenbeck–Yau equation. In order $1/r^4$, we meet the Bianchi identity (16.7.3), which must be supplemented by the equation of motion of the H field:

$$d^* H = 0. \tag{16.7.8}$$

Assuming that $c_2(X) = c_2(T)$, (16.7.3) and (16.7.8) together have a solution, which is unique if (to minimize the energy) we choose H to be orthogonal to the harmonic three forms on K. In order $1/r^4$, we meet for the first time an equation for the dilaton field

$$\Box \phi = \mathrm{tr} F_{ij} F^{ij} - \mathrm{tr} R_{ij} R^{ij}, \tag{16.7.9}$$

as well as a modification of the Einstein equation (16.7.6) that we will not consider here, since (16.7.9) will suffice to illustrate the key concepts.

On the compact manifold K, the equation

$$\Box\phi = J \qquad (16.7.10)$$

does not have a solution for arbitrary J. It is necessary that the source J be orthogonal to the zero modes of the wave operator $\Box$. If indeed ψ is such a zero mode, $\Box\psi = 0$, then (using hermiticity of $\Box$) (16.7.10) implies

$$0 = \int_K \psi\Box\phi = \int_K \psi J \qquad (16.7.11)$$

Vanishing of (16.7.11) for every zero-eigenvalue eigenfunction ψ of $\Box$ is the necessary and sufficient condition for (16.7.10) to have a solution.[*]

The scalar wave operator is particularly simple; the only zero eigenvalue of $\Box$ is the constant function 1. The requirement (16.7.11) that the source should be orthogonal to the zero modes of the wave operator thus reduces in this case to

$$0 = \int_K (\mathrm{tr}F_{ij}F^{ij} - \mathrm{tr}R_{ij}R^{ij}). \qquad (16.7.12)$$

The nonrenormalization theorem for the superpotential, which predicts that it will be possible to readjust the values of the massive fields so as to obtain a vacuum solution order by order in α', predicts that (16.7.12) should be valid. We will now see that indeed (16.7.12) is a consequence of the topological condition $c_2(X) = c_2(T)$ as well as the various conditions that we have imposed in order to obey (16.1.1) in leading order in $1/r$. We note first that

$$\mathrm{tr}F_{ij}F_{ij} = 2g^{a\bar{a}}g^{b\bar{b}}\mathrm{tr}F_{ab}F_{\bar{a}\bar{b}} + 2g^{a\bar{a}}g^{b\bar{b}}\mathrm{tr}F_{a\bar{b}}F_{\bar{a}b}. \qquad (16.7.13)$$

The second term in (16.7.13) can be rewritten using the identity

$$g^{a\bar{a}}g^{b\bar{b}}\mathrm{tr}F_{a\bar{b}}F_{\bar{a}b} = g^{a\bar{a}}g^{b\bar{b}}\mathrm{tr}F_{a\bar{a}}F_{b\bar{b}} - k \wedge \mathrm{tr}F \wedge F, \qquad (16.7.14)$$

where k is the Kähler form. Combining these manipulations gives a formula for the Yang–Mills action:

$$\int_K \mathrm{tr}F_{ij}F^{ij} = \int_K \left[2g^{a\bar{a}}g^{b\bar{b}}\mathrm{tr}F_{ab}F_{\bar{a}\bar{b}} + 2\mathrm{tr}(g^{a\bar{a}}F_{a\bar{a}})^2 - 2k\wedge\mathrm{tr}F\wedge F \right]. \qquad (16.7.15)$$

The last term on the right-hand side of (16.7.15) is a topological invariant; that is, it depends only on the topological class of the Kähler metric on

---

[*] The requisite argument here is equivalent to the one given in §15.6.2 concerning the equation $\Delta\sigma = H$.

K and the topological class of the vector bundle X.[†] (16.7.15) shows that within a given topological class, the solution of the Kähler–Yang–Mills equations predicted by the Donaldson and Uhlenbeck–Yau theorems has the minimum action. Equation (16.7.15) also has a counterpart with F replaced by the Riemann tensor R. Substituting (16.7.15) and its counterpart in (16.7.12), the right-hand side of that equation becomes

$$-2 \int_K k \wedge \mathrm{tr}(F \wedge F - R \wedge R). \qquad (16.7.16)$$

Substituting $dH = \mathrm{tr}(F \wedge F - R \wedge R)$ in (16.7.16), integrating by parts, and using the fact that $dk = 0$, (16.7.16) vanishes. Thus, we have verified that (16.7.9) has a solution.

Let us take stock of what we have done, and compare this to the general discussion following (16.6.21) in the last section. The massive fields X_i in (16.6.21) correspond in our current discussion to the nonzero eigenvalues of the wave operator $\Box$. The massless fields ϕ_k in the earlier discussion correspond to zero eigenvalues ψ of $\Box$. Vanishing of the sources ρ_k in (16.6.22) is analogous to the requirement (16.7.11) that the source terms in (16.7.9) should be orthogonal to the zero modes of the wave operator $\Box$. Vanishing of the right-hand side of (16.7.12) is the condition that ensures that the perturbation $\mathrm{tr}F_{ij}^2 - \mathrm{tr}R_{ij}^2$ does not couple to a massless field coming from a zero eigenvalue of $\Box$; such a coupling would destabilize the vacuum, triggering the spontaneous breaking of supersymmetry.

The general idea, then, is that in expanding the field equations in powers of $1/r$, one will meet in each order equations of the general form

$$\Box \sigma = J, \qquad (16.7.17)$$

where σ is a field, $\Box$ is an elliptic operator with a discrete spectrum, and J is a source term that depends on the vacuum state as constructed in lower orders in $1/r$. The condition for (16.7.17) to have a solution is that J must be orthogonal to all of the zero eigenvalues of $\Box$. The nonrenormalization theorem for the superpotential predicts that this will always be the case, and this has been checked explicitly in low orders, though we have here considered only the simplest case.

[†] Equation (16.7.15) shows that a solution of the Kähler–Yang–Mills equations for which $\mathrm{tr}F \wedge F$ is trivial in cohomology has zero action and therefore must have $F = 0$, as remarked earlier.

16.8 Discussion

In the last subsection, we have argued for the existence, at least to all finite orders in α', of more general states of unbroken supersymmetry. The general starting point is a manifold K of $SU(3)$ holonomy, and a Kähler–Yang–Mills connection on a holomorphic vector bundle with the same second Chern class as the tangent bundle of K. To make all of this a little more specific, let us return to our familiar example, the quintic hypersurface Q in CP^4. Actually, we will discuss the four generation model based on the quotient of such a hypersurface by $Z_5 \times Z_5$.

If we embed the spin connection in the gauge group, we obtain an E_6 model with one $\overline{\mathbf{27}}$ C_x, five $\mathbf{27}$'s $D^x_{(a)}$ and a number of massless E_6 singlets $E_{(m)}$. Here 'x' is an E_6 index, while '(a)' and '(m)' are flavor indices for $\mathbf{27}$'s and singlets, respectively. The E_6 singlets of interest in the present discussion are the ones that arise from $H^1(End\,T)$ when the tangent bundle T is embedded in $E_8 \times E_8$. There are also E_6 singlets coming from deformations in the complex structure and Kähler class of K, but we will not consider them in this section.

Embedding the spin connection in the gauge group corresponds to setting $C = D = E = 0$. It gives a vacuum state in which the superpotential W obeys

$$W = \frac{\partial W}{\partial \phi^k} = 0 \tag{16.8.1}$$

and supersymmetry is unbroken. Let us now ask whether starting from $C = D = E = 0$, there are 'flat directions' in which the superpotential vanishes identically. If so, moving out in these flat directions gives new solutions of (16.8.1), so in this case apart from states obtained by embedding the spin connection in the gauge group, there are other, nearby points in field space with unbroken supersymmetry.

In conventional models of low-energy supersymmetry, such flat directions frequently arise. In the present context, however, looking at things purely from the low-energy point of view, they seem implausible. Let us consider first a flat direction in which just one of the E_6 singlets, E, is excited. Since E is a massless mode corresponding to a nonzero element of $H^1(End\,T)$, there is no E^2 term in the superpotential. However, there is no obvious reason for the superpotential not to have terms of order

$$E^k, \ k = 3, 4, 5, \ldots . \tag{16.8.2}$$

The existence of a flat direction would seemingly require the vanishing of an infinite number of coefficients that have no obvious reason to vanish.

Nevertheless, in example (6) of §15.6.3 we explicitly constructed holomorphic deformations $\tilde{T}$ of the holomorphic tangent bundle T of K. In our discussion in the earlier part of this section, we argued that (at least to all finite orders in α'), we can use *any* holomorphic bundle X with the same second Chern class as T to construct a new vacuum state of unbroken supersymmetry. Being a deformation of T, $\tilde{T}$ certainly has the same second Chern class as T, so it can be used in this construction.

In the example, the number of parameters entering in the construction of the $\tilde{T}$ was the same as the number of massless E_6 singlets $E_{(m)}$, so the new vacuum states constructed in this way are in one-to-one correspondence with possible values of the $E_{(m)}$. Therefore, improbable though it might seem from the low-energy point of view, contributions to the superpotential as in (16.8.2), which couple the E's only, all vanish (to all finite orders in α', at any rate). Thus, we have uncovered our first example of flat directions in the superpotential of the model obtained by compactification on Q. These flat directions leave E_6 unbroken, since we have considered E_6 singlets only.

Of more interest, perhaps, are flat directions that give broken E_6. These must excite the **27**'s and $\overline{\mathbf{27}}$'s $D_{(a)}^x$ and C_x. In this case, there *are* superpotential couplings. Indeed, we have previously given topological formulas for the terms

$$d^{xyz}C_xC_yC_z, \quad d_{xyz}D_{(a)}^x D_{(b)}^y D_{(c)}^z. \tag{16.8.3}$$

Merely having a formula for C^3 or D^3 couplings does not guarantee that they are nonzero. The reader may feel 'once bit, twice shy' after being told that the E^3 couplings, for which we also derived a general topological formula, are all zero in this particular model. However, in this model the C^3 and D^3 couplings are nonzero.[*] At first sight one might think that the existence of C^3 and D^3 superfield couplings would prevent the existence of flat directions in which C or D is excited. This is not so, however; the C^3 and D^3 couplings eliminate most but not all of the flat directions, limiting the possible patterns of E_6 breaking that can arise.

Under a maximal $SO(10) \times U(1)$ subgroup of E_6, the **27** of E_6 decomposes as

$$\mathbf{27} \approx \mathbf{16}^{-1} \oplus \mathbf{10}^2 \oplus \mathbf{1}^{-4}, \tag{16.8.4}$$

---

[*] In the C^3 case this is easily proved. The one $\overline{\mathbf{27}}$ C comes from the one harmonic $(1,1)$ form on Q – the Kähler form. The C^3 superfield coupling is related by our formula to $\int_Q k \wedge k \wedge k$, which is certainly nonzero. As for D^3 couplings, though we constrained them by means of symmetries and pseudosymmetries in §16.6.1 and §16.6.4, the proof that they are nonzero requires more powerful techniques than we have developed.

where the superscript is the $U(1)$ charge. It is easy to see from (16.8.4) that an $SO(10) \times U(1)$ conserving $\mathbf{27}^3$ coupling must have the general form

$$\mathbf{27}^3 \approx \mathbf{16} \cdot \mathbf{16} \cdot \mathbf{10} \oplus \mathbf{10} \cdot \mathbf{10} \cdot \mathbf{1}. \qquad (16.8.5)$$

Notice that (16.8.5) contains no terms cubic or quadratic in the $\mathbf{1}$ of $SO(10)$. If, therefore, C^3 is the only relevant term in the superpotential W, and the $\mathbf{1}$ of $SO(10)$ is the only nonzero component of the C, then $W = \partial W/\partial \phi = 0$. The $\overline{\mathbf{27}}$ C has a decomposition similar to (16.8.4). It again contains an $SO(10)$ singlet component, and the E_6-invariant C^3 coupling is again harmless from the point of view of spoiling a flat direction in which only this component of C is excited.

Let c and d be $SO(10)$ singlet components of C and D, respectively. We have seen that a minimal $C^3 + D^3$ superpotential would leave flat directions in which c and d have vacuum expectation values, and E_6 is therefore broken down to $SO(10)$. Apart from the superpotential, we must also consider the $(D_a)^2$ terms in the potential energy, where a ranges over the Lie algebra of $E_8 \times E_8$, to achieve unbroken supersymmetry. Setting these to zero merely tells us that if c has an expectation value, then d must also.[†] Taking account of the minimal superpotential and the $(D_a)^2$ terms coming from gauge couplings, these are the only possible flat directions.

Do these flat directions breaking E_6 to $SO(10)$ actually exist? This would seem rather implausible from the low-energy point of view. Although the C^3 and D^3 terms permit such flat directions, an infinity of other superfield couplings, like

$$C^x D_x E, \ (C^x D_x)^2, \qquad (16.8.6)$$

etc., would ruin the E_6-breaking flat directions. As in our discussion of (16.8.2), it seems implausible from the low-energy point of view for all of these coefficients to vanish. Nonetheless, this is so (to all finite orders in α'). The requisite reasoning is exactly analogous to what it was in the discussion of (16.8.2). In example (6) of §15.6.3 we constructed rank-four bundles X over Q as holomorphic deformations of $T \oplus L$. The number of parameters in the construction of X was the same as the number of parameters in the flat directions allowed by the minimal superpotential and the $D_a{}^2$ couplings. The existence of these bundles X implies the existence of an enlarged family of possible vacuum states. The Kähler–Yang–Mills connection on X is an $SU(4)$ gauge field whose embedding in

---

[†] α'-dependent corrections might affect the required ratio of c/d, but are otherwise harmless.

E_8 breaks E_8 down to $SO(10)$, just as we were led to expect by thinking about flat directions in superpotentials.

Thus, the nonrenormalization theorem for the superpotential leads to a new mechanism for symmetry breaking; while leaving $N = 1$ super-symmetry unbroken, we can use the above mechanism to break E_6 down to $SO(10)$, which if desired can be further broken by Wilson lines. By using a rank-five holomorphic bundle, and so embedding an $SU(5)$ gauge field in E_8, one can break E_8 down to $SU(5)$; again this can be combined with Wilson lines. We will not construct examples of stable rank-five holomorphic vector bundles over manifolds of $SU(3)$ holonomy, but many examples are possible. The phenomenology of these $SO(10)$ and $SU(5)$ models is quite interesting, though it has been little explored as of yet.

Finally, we point out again that all of these considerations have been justified only to finite orders in α'. There are indications that the non-renormalization theorem for the superpotential breaks down nonpertur-batively, at the level of world-sheet instantons. A good picture of the implications of this has not yet emerged.

16.9   Renormalization of Coupling Constants

Coupling constants at grand unification can be related to coupling con-stants at ordinary energies by means of standard renormalization-group calculations. If we knew the correct scheme for string compactification, we would know all of the input couplings at the unified scale; but we do not know this. Even without knowing all of the microscopic details of a theory, one can often determine some of the input relations among cou-pling constants of a theory on general grounds. One of our main results in §16.4.2 was to show that in the context of Kaluza–Klein theory, with grand-unified symmetry breaking carried out by Wilson lines, tree-level gauge couplings (but *not* Yukawa couplings) obey standard group-theory relations. The Georgi–Quinn–Weinberg calculation of $\sin^2 \theta_W$ and the unification scale is therefore valid. We will discuss its implications here.

Let us denote the strong, weak and electromagnetic coupling constants as α_3, α_2 and α_{em}, respectively.[‡] The standard renormalization-group

---

[‡] Thus, α_3 is the coupling of color $SU(3)$, and α_2 is the coupling of $SU(2)_L$.

formula is

$$\alpha_3^{-1}(\mu) = \alpha_{GUT}^{-1} + \frac{1}{2\pi}b_3 \ln \frac{M_{GUT}}{\mu}$$

$$\alpha_2^{-1}(\mu) = \alpha_{GUT}^{-1} + \frac{1}{2\pi}b_2 \ln \frac{M_{GUT}}{\mu} \qquad (16.9.1)$$

$$\alpha_{em}^{-1}(\mu) = \alpha_{GUT}^{-1} + \frac{1}{2\pi}b_{em} \ln \frac{M_{GUT}}{\mu}.$$

Here M_{GUT} is the mass scale of unification, μ is the scale at which the couplings are observed, α_{GUT} is the coupling at grand unification, and b_3, b_2 and b_{em} are the one-loop beta function coefficients. These can be rearranged as formulas for the unification scale and the weak angle:

$$\ln \frac{M_{GUT}}{\mu} = \frac{\alpha_{em}^{-1} - \frac{8}{3}\alpha_3^{-1}}{(b_{em} - \frac{8}{3}b_3)/2\pi}$$

$$\qquad (16.9.2)$$

$$\sin^2 \theta_W = \alpha_{em}\left(\alpha_3^{-1} + \frac{1}{2\pi}(b_2 - b_3)\ln \frac{M_{GUT}}{\mu}\right).$$

It is convenient to take $\mu = M_W$, in which case the experimental values are roughly $\alpha_{em}^{-1} = 128$, $\alpha_3^{-1} = 9$.

In the case of E_6 models with low-energy supersymmetry,

$$b_3 = 3F + \frac{1}{2}\Delta_3 - 9$$

$$b_2 = 3F + \frac{1}{2}\Delta_2 - 6 \qquad (16.9.3)$$

$$b_{em} = 8F + \Sigma Q^2 - 6C.$$

Here F is the number of complete families or antifamilies (chiral super-multiplets in the **27** or $\overline{\textbf{27}}$ of E_6), Δ_3 is the number of color triplet or antitriplet chiral supermultiplets that do not fit in complete E_6 multi-plets, Δ_2 is the number of weak doublet chiral supermultiplets that do not fit in complete E_6 multiplets, ΣQ^2 is the sum of the squares of the electric charges of chiral superfields that do not fit in complete E_6 multi-plets, and C is the sum of the squares of charges of light gauge bosons of positive electric charge (for the standard model only W^+ contributes, and $C = 1$). We immediately see that for $F \geq 4$, E_6 models with low-energy supersymmetry have $b_3 > 0$ and are plagued by Landau poles. For $F = 4$,

α_3^{-1} goes to zero at $10^{11\pm1}$ GeV. Since experiment seems to require $F \geq 3$,[§] we conclude that E_6 models with low-energy supersymmetry must have precisely three families (and no antifamilies).

Let us first consider the case in which the low-energy gauge group is a group such as $SU(3) \times SU(2) \times U(1) \times U(1)$ in which the standard model is extended only by extra neutral currents. If we assume that the chiral superfields form complete E_6 multiplets, we get $\sin^2 \theta_W = 0.206$, in good agreement with experiment, and $\log_{10}(M_{GUT}/M_W) = 15.8$. The latter value of M_{GUT} is quite reasonable, though this point requires some comment.

In four-dimensional grand-unified models, any value of M_{GUT} between about 10^{15} GeV and the Planck mass would be considered reasonable. In the Kaluza–Klein context, things are quite different. For example, the relation $g^2 \sim \kappa^2/\alpha'$ in the heterotic string means that (as gauge couplings in nature are of order one) the string scale $(\alpha')^{-1/2}$ cannot be much below the Planck mass. If we assume (as we have in this chapter) that compactification and grand-unified symmetry breaking are the same thing, then above M_{GUT} and below the string scale, we are dealing with a ten-dimensional *field* theory. As this is unrenormalizable, the effective couplings will grow rapidly on evolving to higher energies. Noting from observation that $\alpha_{GUT} \sim 1$ and requiring that the string coupling at the string scale not be too much more than one (this is not a strict logical requirement, but something that we certainly hope is true), we conclude that M_{GUT} cannot be too much less than M_{Planck}. The value of M_{GUT} obtained above is probably near the low limit of what we would wish to consider. This means that proton decay is not likely to be mediated by gauge bosons at an observable rate, though it is anybody's guess at what rate proton decay will be mediated by color triplet partners of quarks.

In the above, we assumed $\Delta_2 = \Delta_3 = 0$. To explain quark and lepton masses without inducing proton decay, one may wish to include extra Higgs bosons in incomplete multiplets. In supersymmetric models, two Higgs doublets would be required, one to give mass to up quarks and one to give mass to down quarks. In an E_6 model, Higgs doublets that are to couple to quarks and leptons might plausibly transform as $\mathbf{27}$ of E_6. Fermion partners of such Higgs bosons have triangle anomalies that one might wish to cancel by including additional Higgs doublets coming from

---

[§] It is almost but not quite possible to make a phenomenologically viable model with $F = 2$. In such a model the top quark does not exist and is replaced by an extra b' quark of charge $-1/3$. The b' quark and all known fermions fit neatly in two $\mathbf{27}$'s of E_6, but such models seem to be excluded by observations of b quark decay and of $e^+e^- \rightarrow b\bar{b}$.

a $\overline{27}$. In fact, the mechanism sketched in §16.4.3 for obtaining massless Higgs doublets with superheavy color-triplet partners avoids anomalies in this way. Unfortunately, the predictions for both M_{GUT} and $\sin^2 \theta_W$ are significantly worsened by including Higgs bosons in incomplete E_6 multiplets. For example, if we include two Higgs doublets coming from a **27** of E_6 and two more from a $\overline{27}$, we get $\log_{10}(M_{GUT}/M_W) = 12.9$, and $\sin^2 \theta_W = 0.25$.

So far we have considered models in which the low-energy $SU(3) \times SU(2)_L \times U(1)$ group is extended only by extra neutral currents. There are several other interesting cases to consider. One of the most interesting is the left–right symmetric extension of the standard low-energy model to a group $SU(3) \times SU(2)_L \times SU(2)_R \times U(1) \times U(1)$. In fact, if we take literally the mechanism described in §16.4.3 for getting massless Higgs doublets without color triplet partners, and if we wish to get massless Higgs doublets capable of coupling to up quarks and down quarks alike, then we are forced into this option. Unfortunately, the left–right symmetric extension of the standard model gives unacceptable renormalization-group predictions. In the absence of unaccompanied Higgs bosons, one finds $\sin^2 \theta_W = 0.27$ and $\log_{10}(M_{GUT}/M_W) = 23.7$. Unaccompanied Higgs bosons make M_{GUT} go down, but regrettably $\sin^2 \theta_W$ goes up.

Another option is to use the more general vacuum states described in the last section to break E_6 to $SO(10)$. One of the attractions of this is that in an E_6 generation

$$\mathbf{27} \approx \mathbf{16} \oplus \mathbf{10} \oplus \mathbf{1} \tag{16.9.4}$$

the $\mathbf{10} \oplus \mathbf{1}$ are not observed experimentally so in E_6 models one must in any case assume that those particles gain larger masses than others, perhaps at a TeV scale. Using flat directions to break E_6 to $SO(10)$ might automatically achieve this.

If one breaks E_6 to $SO(10)$ by using flat directions, then the contributions of the $\mathbf{10} \oplus \mathbf{1}$ to the renormalization-group calculation are eliminated. If we drop these contributions, the change in (16.9.3) is $F \to 2F/3$. As a result, $F = 4$ becomes viable and perhaps even $F = 5$. In one-loop approximation, the predictions for $\sin^2 \theta_W$ and M_{GUT} are not changed and are therefore still viable.

Another tempting feature of $SO(10)$ models is that if all charged chiral superfields transform as $\mathbf{16}$ of $SO(10)$, then as a $\mathbf{16}^3$ superfield coupling is forbidden by $SO(10)$, there will be no renormalizable baryon-number-violating interactions. (There may still be dimension-five baryon-number-violating interactions coming from $\mathbf{16}^4$ superfield couplings, and these might be a problem depending on questions such as the value of M_{GUT}.) In this case, however, it is not clear where quark and lepton masses are to come from. If one wishes to obtain quark and lepton masses in the usual way from cubic superfield couplings, then it is necessary to have massless Higgs bosons coming from the **10** of $SO(10)$. If unaccompanied by light color triplets (which can mediate proton decay) they spoil the renormalization-group predictions just as they do in E_6 models. One possibility is that there are unpaired Higgs bosons coming from **10** of $SO(10)$ (unpaired in the sense that their $SO(10)$ partners are superheavy), but the renormalization-group predictions are rescued because the Higgs contributions are balanced by the contributions of extra, unpaired color triplets coming from **16**'s and $\overline{\mathbf{16}}$'s of $SO(10)$. It would take an impressive display of the magic of algebraic geometry and Dolbeault cohomology (or string theory) to bring about this potentially attractive result. Another recent suggestion, little developed as of yet, is that the quark and lepton masses may not come from terms in the superpotential at all, but from soft supersymmetry breaking interactions.

16.10 Orbifolds and Algebraic Geometry

Compactification from ten to four dimensions on a flat six-dimensional torus T is the simplest model of compactification. This compactification scheme, which was discussed in chapter 6 and chapter 9, is far from realistic from the point of view of low-energy phenomenology. The present point of view makes it natural to think of the six-dimensional real torus contemplated in our previous discussions as a three-dimensional complex torus,

$$T \approx C^3/\Gamma, \tag{16.10.1}$$

with Γ being a lattice in three-dimensional complex space C^3. Viewed in this way, T is a complex manifold of complex dimension three with vanishing first Chern class. It admits a flat metric whose holonomy group is a subgroup of $SU(3)$ – in fact, trivial.

In §9.5.2, we explored string propagation on orbifolds as a generalization of string propagation on a flat torus. In this way we were able to obtain a model of compactification that was exactly soluble and almost as simple as compactification on a torus. Our purpose in this section is to explain the connection between orbifolds and manifolds of $SU(3)$ holonomy and at the same time describe a new method for constructing manifolds of $SU(3)$ holonomy.

As in §9.5.2, let z be a complex variable, and let T_0 be the torus obtained by the identification of points $z \approx z + 1 \approx z + e^{2i\pi/3}$ in the complex z plane. The particular torus T_0 admits the action of a group $F_0 \approx Z_3$ generated by

$$z \to \alpha z, \tag{16.10.2}$$

where $\alpha = e^{2i\pi/3}$. Let z_i, $i = 1, 2, 3$ be three complex variables and T_i the torus obtained with the identification $z_i \approx z_i + 1 \approx z_i + e^{2i\pi/3}$ in the complex z_i plane. The complex torus $T = T_1 \times T_2 \times T_3$ then admits the action of a group $F \approx Z_3$ described by

$$z_i \to \alpha z_i, \, i = 1, 2, 3. \tag{16.10.3}$$

As was explained in §9.5.2, the transformation in (16.10.3) has 27 fixed points, and consequently the topological space

$$X = T/F \tag{16.10.4}$$

is not a manifold.

We will now explain how to modify X so as to obtain a complex manifold of vanishing first Chern class. Let P be the collection of 27 fixed points. If we delete these points from T, we get a manifold, which we will call $T - P$, which is a complex but not compact manifold. Because of the additivity of the Euler characteristic, we have

$$\chi(T - P) = \chi(T) - \chi(P) = 0 - 27 = -27. \tag{16.10.5}$$

Here we are using the fact that the Euler characteristic of a point is one, so $\chi(P) = 27$. The group F has three elements and acts freely on $T - P$, so $(T - P)/F$ is a manifold and

$$\chi((T - P)/F) = (-27)/3 = -9. \tag{16.10.6}$$

Since F acts holomorphically on the z_k, the complex structure of $T - P$ is invariant under F, and $(T - P)/F$ is a complex manifold. Also, the everywhere nonzero holomorphic three form

$$\omega = dz^1 \wedge dz^2 \wedge dz^3 \tag{16.10.7}$$

of $T - P$ is evidently invariant under (16.10.3), so it makes sense on $(T - P)/F$, which therefore has vanishing first Chern class.

Of course, $(T - P)/F$ is not compact; it has 27 'holes' where points were originally removed from T. It is, however, possible to fill in these holes with a suitable complex manifold of $c_1 = 0$. The requisite object is a certain three-dimensional complex manifold, which we will call Y. In the terminology of example (5) of §15.6.3, Y is the total space of the bundle $L(-3)$ over CP^2. Y admits a metric of $SU(3)$ holonomy, which (unlike most such metrics) can be described explicitly, though we will not do so. The Euler characteristic of Y is 3. Gluing in a copy of Y at each of the 27 'holes' in $(T - P)/F$, we make a compact and nonsingular manifold $\hat{Z}$, which (because $(T - P)/F$ and Y have these properties) is a complex manifold with $c_1 = 0$. The additivity of the Euler characteristic implies that

$$\chi(\hat{Z}) = \chi((T - P)/F) + 27 \cdot \chi(Y) = -9 + 27 \times 3 = 72. \qquad (16.10.8)$$

Thus, if the spin connection is embedded in the gauge group, compactification on $\hat{Z}$ gives a model with 36 generations of chiral fermions in four dimensions.

The nonzero Hodge numbers of the space Y are $h^{0,0} = h^{1,1} = h^{2,2} = 1$.[*] In particular, the Ricci-flat Kähler metric of Y depends on a single real parameter, its 'radius'. Upon gluing in a copy of Y in each 'hole' in $(T - P)/F$ to make the compact manifold $\hat{Z}$, the radii r_i, $i = 1, \ldots, 27$ of the 27 copies of Y are still free parameters. In the limit of $r_i \rightarrow 0$, we recover the 'orbifold' $Z = T/F$, which was discussed in §9.5.2. Although Z is not a smooth manifold, string propagation on Z seems to make perfect sense, as we indicated in that discussion. The correct number of generations, namely 36, emerges in the orbifold computation provided one includes the contributions of the 'twisted sectors'.

Thus, string propagation on the orbifold Z can be understood as a special limit, $r_i \rightarrow 0$, of string propagation on the smooth manifold $\hat{Z}$ of $SU(3)$ holonomy. A field theorist might consider it necessary to resolve the singularities of Z by replacing the singularities by copies of Y, a process known in algebraic geometry as 'blowing up' the singularities; in field theory the explicit resolution of the singularities enters in computing the values $+72$ for the Euler characteristic and $+36$ for the number of generations. It actually is quite remarkable that the string computation automatically includes the correction term $+3 \times 27$ in (16.10.8) (as contributions from the twisted sectors) even if one does not explicitly resolve the singularities.

---

[*] These coincide with the Hodge numbers of CP^2, because Y, being the total space of the line bundle $L(-3)$ over CP^2, is contractible onto CP^2.

In many ways orbifolds are the simplest examples of manifolds of $SU(3)$ holonomy, the other leading contenders being the complete intersections of hypersurfaces in CP^N. It is quite conceivable that better understanding will uncover physical principles that single out this particularly simple case, and even if this does not occur it is very interesting to have an exactly soluble example.

The answer 36 for the number of generations is not very attractive. However, there are quite a few variants of the above construction that give more reasonable answers. One can replace the torus T and discrete group F by other complex tori and discrete group actions, and generate quite a few analogs of the Z orbifold and its resolution $\hat{Z}$. A few of these examples give four generations if the spin connection is embedded in the gauge group, and by relaxing the assumption of embedding the spin connection in the gauge group there are many other possibilities. Unfortunately, a thorough discussion would take us too far afield.

16.11 Outlook

Elementary-particle phenomena present us with many unsolved problems. If these problems are addressed in the context of compactification of ten-dimensional string theory, then most of the puzzles can be translated into questions about the compact manifold K. For the most part we can write a dictionary translating questions about observed physical phenomena into questions about the compact manifold K. We have tried to give the reader of flavor of what such a dictionary might look like, even though we have omitted many entries (*e.g.*, a discussion of neutrino masses). Undoubtedly, we do not yet understand many entries in this dictionary correctly. There is, however, one conspicuous blank page, or blank chapter, in this dictionary. This is the lack of an understanding of why the cosmological constant vanishes. About this question there are no compelling ideas as of this writing. Very likely, the future development of string theory will depend in large part on success or lack of success in understanding the cosmological constant. Quite possibly, to formulate a proper model of supersymmetry breaking – and most of the phenomena in the low-energy world – requires an understanding of why the cosmological constant vanishes after supersymmetry breaking. The best that we can hope is that the description in this chapter may be an idealization of the picture that might emerge from such an understanding.

Of course, the cosmological constant is just one of the most conspicuous mysteries. The truth is that while much is known about string theory, the roots of this subject lie hidden. We do not know what principles unify the many surprises that make string theory possible. We do not know why propagating strings, or world-sheet path integrals, are a proper starting point for a generalization of nonabelian gauge theory and general relativity. The answers to such questions may lie in directions not yet contemplated.

Bibliography

Chapter 8

The construction of loop amplitudes started very shortly after the formulation of the M-particle tree diagrams of the dual resonance model (before the significance of the critical dimension and the condition on the ground-state mass were understood). First came the paper by Kikkawa, Sakita and Virasoro [337], which was quickly followed by operator constructions by Bardakçi, Halpern and Shapiro [45] and Amati, Bouchiat and Gervais [27] followed by many others. The analysis of the divergences of the planar one-loop amplitude in terms of θ functions was first described by Neveu and Scherk [398,200], and an analysis of all the one-loop open-string diagrams was given by Gross, Neveu, Scherk and Schwarz [263]. The significance of the critical dimension in giving unitary nonplanar amplitudes was first realized by Lovelace [360], who correctly conjectured that there would be an extra set of decoupled states in the critical dimension.

Brink and Olive [77] constructed a projection operator for on-shell physical states, which enabled them to calculate unitary one-loop amplitudes in the critical dimension [78]. Their results can now be obtained more simply by including ghost modes, which is the way we have presented the calculation in §8.1.1.

The systematic factorization of the nonplanar diagrams in terms of the closed-string states was carried out in [112] and [106]. The interpretation of the divergences in planar open-string loop amplitudes in terms of the emission into the vacuum of a tachyon state and a massless dilaton state and their removal by renormalization of the string tension was given by Shapiro [465] and by Ademollo et al. [4]. The closed-string one-loop amplitude was calculated by Shapiro [464], who also realized the significance of modular invariance.

A discussion of the cosmological constant induced by one-loop effects was given by Rohm [429]. It is also discussed in [416]. The disk and projective-plane diagrams with external closed-string states were considered by Green and Schwarz [250].

Chapter 9

There were a few calculations of loop amplitudes in the RNS model
in the days before space-time supersymmetry was appreciated, beginning with Goddard and Waltz [225] and followed by a number of others
[106,234,79,506]. Using the light-cone approach with space-time supersymmetry (the formalism of chapter 5), Green and Schwarz evaluated
four-particle one-loop amplitudes for open superstrings in type I theories
[240], as described in §9.1, and type II closed-string loop amplitudes [242],
as described in §9.1.2. They analyzed and interpreted the divergences of
the open-string loop amplitudes and demonstrated the finiteness of the
closed-string loop amplitudes.

The compactification of the free closed bosonic string theory on a torus
was considered by Cremmer and Scherk [115], who also calculated the
nonplanar open-string loop in the compactified theory, demonstrating the
existence of bound states with nonzero winding number. Toroidal compactification of superstring theories, described in §9.2.2, was considered by
Green, Schwarz and Brink [241], who also considered the limit in which
the compact dimensions have radii that vanish along with the inverse
string tension - leading to a calculation of $N = 4$ Yang-Mills and $N = 8$
supergravity one-loop amplitudes for space-time dimensions $D \leq 10$, the
results of which are summarized in §9.2.3.

The effect of twisting the boundary conditions in order to break supersymmetry was considered in the previously cited paper by Rohm [429],
where the effect of supersymmetry breaking on the cosmological constant
was also considered.

The one-loop heterotic string amplitudes were calculated by Gross,
Harvey, Martinec and Rohm [269] and Yahikozawa [526], where one-loop
modular invariance and finiteness are demonstrated. Even in a modular-
invariant ten-dimensional theory, modular invariance can break down
upon compactification, as shown by Witten [521]. A criterion for modular invariance to survive compactification was formulated in that paper,
where it was also shown that the various closed-superstring models are
modular invariant to all orders.

Orbifolds were introduced by Dixon, Harvey, Vafa and Witten [145].
Modular-invariant ten-dimensional string theories without space-time supersymmetry were constructed in [146,26,454]; the tachyon-free model
described in §9.5.3 was formulated in the papers of Dixon and Harvey
[146] and of Alvarez-Gaumé, Ginsparg, Moore and Vafa [26].

Chapter 10

The literature on anomalies in point-particle theories is enormous, and only a small selection will be referred to here. The subject of anomalies in field theory originated with the papers of Steinberger [468] and Schwinger [452]. The study of the anomalous triangle diagram with one axial current and two vector currents was pioneered by Adler [5], Bell and Jackiw [61] and Bardeen [49], which led to the analysis of anomalies in conservation of gauge currents [265,70,211].

Work on anomalies in higher-dimensional Yang-Mills theory and gravity is particularly relevant to chapter 13 and is referred to in the bibliographic discussion of that chapter. The potential implications of hexagon anomalies for type I superstring theory were discussed in [512], where it was suggested that anomalies would occur in all type I theories. The cancellation of the anomalies in (ten-dimensional) type IIB supergravity was discovered by Alvarez-Gaumé and Witten [21] in the same paper in which gravitational anomalies were first investigated. This work, as well as comments by Friedan and Shenker, prompted the investigation of anomalies in superstring theories. The analysis of anomalies in the hexagon diagram of type I superstring theory, described in §10.2, was carried out by Green and Schwarz [252]. The interpretation of the results in terms of the effective low-energy theory (described in chapter 13) was also given by Green and Schwarz [249]. This analysis indicated the possibility of anomaly cancellation for an $E_8 \times E_8$ superstring theory, which in turn led to the discovery of the heterotic string by Gross, Harvey, Martinec and Rohm [267]. The absence of anomalies in the modular transformations of the heterotic string [268,269] is closely related to the absence of gauge and gravitational anomalies.

Chapter 11

The idea of using path-integral methods in the study of string theory dates back to the earliest days of the subject [300,215]. This approach proved to be quite useful in the construction of bosonic open-string tree diagrams, one-loop diagrams and multiloop diagrams [11,12,320,323,324,325,360]. The resulting form for the amplitudes was in accord with the analog model of Fairlie and Nielsen [171]. While this work was correct for the most part, it was incomplete because it did not take into account the effects of the Fadeev-Popov ghost modes. It was not until the work of Polyakov [417,418] that it was understood how to use covariant functional methods of integrating over string world-sheets to take proper account of the local symmetries. This subject has now blossomed into a field of its own, which would deserve a separate book to do it justice.

A parallel development was the use of the light-cone gauge for functional calculations. It originated with the description of free strings by Goddard, Goldstone, Rebbi and Thorn [226] (discussed in chapter 2) and was applied to string interactions by Mandelstam for the bosonic string in [368]. He developed methods of calculating tree diagrams in the bosonic string theory [365–368] described in §11.3 from which the vertex for coupling three arbitrary open-string states of Ademollo et al. [2] could be recovered. Mandelstam also applied these techniques to the RNS string theory [367]. The method of defining the theory with a lattice cutoff was developed by Giles and Thorn [219,476]. The fact that the divergences cancel between the external wave functions and the vertices in the bosonic theory in the critical dimension is also indicated by the work of Thorn [476].

The Lorentz invariance of the theory is not manifest in the light-cone gauge formalism, although Mandelstam was able to prove it for the bosonic open string in the critical dimension $D = 26$ [369]. The light-cone functional method was also used for one-loop calculations [31,32,368].

The calculation of the measure factors arising from the functional determinant and the Jacobian for the conformal mapping from the string diagram to the upper half plane were given by Mandelstam [370] for open and closed bosonic string tree diagrams, one-loop diagrams and multiloop diagrams. The calculation of the functional determinants is based on the method of McKean and Singer [383] and Alvarez [14] (see also [155]).

The light-cone gauge functional method can be used to define off-shell amplitudes. Amplitudes with off-shell 'point-like' states were studied by Green [235,237-239]. These are closely related to the off-shell current amplitudes of Schwarz [445-447] and Corrigan and Fairlie [111] for open strings and Green and Shapiro [236] for closed strings.

The bosonic three-string vertex was used as the basis of the interacting light-cone field theory by Kaku and Kikkawa [326,327] and Cremmer and Gervais [113,114]. This was also discussed in [293]. Green and Schwarz generalized the construction to the cubic interactions of open superstrings [243] and, together with Brink, to type IIB closed superstrings [244]. The unified light-cone gauge description making use of the $SU(4) \times U(1)$ formalism was developed by Green and Schwarz in [248]. The four-particle superstring amplitudes were calculated in [248], while the M-particle amplitudes were constructed by Mandelstam in [371].

Chapter 12

Coupling of spinors to a gravitational field is described in textbooks on general relativity, e.g. [483], pp. 365-74. A comparatively simple explanation of why a spin structure cannot be defined on an arbitrary Riemannian manifold can be found in [282]. For an interpretation of this in terms of global anomalies in quantum mechanical path integrals, see [194,521]. Some of the elementary facts about spin structures on the string world sheet are explained in somewhat more detail in [454]. Some excellent references on two dimensional world-sheet geometry are [14,15].

Some references for physicists on some of the subjects of this chapter are [159,394]. A very natural extension of ideas developed in this chapter is cohomology with integer coefficients; relatively palatable expositions can be found in [253,69]. References on more detailed aspects of characteristic classes and fiber bundles are [385,306]. For some applications to string theory of cohomology and characteristic classes with integer coefficients, see [490,521,522]. Many references that we have given here are also relevant to chapters 14–16.

Chapter 13

The $N = 1$, $D = 4$ supergravity theory was first constructed in [188,127] and the $N = 1$, $D = 11$ theory was suggested by Nahm [392] and constructed by Cremmer, Julia and Scherk [118]. The $N = 1$, $D = 10$ super Yang–Mills theory of [80,223] is coupled to $N = 1$, $D = 10$ supergravity in [95,62] for the abelian case and [96,99] for the nonabelian one. The $N = 4$ $D = 8$ supergravity theory is given in [119]. For a review of supergravity theories see [481,493,202].

Nahm's paper [392] also suggested the existence of two $N = 2$, $D = 10$ supergravity and superstring theories (now referred to as type IIA and type IIB). The type IIB supergravity was first considered explicitly in [245] and developed further in [449,450,298]. The discussion in §13.1.2 is based on [450]. The possible phenomenological relevance of type IIB supergravity is discussed in [272].

The early history of anomalies was sketched already in the discussion for chapter 10. Important recent developments include the study of gauge anomalies in higher dimensions [258,181,182,183,184,466] and the introduction of gravitational and mixed anomalies [21]. The literature on the subject is enormous. Some modern discussions include [24,25,50,15,59].

The cancellation of anomalies in type IIB supergravity, as well as their noncancellation for type I supergravity, is first demonstrated in [21]. The discovery that string effects can lead to a cancellation for $N = 1$ theories with $SO(32)$ or $E_8 \times E_8$ in terms of an effective field theory analysis was given in [249]. The cancellation mechanism after compactification of some dimensions is discussed in [514,251]. We have not been able to say much about global anomalies. A detailed discussion can be found in [521]. The anomaly cancellation mechanism for the $SO(16) \times SO(16)$ theory is presented in [146,26].

Chapter 14

For references on the early history of Kaluza-Klein theory, and its applications in string theory, see the bibliography to chapter 1 and the review article [435]. A useful collection of old and new papers on Kaluza-Klein theory is the book [28].

Four-dimensional unified theories are an important point of departure in thinking about unification in 4+n-dimensions. An early attempt at quark-lepton unification was [411]. Grand unification in the sense of embedding all known gauge interactions in a unified group was initiated with the $SU(5)$ model in [212], followed by the $SO(10)$ model [213,199] and E_6 [273,274,275,1]. Most of these papers and many more are reprinted in [533].

The connection of massless four-dimensional particles with zero modes of wave operators in $4 + n$ dimensions was proposed in [507]. In that paper it was also suggested that the numbers (and charges) of massless fermions would be derived from topological invariants of a compact manifold. Further work on the latter problem was presented in [98,498,425,519,409,100,348,485]. The problem of fermion quantum numbers was explicitly formulated as an index problem in [498,519]. The fact that one cannot obtain chiral fermions from $(4 + n)$-dimensional theories without elementary gauge fields was proved for odd n in [507], for n divisible by four in [498], and for n of the form $4k + 2$ in [519]. The treatment of the latter case depends on a theorem originally proved in [38]. Theories with elementary gauge fields in $4 + n$ dimensions were investigated in [98,498,425,519,409,100,348,485]. Embedding the spin connection in the gauge group and relating the number of fermions to the Euler characteristic were proposed in [519] and applied in string theory (in a somewhat more specialized context, which we will come to in chapters 15 and 16) in [86]. The closest antecedent of this idea in four-dimensional theories is 'orthogonal family unification' [208,504], or its E_8 analog [55].

Ideas about axions and the strong CP problem were developed in [414, 484,503,341,135]. For the material on axions in §14.3.2 (and additional topics we have omitted here) see [514,489,142].

References on differential geometry on the string world sheet have been given in the bibliography to chapter 12.

There is a connection between index problems and supersymmetric quantum mechanics; it was presented in [510,511]. This connection was used to give a path-integral proof of the Atiyah–Singer index theorem in [194] and [22]. A physics-oriented proof of the character-valued generalization of the index theorem was given in [519] for continuous symmetries and in [229] in the general case.

The question of whether the 'constants' of nature are really time independent is important in string theory, because if the dilaton field is strictly massless in nature, this will result in time variation of the natural 'constants.' This argument is briefly sketched at the end of §14.5, and more detail can be found in [483], pp. 622-631. (The discussion there concerns the Brans-Dicke scalar, which is rather similar to the dilaton of string theory.) The failure to observe changes in time of the natural constants is one of the main reasons to believe that the dilaton is not strictly massless. Time variation of the gravitational constant was first advocated in [143,144] as an approach to what we would now call the hierarchy problem. Direct experimental proof that the gravitational constant is changing more slowly than Dirac proposed came only very recently [287]. The bounds on time variation of other constants are much better; a useful review is [158]. A massless or very light dilaton might also show up in laboratory tests of gravitation. There are two relevant kinds of experiment. If the dilaton is not strictly massless, one anticipates apparent deviations from the $1/r^2$ force law of gravity. The experimental situation is rather complex; see [221] for a review. If the dilaton is strictly massless, it might still show up in tests of the equivalence principle at laboratory distances [351,290].

Chapter 15

An excellent short introduction to complex manifolds is [101]. For more detail, [488] is relatively elementary while [257] is more advanced and comprehensive. Geometry of complex manifolds is just one facet of algebraic geometry, which can be formulated over more general fields; the generalization is important in number theory and may prove to be important in string theory. Some expositions of the more abstract viewpoint about algebraic geometry (in ascending order of difficulty) are [334,460,280]. Some introductions to complex manifolds written for physicists are [17,294]. Many recent papers have initiated a deeper application of holomorphic geometry to world-sheet phenomena [91,68,60,26,374,196]. Among many references on Riemann surfaces we cite [178,391,63].

The relation between unbroken supersymmetry and covariantly constant spinor fields was pointed out in [507]. The application to ten dimensional theories – leading to manifolds of $SU(3)$ holonomy – was developed in [86] and in subsequent papers, which can be found in the bibliography to chapter 16. For the Calabi conjecture see [83]; the proof by Yau is in [530]. For the Donaldson–Uhlenbeck–Yau generalization of the Calabi conjecture to Yang–Mills theory see [149,479].

The examples of complex manifolds and holomorphic vector bundles given in the text are standard except for vector bundles over hypersurfaces in CP^N described as example (6) in §15.6.3. These were constructed in an extension (unpublished) of [523]; see §16.7 for use of these bundles to construct $SO(10)$ models.

Chapter 16

Compactification on manifolds of $SU(3)$ holonomy was first studied in [86], where most of our statements about the spectrum of massless particles were described. We have described just the simplest constructions of examples of manifolds of $SU(3)$ holonomy; for further examples see [531,469]. Implications of symmetry breaking by Wilson lines regarding the symmetry breaking pattern, the low energy couplings, and the fine-tuning problem were explored in [518,455,75]. See [297] for an antecedent of this idea, involving a Wilson line with a continuous range of possible values. Implications of symmetry breaking by Wilson lines for the allowed values of electric and magnetic charge were explored in [490]. Our discussion of global symmetries in compactified models largely follows [518]. The renormalization group calculation presented in §16.9 was done in [137]. Additional phenomenological issues were discussed in [396, 415,140,93,333,126,137,162], along with many more recent papers.

The topological formula for the superpotential (in the field theory limit) was introduced in [469]. There has been much progress in evaluating this formula, both for positive Euler characteristic [470], and negative Euler characteristic [88]. See also [470] for expressions for D terms that are valid in the field-theory limit. The nonrenormalization theorem for the superpotential, showing that the formula derived in the field theory limit has no corrections in finite orders in α', was derived in [523]. That this formula breaks down, under certain conditions, at a nonperturbative level was shown in [142]. We have not explained the latter result in the text, though some of the background can be found in §14.3.2. An argument that the superpotential likewise has no corrections in finite orders in $\hbar$

has been given in [381].

In §16.6.3 and §16.7-8 we present a macroscopic approach [523,142] for predicting the existence of vacuum solutions. Though we do not do so in the text, it is possible to address the same issues microscopically, from the point of view of σ-model calculations of the sort described (for the bosonic theory) in §3.4. The σ models arising in compactification of supersymmetric string theories on manifolds of $SU(3)$ holonomy were discussed in [86,87,84,457,303]. See the bibliography to chapter 3 for additional references on the role of σ models in string theory.

Early calculations of the beta function on σ models based on Ricci-flat Kähler manifolds [17-20] (when interpreted in terms of string theory in the above-cited papers) gave the first suggestion that such manifolds would give solutions of string theory equations. Four-loop σ-model calculations [259] give a nonzero beta function which was analyzed in [419] and [190] and in particular shown to agree with a result that can be extracted from analysis of string scattering amplitudes obtained in [270]. The counterterm in question was shown in [395] to have the interpretation (in terms of a readjustment of the vacuum) that we describe in §16.6.3 and §16.7-8 from the macroscopic viewpoint of [523]. The generalized symmetry breaking scheme sketched in §16.7-8 has been tested by a direct perturbative calculation in [524]. Orbifold calculations [148,278], in addition to being interesting on other grounds, have given a test of a very different sort.

The ultraviolet finiteness of sigma models based on hyper-Kähler manifolds has also been a subject of much study [347,389]. An argument for nonperturbative ultraviolet finiteness of such models was presented in [43].

For other approaches to generalizations of the embedding of the spin connection in the gauge group, see, e.g. [58,471,304]. A detailed discussion of the limiting low-energy supergravity that can emerge from string compactification can be found in [126]. Limits on neutral currents due to light gauge bosons beyond those of the standard model that arise in some compactification schemes are discussed in [124]. General reviews of low-energy phenomenology can be found in [453,310,166]. A recent attempt at detailed realistic models is [256].

REFERENCES

1. Achiman, Y. and Stech, B. (1978), 'Quark-lepton symmetry and mass scales in an E_6 unified gauge model', *Phys. Lett.* **77B**, 389.
2. Ademollo, M., Del Giudice, E., Di Vecchia, P. and Fubini, S. (1974), 'Couplings of three excited particles in the dual-resonance model', *Nuovo Cim.* **19A**, 181.
3. Ademollo, M., D'Adda, A., D'Auria, R., Napolitano, E., Di Vecchia, P., Gliozzi, F. and Sciuto, S. (1974), 'Unified dual model for interacting open and closed strings', *Nucl. Phys.* **B77**, 189.
4. Ademollo, M., D'Adda, A., D'Auria, R., Gliozzi, F., Napolitano, E., Sciuto, S. and Di Vecchia, P. (1975), 'Soft dilatons and scale renormalization in dual theories', *Nucl. Phys.* **B94**, 221.
5. Adler, S.L. (1969), 'Axial-vector vertex in spinor electrodynamics', *Phys. Rev.* **177**, 2426.
6. Adler, S.L. and Bardeen, W.A. (1969), 'Absence of higher-order corrections in the anomalous axial-vector divergence equation', *Phys. Rev.* **182**, 1517.
7. Affleck, I., Dine, M. and Seiberg, N. (1985), 'Dynamical supersymmetry breaking in four dimensions and its phenomenological implications', *Nucl. Phys.* **B256**, 557.
8. Aharonov, Y. and Casher, A. (1986), 'On the origin of the universe in the context of string models', *Phys. Lett.* **166B**, 289.
9. Ahn, Y.J., Breit, J. and Segré, G. (1985), 'The one-loop effective Lagrangian of the superstring', *Phys. Lett.* **162B**, 303.
10. Ahn, Y.J. and Breit, J.D. (1986), 'On one-loop effective Lagrangians and compactified superstrings', *Nucl. Phys.* **B273**, 75.
11. Alessandrini, V. (1971), 'A general approach to dual multiloop diagrams', *Nuovo Cim.* **2A**, 321.
12. Alessandrini, V. and Amati, D. (1971), 'Properties of dual multiloop amplitudes', *Nuovo Cim.* **4A**, 793.
13. Alessandrini, V., Amati, D. and Morel, B. (1972), The asymptotic behavior of the dual Pomeron amplitude', *Nuovo Cim.* **7A**, 797.
14. Alvarez, O. (1983), 'Theory of strings with boundaries. Fluctuations, topology and quantum geometry', *Nucl. Phys.* **B216**, 125.
15. Alvarez, O., Singer, I. and Zumino, B. (1984), 'Gravitational anomalies and the family's index theorem', *Commun. Math. Phys.* **96**, 409.
16. Alvarez, O. (1986), 'Differential geometry in string models', in *Workshop on Unified String Theories, 29 July – 16 August, 1985*, eds. M. Green and D. Gross (World Scientific, Singapore), p. 103.
17. Alvarez-Gaumé, L. and Freedman, D.Z. (1980) in *Unification of the Fundamental Particle Interactions*, eds. S. Ferrara et. al. (Plenum Press).
18. Alvarez-Gaumé, L. and Freedman, D.Z. (1980), 'Kähler geometry and the renormalization of supersymmetric *sigma* models', *Phys. Rev.* **D22**, 846.
19. Alvarez-Gaumé, L. and Freedman, D.Z. (1981), 'Geometrical structure and ultraviolet finiteness in the supersymmetric σ-model', *Commun. Math. Phys.* **80**, 443.
20. Alvarez-Gaumé, Mukhi, S. and D. Z. Freedman, (1981), 'The background field method and the ultraviolet structure of the supersymmetric nonlinear σ-model', *Ann. Phys.* (N.Y.) **134**, 85.

21. Alvarez–Gaumé, L. and Witten, E. (1983), 'Gravitational anomalies', *Nucl. Phys.* **B234**, 269.
22. Alvarez–Gaumé, L. (1983), 'Supersymmetry and the Atiyah–Singer index theorem', *Commun. Math. Phys.* **90**, 161.
23. Alvarez–Gaumé, L. (1983), 'A note on the Atiyah–Singer index theorem', *J. Phys.* **A16**, 4177.
24. Alvarez–Gaumé, L. and Ginsparg, P. (1984), 'The topological meaning of non-Abelian anomalies', *Nucl. Phys.* **B243**, 449.
25. Alvarez–Gaumé, L. and Ginsparg, P. (1985), 'Geometry anomalies', *Nucl. Phys.* **B262**, 439.
26. Alvarez–Gaumé, L, Ginsparg, P., Moore, G. and Vafa, C. (1986), 'An $O(16) \times O(16)$ heterotic string', *Phys. Lett.* **171B**, 155.
27. Amati, D., Bouchiat, C. and Gervais, J.L. (1969), 'On the building of dual diagrams from unitarity', *Nuovo Cim. Lett.* **2**, 399.
28. Appelquist, T., Chodos, A. and Freund, P.G.O. (1987), *Modern Kaluza-Klein Theory and Applications*, (Benjamin-Cummings).
29. Ardalan, F. and Arfaei, H. (1986), 'Critical dimensions from loops in a string sigma model', *Phys. Lett.* **175B**, 164.
30. Aref'eva, I.Y. and Volovich, I.V. (1985), 'Spontaneous compactification of $O(32)$ superstrings', *Phys. Lett.* **158B**, 31.
31. Arfaei, H. (1975), 'Volume element for loop diagram in the string picture of dual models', *Nucl. Phys.* **B85**, 535.
32. Arfaei, H. (1976), 'Theory of closed interacting strings', *Nucl. Phys.* **B112**, 256.
33. Atiyah, M.F. and Singer, I.M. (1968), 'The index of elliptic operators: I', *Ann. of Math.* **87**, 484.
34. Atiyah, M.F. and Segal, G.B. (1968), 'The index of elliptic operators: II', *Ann. of Math.* **87**, 531.
35. Atiyah, M.F. and Singer, I.M. (1968), 'The index of elliptic operators: III', *Ann. of Math.* **87**, 546.
36. Atiyah, M.F. and Singer, I.M. (1968), 'The index of elliptic operators: IV', *Ann. of Math.* **93**, 119.
37. Atiyah, M.F. and Singer, I.M. (1968), 'The index of elliptic operators: V', *Ann. of Math.* **93**, 139.
38. Atiyah, M.F. and Hirzebruch, F. (1970), in *Essays in Topology and Related Subjects*, ed. A. Haefliger and R. Narasimhan (Springer-Verlag, New York).
39. Bailin, D. and Love, A. (1985), 'Compactifications of anomaly-free ten-dimensional supergravity', *Phys. Lett.* **157B**, 375.
40. Bailin, D. and Love, A. (1985), 'Cosmological instability in ten-dimensional supergravity', *Phys. Lett.* **163B**, 135.
41. Bailin, D., Love, A. and Wong, D. (1985), 'Supergravity limit of superstring theory and Friedmann–Robertson–Walker cosmology', *Phys. Lett.* **165B**, 270.
42. Bailin, D., Love, A. and Thomas, S. (1986), 'Dimensional reductions of superstring theory', *Nucl. Phys.* **B273**, 537.
43. Banks, T. and Seiberg, N. (1986), 'Nonperturbative infinities', *Nucl. Phys.* **B273**, 157.
44. Barbieri, R., Cremmer, E. and Ferrara, S. (1985), 'Flat and positive potentials in $N = 1$ supergravity', *Phys. Lett.* **163B**, 143.

45. Bardakçi, K., Halpern, M.B. and Shapiro, J.A. (1969), 'Unitary closed loops in Reggeized Feynman theory', *Phys. Rev.* **185**, 1910.

46. Bardakçi, K. (1974), 'Dual models and spontaneous symmetry breaking', *Nucl. Phys.* **B68**, 331.

47. Bardakçi, K. (1974), 'Dual models and spontaneous symmetry breaking II', *Nucl. Phys.* **B70**, 397.

48. Bardakçi, K. (1978), 'Spontaneous symmetry breakdown in the standard dual string model', *Nucl. Phys.* **B133**, 297.

49. Bardeen, W.A. (1969), 'Anomalous Ward identities in spinor field theories', *Phys. Rev.* **184**, 1848.

50. Bardeen, W.A. and Zumino, B. (1984), 'Consistent and covariant anomalies in gauge and gravitational theories', *Nucl. Phys.* **B244**, 421.

51. Barger, V., Deshpande, N.G. and Whisnant, K. (1986), 'Phenomenological mass limits on extra Z of E_6 superstrings', *Phys. Rev. Lett.* **56**, 30.

52. Barger, V., Deshpande, N.G., Phillips, R.J.N. and Whisnant, K. (1986), 'Extra fermions in E_6 superstrings theories', *Phys. Rev.* **D33**, 1912.

53. Barr, S.M. (1985), 'Harmless axions in superstring theories', *Phys. Lett.* **158B**, 397.

54. Barr, S.M. (1985), 'Effects of extra light Z bosons in unified and superstring models', *Phys. Rev. Lett.* **55**, 2778.

55. Bars, I. and Günaydin, M. (1980), 'Grand unification with the exceptional group E_8', *Phys. Rev. Lett.* **45**, 859.

56. Bars, I. (1985), 'Compactification of superstrings and torsion', *Phys. Rev.* **D33**, 383.

57. Bars, I. and Visser, M. (1985), 'Number of massless fermion families in superstring theory', *Phys. Lett.* **163B**, 118.

58. Bars, I., Nemeschansky, D. and Yankielowicz, S. (1986), 'Torsion in superstrings', in *Workshop on Unified String Theories, 29 July – 16 August, 1985*, eds. M. Green and D. Gross (World Scientific, Singapore), p. 522.

59. Baulieu, L. (1986), 'Anomaly evanescence and the occurrence of mixed Abelian-non-Abelian gauge symmetries', *Phys. Lett.* **167B**, 56.

60. Belavin, A.A. and Knizhnik, V.G. (1986), 'Algebraic geometry and the geometry of quantum strings', *Phys. Lett.* **168B**, 201.

61. Bell, J.S. and Jackiw, R. (1969), 'A PCAC puzzle: $\pi^0 \to \gamma\gamma$ in the σ-model', *Nuovo Cim.* **60A**, 47.

62. Bergshoeff, E., De Roo, M., de Wit, B. and Van Nieuwenhuizen, P. (1982), 'Ten-dimensional Maxwell–Einstein supergravity, its currents, and the issue of its auxiliary fields', *Nucl. Phys.* **B195**, 97.

63. Bers, L. (1972), 'Uniformization, moduli and Kleinian groups', *Bull. London Math. Soc.* **4**, 257.

64. Binétruy, P. and Gaillard, M.K. (1986), 'Radiative corrections in compactified superstring models', *Phys. Lett.* **168B**, 347.

65. Binétruy, P., Dawson, S., Hinchcliffe, I. and Sher, M. (1985), 'Phenomenologically viable models from superstrings', *Nucl. Phys.* **B273**, 501.

66. Bonora, L., Pasti, P. and Tonin, M. (1985), 'Cohomologies and anomalies in supersymmetric theories', *Nucl. Phys.* **B252**, 458.

67. Bonora, L. and Cotta-Ramusino, P. (1986), 'Some remarks on anomaly cancellation in field theories derived from superstrings', *Phys. Lett.* **169B**, 187.

68. Bost, J.B. and Jolicoeur, T. (1986), 'A holomorphy property and critical dimension in string theory from an index theorem', *Phys. Lett.* **174B**, 273.

69. Bott, R. and Tu, L. (1983), *Differential Forms in Algebraic Topology*, (Springer-Verlag).

70. Bouchiat, C, Iliopoulos, J. and Meyer, P. (1972), 'An anomaly-free version of Weinberg's model', *Phys. Lett.* **38B**, 519.

71. Boulware, D.G. and Deser, S. (1985), 'String-generated gravity models', *Phys. Rev. Lett.* **55**, 2656.

72. Bowick, M.J., Smolin, L. and Wijewardhana, L.C.R. (1986), 'Role of string excitations in the last stages of black-hole evaporation', *Phys. Rev. Lett.* **56**, 424.

73. Braden, H.W., Frampton, P.H., Kephart, T.W. and Kshirsagar, A.K. (1986), 'Limitations of heterotic-superstring phenomenology', *Phys. Rev. Lett.* **56**, 2668.

74. Brans, C. and Dicke, R.H. (1961), 'Mach's principle and a relativistic theory of gravity', *Phys. Rev.* **124**, 925.

75. Breit, J.D., Ovrut, B.A. and Segrè, G. (1985), 'E_6 symmetry breaking in the superstring theory', *Phys. Lett.* **158B**, 33.

76. Breit, J.D., Ovrut, B.A. and Segrè, G. (1985), 'The one-loop effective Lagrangian of the superstring', *Phys. Lett.* **162B**, 303.

77. Brink, L., Olive, D. and Scherk, J. (1973), 'The gauge properties of the dual model Pomeron–Reggeon vertex: Their derivation and their consequences', *Nucl. Phys.* **B61**, 173.

78. Brink, L. and Olive, D. (1973), 'Recalculation of the unitary single planar dual loop in the critical dimension of space time', *Nucl. Phys.* **B58**, 237.

79. Brink, L. and Fairlie, D.B. (1974), 'Pomeron singularities in the fermion meson dual model', *Nucl. Phys.* **B74**, 321.

80. Brink, L., Schwarz, J.H. and Scherk, J. (1977), 'Supersymmetric Yang-Mills theories', *Nucl. Phys.* **B121**, 77.

81. Burgess, C., Font, A. and Quevedo, F. (1986), 'Low-energy effective action for the superstring', *Nucl. Phys.* **B272**, 661.

82. Burnett, T.H., Gross, D.J., Neveu, A., Scherk, J. and Schwarz, J.H. (1970), 'Renormalized self-energy operator in the dual-resonance model', *Phys. Lett.* **32B**, 115.

83. Calabi, E. (1955), 'On Kähler manifolds with vanishing canonical class, algebraic geometry and topology', in *Algebraic Geometry and Topology: A Symposium in Honor of S. Lefschetz* (Princeton University Press), p. 78.

84. Callan, C.G., Friedan, D., Martinec, E.J. and Perry, M.J. (1985), 'Strings in background fields', *Nucl. Phys.* **B262**, 593.

85. Campbell, B.A., Ellis, J., Enqvist, K., Nanopoulos, D.V., Hagelin, J.S. and Olive, K.A. (1986), 'Superstring dark matter', *Phys. Lett.* **173B**, 270.

86. Candelas, P., Horowitz, G., Strominger, A. and Witten, E. (1985), 'Vacuum configurations for superstrings', *Nucl. Phys.* **B258**, 46.

87. Candelas, P., Horowitz, G., Strominger, A. and Witten, E. (1985), 'Superstring phenomenology', in *Symp. on Anomalies, Geometry, Topology, March 28 – 30, 1985*, eds. W.A. Bardeen and A.R. White (World Scientific, Singapore), p. 377.

88. Candelas, P. (1985), lecture at the Jerusalem Winter School.

89. Castellani, L. (1986), 'Non-Abelian gauge fields from $10 \rightarrow 4$ compactification of closed superstrings', *Phys. Lett.* **166B**, 54.

90. Castellani, L., D'Auria, R., Gliozzi, F. and Sciuto, S. (1986), 'On the compactification of the closed supersymmetric string', *Phys. Lett.* **168B**, 47.

91. Catenacci, R., Cornalba, M., Martellini, M. and Reina, C. (1986), 'Algebraic geometry and path integrals for closed strings', *Phys. Lett.* **172B**, 328.

92. Cecotti, S., Derendinger, J.P., Ferrara, S., Girardello, L. and Roncadelli, M. (1985), 'Properties of E_6 breaking and superstring theory', *Phys. Lett.* **156B**, 318.

93. Cecotti, S., Ferrara, S., Girardello, L. and Porrati, M. (1985), 'Lorentz Chern–Simons terms in $N = 1$ 4D supergravity consistent with supersymmetry and string compactification', *Phys. Lett.* **164B**, 46.

94. Cecotti, S., Ferrara, S., Girardello, L., Pasquinacci, A. and Porrati, M. (1986), 'Matter couplings in higher derivative supergravity', *Phys. Rev.* **D33**, 2504.

95. Chamseddine, A.H. (1981), '$N = 4$ supergravity coupled to $N = 4$ matter and hidden symmetries', *Nucl. Phys.* **B185**, 403.

96. Chamseddine, A.H. (1981), 'Interacting supergravity in ten dimensions: The role of the six-index gauge field', *Phys. Rev.* **D24**, 3065.

97. Chang, D. and Mohapatra, R.N. (1986), 'A superstring inspired low-energy electro-weak model', *Phys. Lett.* **175B**, 304.

98. Chapline, G.F. and Slansky, R. (1982), 'Dimensional reduction and flavor chirality', *Nucl. Phys.* **B209**, 461.

99. Chapline, G.F. and Manton, N.S. (1983), 'Unification of Yang–Mills theory and supergravity in ten dimensions', *Phys. Lett.* **120B**, 105.

100. Chapline, G.F. and Grossman, B. (1984), 'Dimension reduction and massless chiral fermions', *Phys. Lett.* **135B**, 109.

101. Chern, S.S. (1967), *Complex Manifolds Without Potential Theory* (D. V. Nostrand Co., Princeton).

102. Cho, Y.M. and Freund, P.G.O. (1975), 'Non-Abelian gauge fields as Nambu-Goldstone fields', *Phys. Rev.* **D12**, 1711.

103. Choi, K. and Kim, J.E. (1985), 'Harmful axions in superstring models', *Phys. Lett.* **154B**, 393.

104. Choi, K. and Kim, J.E. (1985), 'Domain walls in superstring models', *Phys. Rev. Lett.* **55**, 2637.

105. Choi, K. and Kim, J.E. (1985), 'Compactification and axions in $E_8 \times E_8'$ superstring models', *Phys. Lett.* **165B**, 71.

106. Clavelli, L. and Shapiro, J.A. (1973), 'Pomeron factorization in general dual models', *Nucl. Phys.* **B57**, 490.

107. Clavelli, L. (1986), 'Proof of one-loop finiteness of type-I SO(32) superstring theory', *Phys. Rev.* **D33**, 1098.

108. Cohen, E., Ellis, J., Gomez, G. and Nanopoulos, D.V. (1985), 'Superstring compactification and supersymmetry breaking', *Phys. Lett.* **160B**, 62.
109. Cohen, E., Ellis, J., Enqvist, K. and Nanopoulos, D.V. (1985), 'Scales in superstring models', *Phys. Lett.* **161B**, 85.
110. Cohen, E., Ellis, H., Enqvist, K. and Nanopoulos, D.V. (1985), 'Experimental predictions from the superstring', *Phys. Lett.* **165B**, 76.
111. Corrigan E.F. and Fairlie, D.B. (1975), 'Off-shell states in dual resonance theory', *Nucl. Phys.* **B91**, 527.
112. Cremmer, E. and Scherk, J. (1972), 'Factorization of the Pomeron sector and currents in the dual resonance model', *Nucl. Phys.* **B50**, 222.
113. Cremmer, E. and Gervais, J.L. (1974), 'Combining and splitting relativistic strings', *Nucl. Phys.* **B76**, 209.
114. Cremmer, E. and Gervais, J.L. (1975), 'Infinite component field theory of interacting relativistic strings and dual theory', *Nucl. Phys.* **B90**, 410.
115. Cremmer, E. and Scherk, J. (1976), 'Dual models in four dimensions with internal symmetries', *Nucl. Phys.* **B103**, 399.
116. Cremmer, E. and Scherk, J. (1976), 'Spontaneous compactification of space in an Einstein–Yang–Mills–Higgs model', *Nucl. Phys.* **B108**, 409.
117. Cremmer, E. and Scherk, J. (1977), 'Spontaneous compactification of extra space dimensions', *Nucl. Phys.* **B118**, 61.
118. Cremmer, E., Julia, B. and Scherk, J. (1978), 'Supergravity theory in 11 dimensions', *Phys. Lett.* **76B**, 409.
119. Cremmer, E. and Julia, B. (1979), 'The SO(8) supergravity', *Nucl. Phys.* **B159**, 141.
120. Cremmer, E., Scherk, J. and Schwarz, J.H. (1979), 'Spontaneously broken $N = 8$ supergravity', *Phys. Lett.* **84B**, 83.
121. Cremmer, E., Julia, B., Scherk, J., Ferrara, S., Girardello, L. and Van Nieuwenhuizen, P. (1979), 'Spontaneous symmetry breaking and Higgs effect in supergravity without cosmological constant', *Nucl. Phys.* **B147**, 105.
122. Daniel, M. and Mavromatos, N.E. (1986), 'The heterotic string and supersymmetric counterparts of the Lorentz Chern–Simons terms', *Phys. Lett.* **173B**, 405.
123. del Aguila, F., Blair, G., Daniel, M. and Ross, G.G. (1986), 'Superstring inspired models', *Nucl. Phys.* **B272**, 413.
124. del Aguila, F., Blair, G., Daniel, M. and Ross, G.G. (1987), 'Analysis of neutral currents in superstring inspired models', *Nucl. Phys.* **B283**, 50.
125. Derendinger, J.P., Ibañez, L.E. and Nilles, H.P. (1985), 'On the low energy $d = 4, N = 1$ supergravity theory extracted from the $d = 10, N = 1$ superstring', *Phys. Lett.* **155B**, 65.
126. Derendinger, J.P., Ibañez, L.E. and Nilles, H.P. (1986), 'On the low-energy limit of superstring theories', *Nucl. Phys.* **B267**, 365.
127. Deser, S. and Zumino, B. (1976), 'Consistent supergravity', *Phys. Lett.* **62B**, 335.
128. de Wit, B. and Freedman, D.Z. (1977), 'On SO(8) extended supergravity', *Nucl. Phys.* **B130**, 105.

129. de Wit, B. and Nicolai, H. (1982), '$N = 8$ supergravity with local $SO(8) \times SU(8)$ invariance', *Phys. Lett.* **108B**, 285.

130. DeWitt, B.S. (1967), 'Quantum theory of gravity. I. The canonical theory', *Phys. Rev.* **160**, 1113.

131. D'Hoker, E. and Phong, D.H. (1986), 'Length-twist parameters in string path integrals', *Phys. Rev. Lett.* **56**, 912.

132. D'Hoker, E. and Phong, D.H. (1986), 'Multiloop amplitudes for the bosonic Polyakov string', *Nucl. Phys.* **B269**, 205.

133. Diamandis, G.A., Ellis, J., Lahanas, A.B. and Nanopoulos, D.V. (1986), 'Vanishing scalar masses in no-scale supergravity', *Phys. Lett.* **173B**, 303.

134. Dimopoulos, S. (1981), 'Softly broken supersymmetry and SU(5)', *Nucl. Phys.* **B193**, 150.

135. Dine, M., Fischler, W. and Srednicki, M. (1981), 'A simple solution to the strong CP problem with a harmless axion', *Phys. Lett.* **104B**, 199.

136. Dine, M., Rohm, R., Seiberg, N. and Witten, E. (1985), 'Gluino condensation in superstring models', *Phys. Lett.* **156B**, 55.

137. Dine, M., Kaplunovsky, V., Mangano, M., Nappi, C.R. and Seiberg, N. (1985), 'Superstring model building', *Nucl. Phys.* **B259**, 549.

138. Dine, M. and Seiberg, N. (1985), 'Couplings and scales in superstring models', *Phys. Rev. Lett.* **55**, 366.

139. Dine, M. and Seiberg, N. (1985), 'Is the superstring weakly coupled?', *Phys. Lett.* **162B**, 299.

140. Dine, M. and Seiberg, N. (1985), 'Is the superstring semiclassical', in *Workshop on Unified String Theories, 29 July – 16 August, 1985*, eds. M. Green and D. Gross (World Scientific, Singapore), p. 678.

141. Dine, M. and Seiberg, N. (1986), 'String theory and the strong CP problem', *Nucl. Phys.* **B273**, 109.

142. Dine, M., Seiberg, N., Wen, X.G. and Witten, E. (1986), 'Nonperturbative effects on the string world sheet', *Nucl. Phys.* **B278**, 769.

143. Dirac, P.A.M. (1937), *Nature* **139**, 323.

144. Dirac, P.A.M. (1938), 'A new basis for cosmology', *Proc. Roy. Soc.* **A**, **165**, 199.

145. Dixon, L., Harvey, J., Vafa, C. and Witten, E. (1985), 'Strings on orbifolds', *Nucl. Phys.* **B261**, 678.

146. Dixon, L. and Harvey, J. (1986), 'String theories in ten dimensions without spacetime supersymmetry', *Nucl. Phys.* **B274**, 93.

147. Dixon, L., Harvey, J., Vafa, C. and Witten, E. (1986), 'Strings on orbifolds II', *Nucl. Phys.* **B274**, 285.

148. Dixon, L., Friedan, D., Martinec, E. and Shenker, S. (1987), 'The conformal field theory of orbifolds', *Nucl. Phys.* **B282**, 13.

149. Donaldson, S. (1983), 'An application of gauge theory to four dimensional topology', *J. Diff. Geom.* **18**, 279.

150. Drees, M., Falck, N.K. and Glück, M. (1986), 'The electroweak sector in superstring models', *Phys. Lett.* **167B**, 187.

151. Duff, M.J., Nilsson, B.E.W. and Pope, C.N. (1985), 'Kaluza–Klein approach to the heterotic string', *Phys. Lett.* **163B**, 343.

152. Duff, M.J., Nilsson, B.E.W., Warner, N.P. and Pope, C.N. (1986), 'Kaluza–Klein approach to the heterotic string II', '*Phys. Lett.* **171B**, 170.

153. Duff, M.J., Nilsson, B.E.W. and Pope, C.N. (1986), 'Gauss–Bonnet from Kaluza–Klein', *Phys. Lett.* **173B**, 69.
154. Duff, M.J. (1986), 'Hidden string symmetries?', *Phys. Lett.* **173B**, 289.
155. Durhuus, B., Nielsen, H.B., Olesen, P. and Petersen, J.L. (1982), 'Dual models as saddle point approximations to Polyakov's quantized string', *Nucl. Phys.* **B196**, 498.
156. Durhuus, B., Olesen, P. and Petersen, J.L. (1982), 'Polyakov's quantized string with boundary terms', *Nucl. Phys.* **198**, 157.
157. Durhuus, B., Olesen, P. and Petersen, J.L. (1982), 'Polyakov's quantized string with boundary terms (II)', *Nucl. Phys.* **201**, 176.
158. Dyson, F.J. (1978), in *Current Trends in the Theory of Fields*, ed. J. E. Lannutti and P. K. Williams, AIP Conference Proceedings No. 48, (Oxford University Press), p. 163.
159. Eguchi, T., Gilkey, P.B. and Hansen, A.J. (1980), 'Gravitation, gauge theories and differential geometry', *Phys. Reports* **66**, 213.
160. Ellis, J., Enqvist, K., Nanopoulos, D.V. and Sarkar, S. (1986), 'Primordial nucleosynthesis, additional neutrinos and neutral currents from the superstring', *Phys. Lett.* **167B**, 457.
161. Ellis, J, Gómez, C. and Nanopoulos, D.V. (1986), 'Axions, dilatons and Wess–Zumino terms in superstring theories', *Phys. Lett.* **168B**, 215.
162. Ellis, J., Gómez, C. and Nanopoulos, D.V. (1986), 'No-scale structure from the superstring', *Phys. Lett.* **171B**, 203.
163. Ellis, J., Enqvist, K., Nanopoulos, D.V. and Zwirner, F. (1986), 'Observables in low-energy superstring models', *Mod. Phys. Lett.* **A1**, 57.
164. Ellis, J., Gómez, C., Nanopoulos, D.V. and Quirós, M. (1986), 'World sheet instanton effects on no-scale structure', *Phys. Lett.* **173B**, 59.
165. Ellis, J., Nanopoulos, D.V. and Quirós, M. (1986), 'On the axion, dilaton, Polonyi, gravitino and shadow matter problems in supergravity and superstring models', *Phys. Lett.* **174B**, 176.
166. Ellis, J. (1986), 'From the Higgs to superstring phenomenology', Proc. of the Lake Louise Winter Institute, p. 225.
167. Ellis, J., Nanopoulos, D.V., Petcov, S.T. and Zwirner, F. (1987), 'Gauginos and Higgs particles in superstring models', *Nucl. Phys.* **B283**, 93.
168. Englert, F. (1982), 'Spontaneous compactification of eleven-dimensional supergravity', *Phys. Lett.* **119B**, 339.
169. Enqvist, K., Nanopoulos, D.V. and Quiros, M. (1986), 'Cosmological difficulties for intermediate scales in superstring models', *Phys. Lett.* **169B**, 343.
170. Evans, M. and Ovrut, B.A. (1986), 'Splitting the superstring vacuum degeneracy', *Phys. Lett.* **174B**, 63.
171. Fairlie, D.B. and Nielsen, H.B. (1970), 'An analogue model for KSV theory', *Nucl. Phys.* **B20**, 637.
172. Fairlie, D.B. and Martin, D. (1974), 'Green's function techniques and dual fermion loops', *Nuovo Cim.* **21A**, 647.
173. Fayet, P. and Iliopoulos, J. (1974), 'Spontaneously broken supergauge symmetries and Goldstone spinors', *Phys. Lett.* **51B**, 46.

174. Fayet, P. (1977), 'Spontaneously broken supersymmetric theories of weak, electromagnetic and strong interactions', *Phys. Lett.* **69B**, 489.

175. Fischler, W. and Susskind, L. (1986), 'Dilaton tadpoles, string condensates and scale invariance', *Phys. Lett.* **171B**, 383.

176. Fischler, W. and Susskind, L. (1986), 'Dilaton tadpoles, string condensates and scale invariance II', *Phys. Lett.* **173B**, 262.

177. Foda, O. and Helayël–Neto, J.A. (1986), 'A coset space compactification of the field theory limit of a heterotic string', *Class. Quant. Grav.* **3**, 607.

178. Ford, L.R. (1951), *Automorphic Functions* (Chelsea, New York).

179. Forgacs, P. and Manton, N.S. (1980), 'Space-time symmetries in gauge theories', *Commun. Math. Phys.* **72**, 15.

180. Frampton, P.H., Goddard, P. and Wray, D. (1971), 'Perturbative unitarity of dual loops', *Nuovo Cim.* **3A**, 755.

181. Frampton, P.H. and Kephart, T.W., (1983), 'Explicit evaluation of anomalies in higher dimensions', *Phys. Rev. Lett.* **50**, 1343.

182. Frampton, P.H. and Kephart, T.W. (1983), 'Consistency conditions for Kaluza-Klein anomalies', *Phys. Rev. Lett.* **50**, 1347.

183. Frampton, P.H. and Kephart, T.W. (1983), 'Analysis of anomalies in higher space-time dimensions', *Phys. Rev.* **D28**, 1010.

184. Frampton, P.H. and Kephart, T.W. (1984), 'Left-right asymmetry from the eight-sphere',*Phys. Rev. Lett.* **53**, 867.

185. Frampton, P.H., van Dam, H. and Yamamoto, K. (1985), 'Chiral fermions from compactification of $O(32)$ and $E(8) \otimes E(8)$ string theories', *Phys. Rev. Lett.* **54**, 1114.

186. Frampton, P.H., Moxhay, P. and Ng, Y.J. (1985), 'One-loop finiteness in $O(32)$ open-superstring theory', *Phys. Rev. Lett.* **55**, 2107.

187. Frampton, P.H., Kikuchi, Y. and Ng, Y.J. (1986), 'Modular invariance in closed superstrings', *Phys. Lett.* **174B**, 262.

188. Freedman, D.Z., Van Nieuwenhuizen, P. and Ferrara, S. (1976), 'Progress toward a theory of supergravity', *Phys. Rev.* **D13**, 3214.

189. Freedman, D.Z., Gibbons, G.W. and West, P.C. (1983), 'Ten into four won't go', *Phys. Lett.* **124B**, 491.

190. Freeman, M.D. and Pope, C.N. (1986), 'Beta-functions and superstring compactifications', *Phys. Lett.* **174B**, 48.

191. Freeman, M.D. and Olive, D.I. (1986), 'The calculation of planar one-loop diagrams in string theory using the BRS formalism', *Phys. Lett.* **175B**, 155.

192. Freund, P.G.O. and Rubin, M.A. (1980), 'Dynamics of dimensional reduction', *Phys. Lett.* **97B**, 233.

193. Freund, P.G.O. and Oh, P. (1985), 'Cosmological solutions with "ten into four" compactification', *Nucl. Phys.* **B255**, 688.

194. Friedan, D. and Windey, P. (1984), 'Supersymmetric derivation of the Atiyah–Singer index and the chiral anomaly', *Nucl. Phys.* **B235** [FS11], 395.

195. Friedan, D., Shenker, S. and Martinec, E. (1985), 'Covariant quantization of superstrings', *Phys. Lett.* **160B**, 55.

196. Friedan, D. and Shenker, S. (1987), 'The analytic geometry of conformal field theory', *Nucl. Phys.* **B281**, 509.

197. Friedan, D. and Shenker, S. (1986), 'The integrable analytic geometry of quantum string', *Phys. Lett.* **175B**, 287.

198. Friedan, D., Martinec, E. and Shenker, S. (1986), 'Conformal invariance, supersymmetry and string theory', *Nucl. Phys.* **B271**, 93.

199. Fritzsch, H. and Minkowski, P. (1975), 'Unified interactions of leptons and hadrons', *Ann. Phys.* **93**, 193.

200. Frye, G. and Susskind, L. (1970), 'Removal of the divergence of a planar dual-symmetric loop', *Phys. Lett.* **31B**, 537.

201. Frye, G. and Susskind, L. (1970), 'Non-planar dual symmetric loop graphs and the Pomeron', *Phys. Lett.* **31B**, 589.

202. Gates, S.J., Grisaru, M., Roček, M. and Siegel, W. (1983), *Superspace or One Thousand and One Lessons in Supersymmetry*, (Benjamin/Cummins).

203. Gates, S.J. and Nishino, H. (1985), 'New $D = 10$, $N = 1$ supergravity coupled to Yang–Mills supermultiplet and anomaly cancellations', *Phys. Lett.* **157B**, 157.

204. Gates, S.J. and Nishino, H. (1986), 'New $D = 10, N = 1$ superspace supergravity and local symmetries of superstrings', *Phys. Lett.* **173B**, 46.

205. Gates, S.J. and Nishino, H. (1986), 'Manifestly supersymmetric $O(\alpha')$ superstring corrections in new $D = 10, N = 1$ supergravity Yang–Mills theory', *Phys. Lett.* **173B**, 52.

206. Gava, E., Iengo, R., Jayaraman, T. and Ramachandran, R. (1986), 'Multiloop divergences in the closed bosonic string theory', *Phys. Lett.* **168B**, 207.

207. Gell-Mann, M., Ramond, P. and Slansky, R. (1978), 'Color embeddings, charge assignments, and proton stability in unified gauge theories', *Rev. Mod. Phys.* **50**, 721.

208. Gell-Mann, M., Ramond, P. and Slansky, R. (1979), 'Complex spinors and unified theories', in *Supergravity*, ed. P. van Nieuwenhuizen et. al. (North-Holland), p. 315.

209. Gell-Mann, M. and Zwiebach, B. (1984), 'Spacetime compactification due to scalars', *Phys. Lett.* **141B**, 333.

210. Gell-Mann, M. and Zwiebach, B. (1985), 'Dimensional reduction of spacetime induced by nonlinear scalar dynamics and noncompact extra dimensions', *Nucl. Phys.* **B260**, 569.

211. Georgi, H. and Glashow, S.L. (1972), 'Gauge theories without anomalies', *Phys. Rev.* **D6**, 429.

212. Georgi, H. and Glashow, S.L. (1974), 'Unity of all elementary-particle forces', *Phys. Rev. Lett.* **32**, 438.

213. Georgi, H. (1974), 'The state of the art – gauge theories', in *Proceedings of the American Institute of Physics #23* ed. C. E. Carlson, p. 575.

214. Georgi, H., Quinn, H.R. and Weinberg, S. (1974), 'Hierarchy of interactions in unified gauge theories', *Phys. Rev. Lett.* **33**, 451.

215. Gervais, J.L. and Sakita, B. (1971), 'Functional-integral approach to dual-resonance theory', *Phys. Rev.* **D4**, 2291.

216. Gervais, J.L. and Sakita, B. (1973), 'Ghost-free string picture of Veneziano model', *Phys. Rev. Lett.* **30**, 716.

217. Gildener, E. and Weinberg, S. (1976), 'Symmetry breaking and scalar bosons', *Phys. Rev.* **D13**, 3333.

218. Gildener, E. (1976), 'Gauge-symmetry hierarchies', *Phys. Rev.* **D14**, 1667.

219. Giles, R. and Thorn, C.B. (1977), 'Lattice approach to string theory', *Phys. Rev.* **D16**, 366.

220. Gilkey, P.B. (1975), 'The spectral geometry of a Riemannian manifold', *J. Diff. Geom.* **10**, 601.

221. Glashow, S. (1986), 'The fifth force', proc. of the 1986 Moriond workshop.

222. Gliozzi, F., Scherk, J. and Olive, D. (1976), 'Supergravity and the spinor dual model', *Phys. Lett.* **65B**, 282.

223. Gliozzi, F., Scherk, J. and Olive, D. (1977), 'Supersymmetry, supergravity theories and the dual spinor model', *Nucl. Phys.* **B122**, 253.

224. Goddard, P. (1971), 'Analytic renormalization of dual one-loop amplitudes', *Nuovo Cim.* **4A**, 349.

225. Goddard, P. and Waltz, R.E. (1971), 'One-loop amplitudes in the model of Neveu and Schwarz', *Nucl. Phys.* **B34**, 99.

226. Goddard, P., Goldstone, J., Rebbi, C. and Thorn, C.B. (1973), 'Quantum dynamics of a massless relativistic string', *Nucl. Phys.* **B56**, 109.

227. Gomez, C. (1986), 'Topologically non-trivial gauge configurations and the heterotic string', *Phys. Lett.* **168B**, 212.

228. Gomez, C. (1986), 'Modular invariance and compactification of the moduli space', *Phys. Lett.* **175B**, 32.

229. Goodman, M. and Witten, E. (1986), 'Global symmetries in four and higher dimensions', *Nucl. Phys.* **B271**, 21.

230. Goodman, M. (1986), 'Proof of character-valued index theorems', Princeton preprint, to appear in *Nucl. Phys.* **B**.

231. Goroff, M.H. and Sagnotti, A. (1985), 'Quantum gravity at two loops', *Phys. Lett.* **160B**, 81.

232. Goroff, M.H. and Sagnotti, A. (1986), 'The ultraviolet behavior of Einstein gravity', *Nucl. Phys.* **B266**, 709.

233. Govindrajan, T.R., Jayraman, T., Mukherjee, A. and Wadia, S.R. (1986), 'Twisted current algebras and gauge symmetry breaking in string theory', *Mod. Phys. Lett.* **A1**, 29.

234. Green, M.B. (1973), 'Cancellation of the leading divergence in dual loops', *Phys. Lett.* **46B**, 392.

235. Green, M.B. (1976), 'Reciprocal space-time and momentum-space singularities in the narrow resonance approximation', *Nucl. Phys.* **B116**, 449.

236. Green, M.B. and Shapiro, J.A. (1976), 'Off-shell states in the dual model', *Phys. Lett.* **64B**, 454.

237. Green, M.B. (1976), 'The structure of dual Green functions', *Phys. Lett.* **65B**, 432.

238. Green M.B. (1977), 'Point-like structure and off-shell dual strings', *Nucl. Phys.* **B124**, 461.

239. Green M.B. (1977), 'Dynamical point-like structure and dual strings', *Phys. Lett.* **69B**, 89.

240. Green, M.B. and Schwarz, J.H. (1982), 'Supersymmetric dual string theory (III). Loops and renormalization', *Nucl. Phys.* **B198**, 441.

241. Green, M.B., Schwarz, J.H. and Brink, L. (1982), '$N = 4$ Yang–Mills and $N = 8$ supergravity as limits of string theories', *Nucl. Phys.* **B198**, 474.

242. Green, M.B. and Schwarz, J.H. (1982), 'Supersymmetrical string theories', *Phys. Lett.* **109B**, 444.
243. Green, M.B. and Schwarz, J.H. (1983), 'Superstring interactions', *Nucl. Phys.* **B218**, 43.
244. Green, M.B., Schwarz, J.H. and Brink, L. (1983), 'Superfield theory of type (II) superstrings', *Nucl. Phys.* **B219**, 437.
245. Green, M.B. and Schwarz, J.H. (1983), 'Extended supergravity in ten dimensions', *Phys. Lett.* **122B**, 143.
246. Green, M.B. (1983), 'Supersymmetrical dual string theories and their field theory limits – a review', *Surveys in High Energy Physics* **3**, 127.
247. Green, M.B. and Schwarz, J.H. (1984), 'Covariant description of superstrings', *Phys. Lett.* **136B**, 367.
248. Green, M.B. and Schwarz, J.H. (1984), 'Superstring field theory', *Nucl. Phys.* **B243**, 475.
249. Green, M.B. and Schwarz, J.H. (1984), 'Anomaly cancellations in supersymmetric $D = 10$ gauge theory and superstring theory', *Phys. Lett.* **149B**, 117.
250. Green, M.B. and Schwarz, J.H. (1985), 'Infinity cancellations in SO(32) superstring theory', *Phys. Lett.* **151B**, 21.
251. Green, M.B., Schwarz, J.H. and West, P.C. (1985), 'Anomaly-free chiral theories in six dimensions', *Nucl. Phys.* **B254**, 327.
252. Green, M.B. and Schwarz, J.H. (1985), 'The hexagon gauge anomaly in type I superstring theory', *Nucl. Phys.* **B255**, 93.
253. Greenberg, M.J. (1967), *Lectures on Algebraic Topology* (Benjamin).
254. Greene, B.R., Kirklin, K.H. and Miron, P.J. (1986), 'Superstring models with SU(5) and SO(10) unifying groups', *Nucl. Phys.* **B274**, 574.
255. Greene, B.R., Kirklin, K.H., Miron, P.J. and Ross, G.G. (1986), 'A three generation superstring model', *Nucl. Phys.* **B278**, 667.
256. Greene, B.R., Kirklin, K.H., Miron, P.J. and Ross, G.G. (1986), 'A superstring inspired standard model', *Phys. Lett.* **180B**, 69.
257. Griffiths, P. and Harris, J. (1978), *Principles of Algebraic Geometry* (Wiley-Interscience).
258. Grimm, R. and Mărculescu, S. (1974), 'The structure of anomalies for arbitrary dimension of the space-time', *Nucl. Phys.* **B68**, 203.
259. Grisaru, M.T., van de Ven, A. and Zanon, D. (1986), 'Four-loop β-function for the $N = 1$ and $N = 2$ supersymmetric non-linear sigma model in two dimensions', *Phys. Lett.* **173B**, 423.
260. Grisaru, M.T., van de Ven, A.E.M. and Zanon, D. (1986), 'Two-dimensional supersymmetric sigma models on Ricci flat Kähler manifolds are not finite', *Nucl. Phys.* **B277**, 388.
261. Grisaru, M.T., van de Ven, A.E.M. and Zanon, D. (1986), 'Four loop divergences for the $N = 1$ supersymmetric nonlinear sigma model in two dimensions', *Nucl. Phys.* **B277**, 409.
262. Gross, D.J., Neveu, A., Scherk, J. and Schwarz, J.H. (1970), 'The primitive graphs of dual-resonance models', *Phys. Lett.* **31B**, 592.
263. Gross, D.J., Neveu, A., Scherk, J. and Schwarz, J.H. (1970), 'Renormalization and unitarity in the dual-resonance model', *Phys. Rev.* **D2**, 697.
264. Gross, D.J. and Schwarz, J.H. (1970), 'Basic operators of the dual-resonance model', *Nucl. Phys.* **B23**, 333.

265. Gross, D.J. and Jackiw, R. (1972), 'Effect of anomalies on quasi-renormalizable theories', *Phys. Rev.* **D6**, 477.

266. Gross, D.J. and Perry, M.J. (1983), 'Magnetic monopoles in Kaluza-Klein theories', *Nucl. Phys.* **B226**, 29.

267. Gross, D.J., Harvey, J.A., Martinec, E. and Rohm, R. (1985), 'Heterotic string', *Phys. Rev. Lett.* **54**, 502.

268. Gross, D.J., Harvey, J.A., Martinec, E. and Rohm, R. (1985), 'Heterotic string theory (I). The free heterotic string', *Nucl. Phys.* **B256**, 253.

269. Gross, D.J., Harvey, J.A., Martinec, E. and Rohm, R. (1986), 'Heterotic string theory (II). The interacting heterotic string', *Nucl. Phys.* **B267**, 75.

270. Gross, D.J. and Witten, E. (1986), 'Superstring modifications of Einstein's equations', *Nucl. Phys.* **B277**, 1.

271. Günaydin, M., Romans, L.J. and Warner, N.P. (1985), 'Gauged $N = 8$ supergravity in five dimensions', *Phys. Lett.* **154B**, 268.

272. Günaydin, M., Romans, L.J. and Warner, N.P. (1985), 'IIB or not IIB: That is the question' *Phys. Lett.* **164B**, 309.

273. Gürsey, F. and Sikivie, P. (1976), 'E_7 as a universal gauge group', *Phys. Rev. Lett.* **36**, 775.

274. Gürsey, F., Ramond, P. and Sikivie, P. (1976), 'A universal gauge theory model based on E_6', *Phys. Lett.* **60B**, 177.

275. Gürsey, F. and Sikivie, P. (1977), 'Quark and lepton assignments in the E_7 model', *Phys. Rev.* **D16**, 816.

276. Guth, A.H. and Tye, S.H. (1980), 'Phase transitions and magnetic monopole production in the very early universe', *Phys. Rev. Lett.* **44**, 631; erratum 963.

277. Guth, A.H. (1981), 'Inflationary universe: A possible solution to the horizon and flatness problems', *Phys. Rev.* **D23**, 347.

278. Hamidi, S. and Vafa, C. (1987), 'Interactions on orbifolds', *Nucl. Phys.* **B279**, 465.

279. Han, C.W., Han, S.K., Jun, J.W., Kim, J.K. and Koh, I.G. (1986), 'Absence of leading divergence in the parity-odd one-loop amplitude of type-I $SO(32)$ superstring theory', *Phys. Rev.* **D34**, 1219.

280. Hartshorne, R. (1977), *Algebraic Geometry* (Springer-Verlag).

281. Harvey, J.A. (1986), 'Twisting the heterotic string', in *Workshop on Unified String Theories, 29 July – 16 August, 1985*, eds. M. Green and D. Gross (World Scientific, Singapore), p. 704.

282. Hawking, S.W. and Pope, C.N. (1978), 'Generalized spin structures in quantum gravity', *Phys. Lett.* **73B**, 42.

283. Hawking, S.W. (1978), 'Spacetime foam', *Nucl. Phys.* **B144**, 349.

284. Hawking, S.W., Page, D.N. and Pope, C.N. (1979), 'The propagation of particles in spacetime foam', *Phys. Lett.* **86B**, 175.

285. Hawking, S.W., Page, D.N. and Pope, C.N. (1980), 'Quantum gravitational bubbles', *Nucl. Phys.* **B170**, 283.

286. Helayël-Neto, J.A. and Smith, A.W. (1986), 'A possible role of gravitino condensates in superstring compactification', *Phys. Lett.* **175B**, 37.

287. Hellings, R.W., Adams, P.J., Anderson, J.D., Keesey, M.S., Lau, E.L. Standish, E.M., Canuto, V.M. and Goldman, I. (1983), 'Experimental

test of the variability of G using Viking lander ranging data', *Phys. Rev. Lett.* **51**, 1609.

288. Henneaux, M. (1986), 'Hamiltonian formulation of $d = 10$ supergravity theories', *Phys. Lett.* **168B**, 233.

289. Hewett, J.L., Rizzo, T.G. and Robinson, J.A. (1986), 'Low-energy phenomenology of some supersymmetric E_6-breaking patterns', *Phys. Rev.* **D33**, 1476.

290. Holding, S.C., Stacey, F.D. and Tuck, G.J. (1986), 'Gravity in mines - an investigation of Newton's law', *Phys. Rev.* **D33**, 3487.

291. Holman, R. and Reiss, D.B. (1986), 'Fermion masses in $E_8 \times E_8'$ superstring theories', *Phys. Lett.* **166B**, 305.

292. Holman, R. and Reiss, D.B. (1986), 'Fermion masses and phenomenology in $SO(10)$ or $SU(5)$ superstring compactifications', *Phys. Lett.* **176B**, 74.

293. Hopkinson, J.F.L., Tucker, R.W. and Collins, P.A. (1975), 'Quantum strings and the functional calculus', *Phys. Rev.* **D12**, 1653.

294. Horowitz, G. (1986), 'What is a Calabi–Yau space?', in *Workshop on Unified String Theories, 29 July – 16 August, 1985*, eds. M. Green and D. Gross (World Scientific, Singapore), p. 635.

295. Horvath, Z. Palla, L., Cremmer, E. and Scherk, J. (1977), 'Grand unified schemes and spontaneous compactification', *Nucl. Phys.* **B127**, 57.

296. Horvath, Z. and Palla, L. (1978), 'Spontaneous compactification and "monopoles" in higher dimensions', *Nucl. Phys.* **B142**, 327.

297. Hosotani, Y. (1983), 'Dynamical gauge symmetry breaking as the Casimir effect', *Phys. Lett.* **129B**, 193.

298. Howe, P.S. and West, P.C. (1984), 'The complete $N = 2, d = 10$ supergravity', *Nucl. Phys.* **B238**, 181.

299. Howe, P.S., Papadopoulos, G. and Stelle, K.S. (1986), 'Quantizing the $N = 2$ super sigma-model in two dimensions', *Phys. Lett.* **174B**, 405.

300. Hsue, C.S., Sakita, B. and Virasoro, M.A. (1970), 'Formulation of dual theory in terms of functional integrations', *Phys. Rev.* **D2**, 2857.

301. Hübsch, T., Nishino, H. and Pati, J.C. (1985), 'Do superstrings lead to quarks or to preons?', *Phys. Lett.* **163B**, 111.

302. Hübsch, T. (1987), 'Calabi-Yau manifolds: motivations and constructions' *Commun. Math. Phys.* **108**, 291.

303. Hull, C.M. and Witten, E. (1985), 'Supersymmetric sigma models and the heterotic string', *Phys. Lett.* **160B**, 398.

304. Hull, C.M. (1986), 'Sigma model beta-functions and string compactifications', *Nucl. Phys.* **B267**, 266.

305. Hull, C.M. (1986), 'Anomalies, ambiguities and superstrings', *Phys. Lett.* **167B**, 51.

306. Husemoller, D. (1966), *Fibre Bundles* (Springer-Verlag).

307. Ibáñez, L.E., López, C. and Muñoz, C. (1985), 'The low-energy supersymmetric spectrum according to $N = 1$ supergravity GUTs', *Nucl. Phys.* **B256**, 218.

308. Ibáñez, L.E. and Nilles, H.P. (1986), 'Low-energy remnants of superstring anomaly cancellation terms', *Phys. Lett.* **169B**, 354.

309. Ibáñez, L.E. (1985), 'Phenomenology from superstrings', in proc. of the First Torino Meeting on Superunification and Extra Dimensions

(World Scientific), p. 189.

310. Ibáñez, L.E. (1986), 'Some topics in the low energy physics from superstrings', CERN preprint Th.4459/86.

311. Ibáñez, L.E., and Mas, J. (1987), 'Low energy supergravity and superstring-inspired models', *Nucl. Phys.* **B286**, 107.

312. Ida, M., Matsumoto, H. and Yazaki, S. (1970), 'Factorization and duality of multiloop diagrams', *Prog. Theor. Phys.* **44**, 456.

313. Imbimbo, C. and Mukhi, S. (1986), 'Chiral fermions and the Witten index for the compactified heterotic string', *Nucl. Phys.* **B263**, 629.

314. Ito, K. (1985), 'Manifestly supersymmetric path integral formulation of the superstring field theories', *Phys. Lett.* **164B**, 301.

315. Itoyama, H. and Leon, J. (1986), 'Some quantum corrections to Calabi–Yau compactification', *Phys. Rev. Lett.* **56**, 2352.

316. Jacob, M. editor. (1974), 'Dual theory', in *Physics Reports, Reprint Volume I*, (North–Holland, Amsterdam, 1974).

317. Jevicki, A. (1986), 'Covariant string theory Feynman amplitudes', *Phys. Lett.* **169B**, 359.

318. Joshipura, A.S. and Sarkar, U. (1986), 'Phenomenologically consistent discrete symmetries in superstrings theories', *Phys. Rev. Lett.* **57**, 33.

319. Kaku, M. and Thorn, C.B. (1970), 'Unitary nonplanar closed loops', *Phys. Rev.* **D1**, 2860.

320. Kaku, M. and Yu, L. (1970), 'The general multi-loop Veneziano amplitude', *Phys. Lett.* **33B**, 166.

321. Kaku, M. and Scherk, J. (1971), 'Divergence of the two-loop planar graph in the dual-resonance model', *Phys. Rev.* **D3**, 430.

322. Kaku, M. and Scherk, J. (1971), 'Divergence of the N-loop planar graph in the dual-resonance model', *Phys. Rev.* **D3**, 2000.

323. Kaku, M. and Yu, L. (1971), 'Unitarization of the dual-resonance amplitude. I. Planar N-loop amplitude', *Phys. Rev.* **D3**, 2992.

324. Kaku, M. and Yu, L. (1971), 'Unitarization of the dual-resonance amplitude. II. The nonplanar N-loop amplitude', *Phys. Rev.* **D3**, 3007.

325. Kaku, M. and Yu, L. (1971), 'Unitarization of the dual-resonance amplitude. III. General rules for the orientable and nonorientable multi-loop amplitudes', *Phys. Rev.* **D3**, 3020.

326. Kaku, M. and Kikkawa, K. (1974), 'Field theory of relativistic strings. I. Trees', *Phys. Rev.* **D10**, 1110.

327. Kaku, M. and Kikkawa, K. (1974), 'Field theory of relativistic strings. II. Loops and pomerons', *Phys. Rev.* **D10**, 1823.

328. Kalb, M. and Ramond, P. (1974), 'Classical direct interstring action', *Phys. Rev.* **D9**, 2273.

329. Kallosh, R.E. (1985), 'Ten-dimensional supersymmetry requires $E_8 \times E_8$ or $SO(32)$', *Phys. Lett.* **159B**, 111.

330. Kallosh, R.E. and Nilsson, B.E.W. (1986), 'Scale invariant $d = 10$ superspace and the heterotic string', *Phys. Lett.* **167B**, 46.

331. Kaluza, Th. (1921), 'On the problem of unity in physics', *Sitz. Preuss. Akad. Wiss.* **K1**, 966.

332. Kalyniak, P. and Sundaresan, M.K. (1986), 'Symmetry-breaking scenarios of Wilson-loop broken E_6', *Phys. Lett.* **167B**, 320.

333. Kaplunovsky, V. (1985), 'Mass scales of the string unification', *Phys. Rev. Lett.* **55**, 1036.

334. Kendig, K. (1977), *Elementary Algebraic Geometry* (Springer-Verlag).
335. Kent, A. (1986), 'Conformal invariance, current algebra and modular invariance', *Phys. Lett.* **173B**, 413.
336. Kephart, T. and Frampton, P. (1983), 'Analysis of anomalies in higher space-time dimensions', *Phys. Rev.* **D28**, 1010.
337. Kikkawa, K., Sakita, B. and Virasoro, M.A. (1969), 'Feynman-like diagrams compatible with duality. I. Planar diagrams', *Phys. Rev.* **184**, 1701.
338. Kikkawa, K. (1969), 'Regge cut from a nonplanar duality amplitude', *Phys. Rev.* **187**, 2249.
339. Kikkawa, K., Klein, S.A., Sakita, B. and Virasoro, M.A. (1970), 'Feynman-like diagrams compatible with duality. II. General discussion including nonplanar diagrams', *Phys. Rev.* **D1**, 3258.
340. Kikuchi, Y., Marzban, C. and Ng, Y. (1986), 'Heterotic string modifications of Einstein's and Yang–Mills' actions', *Phys. Lett.* **176B**, 57.
341. Kim, J.E. (1979), 'Weak-interaction singlet and strong CP invariance', *Phys. Rev. Lett.* **43**, 103.
342. Kim, J.K., Koh, I.G. and Yoon, Y. (1986), 'Calabi–Yau manifolds from arbitrary weighted homogeneous spaces', *Phys. Rev.* **D33**, 2893.
343. Klein, O. (1926), 'Quantentheorie und fünfdimensionale Relativitätstheorie', *Z. Phys.* **37**, 895.
344. Koba, Z. and Nielsen, H.B. (1969), 'Reaction amplitude for n-mesons a generalization of the Veneziano–Bardakçi–Ruegg–Virasoro model', *Nucl. Phys.* **B10**, 633.
345. Koba, Z. and Nielsen, H.B. (1969), 'Manifestly crossing- invariant parametrization of n-meson amplitude', *Nucl. Phys.* **B12**, 517.
346. Kodaira, K. (1985), *Complex Manifolds and Deformation of Complex Structures* (Springer-Verlag).
347. Kogan Ya., Morozov, A. and Perelomov A. (1984), 'Finiteness of $N = 4$ supersymmetry sigma models', *Pis'ma v ZhETF* **40**, 38.
348. Koh, I.G. and Nishino, H., (1985), 'Towards realistic $D = 6, N = 2$ Kaluza–Klein supergravity on coset $E_7/SO(12) \times Sp(1)$ with chiral fermions', *Phys. Lett.* **153B**, 45.
349. Kolb, E.W. and Slansky, R. (1984), 'Dimensional reduction in the early universe: Where have all the massive particles gone?', *Phys. Lett.* **135B**, 378.
350. Kolb, E.W., Perry, M.J. and Walker, T.P. (1986), 'Time variation of fundamental constants, primordial nucleosynthesis, and the size of extra dimensions', *Phys. Rev.* **D33**, 869.
351. Kreuzer, L.B. (1968), 'Experimental measurement of the equivalence of active and passive gravitational mass', *Phys. Rev.* **169**, 1007.
352. Labastida, J.M.F. (1986), 'Equivalence of dual-field theoretical limits of superstring theories', *Phys. Lett.* **171B**, 377.
353. Lam, C.S. and Li, D-X. (1986), 'Modular invariance and one-loop finiteness of five-point amplitudes in type-II and heterotic string theories', *Phys. Rev. Lett.* **56**, 2575.
354. Lang, W. and Louis, J. (1985), '16/16 supergravity coupled to matter: The low energy limit of the superstring', *Phys. Lett.* **158B**, 40.
355. Lazarides, G., Panagiotakopoulos, C. and Shafi, Q. (1986), 'Phenomenology and cosmology with superstrings', *Phys. Rev. Lett.* **56**,

432.

356. Lazarides, G., Panagiotakopoulos, C. and Shafi, Q. (1986), 'Baryon asymmetry, stable proton and $n-\bar{n}$ oscillations in superstring models', *Phys. Lett.* **175B**, 309.

357. Linde, A.D. (1982), 'A new inflationary universe scenario: A possible solution of the horizon, flatness, homogeneity, isotropy and primordial monopole problems', *Phys. Lett.* **108B**, 389.

358. Lorenzo, F.J., Mittelbrunn, J.R., Medrano, M.R. and Sierra, G. (1986), 'Quantum mechanical amplitude for string propagation', *Phys. Lett.* **171B**, 369.

359. Lovelace, C. (1970), 'M-loop generalized Veneziano formula', *Phys. Lett.* **32B**, 703.

360. Lovelace, C. (1971), 'Pomeron form factors and dual Regge cuts', *Phys. Lett.* **34B**, 500.

361. Luciano, J.F. (1978), 'Space-time geometry and symmetry breaking', *Nucl. Phys.* **B135**, 111.

362. Maeda, K. (1986), 'Cosmological solutions with Calabi–Yau compactification', *Phys. Lett.* **166B**, 59.

363. Maeda, K. and Pollock, M.D. (1986), 'On inflation in the heterotic superstring model', *Phys. Lett.* **173**, 251.

364. Mahapatra, S. and Misra, S.P. (1986), 'Fermion condensates and weak symmetry breaking in a superstring-based model', *Phys. Rev.* **D33**, 3464.

365. Mandelstam, S. (1973), 'Interacting-string picture of dual resonance models', *Nucl. Phys.* **B64**, 205.

366. Mandelstam, S. (1973), 'Manifestly dual formulation of the Ramond-model', *Phys. Lett.* **46B**, 447.

367. Mandelstam, S. (1974), 'Interacting-string picture of the Neveu–Schwarz–Ramond model', *Nucl. Phys.* **B69**, 77.

368. Mandelstam, S. (1974), 'Dual-resonance models', *Phys. Reports* **C13**, 259.

369. Mandelstam, S. (1974), 'Lorentz properties of the three-string vertex', *Nucl. Phys.* **B83**, 413.

370. Mandelstam, S. (1986), 'The interacting-string picture and functional integration', in *Workshop on Unified String Theories, 29 July – 16 August, 1985*, eds. M. Green and D. Gross (World Scientific, Singapore), p. 46.

371. Mandelstam, S. (1986), 'Interacting-string picture of the fermionic string', in *Workshop on Unified String Theories, 29 July – 16 August, 1985*, eds. M. Green and D. Gross (World Scientific, Singapore), p. 577.

372. Mangano, M. (1985), 'Low energy aspects of superstring theories', *Z. Phys. C.* **28**, 613.

373. Mani, H.S., Mukherjee, A., Ramachandran, R. and Balachandran, A.P. (1986), 'Embedding of SU(5) GUT in SO(32) superstring theories', *Nucl. Phys.* **B263**, 621.

374. Manin, Yu.I. (1986), 'The partition function of the Polyakov string can be expressed in terms of theta-functions', *Phys. Lett.* **172B**, 184.

375. Manin, Yu.I. (1986), 'Theta-function representation of the partition function of a Polyakov string', *Pis'ma Zh. Eksp. Teor. Fiz.* **43**, 161.

376. Manton, N.S. (1981), 'Fermions and parity violation in dimensional reduction schemes', *Nucl. Phys.* **B193**, 502.
377. Manton, N.S. (1986), 'Dimensional reduction of supergravity', *Ann. Phys.* **167**, 328.
378. Marcus, N. and Sagnotti, A. (1984), 'A test of finiteness predictions for supersymmetric theories', *Phys. Lett.* **135**, 85.
379. Marcus, N. and Sagnotti, A. (1985), 'The ultraviolet behavior of $N = 4$ Yang–Mills and the power counting of extended superspace', *Nucl. Phys.* **B256**, 77.
380. Markushevich, D.G., Ol'shanetskiĭ, M.A. and Perelomov, A.M. (1986), 'Vacuum configurations in superstrings associated with semisimple Lie algebras', *Pis'ma Zh. Eksp. Teor. Fiz.* **43**, 59.
381. Martinec, E. (1986), 'Nonrenormalization theorems and fermionic string finiteness', *Phys. Lett.* **171B**, 189.
382. Matsuoka, T. and Suematsu, D. (1986), 'Gauge hierarchies in the $E_8 \times E_8'$ superstring theory', *Nucl. Phys.* **B274**, 106.
383. McKean, H.P., Jr. and Singer, I.M. (1967), 'Curvature and the eigenvalues of the Laplacian', *J. Diff. Geom.* **1**, 43.
384. Miao, L. (1986), 'The θ-structure in string theories: superstrings', *Phys. Lett.* **175B**, 284.
385. Milnor, J.W. and Stasheff, J.D. (1974), *Characteristic Classes* (Princeton University Press).
386. Mohapatra, R.N. (1986), 'Mechanism for understanding small neutrino mass in superstring theories', *Phys. Rev. Lett.* **56**, 561.
387. Mohapatra, P.K. (1986), 'Realization of the discrete group in E_6 and the possible low-energy gauge groups in superstrings', *Phys. Lett.* **174B**, 51.
388. Moore, G. and Nelson, P. (1986), 'Measure for moduli. The Polyakov string has no local anomalies', *Nucl. Phys.* **B266**, 58.
389. Morozov, A. Yu. and Perelomov, A. M. (1986), 'Hyperkählerian manifolds and exact beta functions of two-dimensional $N = 4$ supersymmetric sigma models', *Nucl. Phys.* **B271**, 620.
390. Müller–Hoissen, F. (1985), 'Spontaneous compactification with quadratic and cubic curvature terms', *Phys. Lett.* **163B**, 106.
391. Mumford, D. (1975), *Curves and Their Jacobians* (University of Michigan Press).
392. Nahm, W. (1978), 'Supersymmetries and their representations', *Nucl. Phys.* **B135**, 149.
393. Nandi, S. and Sarkar, U. (1986), 'Solution to the neutrino-mass problem in superstring E_6 theory', *Phys. Rev. Lett.* **56**, 564.
394. Nash, C. and Sen, S. (1983), *Topology and Geometry for Physicists*, (Academic Press).
395. Nemeschansky, D. and Sen, A. (1986), 'Conformal invariance of supersymmetric σ-models on Calabi–Yau manifolds', *Phys. Lett.* **178B**, 365.
396. Nepomechie, R.I., Wu, Y.S. and Zee, A. (1985), 'New compactifications on Calabi–Yau manifolds', *Phys. Lett.* **158B**, 311.
397. Nepomechie, R.I. (1986), 'Chern–Simons terms and bosonic strings', *Phys. Lett.* **B171**, 195.
398. Neveu, A. and Scherk, J. (1970), 'Parameter-free regularization of one-loop unitary dual diagram', *Phys. Rev.* **D1**, 2355.

399. Neveu, A. and Schwarz, J.H. (1971), 'Factorizable dual model of pions', *Nucl. Phys.* **B31**, 86.

400. Neveu, A. and Schwarz, J.H. (1971), 'Quark model of dual pions', *Phys. Rev.* **D4**, 1109.

401. Neveu, A. and Scherk, J. (1972), 'Gauge invariance and uniqueness of the renormalization of dual models with unit intercept', *Nucl. Phys.* **B36**, 317.

402. Nielsen, H.B. and Olesen, P. (1970), 'A parton view on dual amplitudes', *Phys. Lett.* **32B**, 203.

403. Nilles, H.P. (1984), 'Supersymmetry, supergravity and particle physics', *Phys. Reports* **110**, 1.

404. Nilles, H.P. (1986), 'Supergravity and the low-energy limit of superstring theories', (CERN preprint Th.4444/86).

405. Nilsson, B.E.W. and Tollstén, A.K. (1986), 'Superspace formulation of the ten-dimensional coupled Einstein–Yang–Mills system', *Phys. Lett.* **171B**, 212.

406. Nilsson, B.E.W. and Tollstén, A.K. (1986), 'The geometrical off-shell structure of pure $N = 1; d = 10$ supergravity in superspace', *Phys. Lett.* **169B**, 369.

407. Nilsson, B.E.W. (1986), 'Off-shell $d = 10, N = 1$ Poincaré supergravity and the embeddibility of higher-derivative field theories in superspace', *Phys. Lett.* **175B**, 319.

408. Nishino, H. and Gates, S.J. (1986), 'Dual versions of higher-dimensional supergravities and anomaly cancellations in lower dimensions', *Nucl. Phys.* **B268**, 532.

409. Olive, D. and West, P. (1982), 'The $N = 4$ supersymmetric E_8 gauge theory and coset space dimensional reduction', *Nucl. Phys.* **B217**, 248.

410. Palla, L. (1978), 'Spontaneous compactification', in *Proceedings of the 1978 Tokyo Conference on High Energy Physics*, p. 629.

411. Pati, J.C. and Salam, A. (1973), 'Unified lepton-hadron symmetry and a gauge theory of the basic interactions', *Phys. Rev.* **D8**, 1240.

412. Pati, J.C. and Salam, A. (1974), 'Lepton number as the fourth "color"', *Phys. Rev.* **D10**, 275.

413. Paton, J.E. and Chan H.M. (1969), 'Generalized Veneziano model with isospin', *Nucl. Phys.* **B10**, 516.

414. Peccei, R.D. and Quinn, H. (1977), 'CP conservation in the presence of pseudoparticles', *Phys. Rev. Lett.* **38**, 1440.

415. Pilch, K. and Schellekens, A.N. (1985), 'Fermion spectra from superstrings', *Nucl. Phys.* **B259**, 637.

416. Polchinski, J. (1986), 'Evaluation of the one loop string path integral', *Commun. Math. Phys.* **104**, 37.

417. Polyakov, A.M. (1981), 'Quantum geometry of bosonic strings', *Phys. Lett.* **103B**, 207.

418. Polyakov, A.M. (1981), 'Quantum geometry of fermionic strings', *Phys. Lett.* **103B**, 211.

419. Pope, C.N., Sohnius, M.F. and Stelle, K.S. (1987), 'Counterterm counterexamples', *Nucl. Phys.* **B283**, 192.

420. Preskill, J., Frampton, P.H. and van Dam, H. (1983), 'Anomalies and fermion masses in D dimensions', *Phys. Lett.* **124B**, 209.

421. Quiros, M. (1986), 'On the effective potential and gravitino mass determination in compactified superstring models', *Phys. Lett.* **173B**, 265.

422. Rabin, J.M. (1986), 'Chern–Simons and Wess–Zumino terms in string theory', *Phys. Lett.* **172B**, 333.

423. Raby, S. and Slansky, R. (1986), 'Compactification of closed bosonic strings', *Phys. Rev. Lett.* **56**, 693.

424. Ramond, P. (1971), 'Dual theory for free fermions', *Phys. Rev.* **D3**, 2415.

425. Randjbar-Daemi, S., Salam, A. and Strathdee, S. (1983), 'Spontaneous compactification in six-dimensional Einstein–Maxwell theory', *Nucl. Phys.* **B214**, 491.

426. Randjbar–Daemi, S., Salam, A., Sezgin, E. and Strathdee, J. (1985), 'An anomaly-free model in six dimensions', *Phys. Lett.* **151B**, 351.

427. Rebbi, C. (1974), 'Dual models and relativistic quantum strings', *Phys. Reports* **C12**, 1.

428. Restuccia, A. and Taylor, J.G. (1986), 'On the construction of higher loop closed superstring amplitudes', *Phys. Lett.* **174B**, 56.

429. Rohm, R. (1984), 'Spontaneous supersymmetry breaking in supersymmetric string theories', *Nucl. Phys.* **B237**, 553.

430. Rohm, R. and Witten, E. (1986), 'The antisymmetric tensor field in superstring theory', *Ann. Phys.* **170**, 454.

431. Romans, L.J. (1985), 'New compactifications of chiral $N = 2, d = 10$ supergravity', *Phys. Lett.* **153B**, 392.

432. Romans, L.J. and Warner, N.P. (1986), 'Some supersymmetric counterparts of the Lorentz Chern–Simons term', *Nucl. Phys.* **B273**, 320.

433. Romans, L.J. (1986), 'Massive $N = 2a$ supergravity in ten dimensions', *Phys. Lett.* **169B**, 374.

434. Sakai, N. and Senda, I. (1986), 'Vacuum energies of string compactified on torus', *Prog. Theor. Phys.* **75**, 692.

435. Salam, A. and Strathdee, J. (1982), 'On Kaluza–Klein theory', *Ann. Phys.* **141**, 316.

436. Schellekens, A.N. (1986), 'Anomaly cancellation in ten dimensions and beyond', *Phys. Lett.* **175B**, 41.

437. Scherk, J. (1971), 'Renormalization in the dual resonance model. Its arbitrariness in the general case and for unit intercept', *Nucl. Phys.* **B29**, 357.

438. Scherk, J. (1971), 'Zero-slope limit of the dual resonance model', *Nucl. Phys.* **B31**, 222.

439. Scherk, J. and Schwarz, J.H. (1974), 'Dual models for non-hadrons', *Nucl. Phys.* **B81**, 118.

440. Scherk, J. and Schwarz, J.H. (1974), 'Dual models and the geometry of space-time', *Phys. Lett.* **52B**, 347.

441. Scherk, J. (1975), 'An introduction to the theory of dual models and strings', *Rev. Mod. Phys.* **47**, 123.

442. Scherk, J. and Schwarz, J.H. (1979), 'How to get masses from extra dimensions', *Nucl. Phys.* **B153**, 61.

443. Scherk, J. and Schwarz, J.H. (1979), 'Spontaneous breaking of supersymmetry through dimensional reduction', *Phys. Lett.* **82B**, 60.

444. Schwarz, J.H. (1973), 'Dual resonance theory', *Phys. Reports* **8**, 269.

445. Schwarz, J.H. (1973), 'Off-shell dual amplitudes without ghosts', *Nucl. Phys.* **B65**, 131.

446. Schwarz, J.H. and Wu C.C. (1974), 'Off-mass-shell dual amplitudes (II)', *Nucl. Phys.* **B72**, 397.

447. Schwarz, J.H. (1974), 'Off-mass-shell dual amplitudes III', *Nucl. Phys.* **B76**, 93.

448. Schwarz, J.H. (1982), 'Superstring theory', *Phys. Reports* **89**, 223.

449. Schwarz, J.H. and West, P.C. (1983), 'Symmetries and transformations of chiral $N = 2, D = 10$ supergravity', *Phys. Lett.* **126B**, 301.

450. Schwarz, J.H. (1983), 'Covariant field equations of chiral $N = 2, D = 10$ supergravity', *Nucl. Phys.* **B226**, 269.

451. Schwarz, J.H. (1985), *Superstrings. The First Fifteen Years of Superstring Theory*, in 2 volumes (World Scientific, Singapore).

452. Schwinger, J. (1951), 'On gauge invariance and vacuum polarization', *Phys. Rev.* **82**, 664.

453. Segrè, G. C. (1985), 'Low energy physics from superstrings' (Cargese summer school lectures).

454. Seiberg, N. and Witten, E. (1986), 'Spin structures in string theory', *Nucl. Phys.* **B276**, 272.

455. Sen, A. (1985), 'Heterotic string in an arbitrary background field', *Phys. Rev.* **D32**, 2102.

456. Sen, A. (1986), 'Local gauge and Lorentz invariance of heterotic string theory', *Phys. Lett.* **166B**, 300.

457. Sen, A. (1986), 'σ model approach to the heterotic string theory', in *Workshop on Unified String Theories, 29 July – 16 August, 1985*, eds. M. Green and D. Gross (World Scientific, Singapore), p. 497.

458. Sen, A. (1986), 'Superspace analysis of local Lorentz and gauge anomalies in the heterotic string theory', *Phys. Lett.* **174B**, 277.

459. Sen, A. (1986), 'Central charge of the Virasoro algebra for supersymmetric sigma models on Calabi–Yau manifolds', *Phys. Lett.* **178B**, 370.

460. Shafarevich, I.R. (1974), *Basic Algebraic Geometry* (Springer-Verlag).

461. Shafi, Q. and Wetterich, C. (1983), Cosmology from higher-dimensional gravity', *Phys. Lett.* **129B**, 387.

462. Shapiro, J.A. (1970), 'Electrostatic nalogue for the Virasoro model', *Phys. Lett.* **33B**, 361.

463. Shapiro, J.A. (1971), 'Nonorientable dual loop graphs and isospin', *Phys. Rev.* **D4**, 1249.

464. Shapiro, J.A. (1972), 'Loop graph in the dual-tube model', *Phys. Rev.* **D5**, 1945.

465. Shapiro, J.A. (1975), 'Renormalization of dual models', *Phys. Rev.* **D11**, 2937.

466. Sierra, G. and Townsend, P.K. (1984), 'Chiral anomalies and constraints on the gauge group in higher-dimensional supersymmetric Yang–Mills theories', *Nucl. Phys.* **B222**, 493.

467. Slansky, R. (1981), 'Group theory for unified model building', *Phys. Reports* **79**, 1.

468. Steinberger, J. (1949), 'On the use of subtraction fields and the lifetimes of some types of meson decay', *Phys. Rev.* **76**, 1180.

469. Strominger, A. and Witten, E. (1985), 'New manifolds for superstring compactification', *Commun. Math. Phys.* **101**, 341.

470. Strominger, A. (1985), 'Topology of superstring compactification', in *Workshop on Unified String Theories, 29 July – 16 August, 1985*, eds. M. Green and D. Gross (World Scientific, Singapore), p. 654.

471. Strominger, A. (1986), 'Superstrings with torsion', *Nucl. Phys.* **B274**, 253.

472. Tanii, Y. (1985), 'Absence of the supersymmetry anomaly in heterotic string theory', *Phys. Lett.* **165B**, 275.

473. Taylor, T.R. (1985), 'Hidden sector of superstring models: An effective Lagrangian analysis', *Phys. Lett.* **164B**, 43.

474. Thierry–Mieg, J. (1985), 'Remarks concerning the $E_8 \times E_8$ and D_{16} string theories', *Phys. Lett.* **156B**, 199.

475. Thierry–Mieg, J. (1986), 'Anomaly cancellation and fermionisation in 10-, 18- and 26-dimensional superstrings', *Phys. Lett.* **171B**, 163.

476. Thorn, C.B. (1986), 'The theory of interacting relativistic strings', *Nucl. Phys.* **B263**, 493.

477. Thorn, C.B. (1986), 'Introduction to the theory of relativistic strings', in *Workshop on Unified String Theories, 29 July – 16 August, 1985*, eds. M. Green and D. Gross (World Scientific, Singapore), p. 5.

478. Townsend, P.K. and Sierra, G. (1983), 'Chiral anomalies and constraints on the gauge group in higher-dimensional supersymmetric Yang-Mills theories', *Nucl. Phys.* **B222**, 493.

479. Uhlenbeck, K. and Yau, S.T. (1986), preprint.

480. Vafa, C. (1986), 'Modular invariance and discrete torsion on orbifolds', *Nucl. Phys.* **B273**, 592.

481. Van Nieuwenhuizen, P. (1981), 'Supergravity', *Phys. Reports* **68**, 189.

482. Veneziano, G. (1974), 'An introduction to dual models of strong interactions and their physical motivations', *Phys. Rev.* **C9**, 199.

483. Weinberg, S. (1972), *Gravitation and Cosmology* (Wiley-Interscience).

484. Weinberg, S. (1978), 'A new light boson?', *Phys. Rev. Lett.* **40**, 223.

485. Weinberg, S. (1984), 'Charges from extra dimensions', *Phys. Lett.* **125B**, 265.

486. Weinberg, S. (1984), 'Quasi-Riemannian theories of gravitation in more than four dimensions', *Phys. Lett.* **138B**, 47.

487. Weiss, N. (1986), 'Superstring cosmology: Is it consistent with a matter-dominated universe?', *Phys. Lett.* **172B**, 180.

488. Wells, R.O., Jr. (1980), *Differential Analysis on Complex Manifolds*, (Springer-Verlag).

489. Wen, X.G. and Witten, E. (1986), 'World-sheet instantons and the Peccei–Quinn symmetry', *Phys. Lett.* **166B**, 397.

490. Wen, X.G. and Witten, E. (1985), 'Electric and magnetic charges in superstring models', *Nucl. Phys.* **B261**, 651.

491. Wess, J. and Zumino, B. (1971), 'Consequences of anomalous Ward identities', *Phys. Lett.* **37B**, 95.

492. Wess, J. and Zumino, B. (1974), 'Supergauge transformations in four dimensions', *Nucl. Phys.* **B70**, 39.

493. Wess, J. and Bagger, J. (1983), *Supersymmetry and Supergravity*, (Princeton Univ. Press).

494. Wetterich, C. (1982), '$SO(10)$ unification from higher dimensions', *Phys. Lett.* **110B**, 379.

495. Wetterich, C. (1982), 'Spontaneous compactification in higher dimensional gravity', *Phys. Lett.* **113B**, 377.

496. Wetterich, C. (1983), 'Massless spinors in more than four dimensions', *Nucl. Phys.* **B211**, 177.
497. Wetterich, C. (1983), 'Dimensional reduction of Weyl, Majorana and Majorana–Weyl spinors', *Nucl. Phys.* **B222**, 20.
498. Wetterich, C. (1983), 'Chirality index and dimensional reduction of fermions', *Nucl. Phys.* **B223**, 109.
499. Wetterich, C. (1984), 'Discrete symmetries in Kaluza-Klein theories', *Nucl. Phys.* **B234**, 413.
500. Wetterich, C. (1984), 'Dimensional reduction of fermions in generalized gravity', *Nucl. Phys.* **B242**, 473.
501. Wetterich, C. (1985), 'Spontaneous symmetry breaking and fermion chirality in higher-dimensional gauge theory', *Nucl. Phys.* **B260**, 402.
502. Wetterich, C. (1985), 'Fermion mass predictions from higher dimensions', *Nucl. Phys.* **B261**, 461.
503. Wilczek, F. (1978), 'Problem of strong P and T invariance in the presence of instantons', *Phys. Rev. Lett.* **40**, 279.
504. Wilczek, F. and Zee, A. (1979), Princeton preprint, unpublished.
505. Wilczek, F. and Zee, A. (1982), 'Families from spinors', *Phys. Rev.* **D25**, 553.
506. Winnberg, J.-O. (1975), 'Recalculation of the single planar dual fermion loop', *Nucl. Phys.* **B94**, 205.
507. Witten, E. (1981), 'Search for a realistic Kaluza–Klein theory', *Nucl. Phys.* **B186**, 412.
508. Witten, E. (1981), 'Dynamical breaking of supersymmetry', *Nucl. Phys.* **B188**, 513.
509. Witten, E. (1982), 'Instability of the Kaluza–Klein vacuum', *Nucl. Phys.* **B195**, 481.
510. Witten, E. (1982), 'Constraints on supersymmetry breaking', *Nucl. Phys.* **B202**, 253.
511. Witten, E. (1982) 'Supersymmetry and Morse theory', *J. Diff. Geom.* **17**, 661.
512. Witten, E. (1983), '$D = 10$ superstring theory', in *Fourth Workshop on Grand Unification*, eds. P. Langacker et al. (Birkhauser), p. 395.
513. Witten, E. (1983), 'Global aspects of current algebra', *Nucl. Phys.* **B223**, 422.
514. Witten, E. (1984), 'Some properties of O(32) superstrings', *Phys. Lett.* **149B**, 351.
515. Witten, E. (1985), 'Cosmic superstrings', *Phys. Lett.* **153B**, 243.
516. Witten, E. (1985), 'Dimensional reduction of superstring models', *Phys. Lett.* **155B**, 151.
517. Witten, E. (1985), 'Superconducting strings', *Nucl. Phys.* **B249**, 557.
518. Witten, E. (1985), 'Symmetry breaking patterns in superstring models', *Nucl. Phys.* **B258**, 75.
519. Witten, E. (1985), 'Fermion quantum numbers in Kaluza–Klein theory', in *Shelter Island II: Proceedings of the 1983 Shelter Island Conference on Quantum Field Theory and the Fundamental Problems of Physics*, eds. R. Jackiw et al. (MIT Press, Cambridge, Mass.), p. 227.
520. Witten, E. (1985), 'Global gravitational anomalies', *Commun. Math. Phys.* **100**, 197.

521. Witten, E. (1986), 'Global anomalies in string theory', in *Symposium on Anomalies, Geometry, Topology, March 28-30, 1985*, eds. W.A. Bardeen and A.R. White (World Scientific, Singapore), p. 61.

522. Witten, E. (1986), 'Topological tools in ten dimensional physics', in *Workshop on Unified String Theories, 29 July – 16 August, 1985*, eds. M. Green and D. Gross (World Scientific, Singapore), p. 400. With an appendix by R.E. Stong, 'Calculation of $\Omega_{11}^{Spin}(K(Z,4))$'.

523. Witten, E. (1986), 'New issues in manifolds of SU(3) holonomy', *Nucl. Phys.* **B268**, 79.

524. Witten, L. and Witten, E. (1987), 'Large radius expansion of superstring compactification', *Nucl. Phys.* **B281**, 109.

525. Wu, Y.-S. and Zi Wang (1986), 'The time variation of Newton's gravitational constant in superstring theories', *Phys. Rev. Lett.* **57**, 1978.

526. Yahikozawa, S. (1986), 'Evaluation of the one-loop amplitude in heterotic string theory', *Phys. Lett.* **166B**, 135.

527. Yamamoto, K. (1985), 'Saving the axions in superstring models', *Phys. Lett.* **161B**, 289.

528. Yamamoto, K. (1986), 'The phase transition associated with intermediate gauge symmetry breaking in superstring models', *Phys. Lett.* **168B**, 341.

529. Yasuda, O. (1986), 'Higher derivative terms and zero modes in $D = 10$ supergravity', *Phys. Lett.* **169B**, 64.

530. Yau, S.T. (1977), 'Calabi's conjecture and some new results in algebraic geometry', *Proc. Natl. Acad. Sci.* **74**, 1798.

531. Yau, S.T. (1985), 'Compact three dimensional Kähler manifolds with zero Ricci curvature', in *Symp. on Anomalies, Geometry, Topology, March 28 – 30, 1985*, eds. W.A. Bardeen and A.R. White (World Scientific, Singapore), p. 395.

532. Zee, A. (1972), 'Axial-vector anomalies and the scaling property of field theory', *Phys. Rev. Lett.* **29**, 1198.

533. Zee, A. (1982), *Unity of Forces in The Universe* (World Scientific).

534. Zumino, B. (1975), 'Supersymmetry and the vacuum', *Nucl. Phys.* **B89**, 535.

535. Zumino, B., Wu, Y.S. and Zee, A. (1984), 'Chiral anomalies, higher dimensions, and differential geometry', *Nucl. Phys.* **B239**, 477.

536. Zwiebach, B. (1985), 'Curvature squared terms and string theories', *Phys. Lett.* **156B**, 315.

String Field Theory

String field theory is a very active area of research that we have not presented in these volumes. It concerns the formulation of string theory as a second-quantized field theory based on functional string fields. To do justice to the subject would almost require a third volume. Despite our decision not to develop string field theory in this book, we feel that it would be useful to list some of the papers dealing with this subject.

Most of the early work on string field theory was based on a light-cone gauge formulation. Many of the issues relevant to this approach have been discussed in chapter 11. The basis references in which most of the technology was developed are the papers of Kaku and Kikkawa [40,41] and Cremmer and Gervais [14,15].

The light-cone gauge field theory approach was extended to superstring theory in papers by Green, Schwarz and Brink [29,30,31]. One interesting observation in these papers is that the interaction Hamiltonian is given by the anticommutator of a suitable light-cone component of the supercharge with itself. Moreover, both operators are determined uniquely (given some plausible assumptions) by requiring that the algebra is satisfied.

While a light-cone gauge formulation of the theory may be correct and complete, there there are many questions that it does not elucidate. A covariant formulation with a very large group of gauge invariances is probably required to address many of the deepest questions concerning the geometry of compactification, the occurrence or nonoccurrence of space-time singularities, and so forth. It ought to bring forth the fundamental principles that are the basis for string theory. Whether covariant string field theory will turn out to be the right approach to elucidating the logical structure that underlies string theory remains to be seen. It is certainly the most conservative approach to this problem. A much more radical approach is that of [22].

An early investigation of covariant string field theory in the 'old days' of string theory was [49]. Following the revival of interest in string theory, a first important step toward constructing covariant string field theories was taken by Siegel [67,68], who argued that the BRST operator would play an essential role. By 1985 the subject began to develop rapidly with pioneering studies by Banks and Peskin [6], Siegel and Zwiebach [69] and Neveu and West [52–57], who presented gauge invariant free field theories of strings. Many other contributions to the development of gauge invariant free string field theories are listed in the accompanying references. A gauge invariant nonlinear field theory of open bosonic strings was presented in [77] and interpreted in terms of a noncommutative cohomology ring. Other proposals concerning nonlinear gauge invariant string field theories have been made in [32,33,56]. Many important papers that only exist in preprint form are not listed.

REFERENCES

1. Aratyn, H. and Zimerman, A.H. (1986), 'On covariant formulation of the free Neveu–Schwarz and Ramond string models', *Phys. Lett.* **166B**, 130.
2. Aratyn, H. and Zimerman, A.H. (1986), 'Gauge invariance of the bosonic free field string theory', *Phys. Lett.* **168B**, 75.
3. Aratyn, H. and Zimerman, A.H. (1986), 'Differential form formulation of the Neveu–Schwarz and Ramond free field string theories', *Nucl. Phys.* **B269**, 349.
4. Aulakh, C.S. (1986), 'Consistently truncated open superstring', *Phys. Lett.* **175B**, 297.
5. Awada, M.A. (1986), 'The gauge-covariant formulation of interacting strings and superstrings', *Phys. Lett.* **172B**, 32.
6. Banks, T. and Peskin, M.E. (1986), 'Gauge invariance of string fields', *Nucl. Phys.* **B264**, 513.
7. Banks, T., Friedan, D., Martinec, E., Peskin, M. and Preitschopf, C., (1986), 'All free string theories are theories of forms', *Nucl. Phys.* **B274**, 71.
8. Bardakçi, K. (1986), 'Covariant gauge theory of strings', *Nucl. Phys.* **B217**, 561.
9. Baulieu, L. and Ouvry, S. (1986), 'Quasi–Yang–Mills structure for the open bosonic string', *Phys. Lett.* **171B**, 57.
10. Bengtsson, A.K.H., Brink, L., Cederwall, M. and Ögren, M. (1985), 'Uniqueness of superstring actions', *Nucl. Phys.* **B254**, 625.
11. Bengtsson, I. (1986), 'Hamiltonian treatment of free string field theory', *Phys. Lett.* **172B**, 342.
12. Carson, L. and Hosotani, Y. (1986), 'Line functionals and string field theory', *Phys. Rev. Lett.* **56**, 2144.
13. Chappell, G.J. and Taylor, J.G. (1986), 'On gauge invariant bosonic strings', *Phys. Lett.* **175B**, 159.
14. Cremmer, E. and Gervais, J.L. (1974), 'Combining and splitting relativistic strings', *Nucl. Phys.* **B76**, 209.
15. Cremmer, E. and Gervais, J.L. (1975), 'Infinite component field theory of interacting relativistic strings and dual theory', *Nucl. Phys.* **B90**, 410.
16. Das, S.R. and Rubin, M.A. (1986), 'A Tomonaga–Schwinger–Dirac formulation for string theories', *Phys. Lett.* **169B**, 182.

17. Daté, G.D., Günaydin, M., Pernici, M., Pilch, K. and Van Nieuwenhuizen, P. (1986), 'A minimal covariant action for the free open spinning string field theory', *Phys. Lett.* **171B**, 182.
18. de Alwis, S.P. and Ohta, N. (1986), 'Fully gauge-invariant field theory of free superstrings', *Phys. Lett.* **174B**, 383.
19. de Alwis, S.P. and Ohta, N. (1986), 'All free string theories are theories of BRST cohomology', *Phys. Lett.* **174B**, 388.
20. Floratos, E.G., Kazama, Y. and Tamvakis, K. (1986), 'On the relation between the gauge-covariant formulation of string field theories', *Phys. Lett.* **166B**, 295.
21. Friedan, D. (1985), 'On two-dimensional conformal invariance and the field theory of strings', *Phys. Lett.* **162B**, 102.
22. Friedan, D. and Shenker, S. (1986), 'The integrable analytic geometry of quantum string', *Phys. Lett.* **175B**, 287.
23. Friedan, D. and Shenker, S. (1987), 'The analytic geometry of conformal field theory', *Nucl. Phys.* **B281**, 509.
24. Friedan, D. (1986), 'String field theory', *Nucl. Phys.* **B271**, 540.
25. Gervais, J.-L. (1986), 'Group theoretic approach to the string field theory action', *Nucl. Phys.* **B276**, 339.
26. Giddings, S. (1986), 'The Veneziano amplitude from gauge invariant string field theory', *Nucl. Phys.* **B278**, 242.
27. Giddings, S. and Martinec, E. (1986), 'Conformal geometry and string field theory', *Nucl. Phys.* **B278**, 91.
28. Giddings, S., Martinec, E. and Witten, E. (1986), 'Modular invariance in string field theory', *Phys. Lett.* **176B**, 362.
29. Green, M.B. and Schwarz, J.H. (1983), 'Superstring interactions', *Nucl. Phys.* **B218**, 43.
30. Green, M.B., Schwarz, J.H. and Brink, L. (1983), 'Superfield theory of type (II) superstrings', *Nucl. Phys.* **B219**, 437.
31. Green, M.B. and Schwarz, J.H. (1984), 'Superstring field theory', *Nucl. Phys.* **B243**, 475.
32. Hata, H., Itoh, K., Kugo, T., Kunitomo, H. and Ogawa, K. (1986), 'Manifestly covariant field theory of interacting string I', *Phys. Lett.* **172B**, 186.
33. Hata, H., Itoh, K., Kugo, T., Kunitomo, H. and Ogawa, K. (1986), 'Manifestly covariant field theory of interacting string II', *Phys. Lett.* **172B**, 195.
34. Hata, H., Itoh, K., Kugo, T., Kunitomo, H. and Ogawa, K. (1986), 'Covariant string field theory', *Phys. Rev.* **D34**, 2360.
35. Hata, H., Itoh, K., Kugo, T., Kunitomo, H. and Ogawa, K. (1986), 'Pregeometrical string field theory: creation of space-time and motion', *Phys. Lett.* **175B**, 138.
36. Hopkinson, J.F.L., Tucker, R.W. and Collins, P.A. (1975), 'Quantum strings and the functional calculus', *Phys. Rev.* **D12**, 1653.
37. Horowitz, G.T. and Strominger, A. (1986), 'Origin of gauge invariance in string theory' *Phys. Rev. Lett.* **57**, 519.
38. Horowitz, G.T., Lykken, J., Rohm, R. and Strominger, A. (1986), 'Purely cubic action for string field theory', *Phys. Rev. Lett.* **57**, 283.
39. Itoh, K., Kugo, T., Kunimoto, H. and Ooguri, H. (1986), 'Gauge invariant local action of string field from BRS formalism', *Prog. Theor. Phys.* **75**, 162.

40. Kaku, M. and Kikkawa, K. (1974), 'Field theory of relativistic strings. I. Trees', *Phys. Rev.* **D10**, 1110.
41. Kaku, M. and Kikkawa, K. (1974), 'Field theory of relativistic strings. II. Loops and pomerons', *Phys. Rev.* **D10**, 1823.
42. Kaku, M. (1985), 'Locality in the gauge-covariant field theory of strings', *Phys. Lett.* **162B**, 97.
43. Kaku, M. (1986), 'Gauge field theory of covariant strings', *Nucl. Phys.* **B267**, 125.
44. Kaku, M. and Lykken, J. (1985), 'Supergauge field theory of superstrings', in *Symp. on Anomalies, Geometry, Topology, March 28 – 30, 1985*, eds. W.A. Bardeen and A.R. White (World Scientific, Singapore), p. 360.
45. Kazama, Y., Neveu, A., Nicolai, H. and West, P.C. (1986), 'Symmetry structures of superstring field theories', *Nucl. Phys.* **B276**, 366.
46. LeClair, A. (1986), 'Fermionic string field theory', *Phys. Lett.* **168B**, 53.
47. LeClair, A. and Distler, J. (1986), 'Gauge invariant superstring field theory', *Nucl. Phys.* **B273**, 552.
48. Marcus, N. and Sagnotti, A. (1986), 'String field theory and equations of motion', *Phys. Lett.* **178B**, 343.
49. Marshall, C. and Ramond, P. (1975), 'Field theory of the interacting string: The closed string', *Nucl. Phys.* **B85**, 375.
50. Meurice, Y. (1986), 'About the uniqueness of covariant string field theory', *Phys. Lett.* **173B**, 257.
51. Nakawaki, Y. and Saito, T. (1972), 'Field theory of dual-resonance model', *Prog. Theor. Phys.* **48**, 1324.
52. Neveu, A., Schwarz, J.H. and West, P.C. (1985), 'Gauge symmetries of the free bosonic string field theory', *Phys. Lett.* **164B**, 51.
53. Neveu, A. and West, P.C. (1985), 'Gauge symmetries of the free supersymmetric string field theories', *Phys. Lett.* **165B**, 63.
54. Neveu, A., Nicolai, H. and West, P.C. (1986), 'Gauge covariant local formulation of free strings and superstrings', *Nucl. Phys.* **B264**, 573.
55. Neveu, A., Nicolai, H. and West, P.C. (1986), 'New symmetries and ghost structure of covariant string theories', *Phys. Lett.* **167B**, 307.
56. Neveu, A. and West, P.C. (1986), 'The interacting gauge covariant bosonic string', *Phys. Lett.* **168B**, 192.
57. Neveu, A. and West, P.C. (1986), 'Gauge covariant local formulation of bosonic strings', *Nucl. Phys.* **B268**, 125.
58. Ohta, N. (1986), 'Covariant second quantization of superstrings', *Phys. Rev. Lett.* **56**, 440.
59. Ohta, N. (1986), 'Covariant quantization of superstrings based on Becchi–Rouet–Stora invariance', *Phys. Rev.* **D33**, 1681.
60. Ooguri, H. (1986), 'String field theory with spacetime supersymmetry', *Phys. Lett.* **172B**, 204.
61. Peskin, M.B. and Thorn, C.B. (1986), 'Equivalence of the light-cone formulation and the gauge-invariant formulation of string dynamics', *Nucl. Phys.* **B269**, 509.
62. Pfeffer, D., Ramond, P. and Rodgers, V.G.J. (1985), 'Gauge invariant field theory of free strings', *Nucl. Phys.* **B276**, 131.
63. Raby, S., Slansky, R. and West, G. (1985), 'Toward a covariant string field theory', in proc. of the *Lewes String Theory Workshop*, (World

Scientific), p. 246.

64. Ramond, P. (1986), 'A pedestrian approach to covariant string theory', *Suppl. Prog. Theor. Phys.* **86**, 126.

65. Sciuto, S. (1969), 'The general vertex function in dual resonance models', *Nuovo Cim. Lett.* **2**, 411.

66. Senda, I. (1986), 'Light-cone field theory of closed bosonic strings compactified on a torus', *Phys. Lett.* **174B**, 267.

67. Siegel, W. (1984), 'Covariantly second–quantized string II', *Phys. Lett.* **149B**, 157; **151B**, 391.

68. Siegel, W. (1984), 'Covariantly second–quantized string III', *Phys. Lett.* **149B**, 162; **151B**, 396.

69. Siegel, W. and Zwiebach, B. (1986), 'Gauge string fields', *Nucl. Phys.* **B263**, 105.

70. Taylor, J.G. and Restuccia, A. (1985), European Physical Society meeting, Bari, Italy.

71. Terao, H. and Uehara, S. (1986), 'Covariant second quantization of free superstring', *Phys. Lett.* **168B**, 70.

72. Terao, H. and Uehara, S. (1986), 'Gauge invariant actions and gauge fixed actions of free superstring field theory', *Phys. Lett.* **173B**, 134.

73. Terao, H. and Uehara, S. (1986), 'Gauge invariant actions of free closed superstring field theories', *Phys. Lett.* **173B**, 409.

74. Thorn, C.B. (1985), 'Comments on covariant formulations of string theories', *Phys. Lett.* **159B**, 107.

75. Thorn, C.B. (1986), 'The theory of interacting relativistic strings', *Nucl. Phys.* **B263**, 493.

76. Tseytlin, A.A. (1986), 'Covariant string field theory and effective action', *Phys. Lett.* **168B**, 63.

77. Witten, E. (1986), 'Non-commutative geometry and string field theory', *Nucl. Phys.* **B268**, 253.

78. Witten, E. (1986), 'Interacting field theory of open superstrings', *Nucl. Phys.* **B276**, 291.

79. Yamron, J.P. (1986), 'A gauge invariant action for the free Ramond string', *Phys. Lett.* **174B**, 69.

80. Yoneya, T. (1985), 'Space-time local symmetry of string field theory', *Phys. Rev. Lett.* **55**, 1828.

81. Zwiebach, B. (1985), 'Gauge invariant string actions', in *Workshop on Unified String Theories, 29 July – 16 August, 1985*, eds. M. Green and D. Gross (World Scientific, Singapore), p. 607.

Index